Ecological Studies

Analysis and Synthesis

Volume 239

More information about this series at http://www.springer.com/series/86

Iain J. Gordon • Herbert H. T. Prins
Editors

The Ecology of Browsing and Grazing II

Editors
Iain J. Gordon
Division of Tropical Environment & Societies
James Cook University
Townsville, QLD, Australia

Herbert H. T. Prins
Animal Sciences Group
Wageningen University
Wageningen, The Netherlands

ISSN 0070-8356 ISSN 2196-971X (electronic)
Ecological Studies
ISBN 978-3-030-25867-2 ISBN 978-3-030-25865-8 (eBook)
https://doi.org/10.1007/978-3-030-25865-8

This Springer imprint is published by the registered company Springer Nature Switzerland AG.
The registered company address is: Gewerbestrasse 11, 6330 Cham, Switzerland

Browsers and grazers come from a range of taxa of large mammalian herbivores. Photos Iain J. Gordon (gorilla) and Herbert H. T. Prins (elephant)

Grazers and browsers
Long trotted the ancient lands:
Humankind's lifeblood

Preface

On the 6th of September 2016, the following e-mail came through from Andrea Schlitzberger at 23.46 h. Andrea works at Springer Nature, and Springer published a book we co-edited unassumingly called *The Ecology of Browsing and Grazing*. The message read:

> So far ca. 12,700 chapters of your Ecological Studies volume on "The Ecology of Browsing and Grazing" published in 2008 were downloaded, which is an excellent success and shows the great interest in this topic. Therefore we would like to ask you if you think a new edition would make sense. A new edition should include at least 20% new material. A completely new book on a similar topic would also be possible. What do you think and would you be interested in acting as volume editor?

Now, it took us a couple of days to respond, because we had to liaise as to whether we were interested in taking on the project. We had recently finished three other books for other publishers (Prins and Gordon 2014; Gordon et al. 2017; Prins and Namgail 2017) and were not thinking of jumping into another book project anytime soon. However, both of us are passionate ungulate ecologists and humbled by the interest in the original book. Also, a lot has happened in the world of the ecology of browsing and grazing ungulates in particular and large mammalian herbivores specifically over the past 11 years. For example, the development of new paleontological and genomics techniques, significant improvements in GPS technology and sensors and analytical techniques using artificial intelligence with big data, to name a few. Also, whilst only 11 years had passed since the book was published, it is clear that the plight of many species of wild ungulate species is dire and in need of firm evidence to support conservation and management efforts. So, on the 2nd of October 2016, we went back to Springer to say "yes", we would be interested on four conditions:

1. That this would not be a minor tweak of the original book: we have already mentioned the major new developments that have happened in the field over the past 11 years.
2. Also, with the new book we wanted to have a slight change in the emphasis of the book to provide insights into the impact of large mammalian herbivores on the

ecosystems in which they live. With the growing need to restore dysfunctional ecosystems, keystone herbivore species are often the first that are being brought back into systems and we need evidence to make sure we're doing the right thing and that we preclude or limit perverse outcomes.

3. Rather than go back to the original authors, we try to have first authors for the chapters as women or researchers outside Australia, Europe or the USA and Canada. This would give a voice to those who are not always able to get their research findings and ideas into the mainstream of Anglo-centric thinking.
4. Finally, and this was a sticky one, we wanted to have a share of the sales' value of the book and not an honorarium as is normally the case.

Well, clearly, there was some negotiation required. Surprisingly, we had a bit of knock-back on the second condition as it was deemed that it might affect the "quality" of the book. We hope you will agree, that, whilst we were not 100% successful in our endeavour to have first authors for the chapters as women or researchers outside the Anglo-centric sphere, the book is still a very good summary of the field! The second bone of contention was our fourth condition and we did not manage to reach a compromise in the end: just like chairman Mao Zedong used to say that the last argument comes from the barrel of the gun, Springer informed us kindly but firmly that our wish did not fit in their business model. End of discussion. We accepted because of our love for nature and science that we want to share, and for dissemination, you need publishers. It is as it is, isn't it?

Of course, we wish to thank all of the contributors to the book for their sterling efforts and patience as we have crafted the chapters into a coherent and comprehensive review of the field of the ecology of browsing and grazing. Special thanks, though, to Dr Yvette Williams, who has managed to keep the two of us on point and has ensured that we were apprised of the current state of play of the book and who we needed to chivvy along! We couldn't have done it without her support.

So, there you have it, *The Ecology of Browsing and Grazing II*. Between October 2016 when we were asked to work on the sequel to *The Ecology of Browsing and Grazing*, to the date that we finished our editing work, the number of downloads of the 2008 book has increased from 12,700 to 20,800, showing that this field of science is thriving. We think we will leave you with the last couple of sentences of the Conclusions because these catch our sentiments as both of us love our ecology and are teachers at heart:

> researchers should not forget that these ungulates are made of flesh and blood, that they graze and browse in real landscapes, and that there is a profound need for hard-core ungulate ecologists with a broad set of skills and deep understanding of 'their' animals. As a bonus, we, and all other ungulate ecologists, get to see, feel and understand some of the most beautiful creatures that share our planet.

Townsville, QLD, Australia — Iain J. Gordon
Wageningen, The Netherlands — Herbert H. T. Prins
18th February 2019

References

Gordon IJ, Prins HHT (eds) (2008) The ecology of browsing and grazing. Ecological studies, vol 195. Springer, Berlin

Gordon IJ, Prins HHT, Squire G (eds) (2016) Food production and nature conservation: conflicts and solutions. Routledge, London

Prins HHT, Gordon IJ (eds) (2014) Invasion biology and ecological theory: insights from a continent in transformation. Cambridge University Press, Cambridge

Prins HHT, Namgail T (eds) (2017) Bird migration across the Himalayas: wetland functioning amidst mountains and glaciers. Cambridge University Press, Cambridge

Contents

Chapter 1
The Ecology of Browsing and Grazing II

Iain J. Gordon and Herbert H. T. Prins

Globally, many terrestrial ecosystems have been and are being heavily influenced by human activity, both directly and indirectly. Humanity and our domestic animals (1.4 billion cattle, 1.2 billion sheep and 0.5 billion goats, but only some 120 million horses and 13 million camels; Encyclopedia.com) have now so much impact on global ecosystems that we have entered the Anthropocene (Lewis and Maslin 2015). Wild ruminants number merely 75 million (Hackmann and Spain 2010), and are native to all continents except Antarctica. In such ecosystems extensive grazing and browsing by domestic and wild large mammalian herbivores (hereafter called large herbivores) and, in places, burning have shaped vegetation composition, structure and dynamics. Through their grazing, browsing, trampling and defecation large herbivores not only shape the structure and distribution of the vegetation but also affect nutrient flows and the responses of associated fauna. Consequently, it is the interactions between management or population dynamics of large herbivores and the vegetation they consume that shape the biodiversity, structure and dynamics of these ecosystems, covering vast parts of the globe. Therefore, a knowledge of the determinants of the distribution, movements and activities of herbivores, and how these interact with vegetation composition and dynamics, is required in order to predict the broader impact of these animals, now and into the future.

In 2008 we edited a volume, in Springer's Ecological Studies Series (Gordon and Prins 2008), in which we investigated "*how large herbivores not only influence the structure and distribution of vegetation, but also affect nutrient flows and the response of associated fauna. The mechanisms and processes underlying the herbivores' behaviour, distribution, movement and direct impact on vegetation, and the dynamics of nutrients, plant species, and vegetation* (were) *discussed in detail.*"

I. J. Gordon (✉)
James Cook University, Townsville, QLD, Australia
e-mail: Iain.gordon@jcu.edu.au

H. H. T. Prins
Animal Sciences Group, Wageningen University, Wageningen, The Netherlands

I. J. Gordon, H. H. T. Prins (eds.), *The Ecology of Browsing and Grazing II*,
Ecological Studies 239, https://doi.org/10.1007/978-3-030-25865-8_1

Over 10 years have passed since the publication of the book "*The Ecology of Browsing and Grazing*" and substantial new research has been published in the meantime, improving our understanding of the relationship between and large herbivores and plants how these can be managed to support a range of outcomes including conservation, hunting and livestock production.

In this book we slightly change the emphasis to focus more heavily on the impact of browsing and grazing. Our aim is to draw together the leading research in understanding the mechanisms and processes which underlie the behaviour, distribution, movement of domestic and wild large herbivores, and their direct and indirect impacts on the dynamics of nutrients, plant species and vegetation composition in terrestrial ecosystems, including the impacts of changes in management, environment and the climate. The Chapters in "*The Ecology of Browsing and Grazing II*" encompass fundamental and applied research aimed at cross-cutting issues in palaeontology, ecology, nutrition and management. This research demonstrates how an understanding of the processes of plant/animal interactions allows research to provide practical advice on the management of large herbivores to integrate production and conservation in terrestrial systems.

The focus on "browsing and grazing" relates to the major dichotomy in the form of the foods eaten by large mammalian herbivores. Grasses (i.e., monocotyledonous plants) and browse (i.e., dicotyledonous plants [forbs, bushes and trees]) not only differ in their evolutionary roots but also in their chemistry, morphology, 2D/3D structure and distribution. These differing characteristics of grass and browse plant have led to morphological, physiological and behavioural adaptations of large mammalian herbivores that feed on grass and browse (Gordon and Prins 2008). This in turn has ripple-on effects on the herbivores' ecology, population dynamics and community composition. Whilst we are the first to admit that "browsing and grazing" is a gross simplification (see **Codron et al.** Chap. 4), the classification had provided a useful heuristic allowing researchers to move beyond the complexity of ungulates, and their interactions with ecosystems in which they exist, across the globe, to determine the common processes underlying these interactions. In this book we see what value the classification has had in understanding the ecology and impact of large mammalian herbivores, and other mammalian herbivore taxa (see **Gordon et al.** Chap. 15), over the past 10 years since the publication of *The Ecology of Browsers and Grazers*.

One other issue that we should clarify here, that is, the different definitions of "small", "medium", "large" and "mega-" herbivores in the literature. To our knowledge there is little clarity as to the weight at which vertebrate move from small to medium to large (but see Caughley and Krebs 1983 who state "large mammals and small mammals really do have different ecologies and that both groups of ecologists are right about their animals"). We are attracted to this explanation. We think there may be a break-point in the interval 25–35 kg body weight below which regulation tends to be intrinsic and above which it tends to be extrinsic. However, in the mammalian paleontological literature "mega-herbivores" refers to those having an estimated weight of more than 44 kg (or 100 lbs) (Martin 1967; cf. Martin 1966 where he uses 50 kg); whereas, in the dinosaur palaeontology and mammalian ecology literature mega-herbivores related to those weighing more than 1000 kg (Owen-Smith 1988; Mallon and Anderson 2015). We also note that in the reptilian herbivore literature giant tortoises are classed as mega-herbivores (e.g., Hansen 2015). Given scaling rules there is no reason to assume that the classification is anything other than arbitrary and that the ecology of herbivore species lies on a continuum in relation to weight that may be taxon specific. Being mammalian herbivore ecologists ourselves we have encouraged the authors in the book to adopt definition of mega-herbivores weighing over 1000 kg.

In "*The Ecology of Browsing and Grazing II*" our logical framework for the Chapters is firstly to demonstrate that, since the time of the rise of the first ungulates in the early Eocene (54 M years ago), there have been feedbacks in which these large herbivores have adapted to the plants on which they forage and plants, in turn, have evolved defences against herbivory (**Saarinen** Chap. 2). New techniques, including coprophilous fungi, *Sporormiella*, and dung beetles have allowed us to use paleo-ecological proxies to interpret the fossil record and show that ungulates have had major impacts upon the ecosystems in which they have lived for millennia (**Rowan and Faith** Chap. 3). The dynamics of these ungulate populations and communities still resonates today through the structure and composition of plant communities across the continents (**Smit and Coetsee** Chap. 13).

To understand the impact of modern day assemblages of ungulates on the ecosystems in which they live we firstly need to understand the morphological and physiological adaptations that herbivores have to their diets (**Codron et al.** Chap. 4), and the foraging behaviours of herbivores as they relate to the abundance and distribution of different food types (grass vs browse) (**Venter et al.** Chap. 5).

As resource availability is so important for many large herbivores, the relationship with their plant food types impacts individual large mammalian herbivore species' population dynamics (**Kiffner and Lee** Chap. 6), and their community composition and dynamics (**Mishra et al.** Chap. 7). Obviously, food is not the only determinant of the current size, distribution and dynamics of large mammalian herbivore species and communities, land use change and human induced climate change (**Boone** Chap. 8) are also significant drivers in the Anthropocene.

As a consequence of their keystone role in many of the ecosystems in which large mammalian herbivores exist, they have major impacts on a range of ecological processes, driving the dynamics of soils (**Sitters and Andriuzzi** Chap. 9), plants (**Sabo** Chap. 11) and other fauna that rely on plants for food and/or shelter (**Katona and Coetsee** Chap. 12).

These impacts on individual biological components of ecosystems also has knock on effects for other abiotic system drivers. For example, through their impact on vegetation biomass and distribution large mammalian herbivores create landscapes that are more or less fire prone (**Smit and Coetsee** Chap. 13). These interactions between biotic and abiotic determinants of ecosystem structure and dynamics create positive and negative feedback loops that cascade through the ecosystem (**van Langevelde et al.** Chap. 10). Given this immense importance of humans in determining the future of ungulate species and communities, using our up-to-date understanding of the relationship between large mammalian herbivores and ecological processes will improve the ways in which systems are managed for production and conservation outcomes (**Fynn et al.** Chap. 14), and to meet a range of other ecosystems services (e.g., fire control).

Whilst "*The Ecology of Browsing and Grazing II*" is focused on large herbivorous mammals, primarily wild ungulates and domestic livestock, we sought to include short contributions that highlight how the understanding of the relationship between plants and herbivorous vertebrates has both shaped their evolution and their modern day impacts. As such, we have included a Chapter with sections covering groups from dinosaurs to birds, in which the authors show how the heuristic of classifying plants as grass and browse has improved our understanding across diverse taxa of vertebrate herbivores.

We end this edited volume with a Chapter (**Gordon and Prins** Chap. 16) in which we draw conclusions and flag issues that deserve more research in the vibrant field of plant-animal interactions of vertebrate herbivores.

References

Caughley G, Krebs CJ (1983) Are big mammals simply little mammals writ large? Oecologia 59:7–17

Encyclopedia.com. https://www.encyclopedia.com/science/encyclopedias-almanacs-transcripts-and-maps/livestock. Accessed 14 Jan 2019

Gordon IJ, Prins HHT (eds) (2008) The ecology of grazing and browsing. Springer, Berlin

Hackmann TJ, Spain JN (2010) Invited review: ruminant ecology and evolution: perspectives useful to ruminant livestock research and production. J Dairy Sci 93:1320–1334

Hansen DM (2015) Non-native megaherbivores: the case for novel function to manage plant invasions on islands. AoB Plants 7:1–11

Lewis SL, Maslin MA (2015) Defining the Anthropocene. Nature 519:171–180

Mallon JC, Anderson JS (2015) Jaw mechanics and evolutionary paleoecology of the megaherbivorous dinosaurs from the Dinosaur Park formation (upper Campanian) of Alberta, Canada. J Vertebr Paleontol 35:e904323

Martin PS (1966) Africa and Pleistocene overkill. Nature 212:339–342

Martin PS (1967) Prehistoric overkill. In: Martin PS, Wright HE (eds) Pleistocene extinctions. Yale University Press, New Haven, CT

Owen-Smith RN (1988) Megaherbivores: the influence of very large body size on ecology. Cambridge University Press, Cambridge

Chapter 2
The Palaeontology of Browsing and Grazing

Juha Saarinen

2.1 Introduction

Throughout their evolutionary history herbivorous mammals have encountered drastic climatic and environmental changes, which have led to changes in their diet and adaptations to feeding on various plant materials. The most abundant large herbivorous mammals are the ungulates, large hooved laurasiatherian mammals, which have an extensive fossil record on most continents (except Australia and Antarctica) throughout the Cenozoic era (ca. 65 million years to present). They were the dominant large herbivores in terrestrial ecosystems during the Cenozoic, although some groups, such as pigs (Suidae), include omnivorous forms as well. Many lineages of large herbivorous land mammals have experienced drastic changes in their feeding ecology during the Cenozoic following global and local changes in climate and vegetation (Janis 1993, 2008; Strömberg 2011). The most prominent of these dietary changes is the gradual shift from browsing to various degrees of grazing in many ungulate groups, especially during the latter part of the Cenozoic known as the Neogene (ca. 23 million years to present), following global climatic cooling and the spread of grasslands (Janis 2008). Eventually this led to the evolution of a few lineages of specialised grazing ungulates (mostly among equids and some bovids) during the last few million years (Janis 2008). This was achieved by changes in the digestive system (e.g., Ilius and Gordon 1992; Gordon 2003; Clauss and Rössner 2014) and by increasing wear-resistance of the molar teeth, which are used for grinding food (e.g., Janis and Fortelius 1988). Complex enamel patterns and increased crown heights (hypsodonty) evolved in many large bodied grass-eating mammal lineages (Janis and Fortelius 1988; Fortelius et al. 2002; Janis

J. Saarinen (✉)
Natural History Museum, London, UK

University of Helsinki, Helsinki, Finland
e-mail: juha.saarinen@helsinki.fi

I. J. Gordon, H. H. T. Prins (eds.), *The Ecology of Browsing and Grazing II*,
Ecological Studies 239, https://doi.org/10.1007/978-3-030-25865-8_2

et al. 2002, 2004; Fortelius et al. 2006; Janis 2008; Damuth and Janis 2011). A similar overall pattern occurred in the other major large herbivorous mammal groups, such as proboscideans (relatives of the modern elephants), xenarthrans (sloths and armadillo-like, herbivorous glyptodonts) and large herbivorous marsupials (e.g., Lister 2013; Vizcaíno 2009; Janis et al. 2016).

Browsing refers to feeding on various woody and non-woody dicotyledonous plants, including leaves, shoots and bark of trees, shrubs and dicotyledonous herbs, whereas grazing means eating grass. The most common feeding strategy in extant (and many fossil) ungulates is, however, feeding on various proportions of both browse and grass. This is called mixed-feeding. According to the traditional classification, browsers are defined as having less than 10% grass in their diet, mixed-feeders as having between 10% and 90% grass in diet, and grazers as having more than 90% grass in diet (e.g., Hofmann and Stewart 1972; Fortelius and Solounias 2000). These dietary classes enable simple categorisation of ungulate diets, but there is a continuum of herbivorous species from those utilising almost no grass to those feeding almost exclusively on grass, and this can also vary to some extent between populations of a species according to available vegetation and presence of competing ungulate species (e.g., Rivals et al. 2015; Rivals and Lister 2016; Saarinen et al. 2016; Saarinen and Lister 2016). In this chapter I use the term "frugivore" for mammals feeding mostly on fruits and seeds, whereas the term "folivore" is occasionally used as a synonym of "browser" to specify diets based on leaves and shoots particularly, rather than feeding on fruits (frugivorous diet) or both fruits and leaves (frugivorous-folivorous diet) (see Blondel 2001).

Broadly the trend in the evolution of ungulate faunas since the Mid-Miocene (ca. 16 Ma) has been towards increasing numbers of grazing-adapted species and subsequently a decline in the number of specialized browsing species, but the details are, of course, much more complex depending on local habitats and resources (Janis et al. 2002; Janis 2008). In the past, researchers often used generalized interpretations about diet based on the ecomorphology (e.g., hypsodonty) of herbivorous mammal species. However, diets can vary considerably within species, and many simplified ecomorphological characteristics (such as hypsodonty of teeth) can reflect adaptations to other environmental factors than dietary composition (see further discussion in the next section). This has been clearly demonstrated for many extant and extinct mammals by dietary analyses based on proxy methods such as dental micro- and mesowear and stable isotope analyses (see next section). Dietary adaptations in extant (and Pleistocene Eurasian) ungulate families are variable. Deer (Cervidae) have in general retained relatively unspecialized, browse-based diets required for sufficient nutrient intake for the seasonal growing of antlers (Geist 1998). This is reflected in their relatively unspecialized, mostly low-crowned (brachydont) and simple dentitions (Geist 1998). Also rhinoceroses (Rhinocerotidae) have mostly retained browse-dominated diets and brachydont dentitions, but unlike the deer, some rhinoceros species such as the Pleistocene woolly rhinoceros (*Coelodonta antiquitatis*), the extant African white rhinoceros (*Ceratotherium simum*) and especially the late Neogene elasmotheriines evolved into grazers with relatively hypsodont dentitions (e.g., Janis 2008). Bovids (Bovidae) and derived horses (Equidae, Equinae) thrived in the Pleistocene and recent environments by mostly evolving towards increasingly specialized grazing diets (e.g., by increasing hypsodonty and dental complexity) (e.g., Janis 2008).

2.2 Browsing and Grazing in Fossil Ungulates: What Tooth Wear, Stable Isotopes and Morphological Adaptations Tell us About Diets in the Past

2.2.1 Ecomorphology: Evolutionary Adaptations to Browsing and Grazing

2.2.1.1 Overall Morphology

Many skeletal adaptations of mammals, especially in the skull, reflect their feeding ecology. For example, the skulls of the plesiomorphic browsing ungulates tend to have narrow muzzles, moderate-sized attachment surfaces for the masseter muscles, comparatively large attachment surfaces for the temporalis muscles and relatively shallow jaws holding low-crowned teeth, whereas the more derived grazers usually have wider muzzles, deeper jaws facilitating more high-crowned teeth, posteriorly-located orbits and larger attachment surfaces for the masseter muscles (Janis 1995). Many specialised browsers, such as moose, tapir and many fossil ungulates, have retracted nasals, indicating the presence of a large, flexible upper lip or a proboscis for effective collecting of leaves and other plant parts. Grazing rhinoceroses tend to have backwards-inclined occipital surfaces in their skulls, reflecting a downwards-oriented head posture, whereas in browsing rhinos the occiput is vertical, reflecting a more horizontal head posture (Loose 1975). However, the most prominent clues to dietary adaptations are found in the dentition, as shown in the following two sections.

2.2.1.2 Hypsodonty

Hypsodonty refers to (relative) tooth crown height. It is typically measured as a ratio between tooth crown height and the length or width of the occlusal surface of a tooth, but often it is enough to assign the teeth to three commonly used categories of hypsodonty, which are: (1) brachydont (low-crowned), (2) mesodont (medium-crowned) and (3) hypsodont (high-crowned) (Janis and Fortelius 1988; Fortelius et al. 2002). All herbivorous mammal lineages started as brachydont but many show increase in tooth crown height later during their evolutionary history. An increase in tooth crown height is perhaps the most universal response in herbivorous mammals to increase the durability of their molar teeth in response to increased tooth wear rates caused by abrasive dietary items (Janis and Fortelius 1988). Traditionally it was thought that hypsodonty is specifically an adaptation to grazing, as chewing on tough, fibrous grasses with phytoliths (small mineral particles within grass leaves) abrades the teeth causing increased wear rates (Damuth and Janis 2014). However, another possibility is the general increase of hypsodonty in dry, open environments where exogenous grit on the plant food (following dust accumulation or soil ingestion) could abrade the teeth heavily (e.g., Damuth and Janis 2011; Madden 2015). Several studies indicate that both grazing, exogenous grit especially in dry, open environments, and sometimes even the accumulation of volcanic ash, can accelerate tooth wear and are likely to have contributed to the evolution of hypsodont dentitions in herbivorous mammals (e.g., Fortelius et al. 2002; Damuth and Janis 2011, 2014; Strömberg et al. 2013; Madden 2015). Whatever the ultimate cause, hypsodonty in extant ungulates is nonetheless most common and prominent in grazing forms such as modern horses.

However, the correlation between hypsodonty and grazing only reflects evolutionary adaptation and it does not give more detailed information about dietary variation within species, sometimes even between species. There are several exceptions to the hypsodonty-grazing connection: the extant hippopotamus (*Hippopotamus amphibius*) has brachydont molars despite being a grass-dominated feeder and some hypsodont mammals, such as the extant pronghorn (*Antilocapra americana*) and the takin (*Budorcas taxicolor*) have browse-dominated mixed-feeding diets in their extant populations. Many small mammals developed ever-growing, rootless (hypselodont) dentition to cope with very high tooth wear rates, but in large herbivorous mammals this only happened in a few specialised species, the most intriguing examples being the Plio-Pleistocene rhinoceroses of the genus *Elasmotherium* and many endemic South American notoungulates (Ortiz-Jaureguizar and Cladera 2006). Why hypselodonty evolved only in a few specialised lineages of ungulates is due to the problem of maintaining a complex occlusal morphology once enamel has worn away, unless new enamel is constantly erupting (Janis and Fortelius 1988). A notable exception to the scarcity of hypseolodonty in large herbivorous mammals is the Xenarthra (the major clade of placental mammals including sloths, anteaters and armadillos), all of which have hypselodont dentitions, because, as their teeth lack enamel, their tooth wear rate is high regardless of the abrasiveness of their diet (Vizcaíno 2009).

2.2.1.3 Tooth Morphology and Complexity of Enamel Patterns in Large Herbivorous Mammals

Another way to increase the efficiency of molar teeth is to increase the complexity of enamel patterns on the occlusal surface (Janis and Fortelius 1988). This creates an increasingly complex system of enamel ridges for shearing through tough plants, such as fibrous browse and especially grass.

Mammal teeth are complex in structure and require precise occlusion to function effectively (Ungar 2010). However, this complex system of tooth structure and chewing enables herbivorous mammals to process many kinds of plant foods effectively. The early evolution of mammal dentition during the Mesozoic saw the development of offset upper and lower molar positions (enabling precise occlusion between the upper and lower dentitions), and a triangular construction where three principal cusps in upper molars (protocone, paracone and metacone) form a triangle (trigon) matching a similar (but reverse) construction in the lower molars (trigonid) (Ungar 2010). In more derived tribosphenic molars, a talonid basin was formed in the lower molars by the addition of hypoconid, entoconid and hypoconulid cusps behind the trigonid. Finally, in quadritubercular or eutheromorphic molars, the evolution of another cusp (hypocone) formed the plesiomorphic upper molar morphology of most omnivorous/herbivorous mammals consisting of four principal cusps, while in the lower molars the trigonid and talonid basins became equally sized an often partially fused, also becoming surrounded mainly by four principal cusps (Ungar 2010). This was the starting point of the evolution of herbivorous mammal dentitions during the Cenozoic.

The earliest ungulates, and many other groups of herbivorous mammals such as proboscideans, started their dental evolution from brachydont (low-crowned) molars with the four principal cusps as separate knobs on the tooth surface (bunodont dental morphology) (e.g., Rose 2006; Ungar 2010). These kinds of bunodont dentitions are typical for omnivorous and frugivorous (fruit eating) mammals, and are still seen today (although usually with more complex pattern of cusps) in omnivorous herbivores such as the wild boar (*Sus scrofa*) and many other suids (e.g., Janis 2008; Ungar 2010). Bunodont molars function best in rather unspecialised crushing of various food items (including fruit, seeds and other plant material), whereas the more derived lophodont, selenodont and plagiolophodont molar morphologies of browsing and grazing herbivorous mammals are more optimal for cutting and shearing tough plant material such as branches, leaves and grass (Ungar 2010). When herbivorous mammals started to adapt to consuming tough plant material (dicotyledonous leaves rather than fruit and other softer plant parts), increasingly effective cutting ridges started to evolve between the cusps on the tooth surfaces. This commonly led to the evolution of bilophodont molars where the anterior pair and the posterior pair of principal cusps are connected by two transverse cutting blades, or lophodont (trilophodont or ectolophodont) molars where, in addition to the transverse lophs, there is a prominent outer loph (ectoloph) running longitudinally on the buccal (cheek) side of the tooth (e.g., Fortelius 1985). With the consumption of increasingly tough dietary items, especially grass, these features became more complicated with complex folding and fusion of the lophs, and sometimes added lophs or other ridges (Janis and Fortelius 1988; Janis 2008). These molars with complex folding and fusion of lophs are called plagiolophodont, and they are often further strengthened by extensive dental cement filling gaps between the lophs (Fortelius 1985).

Many artiodactyls, most importantly ruminants and camels, evolved another kind of advanced tooth morphology known as selenodonty (analogous to lophodonty), where the four principal cusps of the molars are elongated into crescent-shaped, longitudinal crests (analogous to lophs), enabling effective processing of plant material (Fortelius 1985). Also selenodont molars often became structurally more complex (becoming convergent to plagiolophodonty) in grazers, such as many grazing bovids (Fortelius 1985). One recently discovered dental adaptation specifically for processing grass-dominated diets in woodland and wetland habitats is the structural fortification of cusps (Žliobaitė et al. 2016, 2018). Molars with structurally fortified cusps are particularly typical in many selenodont ruminants that live in woodland and wetland habitats and feed on fresh grasses, such as water buffalos, reduncine antelopes and axis deer, but similar adaptations are also seen in suids, hippos and other herbivorous mammals (Žliobaitė et al. 2018).

These are the common ways that most large herbivorous mammals, such as ungulates and marsupials, increased the complexity of the occlusal surface of their molar teeth. However, slightly different processes occurred in proboscideans (elephants and their fossil relatives) and herbivorous xenarthrans (sloths and armadillos). The proboscideans started as bunodont, but quite early in their evolution their molars evolved extra pairs of cusps, which were arranged so that they form multiple transverse rows or "lophids" together with smaller accessory cusps, except for deinotheres (Deinotheriidae), which had bilophodont molars similar to some browsing ungulates such as tapirs (Sanders et al. 2010). In mammutid proboscideans the cusp pairs were compressed and formed cutting lophs. Finally, in true elephants, the cusp rows or lophids fused into narrow ridges called lamellae or "plates", which multiplied in number and were bound together with extensive dental cement (loxodont or lamellar tooth morphology). In elephants, an increase in hypsodonty and lamellar count through time also occurred within genera, such as in the mammoth lineage (*Mammuthus*) (Lister et al. 2005). The xenarthrans had peculiar dentitions without the hard, wear-resistant enamel, so their molariform teeth are often simplified in shape (Vizcaíno 2009). Nonetheless, differences between the hardness of differentiated dentine in xenarthran teeth created ridges analogous to the enamel ridges in other placental teeth as the teeth were worn, such as the bilophodont worn shape typical of browsing sloths and the multi-lobed morphology of the glyptodonts (giant herbivorous relatives of armadillos).

2.2.2 Dietary Proxy Methods

2.2.2.1 Microwear

Microwear analysis is the first proxy method for reconstructing diets of fossil mammals based on tooth wear (Walker et al. 1978), by counting the abundance of different kinds of microscopic scratches and pits on tooth enamel caused by chewing different kinds of food objects (Fig. 2.1). The hypothesis behind that method is that during the tooth wear abrasive plant material (such as grass phytoliths) cause long scratches on the worn enamel facets of the teeth, whereas browse does not: in the latter case the microscopic wear pattern is more pitted. The benefit of this method is that it is applicable to virtually all kinds of teeth and it should give consistent results for mammals with very different tooth morphologies. However, microwear analysis

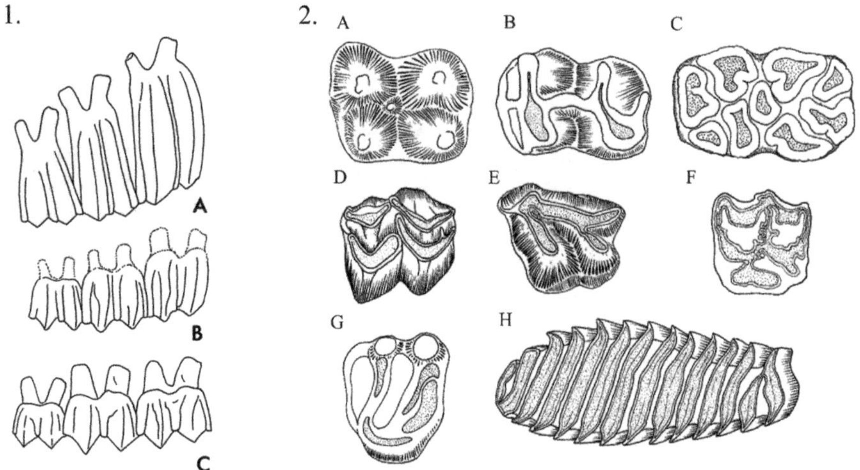

Fig. 2.1 Examples of hypsodonty and molar morphology in herbivorous mammals. (**1**) Hypsodonty categories: A = hypsodont, B = mesodont, C = brachydont. Image from: Fortelius et al. (2002). (**2**) Examples of tooth morphogy types of herbivorous mammals. A. Bunodont (peccary). B. Bilophodont (kangaroo). C. Columnar (warthog). D. Selenodont (deer). E. (Ecto-) Lophodont (rhinoceros). F. Plagiolophodont (horse). G. Bunolophodont (rodent—woodchuck). H. Lamellar (rodent—capybara). Image from: Janis and Fortelius (1988). (Permits to re-use the figures have been received from the original authors and publishers)

only reveals the last few meals of the animal instead of long-term average dietary signal (e.g., Rivals et al. 2010). Furthermore, it has been suggested that other factors than diet, such as external soil material from the environment, may obscure the dietary signal provided by microwear analysis (Rivals et al. 2010). On the other hand, microwear can, for example, reveal feeding on hard seeds, which the other proxy methods do not pick up (Rivals et al. 2012). Ungar et al. (2003) developed a new practical method for analysing microwear surface textures by combining confocal microscopy with scale-sensitive fractal analysis (the dental microwear texture analysis, Scott et al. 2005). This methodological improvement has made dietary analyses based on microscopic tooth wear patterns more objective compared to the original method based on visual counting of scratches and pits.

2.2.2.2 Mesowear

Mesowear analysis (Fortelius and Solounias 2000) is based on the empirical observation that increasing abrasiveness of plant material (especially grasses) will wear the tooth cusps blunter and lower as compared to non-abrasive plants which allow tooth-to-tooth wear (attrition) to maintain high occlusal relief and sharp cusps, in molar teeth that are in a similar state of wear (Fig. 2.2). Why grasses are especially abrasive on tooth enamel is still not totally understood, but whether it is because of high phytolith contents, external grit accumulating on grass leaves or even simply

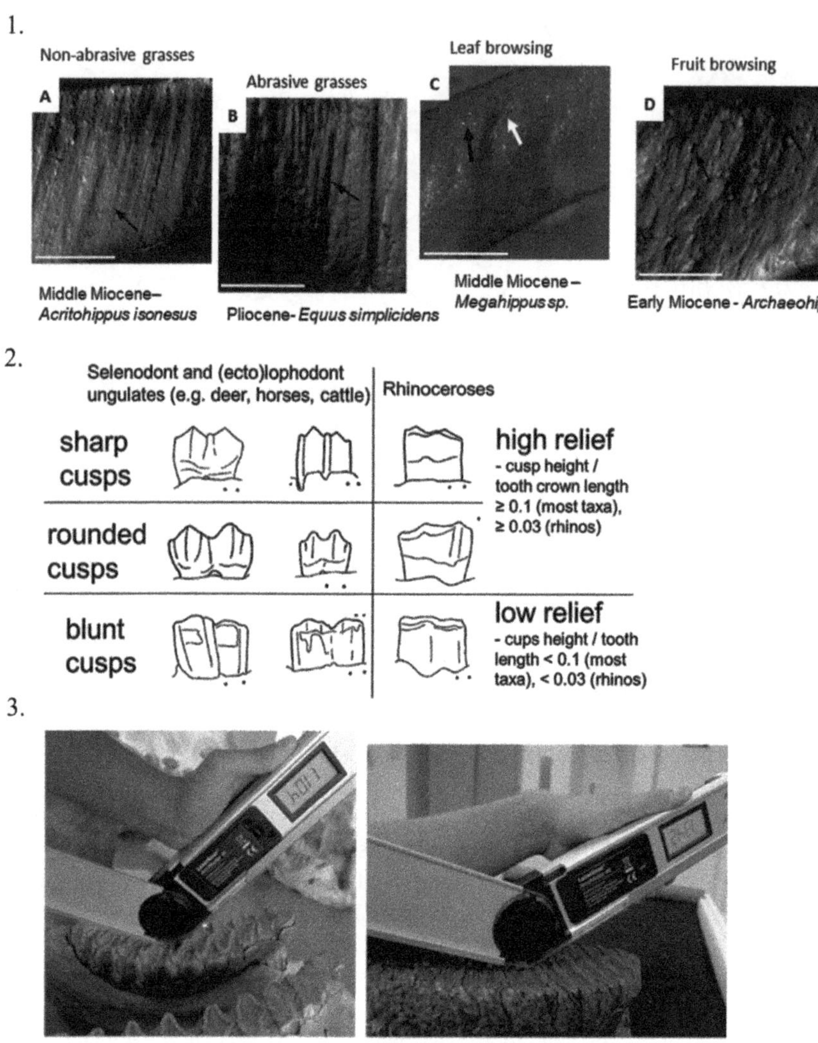

Fig. 2.2 Tooth wear -based dietary analysis methods. (**1**) Examples of microwear: more scratched microwear signal (A and B) indicates grazing and more pitted (C) browsing, whereas fruit eating (D) is indicated by gouges. Image modified from Semprebon et al. (2016a). (**2**) Traditional mesowear scoring. The shape of the cusps is scored as sharp, rounded or blunt, and the relief as high or low. Image modified from Fortelius and Solounias (2000). (**3**) Mesowear angle analysis. The relief of the worn enamel ridges of e.g., proboscidean molars can be measured as angles using a digital angle meter. Sharper angles (as for the modern African elephant, *Loxodonta africana*, molar here) indicate more browsing diet than blunter angles (as for a woolly mammoth, *Mammuthus primigenius*, molar here). Photos: Juha Saarinen, Tsavo Research Station an American Museum of Natural History. (Permits to re-use the figures have been received from the original authors and publishers)

the coarse fibres of plant leaves (e.g., Lucas and Omar 2012; Damuth and Janis 2011; Lucas et al. 2014), the empirical basis of this observation is extensive (e.g., Fortelius and Solounias 2000; Kaiser et al. 2013; Saarinen and Lister 2016). Mesowear analysis is done by visually scoring the cusp shape as sharp, rounded or blunt and the relief (the relative height of the cusps) as high or low. When large samples of ungulate molar teeth are analysed, the mesowear scores reflect the relative amount of abrasive material (mostly grass) in the diet of the ungulate species and local populations/assemblages. The method is easy to use, it can be readily applied to fossil ungulate molars and it gives a robust signal of the abrasiveness of the diet of an herbivorous mammal over a relatively long period of its life. Blind tests have shown that the mesowear analysis based on the traditional visual observing is mostly consistent between observers (Loffredo and DeSantis 2014).

Since the introduction of the mesowear analysis (Fortelius and Solounias 2000), the method has been extended to cover lower molars of rhinoceroses (Hernesniemi et al. 2011), lower molars of ruminants (Fraser et al. 2014), marsupial teeth (Butler et al. 2014) and rodent and lagomorph teeth (Ulbricht et al. 2015). Moreover, a "mesowear ruler" with references to more detailed cusp shape categories has been used (Mihlbachler et al. 2011), and the cusp shape and relief can be transferred into univariate mesowear scores/values to facilitate statistical analyses and comparisons (e.g., Rivals et al. 2007, 2010; Saarinen et al. 2016). A further recent development is the new approach called "mesowear III", where the mesial and distal shape of the inner rather than outer enamel band of buccal cusps/lophs is analysed (Solounias et al. 2014).

2.2.2.3 Extending the Mesowear Method: Mesowear Angle Analysis

The main limitation of the traditional mesowear analysis, based on visually observing the relief and shape of the worn cusps, is that it can only be done for teeth with easily observable outer edge from which those features are easy to observe, such as the selenodont and (ecto-) lophodont molars of perissodactyls and ruminants (Fortelius and Solounias 2000). However, the principle can be extended to cover other types of tooth morphology.

Proboscideans (elephants and their fossil relatives) have specialised transverse-lophed or lamellar molars where the relief and shape of the worn lophids/lophs/lamellae cannot easily be visually observed, and this is further complicated by the specialised fore-aft chewing cycle of the elephants where the lophs/lamellae meet obliquely to the direction of the lophs (Maglio 1973; Saarinen et al. 2015). However, differences in the relief of worn enamel ridges on the molar surface develop in proboscidean molars, and these are related to diet following the principle behind the mesowear analysis: increasingly abrasive foods (usually grass) wear the enamel ridges lower in relation to the softer dentine valleys in the lophs, creating lower relief on the molar surface. This relief can be measured as angles placed at the bottom of the worn dentine valleys on the molar surfaces (Saarinen et al. 2015; Fig. 2.2(3)). The blunter the angles are, the lower is the relief, indicating an increasingly abrasive, grass-based diet. This method has also been successfully utilised for palaeodietary analyses of xenarthrans, which do not have enamel in their teeth, but develop different patterns of occlusal relief on the worn teeth, which are similarly related to grass vs. browse (Saarinen and Karme 2017). As a result of these and other methodological developments, the mesowear analysis can now be used for palaeodietary analyses of practically all herbivorous mammals from ungulates (Fortelius and Solounias 2000; Hernesniemi et al. 2011; Fraser et al. 2014) to proboscideans (Saarinen et al. 2015), xenarthrans (Saarinen and Karme 2017), marsupials (Butler et al. 2014) and small mammals (Ulbricht et al. 2015).

2.2.2.4 Stable Isotopes

Stable isotope fractions in tooth enamel offer proxy methods for dietary analyses of several mammal groups. Values of $\delta^{13}C$ and $\delta^{15}N$ isotopes differ in C3-photosynthetisising plants and C4-photosyhtetisising plants (most of which in the tropics are grasses), so they leave different isotope fractions in the teeth of herbivorous mammals eating them (Lee-Thorp and van der Merwe 1987; Cerling et al. 1997). This methodology can only be used for comparing the fractions of C3 and C4 plants in the diets, but it is useful because strong C4 signal reflects diets based on grasses in tropical and subtropical areas where C4 grasses dominate (Cerling et al. 1997). However, a limitation of this method for palaeodietary analyses is that it only reflects grazing vs. browsing more or less directly in tropical and subtropical areas where most grasses are C4 photosynthesising, and even there only when C4 grasses started to become abundant in the Late Miocene ca. 10 Ma (e.g., Cerling et al. 1997, 2015). Another limitation is that not all grasses, even in the tropics, are C4 photosynthesizing. Thus, tooth wear -based palaeodietary analyses (microwear and mesowear) are needed to evaluate feeding on C3 photosynthesizing grasses.

2.3 The History of Ungulate Diets and Evolution Throughout the Cenozoic Fossil Record

2.3.1 Palaeocene (65–56 Ma): The Time of Archaic Ungulates, or "Condylarths"

Grasses originated long before they became a prominent part of the diets of herbivorous mammals. The earliest fossil record of grasses dates back to Late Cretaceous, and fossilised grass phytoliths have even been discovered in the coprolites (fossil dung) of sauropod dinosaurs from the Late Cretaceous of India (Prasad et al. 2005). However, for a very long period (in most parts of the world until the Miocene, ca. 23–5 Ma), grasses were a relatively minor element of plant communities and they did not form extensive grassland ecosystems (e.g., Strömberg 2011).

After the extinction of large non-avian dinosaurs in the Late Cretaceous mass extinction ca. 65 Ma, mammals started to diversify and fill in the ecological niches of large terrestrial animals (Alroy 1998; Smith et al. 2010; Raia et al. 2013). Global temperatures increased drastically during the Palaeocene, leading to the exceptionally warm "Palaeocene-Eocene Thermal Maximum" at the end of this epoch (Zachos et al. 2001, 2008). Consequently, tropical and subtropical forest environments were the dominant habitats globally (e.g., Strömberg 2011). Among herbivorous mammals, perhaps the most prominent event was the emergence and radiation of early ungulates, often grouped together in a paraphyletic assemblage called the "Condylarthra", which are characterised by brachydont (low-crowned) and bunodont (cusps as separate knobs) dentitions, suggesting adaptation to omnivorous, relatively generalist diets (Rose 2006). The condylarths were globally widespread during the Palaeocene, being the dominant medium- to large-sized mammals in Eurasia, North America and South America, and they ranged from small forms with weasel-like bodies (*Hyopsodus*) to larger, up to about sheep-sized omnivorous forms (e.g., *Arctocyon* and *Phenacodus*). Some of the later condylarths, such as *Meniscotherium*, show the development of increasingly lophodont (cusps elongated and fused into ridges) dentitions, suggesting increasing adaptation to browsing herbivorous diets (Williamson and Lucas 1992).

Other groups of large herbivores, probably unrelated to modern ungulates, emerged too, the ecologically most important being the pantodonts, which evolved dilambdodont (upper molars with a W-shaped ectoloph on the buccal side) or functionally bilophodont (molars with two transverse crests) dentitions suggesting adaptation to frugivorous-folivorous or browsing diets (Rose 2006). The pantodonts originated in Asia and spread to North America where they became particularly diverse, including various species ranging from relatively small (e.g., *Pantolambda*), to large, roughly tapir to bison-sized forms (*Barylambda*, *Coryphodon*, *Titanoides*) (Rose 1981, 2006; Beard 1998). Another group of large, bilophodont herbivores, the Dinocerata, emerged in Asia (*Prodinoceras*), and soon dispersed to North America (Rose 1981, 2006; Beard 1998). In addition, there were a couple of groups with a body build suggesting fossorial lifestyles and sturdy skulls and dentitions suggesting diets based on gritty roots and tubers. These were Tillodonta, which originated in Asia and dispersed to North America, and Taeniodonta, an endemic North American group (Lucas and Schoch 1998; Lucas et al. 1998). In South America, an enigmatic group of ungulate-like mammals called Xenungulata existed during the Palaeocene, and some of them (e.g., *Carodnia*) were quite large, roughly tapir-sized herbivores with bilophodont molars suitable for browsing (Rose 2006).

2.3.2 *Eocene (56–33.9 Ma): The Emergence and Radiation of Modern Mammal Lineages*

The Eocene was the longest epoch of the Cenozoic era, lasting ca. 22 million years, and drastic changes happened in mammal faunas throughout this time. Ungulates had started to diversify during the Late Palaeocene, and during the Early Eocene climatic optimum most of the major modern orders of herbivorous mammals emerged, including the even-toed ungulates (Artiodactyla), the odd-toed ungulates (Perissodactyla) and endemic South American ungulates (Meridiungulata). The proboscideans (Proboscidea) and hyraxes (Hyracoidea) also emerged as a result of a separate radiation of the Afrotherian ungulate-like herbivores (Paenungulata). In addition, several archaic groups of medium- and large-sized herbivorous mammals flourished especially during the Early and Middle Eocene, such as Pantodonta, Dinocerata, Tillodontia, Taeniodonta and the last archaic "condylarth" ungulates. The diversity of mammals was higher on both the species and higher taxonomic levels during the Middle Eocene than ever before or after (Janis 1993; Blondel 2001; Saarinen et al. 2014). The global climate during the Early Eocene was warmer than ever during the Cenozoic, and subtropical forest environments reached very high latitudes (e.g., Erlebe and Greenwood 2012). During the latter part of the Eocene, the global climate started to cool, but still remained warm and humid, much warmer than later during the Cenozoic (Zachos et al. 2001, 2008). Grasses were a rare component of the floras globally throughout the Eocene. No specialised grazers have been discovered from the Eocene. Nevertheless, there was variation in the diets of the herbivorous mammals and in general there was a shift towards folivorous browsing diets throughout the Eocene among many lineages that started as omnivores or frugivores.

2.3.2.1 Eurasia

The climate during the Early and Middle Eocene was warm and humid, and global sea level was high, making Europe a large archipelago which had connections to North America and Asia. The mammal fossil assemblages of this time, such as that from the locality of Abbey Wood, UK, indicate palaeoenvironments essentially similar to modern tropical rainforests (Hooker and Collinson 2012). There was a change in European large herbivore faunas at the beginning of the Eocene when some taxa, especially the large bilophodont pantodont *Coryphodon eocaenus*, arrived from North America (Hooker and Collinson 2012). This species was the largest browser in the Early Eocene European forests. It shared its habitat with some late surviving condylarths, which had bunodont dentitions and probably omnivorous or frugivorous diets, such as *Hyopsodus* and *Phenacodus*, another newcomer from North America. However, the Early Eocene large herbivorous mammal faunas were dominated by early perissodactyls, including early and small horses (Equidae) such as *Pliolophus vulpiceps*, palaeotheres (Eurasian relatives of horses) such as *Hyracotherium leporinum*, large tapiromorphs with bilophodont dentition known as lophiodonts (genus *Lophiodon*) and other smaller tapiromophs such as Hyrachyidae (e.g., Blondel 2001). The dentition of most of these early perissodactyls, together with the less abundant early artiodactyls represented by *Diacodexis* and other forms derived from it, was brachydont and still predominantly bunodont or buno-lophodont, suggesting that they were largely frugivores (Blondel 2001). The key to why the early perissodactyls were more diverse and abundant than the early artiodactyls is probably the difference in the evolution of digestive systems in these major ungulate groups: the perissodactyls developed an enlarged caecum and colon for digesting plant material in the Early Eocene which made them more effective as browsers of leaves than the artiodactyls which probably had not yet developed an effective fermenting system (Janis 1976).

Middle Eocene palaeobotanical evidence suggests that the palaeoenvironments of Europe were similar to modern tropical rainforests (Collinson and Hooker 1987; Collinson 1992) and rich fossil sites, such as the exceptionally preserved assemblage of fossil plants, invertebrates, fish, reptiles, amphibians, birds and mammals from Messel, Germany, support that interpretation. The Middle Eocene large herbivorous mammal faunas from Europe were strongly dominated by what appear to be quite unspecialised frugivorous-folivorous and browsing perissodactyls, such as the buno-lophodont palaeotheriids (e.g., *Propalaeotherium*, *Palaeotherium* and *Plagiolophus*) and the large bilophodont lophiodontids (*Lophiodon*). Also one genus of early true horses (*Eurohippus*) is known from sites such as Messel, Germany. A frugivorous-folivorous diet has been confirmed for the palaeothere *Propalaeotherium* from Messel and Geiseltal by analysis of preserved stomach contents (Sturm 1978; Wilde and Hellmund 2010). However, the diversity of artiodactyls increased drastically as well, with many new lineages emerging, and browsing forms started to dominate over omnivorous and frugivorous forms (Blondel 2001). Among these were the first selenodont ungulates of the families Xiphodontidae and Amphimerycidae.

During the Late Eocene the environments of Europe changed somewhat as the result of the climatic cooling but they were still dominated by evergreen forests, albeit these were subtropical rather than tropical (Collinson and Hooker 1987; Collinson 1992). The herbivorous mammal faunas changed too and there was a shift from frugivorous-folivorous diets to increasingly specialised folivorous browsing in both perissodactyls and artiodactyls. The perissodactyls were dominated by browsing forms (e.g., large palaeotheriids such as *Palaeotherium magnum*), whereas some earlier groups such as the lophiodonts became extinct. True horses (Equidae) also vanished from Eurasia at this time, retuning much later in the Early Miocene with the dispersal of anchitherine equids from North America. The overall diversity of perissodactyls declined and artiodactyls became the dominant ungulates in Europe during the Late Eocene (Blondel 2001). New types of artiodactyls appeared, such as the small-sized selenodont cainotheres (Cainotheriidae) with probably frugivorous-folivorous diets, and the large-sized buno-lophodont anoplotheriids (*Anoplotherium*), which have been interpreted as a specialised, bipedally high-browsing animals somewhat similar in feeding ecology to the later chalicotheriid perissodactyls (Hooker 2007).

The Eocene palaeoenvironments of Eastern and Southern Asia were mostly dominated by forest and woodland palaeoenvironments (e.g., Zaw et al. 2014), and this is reflected in the mammal faunas which were dominated by diverse large browsers, especially perissodactyls (Wang et al. 2007). In Central Asia more arid environments, with drought-adapted shrubs, were present as early as in the Palaeocene, but those were still mixed with forest and woodland vegetation (Jacobs et al. 1999). Several major groups of ungulates originated in Asia during the Early and Middle Eocene and soon migrated to North America, and somewhat later, at the beginning of the Oligocene, to Europe. Among the perissodactyls, these were the amynodonts (Amynodontidae) and hyracodonts (Hyracodontidae), which were early relatives of rhinoceroses, and eomoropids which were early relatives of chalicotheres. True rhinoceroses (Rhinocerotidae) emerged during the Late Eocene in Asia and North America, with either of these continents considered to be the area of origin for that group (Prothero et al. 1989).

Even more important evolutionary radiations occurred among the artiodactyls in Asia during the Middle Eocene, with the emergence of several major groups, including the anthracotheres (Anthracotheriidae), a group of suiform artiodactyls with buno-selenodont dentition, related to later hippopotamids, with typically amphibious lifestyles (Lihoreau and Ducrocq 2007). Their dentition indicates diets based predominantly on unspecialised consumption of fruits, leaves and other plant material (Blondel 2001), but more specific palaeodietary analyses would be needed to further confirm this. Another group of suiforms was the large-sized, fully terrestrial and relatively cursorial entelodonts (Entelodontidae), which were characterised by large skulls and dentitions which comprised carnivore-like canines, strong triangular premolars and bunodont molars (Foss 2007). This kind of dentition indicates omnivorous diet with probably a large component of animal carcasses (Boardman and Secord 2013). The earliest entelodonts was the relatively small *Eoentelodon* from the Middle Eocene of China. Finally among the radiation of suiform artiodactyls was also the ancestral group of pigs (Suidae) and peccaries (Tayassuidae) known as Palaeochoeridae, with bunodont teeth and probably omnivorous diets (Liu 2001).

Perhaps most importantly regarding later Cenozoic mammal faunas, the ruminants originated in Asia during the Middle Eocene, being represented by small species with incipiently selenodont dentitions, such as *Archaeomeryx* and *Archaeotragulus* (Métais et al. 2001, 2005).

2.3.2.2 North America

During the Early Eocene (Wasatchian and Early Bridgerian), North America had a connection with the European archipelago, and, as a result, they shared many similar mammal taxa (Rose 1981; Janis 1993). The global temperatures were extremely high and the environments of North America were dominated by tropical forests and woodlands (Zachos et al. 2008; Strömberg 2011). Even arctic areas were covered by subtropical forest, as shown by the rich fossil assemblage containing, among others, fossil plants, fish, mammals, turtles and alligators from the High Arctic of Canada (Erlebe and Greenwood 2012). The large browsing bilophodont pantodont *Coryphodon* was widely distributed in North America, from where it dispersed to Europe. The ungulate faunas, indicated especially by the rich fossil assemblage from the Green River Formation, Wyoming, were dominated, for the first time, by early horses (Equidae) represented by several small species with very low-crowned, bunolophodont molars (Rose 1981). Their dental morphology indicates a dietary adaptation to feeding on fruit and browse (e.g., Janis 2007), and mesowear analysis indicates a diet that was more abrasive than in later browsing anchitheriine equids, also suggesting the consumption of fruits with hard seeds (Mihlbachler et al. 2011). Originally they were all included in the genus *Hyracotherium*, but actually comprise several genera and species such as *Eohippus*, *Sifrihippus* and *Protorohippus*, possibly including taxa more closely related to the European palaeotheres (Froehlich 2002). The next most abundant taxa, which had been the dominant ungulates in the Late Palaeocene, were the condylarths *Phenacodus*, *Ectocion* and *Hyopsodus*, of which *Phenacodus* and *Hyopsodus* were also present in Europe. The first artiodactyls were represented by the small diacodexeids with unspecialised bunodont dentition, such as *Diacodexis* and *Bunophorus* (Rose 1981). In addition to the abundant early equids, the perissodactylan radiation in North America included early tapiromorphs (e.g., *Homogalax*) and early brontotheres (*Eotitanops borealis*), which were small compared to the later very large forms of their clades and had relatively simple buno-lophodont dentitions and probably rather unspecialised fruit and browse diets. *Eotitanops* was replaced by a somewhat larger (tapir-sized) but otherwise quite similar genus *Palaeosyops* during the latter Early Eocene. As in Europe, most of these early herbivores were probably largely frugivore-folivores. The dinoceratans dispersed from Asia to North America during the Early Eocene and were represented first by the prodinoceratid *Probathyopsis* and later by the larger uintathere *Bathyopsis*. Their bilophodont dentitions indicate browsing diets (Rose 2006). Tillodonts of the genus *Esthonyx* were also present and may have been fossorial animals. Their heavy tooth wear has been interpreted to have resulted from grit associated with roots and tubers and they may have used their enlarged incisors to pull roots from the ground, as indicated by wear marks (Lucas and Schoch 1998).

Middle Eocene large herbivorous mammal faunas from North America were extremely diverse, including many lineages of early perissodactyls (equids, brontotheriids, hyracodonts, amynodonts, eomoropids and various tapiromorphs) and artiodactyls together with archaic groups of roughly ungulate-like herbivores with uncertain phylogenetic relationships, such as Tillodontia and Dinocerata (e.g., Janis 1993; Gunnell et al. 2009). Palaeoenvironmental proxy data indicates the dominance of subtropical-tropical forest environments, although some continental inland areas were starting to change towards drier and more open woodland conditions, although grasses were largely absent in all the environments (Strömberg 2011). The equids were still represented by small archaic "hyracotheres" with buno-lophodont dentitions and frugivorous-folivorous diets, such as *Orohippus* and *Epihippus*, but they were less diverse than in the Early Eocene. In contrast, the brontotheriids radiated into several medium- to large-sized genera, such as *Palaeosyops*, *Telmatherium*, *Mesatirhinus*, *Metarhinus*, *Dolichorhinus* and *Diplacodon*. The brontotheres had brachydont, buno-lophodont teeth with a prominent ectoloph, and typically high and sharp mesowear signals, indicating browsing diets (Mihlbachler 2008; Saarinen, own observations). Other perissodactyls included various tapiromorphs (e.g., *Homogalax*, *Helaletes* and *Hyrachyus*) and early relatives of later chalicotheriids (a group of highly specialised clawed perissodactyls) known as eomoropids (*Eomoropus*). These taxa also had buno-lophdont or incipiently bilophodont molars and were most likely quite unspecialised browsers of fruits and leaves. A new group of perissodactyls, the rhinocerotoid amynodonts arrived from Asia, represented by the genus *Amynodon*, which had more specialised lophodont molars roughly similar to those of true rhinoceroses and were most likely browsers (Wall 1998). However, unlike most early rhinocerotoids, some later amynodonts during the Late Eocene—Early Oligocene (*Metamynodon* in North America and *Cadurcotherium* in Eurasia), evolved more hypsodont molars, but whether this was an adaptation to feeding on gritty or particularly tough plants is, as yet, not well understood (Prothero et al. 1986; Blondel 2001). On the other hand, some late amynodonts, such as *Cadurcodon*, apparently had a tapir-like proboscis, suggesting a specialised browsing rather than grazing diet (Prothero et al. 1986).

Among the artiodactyls, three major new groups endemic to North America emerged: the merycoidodonts (oreodonts) and agriochoerids (Oreodontoidea) represented in the Middle Eocene by *Protoreodon*, the protoceratids (Protoceratidae) represented by *Leptomeryx* and *Leptoreodon* and camels (Camelidae) represented by *Protylopus*. These lineages of selenodont artiodactyls were probably all part of the Tylopoda, of which only camels survive today, and became the dominant groups of large herbivorous mammals in North America during the Oligocene. The last condylarths were represented by the small-sized, squirrel-like genus *Hyopsodus*. The dinoceratans were represented by advanced uintatheriids, which evolved into large-sized taxa such as *Uintatherium* and *Eobasileus*. Some of these were heavily built, megaherbivore-sized (over a ton in body mass) herbivores, and they had robust skulls with bony horns, long and blade-like upper canines and bilophodont molars suitable for leaf browsing (Rose 2006). The presumably root digging tillodonts were represented by the largest, about bear-sized genus *Trogosus*.

Unlike in Europe but similarly to Central Asia, there was a drastic change in the Late Eocene in North America towards substantially more open environments, at least in what is today the western inland of USA (Strömberg 2011). However, the paleoenvironments were still predominantly open woodlands and shrublands, with grasses being present but not yet forming extensive grassland environments (Strömberg 2011). These environments supported diverse herbivore communities with predominantly browsing diets (Boardman and Secord 2013; Mihlbachler and Solounias 2006; Mihlbachler et al. 2011). Probably the best fossil Late Eocene fossil record in North America comes from the famous White River Badlands of western USA, where the deposits range from Late Eocene to Early Oligocene and record a change in climate, vegetation and mammal assemblages across the Eocene-Oligocene boundary. The large herbivorous mammal fauna included the latest and very large-sized brontotheres (*Megacerops*), amynodonts (*Metamynodon*), hyracodonts (*Hyracodon*), early true rhinoceroses (*Trigonias* and *Subhyracodon*) early anchitheriine equids (*Mesohippus*), anthracotheres (*Aepinacodon*, *Elomeryx*), entelodonts (*Archaeotherium*), peccaries (*Perchoerus*), early small camels (*Eotylopus* and *Poebrotherium*), agriochoerids (merycoidodont-like selenodont artiodactyls represented by *Agriochoerus*), merycoidodonts (*Oreonetes*, *Bathygenys* and *Merycoidodon*), protoceratids (*Pseudoprotoceras*) and early traguloid ruminants

(*Hypertragulus* and *Leptomeryx*). Mesowear analysis indicates that the Late Eocene merycoidodonts had somewhat more abrasive diets than the very browse-dominated diets in the abundant Early Oligocene merycoidodonts, reflecting possibly a higher level of mixed-feeding, or feeding on leaves and fruits with seeds (Mihlbachler and Solounias 2006). The protoceratids, a camel-related group of selenodont artiodactyls, had tooth wear typical for browsers (Janis 1982). The small anchitheriine horse *Mesohippus* had more specialised lophodont molars than its "hyracothere" predecessors, and mesowear analysis indicates that it had a diet based largely on dicotyledonous leaves, rather than the more generalised diet of the "hyracotheres" based on leaves and fruit with abrasive seeds. Based on various analyses of dental mesowear, microwear, morphology and stable isotopes, most of the rest of the ungulate fauna were also browsers, but some such as the early rhinoceroses *Subhyracodon* and *Trigonias*, the anthracothere *Aepinacodon* and the early camel *Eotylopus* show some indication of browse-dominated mixed-feeding (Boardman and Secord 2013). Based on microwear analyses, the bunodont peccary *Perchoerus* was a hard object browser and the entelodont *Archaeotherium* was a high-abrasion omnivore (Boardman and Secord 2013). Based on stable isotope analyses, the large browsing brontothere *Megacerops* and the peccary *Perchoerus* were feeding in more closed habitats, and the equid *Mesohippus* and the agriochoerid *Agriochoerus* in more open habitats, than the rest of the cohabiting fauna (Boardman and Secord 2013).

2.3.2.3 Africa

Much less is known about the Eocene mammal faunas and palaeoenvironments of Africa than those of Eurasia and North America. The palaeoenvironments were mostly dominated by forests, woodlands and marshes, although there are indications of more open environments and presence of (possibly wetland) grasses in the Middle Eocene of Western Africa (Jacobs et al. 2010) and Late Eocene of Northern Africa (Strömberg 2011). The latest Palaeocene—earliest Eocene deposits of Ouled Abdoun basin in Morocco comprise enigmatic condylarth-like mammals (*Abdounodus* and *Ocepeia*) and the earliest proboscideans (*Eritherium*, *Phosphatherium* and *Daouitherium*). The El Kohol deposits from Algeria, slightly later Early Eocene in age, contain the larger early proboscidean *Numidotherium* and the basal hyracoid *Seggeurius*. Nothing much is known about the diets of these early African large mammals, but their brachydont, bunodont dentitions indicate omnivorous or frugivorous feeding ecologies, the only exception being *Numidotherium*, which shows bilophodont molars better suited for processing folivorous browse (Rasmussen and Gutiérrez 2010; Sanders et al. 2010).

The best Eocene mammal record from Africa comes from the Late Eocene and Early Oligocene Fayum Beds from Egypt, which will be discussed further in the next section for the Oligocene.

2.3.2.4 South America and Australia

Also South America and Australia remain relatively enigmatic in terms of herbivorous mammal diets during the Eocene. South American palaeoenvironments were dominated by tropical to temperate forests, and grasses were sparse, although fossil phytoliths indicate the presence of grasses in Patagonia (Strömberg 2011). Early and Middle Eocene palaeobotanical records from Australia indicate rainforest and sclerophyllous forest environments in northwestern, central and southeastern parts of the continent (Strömberg 2011). During the Late Cretaceous and earliest Cenozoic there was a land connection between these continents, which enabled some early marsupials to colonise Australia from South America (Woodburne and Case 1996).

The endemic South American ungulates (Meridiungulata) diversified during the Eocene into various species of medium- and large-sized notoungulates and litopterns, which had lophodont dentitions resembling those of perissodactyls (Rose 2006). Some of the notoungulates had already evolved hypsodont cheek teeth earlier than the ungulates on other continents, suggesting perhaps a diet based on relatively fibrous plants, or the accumulation of volcanic ash on the plants they were feeding on, rather than feeding on grasses which were scarce in the still quite heavily forested palaeoenvironments (Janis 1993; Ortiz-Jaureguizar and Cladera 2006; Reguero et al. 2010; Strömberg 2011; Strömberg et al. 2013).

We know next to nothing about the diets and dietary adaptations of Eocene mammals in Australia. A probable condylarth ungulate, *Tingamarra porterorum*, from the Early Eocene of Queensland (Godthelp et al. 1992) had buno-lophodont molars indicating perhaps a frugivorous-folivorous diet, but the fossil mammal record from Australia enables more thorough glimpses to the evolution of herbivorous mammal diets on the continent earliest from Late Oligocene onwards.

2.3.3 Oligocene (33.9–23 Ma): A Time of Global Climatic Cooling

The Oligocene is characterised by a cooling of global climate which started during the Late Eocene, after the globally extremely warm “greenhouse” climate of the Early and Middle Eocene (Zachos et al. 2001, 2008). This cooling trend also resulted in the Antarctic glaciation. However, global temperatures were still much higher than in the late Neogene, especially Pleistocene, and most of the globe was still dominated by subtropical and temperate forest and woodland environments, although more arid and open environments developed e.g., in Central Asia and central North America, and C4 grasses originated, possibly in response to a significant drop in atmospheric CO_2 content (Vicentini et al. 2008; Strömberg 2011). Herbivorous mammal faunas across the globe still consisted predominantly of browsers. However, in the Late Oligocene, some semi-open, grassy woodlands occurred in South America, e.g., in Salla, Bolivia (Strömberg 2011), where some of the earliest grazing ungulates and mylodontid sloths occurred (Croft and Weinstein 2008; Shockey and Anaya 2011).

2.3.3.1 Eurasia

Europe remained predominantly forested throughout the Oligocene, although the vegetation changed towards more temperate conditions (Strömberg 2011). The beginning of the Oligocene in Europe was marked by a turnover of the mammal faunas known as the "Grande Coupure", where new mammal taxa migrated from Asia to Europe as the result of marine regression that opened a land connection between Europe and Asia. There are rich European Early Oligocene mammal assemblages, e.g., from Ronzon and Phosphorites du Quercy, France, and Bouldnor Formation, UK, and Late Oligocene assemblages such as La Milloque, France, and Rickenbach, Switzerland. Among the most prominent newcomers were early ruminants with body sizes similar to modern tragulids and moschids, such as *Gelocus*, *Bachitherium*, *Amphimeryx*, *Lophiomeryx* and *Amphitragulus*, including musk deer -like genera, such as *Dremotherium* and *Bedenomeryx*. Other abundant ungulate groups include anthracotheres (e.g., *Anthracotherium*, *Bothriodon*, *Elomeryx* and *Microbunodon*), entelodonts (*Entelodon*), cainotheres (*Cainotherium*, *Cainomeryx*), xiphodonts (*Xiphodon*), early suids (e.g., *Palaeochoerus*), the last palaeotheres (*Plagiolophus*), late amynodonts and hyracodonts (relatives of rhinos, represented by *Cadurcotherium* and *Egyssodon*, respectively), true rhinoceroses (*Ronzotherium*) and tapirs (*Protapirus*). As far as we know, most of these medium and large-sized herbivorous mammals were browsers and had brachydont dentitions, but some of the early ruminants might already have started to shift towards mixed-feeding, as indicated by tooth wear studies (Novello et al. 2010). Moreover, the amynodont *Cadurcotherium* had evolved more high-crowned molars, but not much is known about its diet, although feeding on plants with grit or phytoliths in more open environments has been suggested (Blondel 2001).

Anthracotheres were a group of eco-morphologically pig or hippo-like ungulates with a tendency for amphibious lifestyles, characterised by elongate skulls and peculiar buno-selenodont molars which were brachydont and had large, separate cusps, but the cusps were elongated into longitudinal, crescent-shaped structures. They had originated in Southern Asia during the Eocene and migrated to Europe during the Eocene-Oligocene transition. Their closest living relatives are whales and hippopotami (Boisserie and Lihoreau 2006). The most likely diet of the European Oligocene anthracotheres has been interpreted as mixed fruit and browse based on the tooth morphology (Blondel 2001), but further palaeodietary analyses would be needed for more precise information about their diets and feeding ecology. The entelodonts, represented in the European Oligocene by the genus *Entelodon*, were by this time already very large (cow to bison-sized), fully terrestrial suiform artiodactyls with brachydont, bunodont molars and strong triangular premolars, indicating an omnivorous/scavenging diet (e.g., Blondel 2001). The early true pigs (Suidae) were small and also had bunodont teeth typical of omnivores.

The early ruminants had brachydont, selenodont molars and they ranged from small forms which probably had rather unspecialised frugivorous-folivorous diets (e.g., Amphimerycidae) to somewhat larger forms with more specialised folivorous browse diets, such as *Bachitherium*, *Bedenomeryx* and *Dremotherium* (Blondel 2001). The browsing ruminants most probably had partially developed foregut-fermenting stomachs (because those are present in all modern ruminants in some form) and were ruminating (Janis 1976; Blondel 1998, 2001). Cainotheres and xiphodonts, the small non-ruminant, tylopod-related selenodont artiodactyls, probably had rather unspecialised fruit and browse-based diets (Blondel 2001). Most of the perissodactyls in the Oligocene of Europe were browsers ranging from frugivorous-folivorous to folivorous, characterised by lophodont, brachydont dentitions (e.g., Blondel 2001; Joomun et al. 2008). Some of them, such as the palaeothere *Plagiolophus* and the amynodont *Cadurcotherium*, had more high-crowned cheek teeth, suggesting a diet based on more abrasive, perhaps fibrous browse (Blondel 2001).

In Central Asia climate became drier at the Eocene-Oligocene transition ca. 34 Ma, as indicated among other lines of evidence by the beginning of Aeolian dust accumulation in the Chinese Loess Plaetau (Sun and Windley 2015). The palaeoenvironments in Central Eastern Asia were a mosaic of wooded and drier, more open environments, where the vegetation probably consisted of abundant shrubs, whereas grasses were still scarce (Métais et al. 2003; Prothero 2013). There was a turnover in the Eastern Asian mammal faunas from the brachydont, perissodactyl-dominated fauna of the Eocene to faunas dominated by hypsodont lagomorphs and rodents in the Oligocene (Wang et al. 2007). This has been interpreted as further evidence for the spread of arid, open habitats. In Southern Asia, the paleoenvironments remained dominated by forests and woodlands, with rare grasses (De Franceschi et al. 2008).

The Asian Oligocene large mammal faunas were largely similar to the European ones in taxonomic composition, comprising anthracotheres (e.g., *Anthracotherium*), entelodonts (*Paraentelodon*), suoids (*Sanitherium*, *Hyotherium*) and early ruminants (e.g., *Gelocus*, *Lophiomeryx*), and abundant perissodactyls comprising hyracodonts (*Egyssodon*, indricotheres), amynodonts (*Cadurcodon*), true rhinoceroses (*Epiaceratherium*) and tapirs (*Colodus*) as well as the clawed chalicotheres (*Schizotherium*) (Wang 1992; Métais et al. 2003, 2009; Orliac et al. 2010; Prothero 2013). They mostly had brachydont dentitions and browsing diets, but a rare hypsodont, small (duiker-sized) bovid-like ruminant (*Palaeohypsodontus*) was present in China and Pakistan, suggesting an adaptation to feeding on abrasive, perhaps particularly fibrous plants (Métais et al. 2003).

One peculiar aspect of the Oligocene Asian faunas are the giant hyracodonts (rhino relatives) known as indricotheres, which evolved into some of the largest land mammals (e.g., *Paraceratherium*, *Indricotherium* and *Dzungariotherium*). They had lophodont, brachydont dentition typical of rhinoceroids and they were browsers (as shown by mesowear and stable isotope results), probably feeding from the canopy level of trees (Prothero 2013). During the Oligocene another major group of perissodactyls that occurred in Eurasia was the chalicotheres (Chalicotheriidae), a group of peculiar tapiromorphs with muscular bodies, elongate forelimbs, claws instead of hooves, relatively small skulls and brachydont, buno-lophodont dentition (Coombs 1978). During the Oligocene the chalicotheres were represented by small to medium-sized species of the genus *Schizotherium*. Studies of tooth mesowear and microwear support the traditional view based on tooth morphology that the chalicotheres (at least during the Miocene) were browsers feeding on leaves, but also included hard objects such as seeds or bark in their diet (Semprebon et al. 2011).

2.3.3.2 North America

The North American Oligocene (Orellan, Whitneyan and early Arikareean) faunas were dominated by the merycoidodonts (*Merycoidodon*, *Miniochoerus*, *Eporeodon*, *Leptauchenia* and *Merycochoerus*), with body sizes ranging from the about mouse deer-sized *Miniochoerus* to the about tapir-sized *Merycochoerus* (Lander 1998; Stevens and Stevens 2007; Janis 2008). All of them had selenodont molars, but some leptaucheniids such as *Sespia* evolved hypsodont molars earlier than most other groups of North American ungulates (Mihlbachler and Solounias 2006). Dental mesowear analysis indicates that the merycoidodonts had predominantly browsing diets in Early Oligocene, but shifted towards increasingly mixed-feeding diets during the Late Oligocene (Mihlbachler and Solounias 2006). The most abundant North American Early Oligocene mammal record comes from the famous White River Badlands, with the fossiliferous sediments spanning from Late Eocene to Early Oligocene, demonstrating some changes in the faunal composition, e.g., the extinction of the large browsing brontotheriid perissodactyls. The paleoenvironment of the Oligocene Badlands have been interpreted as semi-open woodlands and shrublands, roughly similar perhaps to the contemporaneous East Asian ones (Mead and Wall 1998; Strömberg 2011). The camels were represented in the Oligocene by small species such as *Poebrotherium wilsoni*, and the protoceratids by the genus *Protoceras*. The camels were mostly browsers, but *Poebrotherium* was a mixed-feeder of browse and grass, as indicated by dental mesowear and microwear (Semprebon and Rivals 2010). *Protoceras* was a folivorous browser with a long muzzle (Janis 1982; Prothero 1998). Many taxa of medium-sized to large herbivorous mammals were shared between the Late Eocene and Early Oligocene, such as the entelodonts (e.g., *Archaeotherium*), anthracotheres (*Elomeryx*), peccaries (*Perchoerus*) and the agriochoerid *Agriochoerus*, retaining similar diets as during the Late Eocene (Boardman and Secord 2013). However, a new taxon was the tapir *Colodon*, a newcomer from Asia, which was a browser occupying more closed environments in the Early Oligocene Badlands than the other ungulate taxa (Boardman and Secord 2013). Some taxa, such as the early traguloid ruminant *Leptomeryx*, probably occupied more open environments in the Early Oligocene than in the Late Eocene, but retained a browsing diet (Zanazzi and Kohn 2008).

The browsing anchitherine equid *Mesohippus* was also present together with a new, slightly larger sized genus *Miohippus*. Dental mesowear analysis indicates that they had low-abrasive browse diets throughout the Oligocene (Mihlbachler et al.

2011). The hyracodonts of the genus *Hyracodon*, similar in size to early horses, were browsers of tough vegetation in open floodplain and shrubland environments, whereas the true rhinoceroses of the genus *Subhyracodon* were browsers or mixed-feeders in more wooded, riparian environments (Mead and Wall 1998). In addition, the more hypsodont amynodont *Metamynodon* was present (Prothero et al. 1986).

2.3.3.3 Africa

The Oligocene fossil mammal record from Africa is quite scarce and it has large temporal and spatial gaps. The known palaeoenvironments were mostly forests and swamps (Strömberg 2011; Noret et al. 2012; Jacobs et al. 2010). The best record comes from the famous Late Eocene to Early Oligocene deposits of Fayum, Egypt, where the large herbivorous mammal fauna mainly consisted of hyracoids, anthracotheres, embrithopods and proboscideans (Werdelin and Sanders 2010). The hyraxes (Hyracoidea) were the most abundant and diverse group of large herbivores, ranging from relatively small forms (e.g., *Saghatherium*) to long-legged and slender (*Antilohyrax*), to large (*Megalohyrax*) and even giant, megaherbivore-sized (*Titanohyrax*) forms. However, all of these hyracoids had relatively similar brachydont, buno-lophodont or lophodont dentitions resembling those of perissodactyls and they all were probably browsers on various kinds of vegetation (Rasmussen and Gutiérrez 2010). Embrithopods were another group of herbivorous afrotherians, and they were represented in the Early Oligocene of Fayum by the rhinoceros-sized *Arsinoitherium zitteli* characterised by a pair of large bony nasal horns and robust bilophodont molars with relatively thin enamel. *Arsinoitherium* has remained a rather enigmatic taxon in terms of diet, but what little is known indicates that they were browsers feeding perhaps in relatively open habitats (Noret et al. 2012). Artiodactyls were represented in Africa at this time only by the anthracotheres (e.g., *Bothriogenys*), which were roughly hippo-like, amphibious herbivores with brachydont dentition and probably browse-based diets, similarly to their relatives in Eurasia and North America.

Fossil proboscideans from Fayum comprise a diverse group of large herbivores. The largest of the Fayum herbivores was the advanced numidotheriid proboscidean *Barytherium*, an Asian elephant-sized herbivore with a bulky and robust skull indicating the presence of a proboscis, small tusks in upper and lower jaw, and brachydont, bilophodont molar teeth with morphology and tooth wear suggesting a browse-based diet (Sanders et al. 2010). Another group of early proboscideans were moeritheres (*Moeritherium lyonsi* and *M. trigodon*), roughly pig-sized animals with peculiar low and long, barrel-like bodies and small, flat heads with incipiently tusk-like incisors and brachydont, bunodont/bilophodont cheek teeth suitable for browsing (Sanders et al. 2010). Finally palaeomastodonts (*Palaeomastodon* and *Phiomia*), the early relatives of elephantoids (mammutids, gomphotheres and elephants), were present. They were megaherbivore-sized animals similar to, although slightly smaller than, the later gomphotheriid proboscideans (among which the true elephants later originated), with already well-developed trunks, relatively long and low skulls, long mandibular symphyses, well-developed upper and lower tusks and bunodont cheek teeth suitable for processing various kinds of food from fruits and seeds to leaves (Sanders et al. 2010).

The best Late Oligocene African large mammal record comes from Chilga, Ethiopia (Sanders et al. 2004; Noret et al. 2012; Jacobs et al. 2010). It shows a fauna largely similar to that of the Early Oligocene Fayum deposits and consists of hyracoids (*Pachyhyrax*, *Bunohyrax*, *Megalohyrax*) embrithopods (*Arsinoitherium giganteum*) and proboscideans (*Palaeomastodon*, *Phiomia*, *Chilgatherium* and *Gomphotherium*). Stable isotope analyses indicate that all these large herbivores had essentially browse-based diets, but *Arsinoitherium* and the hyracoids were feeding in more open habitats than the proboscideans (Noret et al. 2012). Overall the paleoenvironment has been interpreted as heavily forested (Noret et al. 2012). *Palaeomastodon* and *Phiomia* were essentially similar and probably had similar kinds of diets as their predecessors in Fayum. *Chilgatherium* represents a new kind of proboscidean, the deinotheres (Deinotheriidae), which seem to have evolved from a moerithere-like ancestor (Sanders et al. 2010). The brachydont, bilophodont cheek teeth of *Chilgatherium*, with sharp-angled mesowear also typical of later deinotheres, indicate folivorous browsing diet. The earliest gomphothere (*Gomphotherium* sp.) comes from Chilga, and it had bunodont cheek teeth only slightly more derived than those of the palaeomastodonts. It probably consumed many kinds of dietary items from fruits and seeds to leaves, and would likely have had more diverse diet than the probably mostly folivorous browsing *Chilgatherium*. This dietary niche partitioning between the specialised browsing deinotheres and the often dietary more flexible elephantoids (gomphotheres and later elephants) continued until the extinction of the deinotheres during the Plio-Pleistocene as indicated by dietary proxy studies (Calandra et al. 2008; Cerling et al. 1999).

2.3.3.4 South America

In South America, the Oligocene palaeoenvironments comprised mostly forests and woodlands, but some localities, such as the Late Oligocene deposits from Salla, Bolivia, indicate open, grassy environments (Strömberg 2011; Croft and Weinstein 2008). Based on dental mesowear analysis, the endemic South American ungulates (notoungulates) from Salla were mixed-feeders with a large component of grasses in their diet (*Archaeohyrax* and *Federicoanaya*) or specialised grazers (*Trachytherus*) (Croft and Weinstein 2008). They are, therefore, possibly among some of the earliest evidence of primarily grazing mammals in the fossil record. Interestingly, tooth wear also suggests that the early mylodontid sloth *Paroctodontotherium calleorum* from Salla was a grazer, as were later mylodontids (Shockey and Anaya 2011). However, unlike the Salla notoungulates and sloths which had hypsodont dentitions, the large pyrotheriid ungulate *Pyrotherium* from Salla had brachydont, bilophodont molars and C3-dominated stable carbon isotope values indicating more likely a browsing diet (MacFadden et al. 1994). Also, Late Oligocene leontiniid notoungulates, such as

Fredericoanaya from Salla, still had brachydont dentitions and probably browse-based diets (Croft 2016).

2.3.3.5 Australia

The earliest comprehensive record of a fossil herbivorous mammal fauna from Australia come from the Late Oligocene deposits of Riversleigh in Queensland, and Southern Australia (Archer et al. 1989; Travouillon et al. 2009; Janis et al. 2016). The palaeoenvironments of those faunas have been interpreted as relatively open forests (Travouillon et al. 2009; Strömberg 2011). Similar to modern and Neogene times, this Oligocene herbivorous mammal community already consisted of specialised herbivorous marsupials, including diprotodontid vombatomorphs (*Neohelos*), early koalas (*Litokoala*), potoroid kangaroos (e.g., *Wakiewakie*) and macropodid kangaroos (e.g., *Nambaroo*) (Archer et al. 1989; Travouillon et al. 2009).

Not much is known about the diets of these Oligocene herbivorous marsupials, as detailed palaeodietary analyses based on dental wear and isotopes are largely lacking. However, the brachydont, bilophodont dentition of the diprotodontid *Neohelos*, and the forested palaeoenvironment, indicates a browse-based diet (Archer et al. 1989; Butler et al. 2017). Based on craniodental analysis, the Late Oligocene kangaroos were mostly omnivores, but included browsing forms too, especially in the Riversleigh palaeoenvironment, and they already had specialisations to the hopping locomotion typical of later kangaroos (Janis et al. 2016). The foregut-fermenting digestion of modern macropodid kangaroos, which enabled a few of them to become grazers later during the Neogene, was however probably developing during the Oligocene (Butler et al. 2017). The Oligocene koala (*Litokoala*) had brachydont, selenodont molars similar to the modern koala, suggesting a browse-based diet, but it lacked several of the key cranial adaptations of the modern koala, which gradually evolved during the Neogene as a specialisation to feeding exclusively on *Eucalyptus* leaves (Louys et al. 2009).

2.3.4 Miocene (23–5 Ma): From Forests to Grasslands, and the Evolution of Grazing Mammals

The Miocene was a time of drastic changes in global climate and ecosystems, during which forest and woodland -dominated vegetation was gradually replaced, in many parts of the world, by more open, grassland or savanna-type environments (Cerling et al. 1997; Strömberg 2011). Global climate became cooler and increasingly seasonal, and in mid-latitudes increasingly dry, which favoured the spread of grasslands and eventually C4 photosynthesizing grasses during the later part of the Miocene (Zachos et al. 2001; Strömberg 2011). Land connection formed for the first time between Africa and Eurasia, and new connections between Eurasia and North

America formed too, enabling migrations of major groups of herbivorous mammals between the continents. These changes were reflected in the evolution and dispersal of grazing ungulates and other large herbivorous mammals during the Miocene. Major events include the evolution of hypsodont, mixed-feeding/grazing horses (Equidae, Equinae) in North America and their dispersal to Eurasia and Africa during the Late Miocene, the evolution of deer (Cervidae) and giraffes (Giraffidae) in the Old World with brachydont dentitions and usually browse-dominated diets, the radiation of bovid ruminants (Bovidae) with increasingly hypsodont dentitions and mixed-feeding to grazing diets in the Old World, the dispersal of gomphotheriid, mammutid and deinotheriid proboscideans from Africa to Eurasia and North America during the Early and Middle Miocene, and the evolution of true elephants (Elephantidae) in Africa during the Late Miocene. Since the Miocene, grazing and mixed-feeding mammals have been abundant in open grassland and savanna-like environments, whereas the remaining closed forest environments have still been dominated by browsers and browse-dominated mixed-feeders (Gordon and Prins 2008).

2.3.4.1 Eurasia

Global temperatures had started rising again during the Late Oligocene, and the Early Miocene was relatively warm and humid, and the palaeoenvironments were dominated by subtropical and temperate forests in Europe (Zachos et al. 2001, 2008; Eronen et al. 2010a; Strömberg 2011). The ruminants increased in diversity and body size, and included, among others, early deer and bovids, which by analogue of their extant relatives possibly already hadfully developed four-chambered, foregut-fermenting stomachs (e.g., Janis 1976). Among the ruminants were tragulids (*Dorcatherium*), moschids (e.g., *Pomelomeryx*), palaeomerycids (early deer-like pecorans, such as *Ampelomeryx*, *Triceromeryx* and *Palaeomeryx*), early deer (e.g., *Procervulus*, *Lagomeryx* and *Acteocemas*), and the earliest bovids (*Eotragus* and *Pseudoeotragus*). Interestingly, based on dental mesowear analysis, the earliest Miocene deer seem to have been more mixed-feeding, than the earliest bovids which were more clearly browsers, at least in the early open habitats of Spain (DeMiguel et al. 2008). Later during the Miocene the deer became more specialised browsers whereas the bovids started to diversify and occupy mixed-feeding and grazing niches. The suids were mostly represented by small and medium-sized, bunodont omnivores such as *Hyotherium*, *Conohyus* and *Aureliachoerus*, but included some more specialised forms such as the selenodont *Sanitherium* and the bilophodont *Listriodon* with more specialised browsing diets. In Asia there were even giant, cow-sized listriodontine suids (*Kubanochoerus*) during the Middle Miocene. The Oligocene anthracotheriids went extinct but another kind of anthracothere *Brachyodus* migrated from Southern Asia to Europe and was present there during the Early Miocene. The perissodactyls were represented by abundant rhinoceroses including the hornless aceratherines (e.g., *Plesiaceratherium*, *Hoploaceratherium* and *Alicornops*) and teleoceratines (*Diaceratherium*, *Prosantorhinus* and

Brachypotherium), and the horned rhinocerotines (*Lartetotherium*). All of these rhinos had similar brachydont, lophodont dentitions indicating browse-based diets. However, later during the Early Miocene in Spain there were early open and arid grass-dominated environments (Urban et al. 2010) which had mammal faunas with predominantly hypsodont dentitions, including an early hypsodont elasmotherine rhino *Hispanotherium*. These *Hispanotherium* faunas are the earliest ones in Europe with hypsodontungulates, suggesting open habitat and possibly grazing/mixed-feeding diets in at least some of the species (Fortelius et al. 2002; DeMiguel et al. 2008; Eronen et al. 2010a).

The fruit, bark and browse -feeding chalicotheres (see dietary analysis based on microwear in Semprebon et al. 2011), such as *Anisodon, Chalicotherium* and *Metaschizotherium*, were also present, as well as tapirs (*Paratapirus* and *Tapirus*), which remained browsers throughout their evolutionary history. During the Early Miocene, two important migration events had a significant impact on Eurasian faunas. First, gomphotheriid, mammutid and deinotheriid proboscideans migrated from Africa, facilitated by the opening of the first important land connection between Africa and Eurasia, the so called "*Gomphotherium* land bridge" (Rögl 1998). Analyses of dental microwear (and mesowear, Saarinen pers. obs.) have shown that there were dietary differences between these diverse proboscideans that enabled them to share their environments: e.g., *Gomphotherium* in Central Europe had more mixed-feeding diet than the purely browsing *Deinotherium* (Calandra et al. 2008). Recent results even show the earliest indication of grass-dominated diet among the Eurasian proboscideans in the derived trilophodont gomphothere *Gomphotherium steinheimense* from an early arid grassland environment in the Middle Miocene of Central Asia, as shown by the analysis of phytoliths preserved in dental calculus (Wu et al. 2018). The browsing deinotheres never dispersed to North America, unlike other groups of proboscideans. The "shovel-tusked" gomphothere *Platybelodon* was a browser which used its lower tusks to cut plants (Semprebon et al. 2016b). The second major event was the arrival of equids in Eurasia again after a long gap since the Early Eocene, migrating from North America across Beringia. The Early and Middle Miocene equids in Eurasia were represented by the anchiteriines with brachydont, lophodont molars (*Anchitherium*), which had rather flexible diets ranging from browsing to mixed-feeding according to local conditions, as demonstrated by dental mesowear analyses (Kaiser 2009; Eronen et al. 2010b).

The remarkable climatic cooling and drying of the climate during the Late Miocene is reflected in drastic environmental changes in Eurasia, especially in the mid-latitudes where increasingly dry and seasonal climates led to the spread of extensive woodlands, savannas and grasslands for the first time (Cerling et al. 1997; Strömberg 2011). The most remarkable changes in the herbivorous mammal faunas include the radiation and dispersal of bovids and the dispersal and radiation of the hypsodont, grazing-adapted (but mostly mixed-feeding) hipparionine equids, which arrived from North America and spread across Eurasia ca. 11–10.5 Ma. The European large mammal faunas at this time were diverse and included both archaic forest-adapted and more derived open-adapted species. Central Europe remained wooded throughout the Late Miocene (Kovar-Eder et al. 2008), although there was a shift during the early Late Miocene (Vallesian) from forests to more seasonal semi-open woodland environments (Agustí and Moya-Sola 1990), that supported such large browsers as the proboscideans *Deinotherium* and *Tetralophodon*, the chalicothere *Chalicotherium*, rhinoceroses (e.g., *Aceratherium* and *Dihoplus*), tapirs (*Tapirus*), tragulids (*Dorcatherium*) and deer (e.g., *Procapreolus* and *Lucentia*). However, there were also mixed-feeding bovids (*Miotragocerus*), and the hipparionine horses (*Hippotherium primigenium*) had diets ranging from browsing to grass-dominated feeding (Kaiser 2004). Similarly, South Eastern Asia remained forested and was dominated by browsers, such as chalicotheres (*Anisodon*), tapirs (*Tapirus*), and browsing proboscideans such as gomphotheriids (*Sinomastodon*) and stegodonts (*Stegolophodon*).

However, in large areas over the continent, including the Mediterranean region, Western Asia, Indian Subcontinent and Eastern Asia, environments gradually changed from forests and woodlands to increasingly open savannas and grasslands (Cerling et al. 1997; Strömberg 2011). There was significant aridification in Central Asia already in Middle Miocene, where palaeobotany and mammal ecometrics indicate the presence of open grasslands (e.g., Liu et al. 2009; Tang and Ding 2013). These diverse semi-open environments were home to rich mammal faunas often collectively known as the "*Hipparion*"-fauna. These were characterised by diverse hypsodont three-toed hipparionine horses, ranging from large (about zebra-sized) mixed-feeding forms (e.g., *Hippotherium sp.*) to small grazers (e.g., *Cremohipparion matthewi*) (Solounias et al. 2010, 2013; Bernor et al. 2014). Bovids were diverse, being mostly mesodont and mixed-feeders, such as the abundant *Tragoportax* sp. (Solounias et al. 2013). Many widespread bovid taxa such as *Gazella* sp. and *Urmiatherium* sp. had considerable variation in diet, ranging from browsers to mixed-feeders and sometimes grass-dominated feeders depending on local environments (Bernor et al. 2014; Eronen et al. 2014). Based on tooth wear and isotope studies, early bovines (Bovini) had diets based more heavily on grasses, they lived in more open habitats and had more hypsodont molars than the contemporaneous boselaphine bovids in the Middle Siwaliks beds of the Indian subcontinent (Bibi 2007). The rest of the Late Miocene *Hipparion* faunas comprised abundant rhinoceroses, including species with brachydont or mesodont, lophodont dentition typical of rhinos with browse-based diets (e.g., *Acerorhinus*, *Chilotherium* and *Dihoplus*) and more specialised species with more high-crowned, complex teeth and grazing diets (e.g., *Ceratotherium*, *Iranotherium* and *Sinotherium*), large browsing chalicotheres (*Ancylotherium*), giraffes ranging from browsers (e.g., *Palaeotragus*) to mixed-feeders (e.g., *Samotherium*), large bunodont suids (*Microstonyx*) and proboscideans such as specialised gomphotheres of the genus *Choerolophodon* with grass-dominated diet, as shown by dental microwear and mesowear analyses (Solounias et al. 2010, 2013; Bernor et al. 2014; Eronen et al. 2014; Konidaris et al. 2016). Because of the monsoon climate in Eastern Asia, the palaeoenvironments there were divided into south-eastern areas with more humid and wooded environments and a larger proportion on brachydont, browsing mammals (including even the last archaic anchitheriine equid *Sinohippus*) and south-western inland areas with more dry, seasonal and open environments (Fortelius et al.

2002; Passey et al. 2007). In these environments, the *Hipparion* horses and the hypsodont gazelle *Gazella dorcadoides* consumed some C4 plants (grasses), whereas deer and the mesodont and brachydont gazelles (*G. paotehensis* and *G. gaudryi*) were pure C3 plant (probably browse-dominated) feeders, although the isotopes do not reveal possible feeding on C3 photosynthesizing grasses (Passey et al. 2007). However, mesowear analysis indicates that the C3 feeders were mostly browsers and browse-dominated feeders (Eronen et al. 2014).

2.3.4.2 North America

Drastic environmental changes in North America happened during the Early and Middle Miocene, when savanna and grassland environments started to develop there as a result of the uplift of Rocky Mountains that caused a gradual climatic drying in the western interior of the continent (e.g., Van Devender and McClaran 1995), leading to the evolution of some grazing ungulates (Janis et al. 2002, 2004; Strömberg 2011). A greater change from C3 photosynthesizing to C4 photosynthesising grasslands and from brachydont browser-dominated to hypsodont grazer-dominated ungulate faunas happened from the Middle to the Late Miocene (Janis et al. 2002, 2004; Strömberg and McInerney 2011). The Early and Middle Miocene faunas comprise a wide variety of equids, including browsing anchitheriines (e.g., *Megahippus* and *Hypohippus*), early equine equids with mesodont to hypsodont dentitions (*Merychippus* and *Cormohipparion*), rhinos (*Menoceras*, *Diceratherium*, *Teleoceras*, *Aphelops*), chalicotheres (*Moropus*), large entelodonts (*Dinohyus*), camels ranging from small gazelle-like forms (*Stenomylus*) to larger forms (*Procamelus*, *Oxydactylus*), moschid ruminants (*Machaeromeryx* and *Blastomeryx*), antilocaprid ruminants (*Ramoceros*) dromomerycine palaeomerycid ruminants (e.g., *Aletomeryx*) protoceratids (*Syndyoceras*) and diverse merycoidodonts (e.g., *Merychyus*, *Promerycochoerus*, *Merycochoerus*, *Hypsiops* and *Ustatochoerus*). During the Middle Miocene, proboscideans migrated from Eurasia to North America, being represented by mammutids (*Zygolophodon*) and gomphotheriids (*Gomphotherium*).

Janis et al. (2002 and 2004) noted that there was a higher diversity of browsing ungulates in the Early and Middle Miocene North American semi-open savanna-like environments than in modern savannas and later North American open environments, which indicates that these were high-productivity savanna-/grassland-type environments which do not have a modern analogue. Although the Early and Middle Miocene environments were dominated by browsers in North America, some groups notably show a shift towards increasingly grazing diets, as shown by dental mesowear and adaptations to grazing (especially an increase in hypsodonty). Among these were the first equine horses (e.g., *Merychippus*), which in the late Early Miocene showed a shift to mixed-feeding diets and mesodont/hypsodont dentitions, and existed alongside the (at this time larger) brachydont, browsing anchitheriine horses until the Late Miocene (Mihlbachler et al. 2011). Similarly, mesowear analysis shows that the merycoidodonts, which persisted (but were no longer a significant part of the fauna) in North America until the Late Miocene, shifted their diets towards increasingly grass-dominated mixed-feeding during the Early Miocene, accompanied by increasing hypsodonty in some taxa, as opposed to their completely browse-based diets in the Early Oligocene (Mihlbachler and Solounias 2006).

The Late Miocene palaeoenvironments comprised largely grasslands and savannas with diverse herbivorous mammal faunas of hypsodont equine horses (*Cormohipparion*, *Neohipparion*, *Protohippus*, *Pliohippus*, *Pseudhipparion*, *Nannippus*), camels (*Aepycamelus*, *Megacamelus*), protoceratids (*Synthetoceras*), dromomerycine palaeomerycid ruminants (e.g., *Pediomeryx*, *Cranioceras*), antilocaprid ruminants (*Sphenophalos*, *Texoceros* and *Osbornoceros*), rhinos (*Teleoceras*, *Peraceras*, *Aphelops*), peccaries (*Prosthennops*) and gomphotheriid proboscideans (*Amebelodon*, *Rhynchotherium*, *Tetralophodon*). Following the spread of C4 grasslands in North America during the Late Miocene, the ungulate faunas show a significant decline in browser diversity and the predominance of mixed-feeders and grazers (Janis et al. 2002, 2004; Mihlbachler et al. 2011). However, some of the late protoceratids had long muzzles and retracted nasals, suggesting the presence of a moose-like upper lip and possibly a similar specialised browsing lifestyle (Janis 1982; Prothero 1998). The dominant grazers were the hipparionine horses, which by this time had fully hypsodont dentitions and were all grazers or grass-dominated mixed-feeders in North America, as shown by mesowear analyses (Mihlbachler et al. 2011). Even among the dromomerycines, a group of deer-like ruminants, there was a shift in Late Miocene towards mixed-feeding in the derived cranioceratines such as *Pediomeryx*, as indicated by dental microwear and mesowear, despite their generally browsing morphological adaptations (Semprebon et al. 2004). Microwear and mesowear analyses show that a similar shift to increasingly abrasive, mixed-feeding occurred within the antilocaprids during the Late Miocene and Early Pliocene, accompanied with an increase in the hypsodonty of their dentition (Semprebon and Rivals 2007).

2.3.4.3 Africa

Most of what we know about the Miocene of Africa comes from the rich fossil sites from East Africa, with some important records from the Early and Late Miocene of North and South Africa. During the Early Miocene, the palaeoenvironments in East Africa were dominated by tropical forests and woodlands which largely resembled the modern West African lowland rainforests (Jacobs et al. 2010; Grossman et al. 2014). The North African site of Gebel Zelten (Libya) also indicates a palaeoenvironment dominated by forests and swamps (Grossman et al. 2014). During the Middle Miocene in East Africa the environments started to change locally, and the first C3 grass -dominated grassland environments emerged (e.g., in Maboko and Fort Ternan, Kenya), driven by seasonal changes from dry to waterlogged conditions (Retallack 1992; Retallack et al. 2002). However, more drastic changes in the environments started during the Late Miocene, when C4-photosyntesising grasses spread forming abundant, partially open grassy woodland and savanna environments

(Cerling et al. 1997; Strömberg 2011). However, there were local differences in this pattern, for example, the Lake Turkana region in Eastern Africa got drier earlier than surrounding areas, serving as a "species factory" of mammals adapted to increasingly open grassland environments (Fortelius et al. 2016). In South Africa, the shift to C4 grasses apparently happened later, as stable isotope record indicates at least locally purely C3 conditions throughout the Late Miocene in Langebaanweg (Franz-Odendaal et al. 2002).

The mammal impact of the land connection between Africa and Eurasia resulted in significant changes in the Early Miocene African herbivorous mammal faunas with the arrival of ruminants (e.g., the tragulid *Dorcatherium*, the early pecoran *Walangania* and the giraffid *Palaeotragus*), suids (e.g., *Diamantohyus* and *Nguruwe*), rhinoceroses (*Brachypotherium*, *Turkanatherium*, *Rusingaceros*, *Victoriaceros*) and chalicotheres (*Butleria*). Later, during the Middle Miocene the bovids *Protragocerus* and *Oioceros* arrived, and finally the Late Miocene saw a significant radiation of the African bovids, including such African taxa as *Aepyceros* (impala), cephalophines (duikers), tragelaphines (kudus and relatives), hippotragines (sable antelopes), reduncines (reedbucks and waterbucks) and alcelaphines (relatives of wildebeest and hartebeest), and Eurasian immigrants such as antilopines (e.g., *Gazella*) and "boselaphines" (*Tragoportax*). Meanwhile, the endemic afrotherian herbivores, especially proboscideans, were thriving in Africa during the Miocene. The Late Miocene saw the evolution of true elephants (Elephantidae), which soon started to replace the more archaic gomphotheres. During the Early and Middle Miocene, the large mammal faunas were dominated by browsers (e.g., Uno et al. 2011), although, based on tooth wear data, there is evidence that some gomphotheriid proboscideans were already grass-dominated feeders in the Middle Miocene C3 grasslands of East Africa (Saarinen, personal observation). Several stable isotope-based studies indicate that there was a major shift from predominantly browse-based to largely grass-based diets during the Late Miocene in Africa, following the opening of the environments and the spread of C4 grasses (Cerling 1992; Cerling et al. 1997, 1999; Uno et al. 2011). This happened in many lineages of bovids, derived gomphotheres (*Anancus*), elephants (*Primelephas* and *Stegotetrabelodon*), suids (*Nyanzachoerus australis*), rhinos (*Ceratotherium*), and the hipparionine horses (*Eurygnathohippus*), whereas giraffes and deinotheriid proboscideans retained completely browse-based diets (Uno et al. 2011). Hippos (Hippopotamidae) evolved in Africa from their anthracotheriid ancestors during the Middle Miocene (Boisserie and Lihoreau 2006), and were a prominent part of the herbivore fauna in the Late Miocene (e.g., *Archaeopotamus* and *Hexaprotodon*). Isotope and microwear studies have shown that the Late Miocene hippos already had a tendency for grazing on C4 grasses, in contrast with their browsing anthracotheriid ancestors, although hippos are actually opportunistic feeders rather than obligate grazers and they have retained a variable component of C3 plants in their diet until recent times (Boisserie et al. 2005; Harris et al. 2008).

2.3.4.4 South America

After the initial spread of grasslands and the dominance of grazing notoungulates and sloths in the Late Oligocene of Bolivia, the palaeoenvironments of South America were again predominantly forested over most of the continent during the Miocene, until the Late Miocene (ca. 7 Ma) when they became increasingly dominated by C4 grasslands (Strömberg 2011). The late Miocene spread of grasslands reflects the global climatic cooling, drying and increased seasonality, but was also affected by the uplift of the Andes (Gregory-Wodzicki 2000; Zachos et al. 2001). This is reflected in the herbivorous mammal communities as a shift from predominantly C3 browsing to predominantly C4 grazing during the Late Miocene, similarly to other continents (MacFadden et al. 1996; MacFadden 2000). At this time, South American large herbivorous mammal faunas comprised endemic notoungulates, litopterns and xenarthrans, whereas North American immigrants such as equids, camelids and proboscideans, that became prominent especially in Pleistocene South American faunas, had not yet arrived. Interestingly, most of the notoungulates were already hypsodont in the Early Miocene despite their browse-based diets (e.g., Townsend and Croft 2008), and in fact became largely hypselodont during the Late Miocene (Ortiz-Jaureguizar and Cladera 2006). It has been suggested that this could be explained by the high accumulation of abrasive volcanic ash, which would have created a selection pressure for higher tooth crowns (Strömberg et al. 2013). Among the Xenarthra, the herbivorous sloths and armadillo-related glyptodonts also had hypselodont dentitions, but that is easy to understand as compensation for the lack of enamel in their teeth (Vizcaíno 2009). Microwear analyses of the notoungulates *Nesodon*, *Adinotherium* and *Protypotherium* from the Early Miocene mammal assemblage of Santa Cruz, Argentina, indicate browse-based diets (Townsend and Croft 2008). New mesowear-based dietary analysis indicates browse-based diets also for the sloths (*Eucholoeops* and *Planops*) of the Santa Cruz Formation, and a browse-dominated diet for the early glyptodont *Propalaeohoplophorus* (Saarinen and Karme 2017).

2.3.4.5 Australia

Our knowledge of the Miocene herbivorous marsupial communities of Australia comes from the Miocene records from Riversleigh, Queensland, and Lake Eyre Basin, South Australia. The Early and Middle Miocene palaeoenvironments have been interpreted as rainforests, but during the Late Miocene more open sclerophyllous forests with grassy undergrowth started to be the dominant environments (Travouillon et al. 2009; Strömberg 2011). The browsing diprotodontid *Neohelos* persisted during the Miocene, and new diprotodontids such as *Nimbadon* emerged, as well as the bilophodont palorchestids (*Propalorchestes*) (Archer et al. 1989; Travouillon et al. 2009). Based on craniodental analyses, most of the Miocene kangaroos of Australia were omnivores or browser, and their calcaneal morphology suggests some adaptations to hopping locomotion, although not to the level of modern kangaroos (Janis et al. 2016). However, some kangaroos show an increase in hypsodonty from Middle Miocene onwards and there was a hypselodont species of wombatid in the Late Miocene of Riversleigh, suggesting some adaptation to more abrasive or fibrous diets (Dawson and Dawson 2006). A Middle Miocene species of koala, *Stealokoala riversleighensis*, shows adaptations in dental morphology which indicate the beginning of the specialisation to feeding on fibrous *Eucalyptus* browse, which culminated in the modern koala (*Phascolarctos*) (Black 2016).

2.3.5 *Pliocene (5–2.6 Ma): Towards the Ice Age*

After the drastic climatic cooling and drying of the Late Miocene, the Early Pliocene saw a temporary rise in global temperatures, but during latter part of the Pliocene the temperatures dropped again (Herbert et al. 2016). This gradually led to the onset of the Pleistocene ice ages with cyclic short-term variation between cold glacial stages and warmer interglacials (Zachos et al. 2001, 2008). Atmospheric CO_2 levels declined during the Late Pliocene (Bartoli et al. 2011). Palaeoenvironmental studies indicate a continuum of increasingly open, C4 grass-dominated savanna- and grassland-environments in tropical and mid-latitude regions such as in Eastern Africa and North America, whereas other regions such as Europe and Southeast Asia were dominated by forests and woodlands. Many genera of grazing-adapted mammals originated during the Pliocene, including the monodactyl horses of the genus *Equus*, the Asian elephant lineage (*Elephas*) and the woolly rhinoceroses (*Coelodonta*).

2.3.5.1 Eurasia

During the Pliocene, relatively warm, forested conditions still prevailed in Europe, with such thermophilous trees as redwood (*Sequoia*), bald cypress (*Taxodium*) and hemlocks (*Tsuga*) growing as far north as in the Great Britain (Head 1998; Kovar-Eder et al. 2008; Strömberg 2011). These conditions were reflected by the presence of such "archaic" browsing forest ungulates as tapirs (*Tapirus arvernensis*), and the dominance of deer (Cervidae) such as *Arvernoceros* and *Croitzetoceros* in the ungulate faunas. The deer were mostly browsers, although there were subtle differences in their diets, e.g., in the locality of Saint-Vallier, France, *Croitzetoceros* had a more variable, possibly seasonally changing diet than *Metacervoceros*, as indicated by differences in microwear (Valli and Palombo 2008). They were accompanied by large bovine bovids (*Parabos*, *Alephis*), pigs (e.g., *Sus strotzzii*), brachydont rhinoceroses (*Stephanorhinus*) and the last hipparionine horses (*Plesiohipparion*). Rivals and Lister (2016) studied the diets of the Late Pliocene ungulates from Red Crag Nodule Bed, Great Britain, using mesowear and microwear analyses. Their results show that the tapir (*Tapirus* sp.) and the rhinoceros (*Stephanorhinus etruscus*) were browsers, the tapir also including some fruit in its diet. The hipparionine horse

("*Hipparion*" sp.) from the Red Crag Nodule Bed had a mostly grass-dominated but very flexible diet, ranging from browsing to grazing (Rivals and Lister 2016). The bovid *Parabos* was a browse-dominated mixed-feeder, and the cervids (cf. *Cervus pardinensis* and cf. *Cervus perrieri*) had mixed-feeding and browse-dominated diets, respectively, in the Red Crag Nodule Bed paleoenvironment (Rivals and Lister 2016). The probocideans were represented in Europe by the European mastodon (*Mammut borsoni*), a very large browsing species, and a browse-dominated mixed-feeding gomphothere *Anancus arvernensis* (Rivals et al. 2015; Saarinen and Lister 2016). True elephants dispersed to Eurasia from Africa during the Pliocene and became an ecologically important element. Early mammoths migrated from Northern Africa to Europe through an Eastern Mediterranean route. The earliest species of mammoth in Europe, *Mammuthus rumanus*, had a predominantly browse-based diet, as indicated by microwear analysis of their molar teeth (Rivals et al. 2015).

In Eastern and Southern Asia, the dominant proboscideans were the stegodonts (*Stegodon*), a group of roughly elephant-like forms with brachydont, multi-lophed cheek teeth. Unlike true elephants, stegodonts retained predominantly browsing diets throughout their evolutionary history and were associated with forest and woodland habitats (Sanders et al. 2010; Zhang et al. 2017). The Upper Siwaliks faunas from Indian subcontinent show diverse large mammal assemblages with bovids (e.g., the early bovine *Hemibos*), hippopotami (*Hexaprotodon*), the last surviving anthracotheres (*Merycopotamus*), giant sivatherine giraffes (*Sivatherium*), stegodont proboscideans (*Stegodon*) and early relatives of modern Asian elephant (*Elephas planifrons*). The diets of these large herbivores ranged from grass-dominated (*Hipparion*, *Hemibos*, *Hexaprotodon* and *Sivatherium*) to mixed-feeding (*Merycopotamus*) and browsing (*Stegodon*), based on stable carbon isotope and microwear analyses (Patnaik 2015). Central Asia was dominated by open steppe environments, and the peculiar hypselodont elasmotheriine rhinos of the genus *Elasmotherium* originated there, probably adapting to very abrasive "hypergrazing" grass-based diets (Zhegallo et al. 2005). The Tibetan plateau served as a place of origin for many genera of cold-adapted ungulates with adaptations to grazing in cold steppe environments, such as the early woolly rhinoceros (*Coelodonta thibetana*) (Deng et al. 2011). The Pliocene ugulate fauna of the Tibetan plateau, including early mountain-sheeps/goats, hipparionine horses and deer, were feeding on similar C3 photosynthesizing plants as the modern ungulates in that area, as shown by stable isotope analyses (Wang et al. 2013, 2016). Moreover, the isotopes indicate similar high-altitudes but milder temperature than at present (Wang et al. 2013).

2.3.5.2 North America

As a continuum of the drastic environmental changes that started during the Late Miocene, C4 grass -dominated grasslands were abundant environments over much of the continent (Strömberg and McInerney 2011). The earliest Pliocene in North America was largely a continuum of the Late Miocene (Hemphillian) in term of palaeoenvironments and mammal faunas, but very soon during the next stage of the

Pliocene (Blancan) major changes started to occur (Janis et al. 1998; Morgan and Lucas 2003). One of the most influential events was the formation of a land connection between North and South America, allowing mammals to migrate between the continents (an event known as "the Great American Biotic Interchange") (Woodburne 2010). Although the effect of this faunal interchange was more lasting in South America, it also influenced North America especially through the arrival of megalonychid and mylodontid ground sloths (*Megalonyx*, *Glossotherium*) and the large, herbivorous armadillo-related glyptodonts (*Glyptotherium*).

Horses at this time were represented by the last and small hipparionines (*Nannippus*) and the modern monodactyl genus *Equus*, and based on mesowear analyses they were all grazers, in contrast to the Late Miocene hipparionines which included a wider spectrum of grass-dominated mixed-feeders and grazers (Mihlbachler et al. 2011). The artiodactyls were represented by increasingly modern kinds of relatives of camels, including camelines (*Camelops*) and lamines (*Hemiauchenia*), antilocaprids (*Capromeryx*) and browsing deer (*Odocoileus*, which had migrated to North America from Eurasia) (Morgan and Lucas 2003). The camels, especially *Hemiauchenia*, had browse-dominated mixed-feeding diets (Feranec 2003), and the antilocaprids of this time hade more grass-dominated diets than later during the Pleistocene and Holocene (Semprebon and Rivals 2007). Rhinoceroses went extinct in North America during the Pliocene but the browsing tapirs (*Tapirus*) persisted until the end of Pleistocene in wooded and warm environments. The Gray Fossil site in Tennessee, representing a rare record of a subtropical swamp forest environment with browsing tapirs, *Teleoceras*-rhinos, mastodons and other forest taxa, is now considered to be Early Pliocene rather than Late Miocene in age (Samuels et al. 2018). The Pliocene North American proboscideans included new taxa of derived gomphotheres (*Rhynchotherium*, *Stegomastodon* and *Cuvieronius*) and mammutids (*Mammut sp.*). Of these, based on tooth morphology and tooth wear, *Mammut* sp. was a browser and the gomphotheres more opportunistic mixed-feeders (Rivals et al. 2012; Sánchez et al. 2004).

2.3.5.3 Africa

The Pliocene was a time when the open, C4-grass dominated grassland savannas started to emerge in East Africa after the initial spread of grassy woodland savannas during the Late Miocene (Cerling et al. 1997; Strömberg 2011). However, woodland savannas and associated browser and mixed-feeder dominated herbivorous mammal faunas still prevailed, as indicated by sites such as the Upper Laetoli beds in Tanzania (Bamford 2011a, b; Kaiser 2011). Based on stable carbon isotope studies, the large herbivore faunas of the Turkana Basin in East Africa were dominated by mixed-feeders during the Pliocene, becoming dominated by specialised C4-grass grazers much later, at about the beginning of the Pleistocene ca. 2.5 Ma (Cerling et al. 2015). This also concurs with other palaeoecological and palaeoenvironmental studies, which suggest that the savanna-like environments of the Pliocene were still largely more wooded than in the Pleistocene in East Africa (e.g., Bamford 2011a, b; Fortelius et al. 2016).

The fossil mammal faunas were rich, dominated by Bovidae representing all the modern tribes of African bovids (Tragelaphini, Alcelaphini, Antilopini, Hippotragini, Reduncini and Cephalophini), and pigs (Suidae) represented by such abundant taxa as *Notochoerus* and *Kolpochoerus*. Based on tooth wear and stable isotope studies, both the bovids and the suids included C4 grazers as well as mixed-feeders and browsers (e.g., Uno et al. 2011; Kaiser 2011). Perissodactyls were less diverse and abundant but were still common elements of the mammal faunas, including grazing hipparionine horses (*Eurygnathohippus*), both browsing and grazing rhinoceroses (*Diceros* and *Ceratotherium*, respectively) and rare late-surviving chalicotheres, which would probably have been browsers feeding on leaves and bark of trees and shrubs (*Ancylotherium*) (Uno et al. 2011; Kaiser 2011; see also Semprebon et al. 2011). The proboscideans were diverse and abundant, and they were represented by several often sympatric species including the browsing deinothere *Deinotherium bozasi*, the grazing or mixed-feeding gomphothere *Anancus* and predominantly grazing elephants representing the lineages of modern African and Asian elephants (*Loxodonta adaurora*, *L. exoptata* and *Elephas recki*) (Cerling et al. 1999; Uno et al. 2011; Saarinen et al. 2015).

2.3.5.4 South America

A shift from mostly C3 browsing to predominantly C4 grass-based diets occurred during the Late Miocene in many endemic South American ungulates, but other isotopic evidence suggests that some Pliocene toxodont notoungulates from the Andes retained C3-plant based diets, possibly due to high altitude (MacFadden et al. 1994). Otherwise, detailed palaeodietary analyses seem to be largely lacking for the Pliocene of South America, although Ortiz-Jaureguizar and Cladera (2006) indicate that South American Pliocene faunas were dominated by "grazing types". However, the dietary spectrum of Pliocene South American herbivores was probably much wider, e.g., the common small, burrowing notoungulate *Paedotherium typicum* from the savanna-like paleoenvironment of Chapadmalal, Argentina, was probably a mixed-feeder of low-growing vegetation in various habitats (Croft 2016). Moreover, a peccary (*Platygonus marplatensis*) from Chapadmalal had incipiently bilophodont molars and probably largely browsing/frugivorous diet (Croft 2016). During the Late Pliocene the opening of the land connection between North and South America allowed some new groups of large herbivorous mammals to migrate from North to South America, including some peccaries and camels, but the main event of the Great American interchange and the subsequent faunal changes coincides with the beginning of the Pleistocene ca. 2.6 Ma (MacFadden et al. 1994; Woodburne 2010).

2.3.5.5 Australia

Analysis of the Chinchilla fossil mammal community indicates that the Pliocene palaeoenvironments of Queensland in Australia were mosaic-like with tropical forests, wetlands and grasslands (Montanari et al. 2013). Based on stable isotopes, the kangaroo *Macropus* sp. was a mixed-feeder on both C3 and C4 plants, and the kangaroos *Troposodon* and *Protemnodon*, as well as the large diprotodontid *Euryzygoma* were browse-dominated mixed-feeders with a preference for C3 plants (Montanari et al. 2013). Mesowear analysis of several species of Plio-Pleistocene kangaroos from Australia indicates that most of these species were mixed-feeders (Butler et al. 2014).

2.3.6 Pleistocene (2.6 Ma–10 Ka): The Time of the Ice Ages

The Pleistocene Ice Age started when the global cooling trend had reached a point after which periodical glaciation events started in the Northern hemisphere following cyclic changes in the orbit and axis orientation of the Earth (e.g., Hays et al. 1976; Zachos et al. 2001; Lisiecki and Raymo 2007; Walker and Lowe 2007; Zachos et al. 2008). Continental glaciation of the Antarctic had started much earlier, already in the Oligocene (Zachos et al. 2008). However, at the beginning of the Pleistocene the global cooling intensified the effect of the cyclic changes in the Earth's orbit and axis on climatic changes, causing summer temperatures to drop periodically low enough for the accumulation of large continental ice sheets (glaciers) in northern Europe and in northern North America (Zachos et al. 2008; Lisiecki and Raymo 2007; Walker and Lowe 2007). Since then the climate oscillated between the extremes of cold glacial maxima and warm interglacial stages in Northern hemisphere, and the changes in climate have often been complex and rapid (e.g., Walker and Lowe 2007).

The cold, dry and dramatically oscillating climates of the Pleistocene had a profound effect on all life on earth, shaping vegetation patterns and animal communities, and creating heavy selection and extinction pressures on organisms (e.g., Kurtén 1968, 1972; Kurtén and Anderson 1980; Geist 1998; Guthrie 2001). The cold climatic conditions of the Pleistocene increased plant productivity and quality due to the fertile soils generated by intense glacial erosion and the reduced chemical defences and fibrousness of plants, providing abundant resources for large herbivores, resulting in the evolution of particularly large species with impressive luxury organs (horns, antlers and tusks) (e.g., Zimov et al. 1995; Geist 1998; Saarinen et al. 2014). Atmospheric CO_2 content oscillated following the glacial-interglacial cyclicity, which could also have had an impact on the extent of grassland (especially C4 grass) environments (e.g., Archer et al. 2000). Pleistocene herbivorous mammal communities, across the world, are typically characterized by dietary successions from browsers to grazers, with grazers dominating open grassland environments and browsers dominating forest environments. Mixed-feeders were typical in both kinds of palaeoenvironments. The Pleistocene ended with the mass extinction of megafauna, which led to the impoverished herbivore communities of the Holocene (Barnosky et al. 2004).

2.3.6.1 Eurasia

In Eurasia the cyclic changes in the Pleistocene climate caused especially dramatic periodical changes in biome distributions and environments. There was a general shift during the Early Pleistocene from forests to open savanna-like environments, especially in Southern Europe but also as far north as in England, where semi-open, boreal grassland-heathland environments prevailed during the cool glacial stages and woodland environments during the warm interglacials (e.g., West 1980). Later during the Pleistocene, successions of forest vegetation in Europe and Northern Eurasia were characteristic of the warm interglacial stages (e.g., Stuart 1976). During the glacial stages, especially in the Late Pleistocene, continental ice sheets and mountain ranges blocked moisture from reaching the Eurasian inland areas and caused long-term cold and dry high-pressure climates which resulted in the vast "mammoth steppe" vegetation to spread over the continent (Guthrie 2001). Trampling and grazing by abundant large herbivorous mammals also contributed in maintaining the grass-dominated vegetation of the mammoth steppes (Zimov et al. 1995). South-Eastern Asia remained dominated by tropical and subtropical forests and savannas during the Pleistocene.

In Europe and Northern Eurasia, the constantly changing climatic and environmental conditions of the Early and Middle Pleistocene favoured large herbivorous mammal species which were able to adapt to these changes and tended to have rather opportunistic dietary strategies, being able to change their diets based on environmental changes despite their adaptations to either grazing or browsing. Many "archaic" mammals which were adapted to the warm forest environments of the Pliocene and had browsing diets, such as tapirs (*Tapirus*) and the European mastodon (*Mammut borsoni*) went extinct in Europe at the transition from Pliocene to Pleistocene. Also the last hipparionine horses went extinct and were replaced by the modern horses (*Equus*), which migrated to Eurasia (and somewhat later to Africa) at the beginning of the Pleistocene. The large herbivorous mammal communities of the Pleistocene typically show a continuum of dietary strategies from browsers to mixed-feeders and grazers. For example in the Early Pleistocene community of Coste San Giacomo from Italy where the deer (*Axis*, *Eucladoceros* and *Croitzetoceros*) were browsers, the bovids (*Gallogoral*, *Gazella* and *Leptobos*) were predominantly mixed-feeders, and the horse (*Equus stenonis*) was a grazer, as shown by dental mesowear analysis (Strani et al. 2015). However, the flexibility of dietary strategies, which enabled mammals to cope with the variable and changing climatic and environmental conditions, are clearly indicated by dental mesowear and microwear analyses: e.g., the rhinoceros *Stephanorhinus hudsheimensis* was able to shift its diets from browsing in a forest environment to mixed-feeding in an open environment (Kahlke and Kaiser 2011; Van Asperen and Kahlke 2015). Similarly, the gomphotheriid proboscidean *Anancus arvernensis* and the ancestral mammoth *Mammuthus meridionalis* were able to shift their diets according to environmental conditions, despite *A. arvernensis* being more adapted to browsing and *M. meridionalis* to grazing (Rivals et al. 2015; Saarinen and Lister 2016).

During the latter part of the Pleistocene, the glacial-interglacial cyclicity intensified, and the warm forest-adapted interglacial faunas (the so-called *Palaeoloxodon-Stephanorhinus*—chronofaunas) alternated with the cold steppe-adapted *Mammuthus-Coelodonta* (woolly mammoth—woolly rhino) chronofaunas locally over much of the Northern Eurasia (Kahlke 1999; Pushkina 2007). Also these Middle and Late Pleistocene mammal communities demonstrate a dietary spectrum from browsers (moose (*Alces alces*), roe deer (*Capreolus capreolus*) and Merck's rhinoceros (*Stephanorhinus kirchbergensis*)) to mixed-feeders (giant deer (*Megaloceros giganteus*), red deer (*Cervus elaphus*), fallow deer (*Dama dama*), reindeer (*Rangifer tarandus*) and the narrow-nosed rhinoceros (*Stephanorhinus hemitoechus*)) to grass-dominated mixed-feeders (steppe bison (*Bison priscus*), aurochs (*Bos primigenius*)), to grazers (wild horse (*Equus ferus*) and woolly rhinoceros (*Coelodonta antiquitatis*)) (Rivals et al. 2010; Rivals and Lister 2016; Saarinen et al. 2016). However, again dental mesowear and microwear analyses show that there was typically local variation in the diets of these species, following changes in vegetation (Rivals and Lister 2016; Saarinen et al. 2016). The diets of mountain-adapted caprine bovids ranged from mixed-feeding (*Rupicapra pyrenaica* and *Capra caucasica* in the Late Pleistocene of France) to grazing (Middle Pleistocene argali-like sheep (*Ovis ammon antiqua*) from Caune de l'Arago, France) in high-altitude meadows, as indicated by dental microwear analyses (Rivals and Deniaux 2003, 2005). The ancestral mammoth (*Mammuthus meridionalis*) was replaced in the Middle Pleistocene by a more specialised woodland species with browse-dominated mixed-feeding diet, the straight-tusked elephant (*Palaeoloxodon antiquus*), and a more specialised open adapted grazing species, the steppe mammoth (*Mammuthus trogontherii*) (Saarinen and Lister 2016). Finally, the cold adapted woolly mammoth (*Mammuthus primigenius*) became the only species of elephant in Europe and Northern Eurasia during the last glaciation. Based on dental microwear and mesowear analyses, all of these species of elephant, even the woolly mammoth with the strongest adaptation to grazing (the most hypsodont molars with the highest number of enamel ridges), had quite large variation in their diets following again changes in environments and vegetation structure (Rivals and Lister 2016; Saarinen and Lister 2016).

2.3.6.2 North America

In North America, a roughly similar pattern occurred as in Northern Eurasia, except that throughout the Pleistocene the environmental differences were not as great between glacial and interglacial stages but were greater between the heavily forested eastern part of the continent and the dry, grassland-dominated western part

(Williams et al. 1998). However, a much larger area of North America was covered by ice sheets during the glacial stages than in Eurasia, and north of the ice sheet the mammal fauna was similar to the Siberian mammoth steppes, whereas south of the ice there were forests and grassland habitats throughout the Pleistocene (Williams et al. 1998; Guthrie 2001).

The diets of Pleistocene North American large herbivores were characteristically diverse, ranging from browsers (e.g., moose (*Alces alces*), white-tailed deer and mule deer (*Odocoileus*), American mastodon (*Mammut americanum*) and Jefferson's ground sloth (*Megalonyx jeffersoni*)) to abundant, often grass-dominated mixed-feeders (e.g., antilocaprids, bisons (*Bison*) and the Columbian mammoth (*Mammuthus columbi*) and obligate grazers (e.g., horses (*Equus*) and Harlan's ground sloth (*Paramylodon harlani*)) (Rivals et al. 2007; Semprebon and Rivals 2007; Mihlbachler et al. 2011; Rivals et al. 2012; Saarinen et al. 2015; Saarinen and Karme 2017). Dental mesowear and microwear analyses indicate that the diets of the Pleistocene North America bisons (*Bison* sp.) were less heavily abrasive than in the modern plains bison (*Bison bison*), which is a "hyper-grazer" among bovids (Rivals et al. 2007). The camels in North America during the Pleistocene, including *Camelops*, *Hemiauchenia* and *Palaeolama* were all browse-dominated feeders, based on dental mesowear and microwear analyses (Semprebon and Rivals 2010). The Pleistocene antilocaprids (*Tetrameryx*, *Stockoceros*, *Capromeryx* and *Antilocapra*) had less abrasive mesowear signals and more C3 dominated diets than their Pliocene predecessors, but their microwear analyses still suggest grazing, indicating perhaps specialisation to utilising relatively non-abrasive C3 photosynthesising grasses (Semprebon and Rivals 2007). The mesowear based analysis of North American horses through time (Mihlbachler et al. 2011) indicates that the *Equus*-horses in the Pleistocene of North America were all grazers. Microwear analysis indicates that the flat-headed peccary (*Platygonus compressus*) had predominantly omnivorous/browsing diet (Schmidt 2008). Within the proboscideans, the American mastodon (*Mammut americanum*) with brachydont, lophodont molars was a dedicated browser and is typically associated with forest environments, whereas the Columbian mammoth (*Mammuthus columbi*) occupied more open habitats, had specialised hypsodont dentition and was clearly adapted to grazing, but had quite variable diet ranging from browse-dominated to grazing (Rivals et al. 2012; Saarinen et al. 2015). Dental mesowear and microwear analyses of the Late Pleistocene ungulate community of Alaska indicate unusually heavily grass-dominated dietary spectrum, with even typically browsing or mixed-feeding deer such as *Rangifer* and *Cervus* being grazers despite their brachydont teeth (Rivals et al. 2010).

2.3.6.3 Africa

The palaeoenvironments of East Africa became increasingly dominated by open C4 grasslands during the Pleistocene (e.g., Cerling et al. 1997; Jacobs et al. 2010). The ungulate communities were by now clearly dominated by abundant and diverse bovids. Based on stable isotope studies, the Pleistocene large herbivorous mammal faunas became dominated by C4 grazers, including alcelaphine (e.g., *Megalotragus*, *Connochaetes*, *Alcelaphus*), bovine (e.g., *Pelorovis*) and reduncine (e.g., *Kobus*) bovids, zebras (*Equus*), suids of the *Metridiochoerus-Phacochoerus* lineage, the white rhinoceros (*Ceratotherium*), elephants (*Elephas* and *Loxodonta*) and even sivatherine giraffes (*Sivatherium*) (Cerling et al. 2015). Also hippos (*Hippopotamus*) were locally C4 grazers in some regions such as Turkana and Nakuru. The less diverse browsers included the black rhinoceros (*Diceros*), giraffe (*Giraffa*), cephalophine bovids and the last surviving deinotheriid proboscidean *Deinotherium bozasi* (Cerling et al. 2015). Some lineages of large herbivores, such as the African elephant (*Loxodonta africana*) and suids of the *Kolpochoerus-Hylochoerus* lineage, shifted their diet from C4-grass dominated grazing to C3-dominated browsing from the Pleistocene to the Holocene (Cerling et al. 2015). Also, dental mesowear analysis of the Middle Pleistocene to Holocene bovids from the Kibish Formation, Ethiopia, shows that all of the taxa shifted their diets to less abrasive (more browse-dominated) from the Pleistocene to the Holocene (Rowan et al. 2015).

2.3.6.4 South America

During the Early Pleistocene, the Great American Biotic Interchange between North and South America resulted in the introduction of many new groups of large herbivorous mammals from North to South America, such as gomphotheriid proboscideans (*Stegomastodon*, *Cuvieronius*), lamine camels (e.g., *Palaeolama*), horses (*Onohippidion*, *Hippidion*, *Equus*), deer (which radiated into many endemic South American genera such as *Ozotoceros*, *Hippocamelus* and *Mazama*) and tapirs (*Tapirus*) (Woodburne 2010). Of these, proboscideans and horses went extinct in South America at the end of the Pleistocene. The new immigrants formed diverse mammal communities together with the endemic South American mammals, such as notoungulates (e.g., *Toxodon*), litopterns (e.g., *Macrauchenia*) and xenarthans, including diverse megatheriid, megalonychid and mylodontid ground sloths (e.g., *Megatherium*, *Nothrotherium*, *Glossotherium*, *Lestodon* and *Scelidotherium*) and herbivorous armadillo-related glyptodonts (e.g., *Glyptodon*, *Neosclerocalyptus*, *Doedicurus*) and pampatheres (e.g., *Holmesina*).

MacFadden et al. (1994) analysed the dietary composition of the typical Pleistocene South American ungulates. Their results indicate a continuum of diets from browsing to grazing, similar to the other continents during the Pleistocene: among the horses, *Hippidion* had C3-plant dominated (browsing) diets, *Onohippidion* had mixed C3/C4 diets and *Equus* had C4 grass -dominated diets, and similarly among the lamine camels, *Palaeolama* had C3-dominated and *Lama* C4 grass -dominated diets. The gomphothere *Cuvieronius* had highly variable isotope signals indicating a diet based on wide range of plants. Finally, based on the isotope analyses, the large, hypselodont notoungulate *Toxodon* was a C4 grazer, whereas the litoptern *Macrauchenia* was a browser on C3 photosynthesising vegetation. New stable isotope analyses of South American Pleistocene gomphotheres indicate that most of them were mixed-feeders, but their diets varied considerably according to environmental conditions (Sánchez et al. 2004). For example, the population of *Cuvieronius* from Chile had exclusive C3 diets whereas the populations from Bolivia and Ecuador had mixed C3/C4 diets. The diets of *Stegomastodon* varied even more, ranging from C3 browsing in Argentina to C4 grazing in Ecuador. This kind of flexibility in diet is typical for Pleistocene herbivorous mammals on other continents too, and is particularly pronounced in proboscideans (Rivals et al. 2012, 2015, Rivals and Lister 2016; Saarinen et al. 2016; Saarinen and Lister 2016).

Finally large herbivorous xenarthrans, including several species of giant ground sloths, glyptodonts and pampatheres, were abundant, adding to the already diverse large herbivore faunas (Fariña 1996; Vizcaíno 2009). In fact, it has been puzzling how such abundant and diverse megaherbivore communities could have coexisted in the Pleistocene South American palaeoenvironments (see Fariña 1996). However, Saarinen and Karme (2017) noted, based on new dental mesowear analysis, that there were significant differences between the diets of the Pleistocene xenarthrans: among the ground sloths *Megatherium* and *Nothrotherium* were specialised browsers, whereas the various mylodontid sloths had diets ranging from mixed-feeding (*Glossotherium*) to grazing (*Scelidotherium* and *Lestodon*). Among the glyptodonts, *Neosclerocalyptus* and *Glyptodon* were grass-dominated mixed-feeders, whereas the pampatheres, such as *Holmesina*, were most likely specialised grazers. These new dietary observations are in agreement with previous results based on ecomorphological, stable isotope and microwear analyses (e.g., Bargo et al. 2006; Vizcaíno 2009; Domingo et al. 2012; Green and Resar 2012), but revealed more diversity in the diets than had been anticipated.

2.3.6.5 Australia

The Pleistocene herbivorous marsupial faunas of Australia were diverse and included species with highly varied diets, paralleling the patterns seen on other continents. Based on stable isotope and dental microwear analyses, DeSantis et al. (2017) were able to resolve the diets of the Middle Pleistocene megafauna from Cuddie Springs, and found them to be largely browse-dominated. Among kangaroos, the "short-faced" kangaroo *Sthenurus* was a browser in more dense-canopy environments than the rest of the taxa, *Protemnodon* appears to have been a browser consuming C3 browse but also C4 photosynthesising shrubs, and *Macropus* a mixed-feeder feeding more heavily on C4 grasses and shrubs. Stable isotope and dental microwear analyses indicate that the giant Pleistocene short-faced kangaroo, *Procoptodon goliath*, was a specialised browser of chenopods and C4 photosynthesising saltbushes (Prideaux et al. 2009). All the Pleistocene kangaroos seem to have been more browse-dominated feeders than the modern grazing kangaroos. The diprotodontids, including the giant, megaherbivore-sized *Diprotodon* and the somewhat smaller *Zygomaturus*, were browsers or mixed-feeders with seasonally varying diets. The wombats *Phascolonus* and *Vombatus* were the most heavily C4-dominated grazers, as wombats are today (DeSantis et al. 2017). There was a considerable drying and desertification of environments from the Middle to Late Pleistocene in Australia, which led to a restriction in the diets of the herbivorous marsupials from C4 plants to increasingly C3-dominated vegetation (DeSantis et al. 2017).

2.4 Discussion

The evolutionary history of browsing and grazing and be summarized as follows:

- Many lineages of frugivorous-folivorous mammals evolved from omnivorous ancestors during the Palaeocene and early Eocene, including early artiodactyls, perissodactyls and proboscideans.
- Specialised folivorous browsing mammals evolved and prevailed during the Eocene and the Oligocene, and continued to be the dominant elements in most parts of the world until the Late Miocene.
- Following the spread of grassland environments, the major shift from browsing to grazing happened in many major lineages of large herbivorous mammals (e.g., bovids and elephants) during the Late Miocene. However, following the earlier emergence of grassland environments, the earliest grazers evolved during the Oligocene in South America among notoungulates and sloths, and in North America during the mid-Miocene among equine horses. Morphological adaptations to grazing in herbivorous mammals follows the emergence of new (abundant) resources, first the increase in C3 grasses, and then that of C4 grasses. This emergence of resources is a reaction to changing rainfall patterns, temperature and CO_2-levels, all of which react to tectonic movements and tilt of the earth's axis.
- During the Pliocene and Pleistocene, and to a lesser degree in Holocene and modern times, diverse herbivore faunas with a dietary spectrum ranging from browsers to grazers have been typical in all continents. The dietary spectrum of the species, and dietary variation within species, reflect environmental and climatic variations locally and through time.

I have deliberately concentrated in discussing the dietary history of specialised large terrestrial herbivorous mammals such as ungulates, proboscideans and herbivorous xenarthrans and marsupials, and left out primates (which have specialised, often arboreal lifestyles) and small mammals (especially rodents, which are predominantly herbivorous but are able to utilise micro-niches in their environments). However, primates and rodents do, in fact, show roughly similar dietary histories as well, starting out as omnivores and frugivores, and later specialising at browsing and in some cases grazing. For example, many cercopithecid monkeys, which were the dominant primates in African forests from the Miocene until today, were specialised leaf browsers, whereas giant baboons (*Theropithecus oswaldi*) occupying East African savannas in the Plio-Pleistocene were grazers (Cerling et al. 2013). In fact, even some early human relatives (*Paranthropus boisei*) could have been feeding on C4 grasses (or their seeds) in East Africa during the Early Pleistocene (Cerling et al. 2011).

The various proxy methods (stable isotope analyses and dental microwear and mesowear) together provide us with an increasingly comprehensive set of tools for examining the diets of fossil herbivorous mammals, especially browsing vs. grazing. Whereas the ecomorphology of fossil mammals such as tooth crown height and complexity of their enamel patterns give valuable clues to what they were adapted to eat, the tooth wear and isotope proxies enable us to see variation in their diets at the level of ancient populations and even individuals. Thus we are now provided with fascinatingly detailed information about past herbivores' diets. For example, recent studies of dental mesowear have revealed a roughly similar evolution of dietary strategies in many groups throughout their evolutionary history, for example the horses (Equidae): they started as relatively generalistic herbivores feeding on leaves, fruit and seeds in tropical and subtropical forests, shifted to browsing in increasingly seasonal forests and woodlands, and finally shifted towards increasingly grass-based diets when grassland environments started to spread (Mihlbachler et al. 2011). Despite adaptations to utilising different diets, many large herbivorous mammals kept a degree of flexibility in their diet, being able to shift it according to changing environmental conditions (e.g., Rivals and Lister 2016; Saarinen et al. 2016; Saarinen and Lister 2016).

Acknowledgements I thank Jenny and Antti Wihuri Foundation for funding my work as a post doc researcher in the Natural History Museum of London during this work. I would also like to thank Christine Janis and Iain Gordon for their constructive suggestions which helped me improve this chapter.

Glossary

Cenozoic the last ca. 66 million years that started from the end-Cretaceous mass extinction and continues today. The Cenozoic is characterized by a warm and humid beginning followed by mostly cooling and drying global climate, which led to the spread of open, arid environments and grasslands. It is also the time when mammals diversified and filled the ecological niches of large terrestrial herbivores.

Neogene ca. 23–2.6 million years ago, during which climatic cooling and drying led to increasing coverage of grass-dominated open habitats in many parts of the world, driving the evolution of adaptations to grazing in many lineages of herbivorous mammals.

Pleistocene ca. 2.6 million years to ca. 11,000 years ago, until the present warm-climatic stage. This was the time of the ice ages characterized by strong cyclic variations in climate, environments and the distribution of plants and animals in the northern hemisphere.

Mesowear dietary analysis method applicable to present and fossil mammals, based on the wear-induced shape of the occlusal surface of molar teeth. It indicates the amount of grass (in relation to browse) in diet.

Microwear dietary analysis method based on microscopic wear marks on tooth enamel, which reflect the relative amounts of grass, browse, seeds and other dietary items during the last days of an animal's life.

Bunodont tooth morphology type where the cusps are separate and not fused or connected by elongated ridges.

Lophodont tooth morphology type where the cusps are elongated and connected into long cutting ridges (lophs).

Plagiolophodont derived lophodont tooth morphology where the lophs are folded and fused to form a flat occlusal surface with shearing enamel edges, often supported by extensive dental cement that covers the tooth crown.

Selenodont tooth morphology type where the cusps have been elongated into crescent-shaped cutting blades. This is the typical tooth morphology of ruminants and camels.

Bilophodont tooth morphology type where anterior and posterior cusp pairs have been fused into two transverse cutting lophs.

Loxodont tooth morphology type where the amount of transverse cutting lophs has been multiplied to form an efficient shearing surface with multiple enamel ridges, often supported by extensive dental cement between the lamellae.

Hypsodont a relatively high tooth crown, as opposed to Brachydont which refers to a relatively short crown.

Perissodactyla odd-toed ungulates, including horses (Equidae), rhinoceroses (Rhinocerotidae), tapirs (Tapiridae) and many extinct families such as chalicotheres (Chalicotheriidae), brontotheres (Brontotheriidae), paleotheres (Palaeotheriidae), hyracodonts (Hyracodontidae) and amynodonts (Amynodontidae).

Artiodactyla even-toed ungulates, including ruminants (Ruminantia), camels (Camelidae), pigs (Suidae), peccaries (Tayassuidae), hippopotami (Hippopotamidae) and many extinct families such as anthracotheres (Anthracotheriidae), entelodont (Entelodontidae) and oreodonts (Merycoidodontidae).

Proboscidea the mammal order that comprises elephants and their fossil relatives

Xenarthra the mammal order containing sloths, armadillos, anteaters and their fossil relatives such as glyptodonts and ground sloths.

Notoungulata an extinct order of endemic South American ungulates.

Litopterna an extinct order of endemic South American ungulates.

References

Agustí J, Moya-Sola S (1990) Mammal extinctions in the Vallesian (Upper Miocene). In: Kauffman EG (ed) Extinction events in earth history, vol IV. Springer, Berlin, pp 425–432

Alroy J (1998) Cope's rule and the dynamics of body mass evolution in North American fossil mammals. Science 280:731–734

Archer M, Godthelp H, Hand SJ, Megirian D (1989) Fossil mammals of Riversleigh, Northwestern Queensland: preliminary overview of biostratigraphy, correlation and environmental change. Aust Zool 25:29–65

Archer D, Winguth A, Lea D, Mahowald N (2000) What caused the glacial/interglacial atmospheric pCO_2 cycles? Rev Geophys 38:159–189
Bamford MK (2011a) Fossil woods. In: Harrison T (ed) Paleontology and geology of Laetoli: human evolution in context: volume 1: geology, geochronology, paleoecology and paleoenvironment. Springer, Amsterdam, pp 217–134
Bamford MK (2011b) Fossil leaves, fruits and seeds. In: Harrison T (ed) Paleontology and geology of Laetoli: human evolution in context: volume 1: geology, geochronology, paleoecology and paleoenvironment. Springer, Amsterdam, pp 235–252
Bargo MS, Toledo N, Vizcaíno SF (2006) Muzzle of South American Pleistocene ground sloths (Xenarthra, Tardigrada). J Morphol 267:248–263
Barnosky AD, Koch PL, Feranec RS, Wing SL, Shabel AB (2004) Assessing the causes of Late Pleistocene extinctions on the continents. Science 306:70–75
Bartoli G, Hönisch B, Zeebe RE (2011) Atmospheric CO2 decline during the Pliocene intensification of Northern Hemisphere glaciations. Paleoceanography 26:PA4213. https://doi.org/10.1029/2010PA002055
Beard C (1998) East of Eden: Asia as an important center of taxonomic origination in mammalian evolution. Bull Carnegie Mus Nat Hist 34:5–39
Bernor RL, Semprebon G, Damuth J (2014) Maragheh ungulate mesowear: interpreting paleodiet and paleoecology from a diverse fauna with restricted sample. Ann Zool Fenn 51:201–208
Bibi F (2007) Dietary niche partitioning among fossil bovids in late Miocene C3 habitats: consilience of functional morphology and stable isotope analysis. Palaeogeogr Palaeoclimatol Palaeoecol 253:529–538
Black KH (2016) Middle Miocene origins for tough-browse dietary specialisations in the koala (Marsupialia, Phascolarctidae) evolutionary tree: description of a new genus and species from the Riversleigh World Heritage Area. Mem Mus Vic 74:255–262
Blondel C (1998) Etude morphologique du squelette appendiculaire des ruminants de l'OligoceÁne d'Europe occidentale; implications environnementales. C R Acad Sci 326:527–532
Blondel C (2001) The Eocene-Oligocene ungulates from Western Europe and their environment. Palaeogeogr Palaeoclimatol Palaeoecol 168:125–139
Boardman GS, Secord R (2013) Stable isotope paleoecology of White River ungulates during the Eocene–Oligocene climate transition in northwestern Nebraska. Palaeogeogr Palaeoclimatol Palaeoecol 375:38–49
Boisserie J-R, Lihoreau F (2006) Emergence of Hippopotamidae: new scenarios. C R Palevol 5:749–756
Boisserie J-R, Zazzo A, Merceron G, Blondel C, Vignaud P, Likius A, Mackaye HT, Brunet M (2005) Diets of modern and late Miocene hippopotamids: evidence from carbon isotope composition and micro-wear of tooth enamel. Palaeogeogr Palaeoclimatol Palaeoecol 221:153–174
Butler K, Louys J, Travouillon K (2014) Extending dental mesowear analyses to Australian marsupials, with applications to six Plio-Pleistocene kangaroos from Southeast Queensland. Palaeogeogr Palaeoclimatol Palaeoecol 408:11–25
Butler K, Travouillon KJ, Price GJ, Archer M, Hand SJ (2017) Species abundance, richness and body size evolution of kangaroos (Marsupialia: Macropodiformes) throughout the Oligo-Miocene of Australia. Palaeogeography, Palaeoclimatology, Palaeoecology. https://doi.org/10.1016/j.palaeo.2017.08.016
Calandra I, Göhlich UB, Merceron G (2008) How could sympatric megaherbivores coexist? Example of niche partitioning within a proboscidean community from the Miocene of Europe. Naturwissenschaften 95:831–838
Cerling TE (1992) Development of grasslands and savannas in East Africa during the Neogene. Palaeogeogr Palaeoclimatol Palaeoecol 97:241–247
Cerling TE, Harris JM, MacFadden BJ, Leakey MG, Quade J et al (1997) Global change through the Miocene/Pliocene boundary. Nature 389:153–158
Cerling TE, Harris JM, Leakey MG (1999) Browsing and grazing in elephants: the isotope record of modern and fossil proboscideans. Oecologia 120:364–374

Cerling TE, Mbua E, Kirera FM, Manthi FK, Grine FE, Leakey MG, Sponheimer M, Uno KT (2011) Diet of *Paranthropus boisei* in the early Pleistocene of East Africa. PNAS 108:9337–9341

Cerling TE, Chritz KL, Jablonski NG, Leakey MG, Manthi FK (2013) Diet of *Theropithecus* from 4 to 1 Ma in Kenya. PNAS 110:10507–10512

Cerling TE, Andanje SA, Blumenthal SA, Brown FH, Chritz KL, Harris JM, Hart JA, Kirera FM, Kaleme P, Leakey LN, Leakey MG, Levin NE, Manthi FK, Passey PK, Uno KT (2015) Dietary changes of large herbivores in the Turkana Basin, Kenya from 4 to 1 Ma. PNAS 112:11467–11472

Clauss M, Rössner GE (2014) Old world ruminant morphophysiology, life history, and fossil record: exploring key innovations of a diversification sequence. Ann Zool Fenn 51:80–94

Collinson ME (1992) Chapter 22: Vegetational and foristic changes around the Eocene/Oligocene boundary in Western and Central Europe. In: Prothero DR, Berggren WA (eds) Eocene-Oligocene climatic and biotic evolution. Princeton University Press, Oxford, pp 437–450

Collinson ME, Hooker JJ (1987) Vegetational and mammalian faunal changes in the early tertiary of Southern England. In: Chaloner WG, Crane PR, Friis EM (eds) The origin of angiosperms and their biological consequences. Cambridge University Press, Cambridge, pp 259–304

Coombs MC (1978) Additional *Schizotherium* material from China, and a review of *Schizotherium* dentitions (Perissodactyla, Chalicotheriidae). Am Mus Novit 2647:1–18

Croft DA (2016) Horned armadillos and rafting monkeys – the fascinating fossil mammals of South America. Indiana University Press, Bloomington, IN

Croft DA, Weinstein D (2008) The first application of mesowear method to endemic South American ungulates (Notoungulata). Palaeogeogr Palaeoclimatol Palaeoecol 269:103–114

Damuth J, Janis CM (2011) On the relationship between hypsodonty and feeding ecology in ungulate mammals, and its utility in palaeoecology. Biol Rev 86:733–758

Damuth J, Janis CM (2014) A comparison of observed molar wear rates in extant herbivorous mammals. Ann Zool Fenn 51:188–200

Dawson TJ, Dawson L (2006) Evolution of arid Australia and consequences for vertebrates. In: Merrick JR, Archer M, Hickey GM, Lee MSY (eds) Evolution and biogeography of Australasian vertebrates. Auscipub, Sydney, pp 51–70

De Franceschi D, Hoorn C, Antoine PO, Cheema IU, Flynn LJ, Lindsay EH, Marivaux L, Metais G, Rajpar R, Welcomme J-L (2008) Floral data from themid-Cenozoicof Central Pakistan. Rev Palaeobot Palynol 150:115–129

DeMiguel D, Fortelius M, Azanza B, Morales J (2008) Ancestral feeding state of ruminants reconsidered: earliest grazing adaptation claims a mixed condition for Cervidae. BMC Evol Biol 8:1–13

Deng T, Wang X, Fortelius M, Li Q, Wang Y, Tseng ZJ, Takeuchi GT, Sylor JE, Säilä LK, Xie G (2011) Out of Tibet: Pliocene woolly rhino suggests high-plateau origin of Ice Age megaherbivores. Science 333:1285–1288

DeSantis LRG, Field JH, Wroe S, Dodson JR (2017) Dietary responses of Sahul (Pleistocene Australia–New Guinea) megafauna to climate and environmental change. Paleobiology 43:181–195

Domingo L, Prado JL, Alberdi MT (2012) The effect of paleoecology and paleobiogeography on stable isotopes of Quaternary mammals from South America. Quat Sci Rev 55:103–113

Erlebe JJ, Greenwood DR (2012) Life at the top of the greenhouse Eocene world - a review of the Eocene flora and vertebrate fauna from Canada's High Arctic. Geol Soc Am Bull 124:3–23

Eronen JT, Puolamaki K, Liu L, Lintulaakso K, Damuth J et al (2010a) Precipitation and large herbivorous mammals II: application to fossil data. Evol Ecol Res 12:235–248

Eronen JT, Evans AR, Fortelius M, Jernvall J (2010b) The impact of regional climate to the evolution of mammals: a case study using fossil horses. Evolution 64:398–408

Eronen JT, Kaakinen A, Liu L-P, Passey BH, Tang H, Zhang Z-Q (2014) Here be dragons: Mesowear and tooth enamel isotopes of the classic Chinese "*Hipparion*" faunas from Baode, Shanxi Province, China. Ann Zool Fenn 51:227–244

Fariña RA (1996) Trophic relationships among Lujanian mammals. Evol Theory 11:125–134
Feranec RS (2003) Stable isotopes, hypsodonty, and the paleodiet of *Hemiauchenia* (Mammalia: Camelidae): a morphological specialization creating ecological generalization. Paleobiology 29:230–242
Fortelius M (1985) Ungulate cheek teeth: developmental, functional and evolutionary interrelations. Acta Zool Fenn 180:1–76
Fortelius M, Solounias N (2000) Functional characterization of ungulate molars using the abrasion-attrition wear gradient: a new method for reconstructing paleodiets. Am Mus Novit 3301:1–35
Fortelius M, Eronen JT, Jernvall J, Liu L, Pushkina D et al (2002) Fossil mammals resolve regional patterns of Eurasian climate change during 20 million years. Evol Ecol Res 4:1005–1016
Fortelius M, Eronen J, Liu L, Pushkina D, Tesakov A, Vislobokova I, Zhang Z (2006) Late Miocene and Pliocene large land mammals and climatic changes in Eurasia. Palaeogeogr Palaeoclimatol Palaeoecol 238:219–227
Fortelius M, Zliobaité I, Kaya F, Bibi F, Bobe R, Leakey L, Leakey M, Patterson D, Rannikko J, Werdelin L (2016) An ecometric analysis of the fossil mammal record of the Turkana Basin. Philos Trans R Soc B 371:1–13
Foss SE (2007) Family Entelodontidae. In: Prothero DR, Foss SE (eds) The evolution of artiodactyls. Johns Hopkins University Press, Baltimore, MD, pp 120–129
Franz-Odendaal TA, Lee-Thorp JA, Chinsamy A (2002) New evidence for the lack of C4 grassland expansions during the early Pliocene at Langebaanweg, South Africa. Paleobiology 28:378–388
Fraser D, Zybutz T, Lightner E, Theodor J (2014) Ruminant mandibular tooth mesowear: a new scheme for increasing paleoecological sample sizes. J Zool 294:41–48
Froehlich DJ (2002) Quo vadis eohippus? The systematics and taxonomy of the early Eocene equids (Perissodactyla). Zool J Linnean Soc 134:141–256
Geist V (1998) Deer of the world – their evolution, behaviour, and ecology. Stackpole Books, Mechanicsburg, Pennsylvania
Godthelp H, Archer M, Cifelli R, Hand SJ, Gilkeson CF (1992) Earliest known Australian tertiary mammal fauna. Nature 356:514–516
Gordon IJ (2003) Browsing and grazing ruminants: are they different beasts? For Ecol Manag 181:13–21
Gordon IJ, Prins HHT (2008) The ecology of browsing and grazing, Ecological Studies 195. Springer, Berlin
Green JL, Resar NA (2012) The link between dental microwear and feeding ecology in tree sloths and armadillos (Mammalia: Xenarthra). Biol J Linn Soc 107:277–294
Gregory-Wodzicki KM (2000) Uplift history of the Central and Northern Andes: a review. Geol Soc Am Bull 112:1091–1105
Grossman A, Liutkus-Pierce C, Kyongo B, M'Kirera F (2014) New fauna from Loperot contributes to the understanding of early Miocene catarrhine communities. Int J Primatol 35:1253–1274
Gunnell GF, Murphey PC, Stucky RK, Townsend KEB, Robinson B, Zonneveld J-P, Bartels WS (2009) Biostratigraphy and biochronology of the latest Wasatchian, Bridgerian, and Uintan North American land mammal "ages". Mus North Ariz Bull 65:279–330
Guthrie DR (2001) Origin and causes of the mammoth steppe: a story of cloud cover, woolly mammal tooth pits, buckles, and inside-out Beringia. Quat Sci Rev 20:549–574
Harris JM, Cerling TE, Leakey MG, Passey BH (2008) Stable isotope ecology of fossil hippopotamids from the Lake Turkana Basin of East Africa. J Zool 275:323–331
Hays JD, Imbrie J, Shackleton NJ (1976) Variations in the earth's orbit: pacemaker of the Ice Ages. Science 194:1121–1132
Head MJ (1998) Pollen and dinoflagellates from the Red Crag at Walton-on-the-Naze, Essex: evidence for a mild climatic phase during the early late Pliocene of eastern England. Geol Mag 135:803–817
Herbert TD, Lawrence KT, Tzanova A, Cleaveland-Patterson L, Caballero-Gill R, Kelly CS (2016) Late Miocene global cooling and the rise of modern ecosystems. Nat Geosci 9:843–847

Hernesniemi E, Blomstedt K, Fortelius M (2011) Multi-view stereo three-dimensional reconstruction of lower molars of recent and Pleistocene rhinoceroses for mesowear analysis. Palaeontol Electron 14:1–15
Hofmann RR, Stewart DRM (1972) Grazer or browser: a classification based on the stomach-structure and feeding habits of East African ruminants. Mammalia 36:226–240
Hooker JJ (2007) Bipedal browsing adaptations of the unusual Late Eocene–earliest Oligocene tylopod *Anoplotherium* (Artiodactyla, Mammalia). Zool J Linnean Soc 151:609–659
Hooker JJ, Collinson ME (2012) Mammalian faunal turnover across the Palaeocene-Eocene boundary in NW Europe: the roles of displacement, community evolution and environment. Austrian J Earth Sci 105:17–28
Ilius AW, Gordon IJ (1992) Modelling the nutritional ecology of ungulate herbivores: evolution of body size and competitive interactions. Oecologia 89:428–434
Jacobs BF, Kingston JD, Jacobs LL (1999) The origin of grass-dominated ecosystems. Ann Mo Bot Gard 86:590–643
Jacobs BF, Pan AD, Scotese CR (2010) A review of the Cenozoic vegetation history of Africa. In: Werdelin L, Sanders WJ (eds) Cenozoic mammals of Africa. University of California Press, Berkeley, CA, pp 57–72
Janis CM (1976) The evolutionary strategy of the Equidae and the origins of rumen and cecal digestion. Evolution 30:757–774
Janis CM (1982) Evolution of horns in ungulates: ecology and paleoecology. Biol Rev 57:261–318
Janis CM (1993) Tertiary mammal evolution in the context of changing climates, vegetation, and tectonic events. Annu Rev Ecol Syst 24:467–500
Janis CM (1995) Correlation between craniodental morphology and feeding behavior in ungulates: reciprocal illumination between living and fossil taxa. In: Thomason JJ (ed) Functional morphology in vertebrate paleontology. Cambridge University Press, Cambridge, pp 76–98
Janis CM (2007) The horse series. In: Regal B (ed) Icons of evolution. Greenwood Press, West-port, CT, pp 257–280
Janis CM (2008) An evolutionary history of browsing and grazing ungulates. In: Gordon IJ, Prins HHT (eds) The ecology of browsing and grazing, Ecological Studies 195. Springer, Berlin, pp 21–45
Janis CM, Fortelius M (1988) On the means whereby mammals achieve increased functional durability of their dentitions, with special reference to limiting factors. Biol Rev 63:197–230
Janis CM, Scott KM, Jacobs LL (1998) Evolution of tertiary mammals of North America: volume 1, terrestrial carnivores, ungulates and ungulatelike mammals. Cambridge University Press, Cambridge
Janis CM, Damuth J, Theodor JM (2002) The origins and evolution of the North American grassland biome: the story from the hoofed mammals. Palaeogeogr Palaeoclimatol Palaeoecol 177:183–198
Janis CM, Damuth J, Theodor JM (2004) The species richness of Miocene browsers, and implications for habitat type and primary productivity in the North American grassland biome. Palaeogeogr Palaeoclimatol Palaeoecol 207:371–398
Janis CM, Damuth J, Travouillon KJ, Figueirido B, Hand SJ, Archer M (2016) Palaeoecology of Oligo-Miocene macropodoids determined from craniodental and calcaneal data. Mem Mus Vic 74:209–232
Joomun SC, Hooker JJ, Collinson ME (2008) Dental wear variation and implications for diet: an example from Eocene perissodactyls (Mammalia). Palaeogeogr Palaeoclimatol Palaeoecol 263:92–106
Kahlke R-D (1999) The history of the origin, evolution and dispersal of the late pleistocene *Mammuthus-Coelodonta* faunal complex in Eurasia (Large Mammals). Rotterdam, the Netherlands
Kahlke R-D, Kaiser TM (2011) Generalism as a subsistence strategy: advantages and limitations of the highly flexible feeding traits of Pleistocene *Stephanorhinus hundsheimensis* (Rhinocerotidae, Mammalia). Quat Sci Rev 30:2250–2261

Kaiser TM (2004) The dietary regimes of two contemporaneous populations of *Hippotherium primigenium* (Perissodactyla, Equidae) from the Vallesian (Upper Miocene) of Southern Germany. Palaeogeogr Palaeoclimatol Palaeoecol 198:381–402

Kaiser TM (2009) *Anchitherium aurelianense* (Equidae, Mammalia): a brachydont "dirty browser" in the community of herbivorous large mammals from Sandelzhausen (Miocene, Germany). Paläontol Z 83:131–140

Kaiser TM (2011) Feeding ecology and niche partitioning of Laetoli ungulate faunas. In: Harrison T (ed) Paleontology and geology of Laetoli: human evolution in context: volume 1: geology, geochronology, paleoecology and paleoenvironment. Springer, Dordrecht, pp 329–354

Kaiser TM, Müller DWH, Fortelius M, Schulz E, Codron D, Clauss M (2013) Hypsodonty and tooth facet development in relation to diet and habitat in herbivorous ungulates: implications for understanding tooth wear. Mammal Rev 43:34–46

Konidaris GE, Koufos GD, Kostopoulos DS, Merceron G (2016) Taxonomy, biostratigraphy and palaeoecology of *Choerolophodon* (Proboscidea, Mammalia) in the Miocene of SE Europe – SW Asia: implications for phylogeny and biogeography. J Syst Palaeontol 14:1–27

Kovar-Eder J, Jechorek H, Kvaček Z, Parashiv V (2008) The integrated plant record: an essential tool for reconstructing Neogene zonal vegetation in Europe. PALAIOS 23:97–111

Kurtén B (1968) Pleistocene mammals of Europe. Aldine, Chicago

Kurtén B (1972) The ice age. Hart-Davis, London

Kurtén B, Anderson E (1980) Pleistocene mammals of North America. Columbia University Press, New York

Lander B (1998) Oreodontoidea. In: Janis CM, Scott KM, Jacobs LL (eds) Evolution of tertiary mammals of North America: volume 1, terrestrial carnivores, ungulates and ungulatelike mammals. Cambridge University Press, Cambridge, pp 402–425

Lee-Thorp J, van der Merwe NJ (1987) Carbon isotope analysis of fossil bone apatite. S Afr J Sci 83:712–715

Lihoreau F, Ducrocq S (2007) Family Anthracotheriidae. In: Prothero DR, Foss SE (eds) The evolution of artiodactyls. Johns Hopkins University Press, Baltimore, MD, pp 89–105

Lisiecki LE, Raymo ME (2007) Plio-Pleistocene climate evolution: trends and transitions in glacial cycle dynamics. Quat Sci Rev 26:56–69

Lister AM (2013) The role of behaviour in adaptive morphological evolution of African proboscideans. Nature 500:331–334

Lister AM, Sher AV, van Essen H, Wei G (2005) The pattern and process of mammoth evolution in Eurasia. Quat Int 126–128:49–64

Liu L-P (2001) Eocene suoids (Artiodactyla, Mammalia) from Bose and Yongle basins, China and the classification and evolution of the Paleogene suoids. Vertebrata Pal Asiatica 39:115–128

Liu L-P, Eronen JT, Fortelius M (2009) Significant mid-latitude aridity in the middle Miocene of East Asia. Palaeogeogr Palaeoclimatol Palaeoecol 279:201–206

Loffredo LF, DeSantis LRG (2014) Cautionary lessons from assessing dental mesowear observer variability and integrating paleoecological proxies of an extreme generalist *Cormohipparion emsliei*. Palaeogeogr Palaeoclimatol Palaeoecol 395:42–52

Loose HK (1975) Pleistocene Rhinocerotidae of W. Europe with reference to the recent two-horned species of Africa and S.E. Asia. Scr Geol 33:1–59

Louys J, Aplin K, Beck RMD, Archer M (2009) Cranial anatomy of Oligo-Miocene koalas (Diprotodontia: Phascolarctidae): stages in the evolution of an extreme leaf-eating specialization. J Vertebr Paleontol 29:981–992

Lucas PW, Omar R (2012) New perspectives of tooth wear. Int J Dent 2012. https://doi.org/10.1155/2012/287573

Lucas SG, Schoch RM (1998) Tillodontia. In: Janis CM, Scott KM, Jacobs LL (eds) Evolution of tertiary mammals of North America: volume 1, terrestrial carnivores, ungulates and ungulatelike mammals. Cambridge University Press, Cambridge, pp 268–273

Lucas SG, Schoch RM, Williamson TE (1998) Taeniodonta. In: Janis CM, Scott KM, Jacobs LL (eds) Evolution of tertiary mammals of North America: volume 1, terrestrial carnivores, ungulates and ungulatelike mammals. Cambridge University Press, Cambridge, pp 260–267

Lucas PW, van Casteren A, Al-Fadhalan K, Almusallam AS, Henry AG, Michael S, Watzke J, Reed DA, Diekwisch TGH, Strait DS, Atkins AG (2014) The role of dust, grit and phytoliths in tooth wear. Ann Zool Fenn 51:143–152
MacFadden BJ (2000) Cenozoic mammalian herbivores from the Americas: reconstructing ancient diets and terrestrial communities. Annu Rev Ecol Syst 31:33–59
MacFadden BJ, Wang Y, Cerling TE, Anaya F (1994) South American fossil mammals and carbon isotopes: a 25 million-year sequence from the Bolivian Andes. Palaeogeogr Palaeoclimatol Palaeoecol 107:257–268
MacFadden BJ, Cerling TE, Prado J (1996) Cenozoic terrestrial ecosystem evolution in Argentina: evidence from carbon isotopes of fossil mammal teeth. PALAIOS 11:319–327
Madden RH (2015) Hypsodonty in mammals. Cambridge University Press, Cambridge
Maglio VJ (1973) Origin and evolution of the Elephantidae. Trans Am Philos Soc 63:1–149
Mead AJ, Wall WP (1998) Dietary implications of jaw mechanics in the rhinocerotoids *Hyracodon* and *Subhyracodon* from Badlands National Park, South Dakota. In: McClelland L (ed) Santucci VL. National Park Service Paleontological Research, National Park Service, pp 18–23
Métais G, Chaimanee Y, Jaeger J-J, Ducrocq S (2001) New remains of primitive ruminants from Thailand: evidence of the early evolution of the Ruminantia in Asia. Zool Scr 30:231–248
Métais G, Antoine P-O, Marivaux L, Welcomme J-L, Ducrocq S (2003) New artiodactyl ruminant mammal from the late Oligocene of Pakistan. Acta Palaeontol Pol 48:375–382
Métais G, Qi T, Guo J, Beard KC (2005) A new bunoselenodont artiodactyl from the Middle Eocene of China and the early record of selenodont artiodactyls in Asia. J Vertebr Paleontol 25:994–997
Métais G, Welcomme J-L, Ducrocq S (2009) New lophiomerycid ruminants from the Oligocene of the Bugti Hills (Balochistan, Pakistan). J Vertebr Paleontol 29:231–241
Mihlbachler MC (2008) Species taxonomy, phylogeny, and biogeography of the Brontotheriidae (Mammalia: Perissodactyla). Bull Am Mus Nat Hist 311:1–475
Mihlbachler MC, Solounias M (2006) Coevolution of tooth crown height and diet in oreodonts (Merycoidodontidae, Artiodactyla) examined with phylogenetically independent contrasts. J Mamm Evol 13:11–36
Mihlbachler MC, Rivals F, Solounias N, Semprebon GM (2011) Dietary change and evolution of horses in North America. Science 331:1178–1181
Montanari S, Louys J, Price GJ (2013) Pliocene paleoenvironments of southeastern Queensland, Australia, inferred from stable isotopes of marsupial tooth enamel. PLoS One 8:e66221
Morgan GS, Lucas SG (2003) Mammalian biochronology of Blancan and Irvingtonian (Pliocene and early Pleistocene) faunas from New Mexico. Bull Am Mus Nat Hist 279:269–320
Noret J, Tabor NI, Jacobs BF, Sanders WJ, Kappelman J (2012) Stable isotope data from the Chilga Basin, Ethiopia, and their implications for resource partitioning among late Paleogene African endemic mammals. J Vertebr Paleontol 32(Suppl 2):150
Novello A, Blondel C, Brunet M (2010) Feeding behavior and ecology of the late Oligocene Moschidae (Mammalia, Ruminantia) from La Milloque (France): evidence from dental microwear analysis. C R Palevol 9:471–478
Orliac MJ, Antoine P-O, Roohi G, Welcomme J-L (2010) Suoidea (Mammalia, Cetartiodactyla) from the early Oligocene of the Bugti Hills, Balochistan, Pakistan. J Vertebr Paleontol 30:1300–1305
Ortiz-Jaureguizar E, Cladera GA (2006) Paleoenvironmental evolution of southern South America during the Cenozoic. J Arid Environ 66:498–532
Passey BH, Eronen JT, Fortelius M (2007) Paleodiets and paleoenvironments of late Miocene gazelles from North China: evidence from stable carbon isotopes. Vertebrata Pal Asiatica 45:118–127
Patnaik R (2015) Diet and habitat changes among Siwalik herbivorous mammals in response to Neogene and quaternary climate changes: an appraisal in the light of new data. Quat Int 371:232–243

Prasad V, Strömberg CAE, Alimohammadian H, Sahni A (2005) Dinosaur coprolites and the early evolution of grasses and grazers. Science 310:1177–1180

Prideaux GJ, Ayliffe LK, DeSantis LGR, Schubert BW, Murray PF, Gagan MK, Cerling TE (2009) Extinction implications of a chenopod browse diet for a giant Pleistocene kangaroo. PNAS 106:11646–11650

Prothero DR (1998) Protoceratidae. In: Janis CM, Scott KM, Jacobs LL (eds) Evolution of tertiary mammals of North America: volume 1, terrestrial carnivores, ungulates and ungulatelike mammals. Cambridge University Press, Cambridge, pp 431–438

Prothero DR (2013) Rhinoceros giants: the paleobiology of indricotheres. Indiana University Press, Bloomington, IN

Prothero DR, Manning E, Hanson CB (1986) The phylogeny of the Rhinocerotoidea (Mammalia, Perissodactyla). Zool J Linnean Soc 87:341–366

Prothero DR, Guérin C, Manning E (1989) The history of the Rhinocerotoidea. In: Prothero DR, Schoch RM (eds) The evolution of Perissodactyls. Oxford University Press, New York, pp 321–340

Pushkina D (2007) The Pleistocene easternmost distribution in Eurasia of the species associated with the Eemian *Palaeoloxodon antiquus* assemblage. Mammal Rev 37:224–245

Raia P, Carotenuto F, Passaro F, Piras P, Fulgione D, Werdelin L, Saarinen J, Fortelius M (2013) Rapid action in the Palaeogene, the relationship between phenotypic and taxonomic diversification in Cenozoic mammals. Proc R Soc B Biol Sci 280:2012–2244

Rasmussen DT, Gutiérrez M (2010) Hyracoidea. In: Werdelin L, Sanders WJ (eds) Cenozoic mammals of Africa. University of California Press, Berkeley, CA, pp 124–146

Reguero MA, Candela AM, Cassini GH (2010) Hypsodonty and body size in rodent-like notoungulates. In: Madden RH, Carlini AA, Vucetich MG, Kay RF (eds) The paleontology of gran Barranca: evolution and environmental change through the middle Cenozoic of Patagonia. Cambridge University Press, Cambridge, pp 362–371

Retallack GJ (1992) Middle Miocene fossil plants from fort Ternan (Kenya) and evolution of African grasslands. Paleobiology 18:383–400

Retallack GJ, Wynn JG, Benefit BR, Mccrossin ML (2002) Paleosols and paleoenvironments of the middle Miocene, Maboko formation, Kenya. J Hum Evol 42:659–703

Rivals F, Deniaux B (2003) Dental microwear analysis for investigating the diet of an argali population (*Ovis ammon antiqua*) of mid-Pleistocene age, Caune de l'Arago cave, eastern Pyrenees, France. Palaeogeogr Palaeoclimatol Palaeoecol 193:443–455

Rivals F, Deniaux B (2005) Investigation of human hunting seasonality through dental microwear analysis of two Caprinae in late Pleistocene localities in Southern France. J Archaeol Sci 32:1603–1612

Rivals F, Lister AM (2016) Dietary flexibility and niche partitioning of large herbivores through the Pleistocene of Britain. Quat Sci Rev 146:116–133

Rivals F, Solounias N, Mihlbachler MC (2007) Evidence for geographic variation in the diets of late Pleistocene and early Holocene *Bison* in North America, and differences from the diets of recent *Bison*. Quat Res:68338–68346

Rivals F, Mihlbachler MC, Solounias N, Mol D, Semprebon GM, de Vos J, Kalthoff DC (2010) Palaeoecology of the Mammoth Steppe fauna from the late Pleistocene of the North Sea and Alaska: separating species preferences from geographic influence in paleoecological dental wear analysis. Palaeogeogr Palaeoclimatol Palaeoecol 286:42–54

Rivals F, Semprebon G, Lister A (2012) An examination of dietary diversity patterns in Pleistocene proboscideans (*Mammuthus*, *Palaeoloxodon*, and *Mammut*) from Europe and North America as revealed by dental microwear. Quat Int 255:188–195

Rivals F, Mol D, Lacombat F, Lister AM, Semprebon GM (2015) Resource partitioning and niche separation between mammoths (*Mammuthus rumanus* and *Mammuthus meridionalis*) and gomphotheres (*Anancus arvernensis*) in the early Pleistocene of Europe. Quat Int 379:164–170

Rögl F (1998) Palaeogeographic considerations for Mediterranean and Paratethys seaways (Oligocene to Miocene). Annalen des Naturhistorischen Museums in Wien 99A:279–310

Rose KD (1981) Composition and species diversity in Paleocene and Eocene mammal assemblages: an empirical study. J Vertebr Paleontol 1:367–388

Rose KD (2006) The beginning of the age of mammals. Johns Hopkins University Press, Baltimore, MD

Rowan J, Faith JT, Gebru Y, Feagle JG (2015) Taxonomy and paleoecology of fossil Bovidae (Mammalia, Artiodactyla) from the Kibish formation, southern Ethiopia: implications for dietary change, biogeography, and the structure of the living bovid faunas of East Africa. Palaeogeogr Palaeoclimatol Palaeoecol 420:210–222

Saarinen J, Karme A (2017) Tooth wear and diets of extant and fossil xenarthrans (Mammalia, Xenarthra) – applying a new mesowear approach. Palaeogeogr Palaeoclimatol Palaeoecol 476:42–54

Saarinen J, Lister AM (2016) Dental mesowear reflects local vegetation and niche separation in Pleistocene proboscideans from Britain. J Quat Sci 31:799–808

Saarinen J, Boyer AG, Brown JH, Costa DB, Ernest SKM, Evans AR, Fortelius M, Gittleman JL, Hamilton MJ, Harding LE, Lintulaakso K, Lyons SK, Okie JG, Sibly RM, Stephens PR, Theodor J, Uhen MD, Smith FA (2014) Patterns of body size evolution in Cenozoic land mammals: intrinsic biological processes and extrinsic forcing. Proc R Soc B Biol Sci 281:20132049. https://doi.org/10.1098/rspb.2013.2049

Saarinen J, Karme A, Cerling T, Uno K, Säilä L, Kasiki S, Ngene S, Obari T, Mbua E, Manthi FK, Fortelius M (2015) A new tooth wear -based dietary analysis method for Proboscidea (Mammalia). J Vertebr Paleontol 35. https://doi.org/10.1080/02724634.2014.918546

Saarinen J, Eronen J, Fortelius M, Seppä H, Lister AM (2016) Patterns of diet and body mass of large ungulates from the Pleistocene of Western Europe, and their relation to vegetation. Palaeontol Electron 19.3.32A:1–58. palaeo-electronica.org/content/2016/1567-pleistocene-mammal-ecometrics

Samuels JX, Bredehoeft KE, Wallace SC (2018) A new species of *Gulo* from the Early Pliocene Gray Fossil Site (Eastern United States); rethinking the evolution of wolverines. PeerJ 6:e4648. https://doi.org/10.7717/peerj.4648

Sánchez B, Prado JL, Alberdi MT (2004) Feeding ecology, dispersal, and extinction of South American Pleistocene gomphotheres (Gomphotheriidae, Proboscidea). Paleobiology 30:146–161

Sanders WJ, Kappelman J, Rasmussen TD (2004) New large bodied mammals from the late Oligocene site of Chilga, Ethiopia. Acta Palaeontol Pol 49:365–392

Sanders WJ, Gheerbrant E, Harris JM, Saegusa H, Delmer C (2010) Proboscidea. In: Werdelin L, Sanders WJ (eds) Cenozoic mammals of Africa. University of California Press, Berkeley, CA, pp 124–146

Schmidt CW (2008) Dental microwear analysis of extinct flat-headed peccary (*Platygonus compressus*) from Southern Indiana. Proc Indiana Acad Sci 117:95–106

Scott RS, Ungar PS, Bergstrom TS, Brown CA, Grine FE, Teaford MF, Walker A (2005) Dental microwear texture analysis shows within-species diet variability in fossil hominins. Nature 436:693–695

Semprebon GM, Rivals F (2007) Was grass more prevalent in the pronghorn past? An assessment of the dietary adaptations of Miocene to recent Antilocapridae (Mammalia: Artiodactyla). Palaeogeogr Palaeoclimatol Palaeoecol 253:332–347

Semprebon GM, Rivals F (2010) Trends in the paleodietary habits of fossil camels from the tertiary and quaternary of North America. Palaeogeogr Palaeoclimatol Palaeoecol 295:131–145

Semprebon GM, Janis CM, Solounias N (2004) The diets of the Dromomerycidae (Mammalia: Artiodactyla) and their response to Miocene vegetational change. J Vertebr Paleontol 24:427–444

Semprebon GM, Sise PJ, Coombs MC (2011) Potential bark and fruit browsing as revealed by stereomicrowear analysis of the peculiar clawed herbivores known as chalicotheres (Perissodactyla, Chalicotherioidea). J Mamm Evol 18:33–55

Semprebon GM, Rivals F, Solounias N, Hulbert RC Jr (2016a) Paleodietary reconstruction of fossil horses from the Eocene through Pleistocene of North America. Palaeogeogr Palaeoclimatol Palaeoecol 442:110–127

Semprebon GM, Tao D, Hasjanova J, Solounias N (2016b) An examination of the dietary habits of *Platybelodon grangeri* from the Linxia Basin of China: evidence from dental microwear of molar teeth and tusks. Palaeogeogr Palaeoclimatol Palaeoecol 457:109–116

Shockey BJ, Anaya F (2011) Grazing in a new late Oligocene mylodontid sloth and a mylodontid radiation as a component of the Eocene-Oligocene faunal turnover and the early spread of grasslands/savannas in South America. J Mamm Evol 18:101–115

Smith FA, Boyer AG, Brown JH, Costa DP, Dayan T, Ernest SKM, Evans AR, Fortelius M, Gittleman J, Hamilton MJ, Harding LE, Lintulaakso K, Lyons SK, McCain C, Okie JK, Saarinen J, Sibly RM, Stephens PR, Theodor J, Uhen MD (2010) The evolution of maximum body size of terrestrial mammals. Science 330:1216–1219

Solounias N, Rivals F, Semprebon GM (2010) Dietary interpretation and paleoecology of herbivores from Pikermi and Samos (Late Miocene of Greece). Paleobiology 36:113–136

Solounias N, Semprebon GM, Mihlbachler MC, Rivals F (2013) Paleodietary comparisons of ungulates between the late Miocene of China, and Pikermi and Samos in Greece. In: Wang X, Flynn LJ, Fortelius M (eds) Fossil mammals of Asia: Neogene biostratigraphy and chronology, Columbia University Press, New York, p 676–692

Solounias N, Tariq M, Hou S, Danowitz M, Harrison M (2014) A new method of tooth mesowear and a test of it on domestic goats. Ann Zool Fenn 51:111–118

Stevens MS, Stevens JB (2007) Family Merycoidodontidae. In: Prothero DR, Foss SE (eds) The evolution of artiodactyls. Johns Hopkins University Press, Baltimore, pp 157–168

Strani F, DeMiguel D, Sardella R, Bellucci L (2015) Paleoenvironments and climatic changes in the Italian Peninsula during the early Pleistocene: evidence from dental wear patterns of the ungulate community of Coste San Giacomo. Quat Sci Rev 121:28–35

Strömberg CAE (2011) Evolution of grasses and grassland ecosystems. Annu Rev Earth Planet Sci 39:517–544

Strömberg CAE, McInerney FA (2011) The Neogene transition from C3 to C4 grasslands in North America: assemblage analysis of fossil phytoliths. Paleobiology 37:50–71

Strömberg CAE, Dunn RE, Madden RH, Kohn MJ, Carlini AA (2013) Decoupling the spread of grasslands from the evolution of grazer-type herbivores in South America. Nat Commun 4 (1478):1–8

Stuart AJ (1976) The history of the mammal fauna during the Ipswichian/last interglacial in England. Philos Trans R Soc B 276:221–250

Sturm M (1978) Maw contents of an Eocene horse (Propalaeotherium) out of the oil shale of Messel near Darmstadt. In: Kvacek Z, Schaarschmidt F (eds) Advances in angiosperm Palaeobotany, vol 30. Courier Forschungsinstitut, Senckenberg, pp 2–120

Sun J, Windley BF (2015) Onset of aridification by 34 Ma across the Eocene-Oligocene transition in Central Asia. Geology 43:1015–1018

Tang Z-H, Ding Z-L (2013) A palynological insight into the Miocene aridification in the Eurasian interior. Palaeoworld 22:77–85

Townsend KEB, Croft DA (2008) Diets of notoungulates from the Santa Cruz formation, Argentina: new evidence from enamel microwear. J Vertebr Paleontol 28:217–230

Travouillon KJ, Legendre S, Archer M, Hand SJ (2009) Palaeoecological analyses of Riversleigh's Oligo-Miocene sites: implications for Oligo-Miocene climate change in Australia. Palaeogeogr Palaeoclimatol Palaeoecol 276:24–37

Ulbricht A, Maul LC, Schulz E (2015) Can mesowear analysis be applied to small mammals? A pilot-study on leporines and murines. Mamm Biol 80:14–20

Ungar PS (2010) Mammal teeth: origin, evolution and diversity. Johns Hopkins University Press, Baltimore, MD

Ungar PS, Brown CA, Bergstrom TS, Walker A (2003) A quantification of dental microwear by tandem scanning confocal microscopy and scale-sensitive fractal analyses. Scanning 25:189–193

Uno KT, Cerling TE, Harris JM, Kunimatsu Y, Leakey MG, Nakatsukasa M, Nakaya H (2011) Late Miocene to Pliocene carbon isotope record of differential diet change among East African herbivores. PNAS 108:6509–6514

Urban MA, Nelson DM, Jiménez-Moreno G, Châteauneuf J-J, Pearson A, Hu FS (2010) Isotopic evidence of C4 grasses in southwestern Europe during the Early Oligocene–Middle Miocene. Geology 38:1091–1094

Valli AF, Palombo MR (2008) Feeding behaviour of middle-size deer from the Upper Pliocene site of Saint-Vallier (France) inferred by morphological and micro/mesowear analysis. Palaeogeogr Palaeoclimatol Palaeoecol 257:106–122

Van Asperen E, Kahlke R-D (2015) Dietary variation and overlap in Central and Northwest European *Stephanorhinus kirchbergensis* and *S. hemitoechus* (Rhinocerotidae, Mammalia) influenced by habitat diversity: "You'll have to take pot luck!" (proverb). Quat Sci Rev 107:47–61

Van Devender TR, McClaran MP (1995) Desert grassland history. In: McClaran MP, Van Devender TR (eds) The desert grassland. The University of Arizona Press, Tucson, Arizona, pp 68–99

Vicentini A, Barber JC, Aliscioni SS, Giussani LM, Kellogg EA (2008) The age of the grasses and clusters of origins of C4 photosynthesis. Glob Chang Biol 14:2963–2977

Vizcaíno SF (2009) The teeth of the "toothless": novelties and key innovations in the evolution of xenarthrans (Mammalia, Xenarthra). Palaeobiology 35:343–366

Walker M, Lowe J (2007) Quaternary science 2007: a 50-year retrospective. J Geol Soc 164:1073–1092

Walker A, Hoeck HN, Perez L (1978) Microwear of mammalian teeth as an indicator of diet. Science 201:908–910

Wall WP (1998) Amynodontidae. In: Janis CM, Scott KM, Jacobs LL (eds) Evolution of tertiary mammals of North America: volume 1, terrestrial carnivores, ungulates and ungulatelike mammals. Cambridge University Press, Cambridge, pp 583–588

Wang B (1992) The Chinese Oligocene: a preliminary review of mammalian localities and local faunas. In: Prothero DR, Berggren WA (eds) Eocene-Oligocene climatic and biotic evolution. Princeton University Press, Princeton, pp 529–547

Wang Y, Meng J, Ni X, Li C (2007) Major events in Paleogene mammal radiation in China. Geol J 42:415–430

Wang Y, Xu Y, Khawaja S, Passey BH, Zhang C, Wang X, Li Q, Tseng ZJ, Takeuchi GT, Deng T, Xie G (2013) Diet and environment of a mid-Pliocene fauna from southwestern Himalaya: Paleo-elevation implications. Earth Planet Sci Lett 376:43–53

Wang X, Li Q, Takeuchi GT (2016) Out of Tibet: an early sheep from the Pliocene of Tibet, Protovis himalayensis, genus and species nov. (Bovidae, Caprini), and origin of Ice Age mountain sheep. J Vertebr Paleontol 36:5, e1169190. https://doi.org/10.1080/02724634.2016.1169190

Werdelin L, Sanders WJ (2010) Cenozoic mammals of Africa. University of California Press, Berkeley, CA

West RG (1980) The pre-glacial Pleistocene of the Suffolk and Norfolk Coasts. Cambridge University Press, Cambridge

Wilde V, Hellmund M (2010) First record of gut contents from a middle Eocene equid from the Geiseltal near Halle (Saale), Sachsen-Anhalt, Central Germany. Palaeobiodivers Palaeoenviron 90:153–162

Williams M, Dunkerley D, De Deckker P, Kershaw P, Chappell J (1998) Quaternary environments. Arnold, London

Williamson TE, Lucas SG (1992) *Meniscotherium* (Mammalia, "Condylarthra") from the Palaeocene-Eocene of Western North America. BullMexico Mus Nat Hist Sci 1:1–75

Woodburne MO (2010) The great American Biotic Interchange: dispersals, tectonics, climate, sea level and holding pens. J Mamm Evol 17:245–264

Woodburne MO, Case JA (1996) Dispersal, vicariance, and the Late Cretaceous to early tertiary land mammal biogeography from South America to Australia. J Mamm Evol 3:121–161

Wu Y, Deng T, Ma J, Zhou X, Mao L, Zhang H, Ye J, Wang S-Q (2018) A grazing *Gomphotherium* in Middle Miocene Central Asia, 10 million years prior to the origin of the Elephantidae. Sci Rep 8:7640. https://doi.org/10.1038/s41598-018-25909-4

Zachos JC, Pagani M, Sloan L, Thomas E, Billups K (2001) Trends, rhythms and aberrations in global climate 65 Ma to present. Science 292:686–693

Zachos JC, Dickens GR, Zeebe RE (2008) An early Cenozoic perspective on greenhouse warming and carbon-cycle dynamics. Nature 451:279–283

Zanazzi A, Kohn MJ (2008) Ecology and physiology of White River mammals based on stable isotope ratios of teeth. Palaeogeogr Palaeoclimatol Palaeoecol 257:22–37

Zaw K, Meffre S, Takai M, Suzuki H, Burret C, Htike T, Thein ZMM, Tsubamoto T, Egi N, Maung M (2014) The oldest anthropoid primates in SE Asia: Evidence from LA-ICP-MS U–Pb zircon age in the Late Middle Eocene Pondaung formation, Myanmar. Gondwana Res 26:122–131

Zhang H, Wang Y, Janis CM, Goodall RH, Purnell MA (2017) An examination of feeding ecology in Pleistocene proboscideans from southern China (*Sinomastodon*, *Stegodon*, *Elephas*), by means of dental microwear texture analysis. Quat Int 445:60–70

Zhegallo V, Kalandadze N, Shapovalov A, Bessudnova Z, Noskova N, Tesakova E (2005) On the fossil rhinoceros *Elasmotherium* (including the collections of the Russian Academy of Sciences). Cranium 22:17–40

Zimov SA, Chuprynin VI, Oreshko AP, Chapin FS III, Reynolds JF, Chapin MC (1995) Steppe-tundra transition: a herbivore-driven biome shift at the end of the Pleistocene. Am Nat 146:765–794

Žliobaitė I, Rinne J, Tóth AB, Mechenich M, Liu L, Behrensmeyer AK, Fortelius M (2016) Herbivore teeth predict climatic limits in Kenyan ecosystems. PNAS 45:12751–12756

Žliobaitė I, Tang H, Saarinen J, Fortelius M, Rinne J, Rannikko J (2018) Dental ecometrics of tropical Africa: linking vegetation types and communities of large plant-eating mammals. Evol Ecol Res 19:127–147

Chapter 3
The Paleoecological Impact of Grazing and Browsing: Consequences of the Late Quaternary Large Herbivore Extinctions

John Rowan and J. T. Faith

3.1 Introduction

Extant species of large herbivorous mammals, considered here as taxa ≥44 kg, are the primary source of the world's remaining large mammal diversity. At the family-level, this includes Antilocapridae (pronghorn), Bovidae (antelopes), Camelidae (camels), Caviidae (cavys), Cervidae (deer), Elephantidae (elephants), Giraffidae (giraffes and okapi), Hippopotamidae (hippopotamuses), Rhinocerotidae (rhinoceroses), Suidae (pigs), Tapiridae (tapirs), and Tayassuidae (peccaries) (Wilson and Reeder 2005). By virtue of their diversity, abundance, and body size, these herbivores have massive impacts on their local environments and play important roles in maintaining ecosystem functionality and diversity (see Chapters in this Book). Such roles include, but are not limited to, altering fire regimes, facilitating nutrient cycling, promoting food web diversity, and modifying vegetation structure to the benefit of other species (Ripple et al. 2015; Malhi et al. 2016). For example, African savanna elephants (*Loxodonta africana*), the largest extant terrestrial herbivores, can disperse seeds up to 65 km from their source, while African forest elephants (*Loxodonta cyclotis*) are in many cases the sole disperser of the largest seeds (e.g., *Cola* spp.) of rainforest trees (Bunney et al. 2017). Grazing by white rhinoceros (*Ceratotherium simum*) maintains short-grass swards that are favored by smaller-

J. Rowan (✉)
Organismic and Evolutionary Biology, University of Massachusetts Amherst, Amherst, MA, USA

Department of Anthropology, University of Massachusetts Amherst, Amherst, MA, USA
e-mail: jjrowan@umass.edu

J. T. Faith
Natural History Museum of Utah, University of Utah, Salt Lake City, UT, USA

Department of Anthropology, University of Utah, Salt Lake City, UT, USA
e-mail: jfaith@nhmu.utah.edu

I. J. Gordon, H. H. T. Prins (eds.), *The Ecology of Browsing and Grazing II*,
Ecological Studies 239, https://doi.org/10.1007/978-3-030-25865-8_3

bodied herbivores (e.g., blue wildebeest, *Connochaetes taurinus*) and reduces the strength and spread of wildfires (Waldram et al. 2008). Likewise, at a macroecological scale, the synergistic relationship between fire and herbivory regimes has been shown to delimit the major vegetation biomes across Africa (Hempson et al. 2015).

As implied by the examples above, the vast majority of our knowledge of large mammal functional ecology and its impacts comes from where they survive in greatest diversity today—sub-Saharan Africa and, to a lesser extent, southeast Asia (Olff et al. 2002; Ripple et al. 2015). However, as recently as ~50,000 years ago, large herbivore diversity paralleling present-day Africa could be found within virtually all of Earth's major terrestrial regions (Fig. 3.1). This lost diversity includes familiar components of high-latitude Ice Age faunas—wooly mammoths (*Mammuthus primigenius*), wooly rhinos (*Coelodonta antiquitatis*), and long-horned bison (*Bison latifrons*)—as well as lesser-known taxa, such as the bizarre South American liptotern *Macrauchenia*, and the giant Australian marsupial

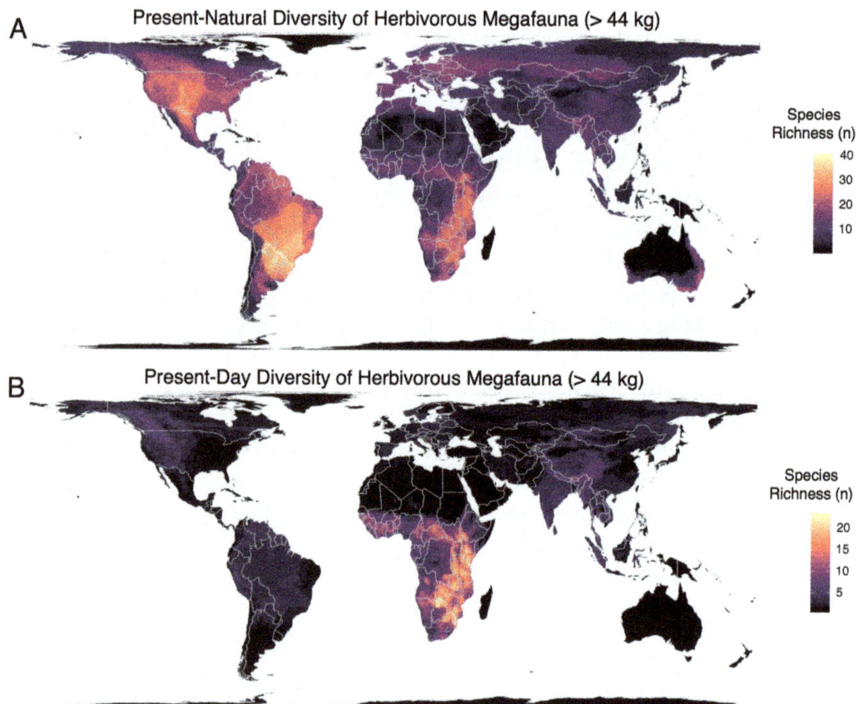

Fig. 3.1 'Present-natural' (**a**) versus present-day (**b**) maps of global species richness for large herbivores (species >44 kg). Note that the present-natural map does not reflect literal large herbivore distribution patterns during the late Quaternary, but instead estimates what the world might have looked like today if late Quaternary extinctions and extirpations never occurred [see Faurby and Svenning (2015) for more information]. Data from Faurby and Svenning (2015), Faurby et al. (2018), and the International Union for Conservation of Nature (IUCN) Red List

Diprotodon (Stuart 2015). Recent estimates suggest that, in total, two-thirds of the world's large mammal genera and over half of its species went extinct towards the end of the Pleistocene, the bulk of these being herbivores (Barnosky 2008). Based on our knowledge of present-day herbivore ecology, it follows that the abrupt loss of so many taxa within a relatively short span of time had significant, far-reaching impacts on terrestrial ecosystems. Moreover, given their recency, the ecological impacts of late Quaternary extinctions almost certainly linger to the present-day and are, therefore, critical to understanding the factors shaping current patterns of biodiversity and its future (Corlett 2013; Galetti et al. 2018).

Here, we review global late Quaternary extinctions of large herbivores, emphasizing what is known about their consequences for the structure and functioning of ecosystems. In the literature on late Quaternary extinctions, most scholars follow Martin (1967) in describing mammalian species ≥44 kg as 'megafauna.' We use the terms 'large mammals' or 'large herbivores' instead to avoid potential confusion with the term 'megaherbivore,' which is often used for animals ≥1000 kg (after Owen-Smith 1988). Though we acknowledge the importance of many other extinctions that occurred during older time periods of the Cenozoic (e.g., the Eocene-Oligocene 'Grande Coupure'), we choose to focus this Chapter on the late Quaternary for three primary reasons. First, this interval is data-rich: late Quaternary extinctions are documented from well-constrained and highly fossiliferous assemblages across much of the world, providing a temporally detailed and geographically extensive sample (**Saarinen** Chap. 2). Second, there is a wealth of paleobiological and paleoecological data for extinct late Quaternary taxa, including body mass estimates (Smith et al. 2003, 2018) and dietary reconstructions from stable carbon isotopes (e.g., Koch et al. 1998; Coltrain et al. 2004) and other proxies (e.g., Rivals et al. 2007; Prideaux et al. 2009; González-Guarda et al. 2018). These data can then be compared to several late Quaternary paleoenvironmental records (e.g., charcoal, pollen), permitting analyses of the relationship between extinctions and ecosystem change through time. Third, because the late Quaternary extinctions preferentially affected the very largest mammal species (Lyons et al. 2004a; Smith et al. 2018), this event provides an important comparison for present-day biodiversity loss. It is now widely acknowledged that human activity since the Industrial Revolution (e.g., climate and land-use change) has fundamentally reshaped global biodiversity patterns (Barnosky et al. 2011; Ceballos et al. 2015). In many cases, this has caused extirpation and near-extinction among large-bodied mammals (e.g., Ripple et al. 2014, 2015), and it is likely that the extinction of many of these species, at least in the wild, will be a hallmark of anthropogenic activity in the twenty-first century (Barnosky et al. 2011; Ceballos et al. 2015). Thus, late Quaternary extinctions provide the best 'natural experiment' with which we can contextualize and compare present-day large mammal diversity loss and its downstream ecological consequences.

The structure of this Chapter is as follows: first, because our review is heavily reliant on paleoecological proxy data, we provide a brief overview of recent methods, focusing on advances using dung beetle and dung fungi proxies. This is followed by an introduction to the topic of late Quaternary extinctions and a brief overview of the major hypotheses for why they occurred. Regional summaries of extinctions and their impacts are then presented for Australia, North America, and northern Eurasia, which form the bulk of studies to date. Finally, we conclude our Chapter by discussing the importance of late Quaternary extinctions for guiding conservation efforts aimed at preserving large mammal diversity today and mitigating losses of ecosystem functionality.

3.2 Paleoecological Proxies

Traditional paleontological approaches for studying fossil mammals, namely those founded on the analysis of their hard-tissue skeletal remains, do not readily lend themselves to assessments of the paleoecological impacts of herbivory. This is because the vertebrate fossil record is discontinuous and patchy in both space and time, meaning that empirical fossil evidence for the appearance or disappearance of a species can often be off-the-mark by a considerable length of time. Likewise, most species are so infrequent in the vertebrate fossil record that it can be very difficult to assess temporal trends in their population histories (but see Boulanger and Lyman 2014; Broughton and Weitzel 2018). Together with chronological uncertainties typically in excess of 100 years (in the case of Late Pleistocene mammals), establishing causal linkages between the vertebrate fossil record and other indicators of paleoecological change has proven exceptionally difficult. However, analysis of other, non-vertebrate fossil proxies in recent years has allowed for the generation of exciting new insights into the environmental and ecological impacts of late Quaternary extinctions. We highlight two relatively new proxies related to large herbivore dung as examples.

3.2.1 Dung Fungi

The coprophilous fungi *Sporormiella* is often found in lake and bog sediments that also preserve ancient pollen (a record of vegetation history), charcoal (a record of fire history), and other paleoecological proxies, and is increasingly used as a measure of large herbivore biomass on Quaternary landscapes (e.g., Gill et al. 2009, 2012). Its use as a biomass proxy stems from the fact that the spores of *Sporormiella* must pass through the digestive system of mammalian herbivores in order to complete their life cycle: *Sporormiella* germinates on dung and releases spores that, under certain situations, become incorporated into sedimentary sequences (see review in Baker et al. 2013). Thus, in much the same way that palynologists infer vegetation change through the shifting abundance of different pollen taxa in a sedimentary sequence, it is reasoned that changes in the abundance of *Sporormiella* provide insight into the abundance of large herbivores on ancient landscapes. In an early and influential study, Davis (1987) observed that *Sporormiella* tends to be rare during much of the Holocene in the western United States, but are abundant prior to extinction of

Pleistocene mammals, and after the historic introduction of grazing livestock, leading to the inference that spores could be used to track herbivore biomass. Though well-recognized taphonomic problems can complicate this situation (e.g., Feranec et al. 2011; Johnson et al. 2015), a growing number of studies highlight the utility of *Sporormiella* as a reliable paleoecological proxy (e.g., Baker et al. 2013; Gill et al. 2013). For Quaternary scientists, this means that it is possible to combine *Sporormiella* abundance with other records and explore the relationship between large herbivore biomass, vegetation, and fire history through time (Gill et al. 2013).

3.2.2 Dung Beetles

So-called 'dung beetles' include several lineages of the superfamily Scarabaeoidea (Order Coleoptera) that have evolved into specialized niches revolving around the consumption of vertebrate faeces, with the most dedicated coprophagous taxa falling into the subfamilies Scarabaeinae and Aphodiinae. Though dung beetles originated alongside the earliest truly massive herbivores—Mesozoic dinosaurs—nearly all lineages surviving the Cretaceous-Paleogene boundary (~66 million years ago) had switched to mammalian dung sources, and subsequently diversified and co-evolved with the radiation of mammals during the Cenozoic (Gunter et al. 2016).

The correlated diversification of dung beetles and their mammalian dung sources throughout the Cenozoic implies that the two have a tight ecological association. This is confirmed by studies of both the fossil record and modern ecosystems. For example, the flightless dung beetle (*Circellium bacchus*), a species endemic to southern Africa, preferentially consumes elephant faeces (Galetti et al. 2018). Though formerly widespread, this species is now restricted to a handful of areas retaining high elephant densities and is considered to be a species of conservation concern (Chown et al. 1995; Kryger et al. 2006; Galetti et al. 2018). Likewise, a study by Schweiger and Svenning (2018) showed that the mean body size of dung beetle assemblages was correlated with declines in large mammal diversity over the last ~50,000 years. Such studies imply that late Quaternary dung beetle assemblages provide a proxy for large mammal biomass and diversity on ancient landscapes.

3.3 Late Quaternary Extinctions and Their Ecological Consequences

Present-day large mammal communities are depauperate when compared to those of the Late Pleistocene. Between ~50,000 and 12,000 years ago, the majority of the world's large mammals disappeared, either in a series of cascading extinctions over thousands of years, or within a comparatively shorter amount of time, depending on geography and other factors (Barnosky et al. 2004; Koch and Barnosky 2006; Stuart

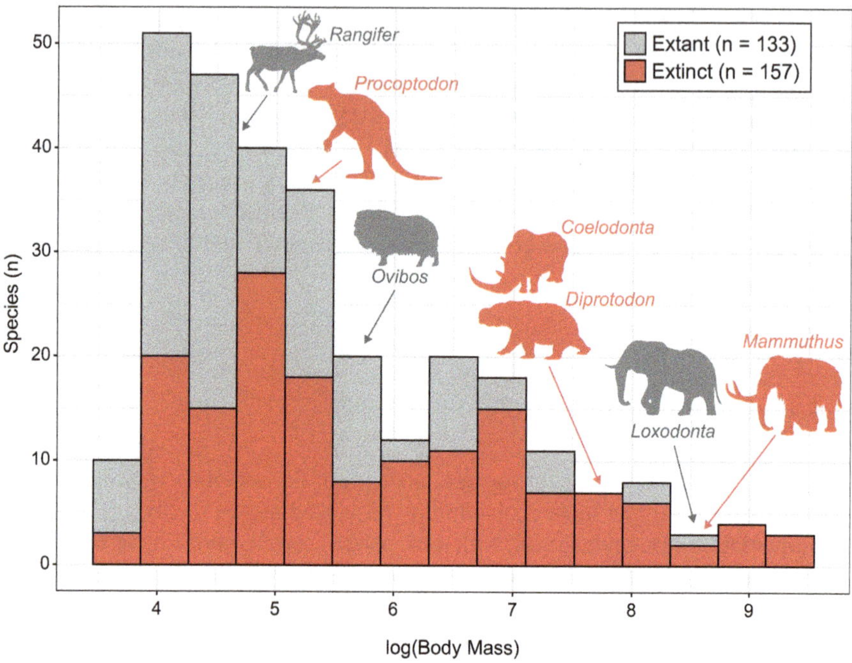

Fig. 3.2 Body mass histogram (log g) for large herbivore species that survived and did not survive the late Quaternary extinctions. Silhouettes represent a handful of species mentioned in the text as examples. From left to right: reindeer (*Rangifer tarandus*), giant kangaroo (*Procoptodon goliah*), muskox (*Ovibos moschatus*), wooly rhino (*Coelodonta antiquitatis*), diprotodon (*Diprotodon optatum*), African savanna elephant (*Loxodonta africana*), wooly mammoth (*Mammuthus primigenius*). Data from Faurby et al. (2018)

2015). Despite these regional differences, the most striking ecological characteristic of the late Quaternary extinctions as a whole was their selectivity for taxa of large body sizes (Fig. 3.2; Lyons et al. 2004a; Smith et al. 2018). Large body size is a trait correlated with low population densities and a slow life history profile (e.g., delayed reproduction, long inter-birth intervals) that, in turn, confers significantly greater extinction risk (Johnson 2002; Cardillo et al. 2005). However, several studies have demonstrated that the late Quaternary extinction is unique in comparison to body size selectivity in previous Cenozoic extinctions (e.g., Alroy 1999; Smith et al. 2018), a fact that has fostered ongoing debate over their underlying driver(s) (e.g., Koch and Barnosky 2006).

Late Quaternary extinctions have long captured the attention of paleontologists and archaeologists alike, mainly because humans may have played a decisive role. Besides the unusual size selectivity of the extinctions, perhaps the most important reason why human impacts are invoked is that the timing of large mammal disappearances across continents roughly tracks the chronology of human dispersals out of Africa and across the globe during the Late Pleistocene (e.g., Martin 1984; Martin and Steadman 1999; Sandom et al. 2014). The alternative hypothesis to human impacts is that the late Quaternary extinctions were the outcome of climate-driven environmental change, though many hypotheses rely on some combination of climatic or environmental drivers (e.g., Burney and Flannery 2005; Koch and Barnosky 2006). The influence of one driver over the other may have also varied by continent; for example, Barnosky et al. (2004) suggested that human impacts were mostly responsible for Australian and North American extinctions, whereas climate change was a more likely scenario for Eurasian extinctions. A consensus is emerging that it is probable that both human hunting and climate change acted in tandem, with glacial-interglacial transitions forcing the contraction and fragmentation of herbivore ranges, making populations more susceptible to the effects of human hunting (Barnosky et al. 2004; Koch and Barnosky 2006; Stuart 2015). Other hypotheses (e.g., comet impacts, hyperdisease) for the late Quaternary extinctions tend to be controversial and not widely accepted (Lyons et al. 2004b; Pinter et al. 2011; Holliday et al. 2014).

As discussed above, most research on late Quaternary extinctions has focused on the hypothesized extinction drivers (e.g., the relative roles of humans vs. climate change) and the magnitude of taxonomic losses across continents (Barnosky et al. 2004; Koch and Barnosky 2006). Recently, however, studies have turned their attention towards the downstream ecological impacts of late Quaternary extinctions for the structure and functioning of ecological communities both in the past and today (Malhi et al. 2016; Smith et al. 2016). Here, we synthesize the available paleobiological and ecological datasets bearing on large herbivore extinctions and corresponding ecosystem change from 50,000 years ago to the present-day in Australia, North America, and northern Eurasia, keeping in mind that the consequences of these extinctions can provide powerful insights into understanding the paleoecological impacts of browsing and grazing and the predicted consequences of further extinctions due to human impacts in the Anthropocene.

3.3.1 Australia

The chronology of late Quaternary extinctions in Australia has been heavily debated, with some studies favoring an early (>46 ka), prolonged, and climate-caused disappearance of the continent's large mammals (e.g., Wroe et al. 2013), whereas others argue that most extinctions occurred rapidly between 50 and 40 ka (e.g., Roberts et al. 2001; Rule et al. 2012) and relate to the appearance of humans on the continent (van der Kaars et al. 2017). It is generally accepted that by ~40 ka, most, if not all, extinctions had occurred (Roberts et al. 2001; Gillespie et al. 2006), with potential sites documenting late survival of extinct large mammals (e.g., the Cuddie Springs site) being highly controversial (Gillespie and Brook 2006). Regardless of its underlying drivers and exact chronology, the late Quaternary Australian extinction ranks among one of the most severe in the world. Although exact counts vary, it is generally estimated that this region lost ~88% of genera (Koch and

Barnosky 2006) and ~91% of species of large mammals, including all taxa >100 kg (Stuart 2015).

Extinct Australian herbivorous large mammals include a diversity of forms, all of which evolved from endemic marsupial lineages (Long et al. 2002). This includes several species with no modern analogs, either in Australia or globally. A prime example is the giant short-faced kangaroo (*Procoptodon goliah*), which is estimated to have weighed 230 kg, and stood up to two meters tall (Helgen et al. 2006). Multiproxy evidence from functional morphology, dental wear, and stable carbon isotopes suggests that this species was a specialized consumer of saltbushes (*Atriplex* spp.), a rare C_4 dicot (most C_4 plants are tropical grasses). If this dietary inference is correct, *P. goliah* would be the only known large mammal species to specialize on C_4 browse (Prideaux et al. 2009). Other enigmatic taxa of the Australian Quaternary include the largest marsupial ever to exist, *Diprotodon optatum*, which weighed nearly 2800 kg and had a shoulder height of 1.8 m (Wroe et al. 2004). In terms of diet, *D. optatum* has been reconstructed as either a browser (DeSantis et al. 2017), or a mixed-feeder preferring browse (Price et al. 2017), which invites ecological parallels in terms of both body size and diet with extant black rhinoceroses (*Diceros bicornis*) in Africa. A recent study by Price et al. (2017), used strontium isotope ratios ($^{87}Sr/^{86}Sr$) to show that *D. optatum* underwent roundtrip seasonal migrations of up to 200 km, rivaling those of some eastern Africa ungulates in the Serengeti today (Estes 2014).

Sediment cores from paleolakes yielding pollen, charcoal, and *Sporormiella* have greatly improved our knowledge of the ecological impacts of late Quaternary extinctions. Rule et al. (2012) analyzed the relationship between inferred herbivore biomass and ecosystem change using a 130,000-year paleoenvironmental record based on two sediment cores from Lynch's Crater paleolake (northeast Australia). Collectively, the Lynch's Crater records show three major, stepwise vegetation shifts over the last 130 ka. The first two occurred at ~125 ka and ~75 ka and correspond to the termination of the Last Interglacial (Marine Isotope Stage 5e) and the onset of Marine Isotope Stage 4, respectively, both of which witnessed cooling and aridification of Australia. These climate changes led to the replacement of rainforest angiosperms by rainforest gymnosperms and sclerophyll taxa. At ~41 ka, the third major vegetation change has no climatic explanation but instead follows, within a few hundred years, the loss of large-bodied herbivores as inferred from the abrupt drop in *Sporormiella* abundance to near-zero. At the same time, charcoal increased in abundance, and sclerophyll taxa and grasses quickly came to dominate local plant communities. Rule et al. (2012) proposed that the sudden disappearance of large herbivores and relaxed herbivory ~41 ka created larger fuel loads which, in turn, lead to larger and more intense fires, favoring the replacement of rainforest vegetation by fire-tolerant sclerophyllous plant communities and grasses. Though the *Sporormiella* proxy tells us nothing about which species were involved, and there is little direct large mammal fossil evidence nearby, such changes are fully consistent with contemporary observations highlighting the influence of herbivory on fire regimes (e.g., Waldram et al. 2008).

A similar study by Johnson et al. (2016) on a 135,000-year sediment core from Caledonia Fen suggests that herbivore extinctions may have had varying impacts on ecosystems depending on geography and primary productivity. Caledonia Fen is a high-elevation site in southeast Australia that was covered by low grasslands and shrub steppe for much of the Pleistocene, especially during glacial phases with reduced temperature, precipitation, and atmospheric CO_2 levels (Kershaw et al. 2007). Like Lynch's Crater, this site suggests an abrupt loss of large-bodied herbivores midway through the Late Pleistocene ~52–45 ka based on *Sporormiella* abundance. Unlike Lynch's Crater, however, the loss of herbivorous large mammals at Caledonia Fen occurred independent of any significant shift in fire regimes or vegetation structure. Johnson et al. (2016) propose that climatic constraints on plant growth and primary productivity likely limited the ability of vegetation to respond to relaxed herbivory pressures. For shrubby and grassy landscapes, like those of glacial-phase Caledonia Fen, such responses might have been an increase in woody plant cover and an overall increase in vegetation density if fires remained infrequent, as indicated by charcoal abundance. The record from Caledonia Fen thus highlights the role of climate and primary productivity in modulating the magnitude of large herbivore loss on ecosystems.

Outside data from paleolake cores, evidence of large herbivore extinction impacts in Australia are detectable in the morphology of extant plants, as discussed in Weber's (2013) recent review of the co-evolution of plants with extinct herbivores. For example, many of Australia's acacias are heavily equipped with large, sharp thorns. In Africa, such defenses prevent or lessen consumption of acacias by giraffes (*Giraffa camelopardalis*) and other large-bodied browsers (Tomlinson et al. 2016), but no such animal exists in Australia today—the largest browsers are wallabies, no more than a meter tall (Menkhorst and Knight 2011). When looking to the late Quaternary record, however, it is clear that the thorny spines of Australia's acacias likely evolved to deter consumption by several extinct large-bodied browsers, such as the massive *Diprotodon optatum* mentioned above. A similar case of anachronistic plant morphologies involves fruits of the rainforest vine *Omphalea*, which can be up to 12.5 cm wide and are, therefore, too large to be dispersed by any living mammalian herbivore. They are not consumed by the large rainforest bird, the southern cassowary (*Casuarius casuarius*), which is Australia's largest fruit-dispersing species today, potentially suggesting their disperser was likely a mammalian casualty of the late Quaternary extinctions, but this remains to be shown.

3.3.2 North America

Like the extinctions in Australia, those in North America were dramatic in terms of both magnitude and the breadth of taxonomic and ecological diversity that disappeared. In total, at least 37 genera went extinct (tally from Grayson and Meltzer 2015), as did a handful of species with surviving North American congeners (*Oreamnos harringtoni*, *Dasypus bellus*, *Panthera atrox*, *Canis dirus*). Though many of North America's extinct large mammals are familiar, such as mammoths (*Mammuthus*) and mastodons (*Mammut*), a substantial number are decidedly not. The giant ground sloth *Eremotherium*, for example, overlapped in body mass with African elephants (~3500 kg), whereas the glyptodont *Glyptotherium*, broadly resembling a heavily armored armadillo, was about the size of small car (~1100 kg). Altogether, North America lost at least 73% of its large mammal genera (tally updated from Koch and Barnosky 2006 to include *Cuvieronius* and *Mixotoxodon*), including all of those weighing >1000 kg (*Camelops*, *Mammuthus*, *Mammut*, *Cuvieronius*, *Glyptotherium*, *Eremotherium*, *Glossotherium*, *Mixotoxodon*).

Of the 37 extinct genera, 17 are securely dated to the terminal Pleistocene, between 12,000 and 10,000 radiocarbon years ago (= ~14,000 and 11,500 calendar years ago). The others tend to be rare in the fossil record and are poorly dated, which means that a well-known sampling phenomenon (the Signor-Lipps effect) could very well explain their apparent absence from the terminal Pleistocene. Indeed, quantitative analysis of the existing radiocarbon chronology are consistent with a more or less synchronous loss of all genera at this time (Faith and Surovell 2009). The timing of the extinctions shortly follows (at least over geological timescales) the arrival of the earliest *Homo sapiens* in North America, a temporal correspondence that played a key role Paul Martin's (1967, 1984, 2005) formulation of the overkill hypothesis, which proposes that human hunting was directly responsible for extinctions in North America. Similar parallels between the timing of extinctions and human arrival elsewhere led him to extend this hypothesis to account for the demise of the planet's large mammals across the globe. Though, as noted above, the overkill hypothesis is hotly contested, particularly among archaeologists (e.g., Grayson and Meltzer 2015; Meltzer 2015; Nagaoka et al. 2018).

Regardless of the cause(s) of North America's large mammal extinctions, given what is known about the ecological roles of large-bodied herbivores (Owen-Smith 1988; Ripple et al. 2015; Malhi et al. 2016), we should expect that the disappearance of so many large-bodied species would have had tremendous downstream effects (Johnson 2009; Gill 2014). A growing body of evidence suggests that this is indeed the case (Robinson et al. 2005; Gill et al. 2009, 2012). Perhaps the most compelling evidence comes from late Quaternary pollen, charcoal, and *Sporormiella* records from two sites in the Midwestern United States, Appleman Lake in Indiana (Gill et al. 2009) and Silver Lake in Ohio (Gill et al. 2012). Both records show that spores of *Sporormiella* decline in abundance shortly after ~14,000 years ago, roughly coincident with human arrival in the Americas (e.g., Halligan et al. 2016). This decline precedes the youngest-dated fossils of extinct species in the region (e.g., Faith and Surovell 2009; Woodman and Athfield 2009), so it is interpreted as representing population collapse and local functional extinction.

At both Appleman Lake and Silver Lake, the collapse of large herbivore populations is followed by vegetation change and altered fire regimes. Broad-leaved deciduous trees, namely *Fraxinus* (ash) and *Ostrya* (hophornbeam), rapidly increase in abundance. Such species are known to be suppressed by browsing pressure from extant large-bodied herbivores, leading to the inference that their increase is related to herbivory release stemming from the functional loss of large herbivores. The temporal resolution at Silver Lake is sufficiently high to show that this phenomenon occurred within 20 years of the *Sporormiella*-inferred herbivore decline. As was also demonstrated at Lynch's Crater in Australia (Rule et al. 2012), the decline of large herbivores is also linked to charcoal peaks that imply enhanced fire regimes relative to preceding intervals, likely due to the accumulation of fuel loads in the absence of large mammal herbivory. Paleoecological archives from southeastern New York State show a similar pattern of enhanced fire regimes following the decline of *Sporormiella* at the end of the Pleistocene ~14,000 to 11,500 years ago (Robinson et al. 2005).

A common consequence of large mammal extinctions is the loss of seed dispersers for plant species that bear large fruits (Janzen and Martin 1982; Galetti et al. 2018; Pires et al. 2018). In many cases, this has led to range contractions and near-extinctions of mammal-dispersed plant taxa, as was likely the case for *Omphalea* in Australia. In North America, Kistler et al. (2015) showed using multiple lines of evidence that squashes, pumpkins, and gourds of the genus *Cucurbita* were heavily impacted by late Quaternary extinctions. For example, though many species are rare or extinct in the wild today, *Cucurbita* is common and widespread in Late Pleistocene archaeological and paleontological assemblages, revealing a greater diversity and broader distribution in the past. These assemblages include fossilized *Cucurbita* seeds in the dung of American mastodons (*Mammut americanum*), suggesting that the fruits of this genus co-evolved with and were mostly reliant upon megaherbivores (species >1000 kg) for their dispersal (Newsom and Mihlbachler 2006). Thus, their survival from the early Holocene to today has solely relied on several independent cases of domestication by early agriculturalists following the late Quaternary extinctions (Kistler et al. 2015).

Finally, we can consider how the ecological impacts of late Quaternary extinctions of plant-consuming species might extend beyond the herbivore guild. Though data from modern ecosystems indicate that megaherbivores have relatively low predation risk (Sinclair et al. 2003), this might not have been the case in the past. Late Pleistocene ecosystems in North America contained a diversity of extinct carnivores, such as saber-toothed felids (e.g., *Smilodon* and *Homotherium*) and large relatives of extant lions (e.g., *Panthera atrox*) and gray wolves (e.g., *Canis dirus*). There is some evidence suggesting that these extinct carnivores might have favored, and were capable of, hunting megaherbivore prey in the past. For example, Marean and Ehrhardt (1995) used zooarchaeological data to show that a Late Pleistocene den of *Homotherium serum* from Texas (Friesenhahn Cave) was dominated by juvenile remains of Columbian mammoth (*Mammuthus columbi*), implying that large saber-toothed felids may have specialized on megaherbivore prey. Likewise, Van Valkenburgh et al. (2016) used comparative analyses of predator-prey relationships to argue that Late Pleistocene carnivores could effectively hunt subadult proboscideans, including individuals >1000 kg. Collectively, these studies suggest that the extinction of large herbivores had cascading effects into the carnivore guild and may be, at least in part, responsible for the extinction of many carnivore taxa during the late Quaternary (see Faith et al. 2018 for an example of megaherbivores in African ecosystems).

3.3.3 Northern Eurasia

Unlike the relatively abrupt late Quaternary extinction of North America, Quaternary extinctions in Eurasia occurred in a series of waves. Stuart (2015) considered four major extinction waves to have occurred: one during the early Last Glacial (~117–40 ka), one near the onset of the Last Glacial Maximum (~30–27 ka), one spanning the Pleistocene-Holocene boundary (~15–4 ka), and another one beginning in the latest Holocene and continuing today (~4 ka to the present-day). Among large mammals, Eurasia is estimated to have lost ~35% of its genera and ~37% of its species, revealing a much more limited extinction than that of Australia and North America (Koch and Barnosky 2006; Stuart 2015). It is often proposed that limited Eurasian extinctions relate to the fact that human ancestors have been present in the region since at least ~500,000 years ago in the form of Paleolithic tool-bearing species like *Homo heidelbergensis* and Neanderthals, perhaps in some way buffering these prey species from the more recent appearance of *Homo sapiens* (e.g., Sandom et al. 2014).

Eurasian extinctions are instead often explained in relation to interglacial-glacial climate change and corresponding vegetation shifts (Prins 1998; Barnosky et al. 2004; Koch and Barnosky 2006). During interglacial periods, such as warm and wet Marine Isotope Stage 5e (~124,000–119,000 years ago), temperate and Mediterranean forests were widespread across western Europe and correspond to broad distributions of the extinct straight-tusked elephant (*Palaeoloxodon antiquus*) and narrow-nosed rhino (*Stephanorhinus hemitoechus*) (Stuart 2015). Marine Isotope Stage 5e climates in Eurasia were so amicable that today what we consider a distinctly tropical species, *Hippopotamus amphibius*, could be found in Great Britain and was somewhat larger than its extant African relatives (Stuart 1982). The population decline and eventual extinction of these common interglacial species shows clear climatic patterning, such as cyclical north-south range shifts through time. In terms of vegetation, their extinctions seem to be related to the widespread replacement of Mediterranean deciduous forest and woodland by open and herb-dominated steppe tundra—so called 'mammoth steppe'—during glacial periods (Allen et al. 2010; Stuart 2015).

The mammoth steppe, stretching from western Europe, northern Asia, and Beringia to the Yukon (Alaska), was the world's largest vegetation biome for much of the late Quaternary and hosted the period's most iconic fauna. This includes its namesake, the wooly mammoth (*Mammuthus primigenius*), extinct steppe bison (*Bison priscus*), and wooly rhino (*Coelodonta antiquitatis*), as well as a handful of extant species (e.g., muskox *Ovibos moschatus*, saiga *Saiga tatarica*, reindeer *Rangifer tarandus*, and onager *Equus hemionus*). Many of the largest species were specialized grazers, and it is believed that grazing pressure by megaherbivores, especially mammoth and rhino, may have helped to maintain open steppe environments during glacial phases (e.g., Guthrie 2001; Zimov 2005).

Yeakel et al. (2013) used stable isotope ratios and bipartite network analysis to show that population fluctuations of herbivore species before, during, and after the Last Glacial Maximum (~26,000–19,000 years ago) cascaded upwards into the large carnivore guild. For example, large felids in Europe (cave lion *Panthera spelea*, and saber-tooth felid *Homotherium serum*) show shifts towards reindeer-dominated diets after the Last Glacial Maximum in response to declining availability of other prey items. Yeakel et al. (2013) suggest that this post-glacial dietary specialization may explain the eventual disappearance of large felids from Eurasia, assuming that niche specialization makes species more prone to extinction. This also explains why generalist carnivores (*Canis* and *Ursus*) survived the late Quaternary and are still found in the region today. Prey-loss as an extinction driver of cave lions was also discussed by Stuart and Lister (2011) in their review of this taxon.

Other studies show how the collapse of the mammoth steppe fauna could also cascade downwards, all the way to invertebrate and vegetation communities. An interesting study by Sandom et al. (2015) used fossilized beetle remains from Great Britain to test the contested 'wood-pasture hypothesis', which proposes that a high diversity and density of large-bodied herbivores during interglacial phases of the late Quaternary was associated with heterogeneous vegetation. They found that beetle assemblages from the Last Interglacial (~132,000–110,000 years ago) document a mixture of semi-open vegetation consisting mainly of grassy woodlands, and that beetles reliant on herbivore dung sources comprise over 50% of assemblages, speaking to a diverse and abundant large herbivore community. On the other hand, beetle assemblages from the early Holocene (~10,000–5000 years ago) contained taxa mainly indicative of closed forest conditions, with dung-consuming beetles comprising only 29% of fossil assemblages. These results suggest that large mammals were sufficiently abundant during past interglacial periods to prevent the formation of large stands of closed forest. The prevalence of forests during the early Holocene is likely related to the fact that this period post-dates the extinctions in Eurasia, thus having a significantly lowered diversity and abundance of large herbivores to exert varied browsing and grazing pressures, including limiting tree recruitment (Sandom et al. 2015).

A study by Schweiger and Svenning (2018) used a similar record of Eurasian dung beetle assemblages to ask whether downsizing of large herbivore communities during the late Quaternary reverberated into invertebrate communities. They showed that over the last ~50,000 years, the mean body size of fossil dung beetle assemblages gradually declined before abruptly crashing after the Pleistocene-Holocene boundary, which corresponds to the termination of prehistoric mammal extinctions.

3.4 Synthesis, Conclusions, and the Future

Our review suggests that the disappearance of the majority of Earth's large-bodied herbivores during the late Quaternary fundamentally altered the structure and functioning of ecosystems. Though the underlying drivers and exact chronology of extinctions varied between Australia, North America, and northern Eurasia, many of the ecological outcomes of extinctions are shared between them, which speaks to some degree of universality of the impacts of browsing and grazing mammals. First, the disappearance of large herbivores fundamentally alters fire regimes and vegetation communities, keeping in mind that there were important feedbacks between the two (e.g., Gill et al. 2012), as was seen in the late Quaternary of both Australia and North America. Second, high herbivore diversity and abundance can maintain heterogeneous vegetation across landscapes, whereas the decline and subsequent disappearance of large mammals may have led to homogenization of plant communities. Third, herbivore losses have cascading effects on both higher and lower trophic levels. Losses of herbivorous large mammals are clearly linked to the collapse of large carnivore guilds in North America and Eurasia, but also to invertebrate communities that have mutualistic relationships with them (e.g., dung beetles). Such effects are likely to cascade through food webs as large carnivores provide important scavenging resources for smaller carnivorous species, whereas dung beetles play important roles in nutrient cycling and secondary seed dispersal. Fourth, as was seen in both Australia and North America, the disappearance of large herbivores severely hinders the ability of many large fruit-bearing plant species to disperse, leading to lower recruitment and in some cases near-extinction in the wild. Though not discussed here, this anachronism is well-known in South America where extinct proboscideans once roamed, and many large fruits lack dispersers today (Janzen and Martin 1982; Pires et al. 2018).

The examples from global late Quaternary extinctions reviewed here underscore the important role of large herbivores in maintaining ecosystem diversity, functionality, and stability. Such late Quaternary 'natural experiments' assume even greater relevance today considering that many extant large herbivores are declining and at risk of extinction. When plotting the body mass of present-day herbivorous large mammals by their International Union for Conservation of Nature (IUCN) Red List status (Fig. 3.3), we see that many of the very largest species are of major conservation concern (i.e., listed by the IUCN as Vulnerable, Endangered, or Critically Endangered). Across the globe, these species are being decimated by human populations via bushmeat hunting, deforestation, and the establishment of agricultural lands and associated with habitat change and competition with domestic livestock (Ripple et al. 2015). Indeed, it is now generally accepted that the decline and disappearance of the world's remaining large mammals will likely be a hallmark of Earth's human-induced sixth mass extinction over the next century (Barnosky et al. 2011; Ceballos et al. 2015).

Though mitigating current and future extinctions are of high conservation significance in terms of both taxonomic and phylogenetic diversity (e.g., Davis et al. 2018), of greater importance is preserving the ecological roles and interactions of

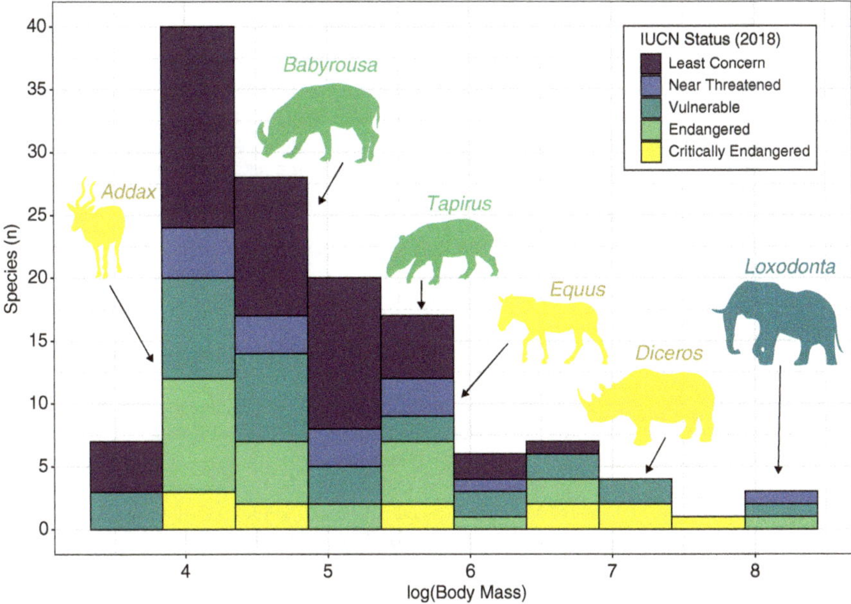

Fig. 3.3 Body mass histogram (log g) for extant large herbivores, colour-coded by their conservation status according to the International Union for Conservation of Nature (IUCN) Red List. Silhouettes represent a handful of threatened species. From left to right: addax (*Addax nasomaculatus*), babyrousa (*Babyrousa togeanensis*), Baird's tapir (*Tapirus bairdii*), wild ass (*Equus africanus*), black rhino (*Diceros bicornis*), and African savanna elephant (*Loxodonta africana*). Data from Faurby et al. (2018)

these species (Estes et al. 2011). A newly emerging subdiscipline of conservation biology and restoration ecology known as 'rewilding' seeks to restore the lost ecological functions of taxa that have been extirpated or driven to extinction by human activity. Though there is vigorous debate about what baseline we should use to rewild landscapes (e.g., Pleistocene vs. pre-Industrial Revolution baselines), the potential benefits of large herbivore reintroductions are clear (Svenning et al. 2016). This has sometimes been done through the reintroduction of formerly native species (e.g., *Bison bonasus* in European forests), and conservationists are now increasingly open to the idea of using introduced (non-native) species for restoring vacated ecological roles and restoring lost ecosystem functionality (Wallach et al. 2018). It is well understood that we do not yet know enough about the potential ecological function of such introduced large mammals, and that future research in modern systems is essential (Lundgren et al. 2018). We agree, and note that a better understanding of ancient systems can also help in our efforts to sustain biodiversity into the future.

Acknowledgements Thanks to editors I. Gordon and H. Prins for inviting us to contribute a chapter on the paleoecological impacts of browsing and grazing, and for their careful editing of this chapter. We acknowledge the contributions of Søren Faurby and colleagues and the PHYLACINE dataset, which we heavily relied upon in this chapter.

References

Allen JR, Hickler T, Singarayer JS et al (2010) Last glacial vegetation of northern Eurasia. Quat Sci Rev 29:2604–2618

Alroy J (1999) Putting North America's end-Pleistocene megafaunal extinction in context. In: MacPhee RDE, Sues HD (eds) Extinctions in near time. Springer, Boston, pp 105–143

Baker AG, Bhagwat SA, Willis KJ (2013) Do dung fungal spores make a good proxy for past distribution of large herbivores? Quat Sci Rev 62:21–31

Barnosky AD (2008) Megafauna biomass tradeoff as a driver of quaternary and future extinctions. Proc Natl Acad Sci USA 105:11543–11548

Barnosky AD, Koch PL, Feranec RS et al (2004) Assessing the causes of late Pleistocene extinctions on the continents. Science 306:70–75

Barnosky AD, Matzke N, Tomiya S et al (2011) Has earth's sixth mass extinction already arrived? Nature 471:51–57

Boulanger MT, Lyman RL (2014) Northeastern North American Pleistocene megafauna chronologically overlapped minimally with Paleoindians. Quat Sci Rev 85:35–46

Broughton JM, Weitzel EM (2018) Population reconstructions for humans and megafauna suggest mixed causes for North American Pleistocene extinctions. Nat Commun 9:5441

Bunney K, Bond WJ, Henley M (2017) Seed dispersal kernel of the largest surviving megaherbivore—the African savanna elephant. Biotropica 49:395–401

Burney DA, Flannery TF (2005) Fifty millennia of catastrophic extinctions after human contact. Trends Ecol Evol 20:395–401

Cardillo M, Mace GM, Jones KE et al (2005) Multiple causes of high extinction risk in large mammal species. Science 309:1239–1241

Ceballos G, Ehrlich PR, Barnosky AD et al (2015) Accelerated modern human–induced species losses: entering the sixth mass extinction. Sci Adv 1:e1400253

Chown SL, Scholtz CH, Klok CJ et al (1995) Ecophysiology, range contraction and survival of a geographically restricted African dung beetle (Coleoptera: Scarabaeidae). Funct Ecol 9:30–39

Coltrain JB, Harris JM, Cerling TE (2004) Rancho La Brea stable isotope biogeochemistry and its implications for the palaeoecology of late Pleistocene, coastal southern California. Palaeogeogr Palaeoclimatol Palaeoecol 205:199–219

Corlett RT (2013) The shifted baseline: prehistoric defaunation in the tropics and its consequences for biodiversity conservation. Biol Conserv 163:13–21

Davis OK (1987) Spores of the dung fungus *Sporormiella*: increased abundance in historic sediments and before Pleistocene megafaunal extinction. Quat Res 28:290–294

Davis M, Faurby S, Svenning JC (2018) Mammal diversity will take millions of years to recover from the current biodiversity crisis. Proc Natl Acad Sci USA 115:11262–11267

DeSantis LR, Field JH, Wroe S et al (2017) Dietary responses of Sahul (Pleistocene Australia–New Guinea) megafauna to climate and environmental change. Paleobiology 43:181–195

Estes R (2014) The Gnu's world: Serengeti wildebeest ecology and life history. University of California Press, Berkeley

Estes JA, Terborgh J, Brashares JS et al (2011) Trophic downgrading of planet earth. Science 333:301–306

Faith JT, Surovell TA (2009) Synchronous extinction of North America's Pleistocene mammals. Proc Natl Acad Sci USA 106:20641–20645

Faith JT, Rowan J, Du A, Koch PL (2018) Plio-Pleistocene decline of African megaherbivores: no evidence for ancient hominin impacts. Science 362:938–941

Faurby S, Svenning JC (2015) Historic and prehistoric human-driven extinctions have reshaped global mammal diversity patterns. Divers Distrib 21:1155–1166

Faurby S, Davis M, Pedersen RØ et al (2018) PHYLACINE 1.2: the phylogenetic atlas of mammal macroecology. Ecology 99:2626–2626

Feranec RS, Miller NG, Lothrop JC et al (2011) The *Sporormiella* proxy and end-Pleistocene megafaunal extinction: a perspective. Quat Int 245:333–338

Galetti M, Moleón M, Jordano P et al (2018) Ecological and evolutionary legacy of megafauna extinctions. Biol Rev 93:845–862
Gill JL (2014) Ecological impact of late Quaternary megaherbivore extinctions. New Phytol 201:1163–1169
Gill JL, Williams JW, Jackson ST et al (2009) Pleistocene megafaunal collapse, novel plant communities, and enhanced fire regimes in North America. Science 326:1100–1103
Gill JL, Williams JW, Jackson ST et al (2012) Climatic and megaherbivory controls on late-glacial vegetation dynamics: a new, high-resolution, multi-proxy record from Silver Lake, Ohio. Quat Sci Rev 34:66–80
Gill JL, McLauchlan KK, Skibbe AM et al (2013) Linking abundances of the dung fungus *Sporormiella* to the density of bison: implications for assessing grazing by megaherbivores in paleorecords. J Ecol 101:1125–1136
Gillespie R, Brook BW (2006) Is there a Pleistocene archaeological site at Cuddie Springs? Archaeol Ocean 41:1–11
Gillespie R, Brook BW, Baynes A (2006) Short overlap of humans and megafauna in Pleistocene Australia. Alcheringa 30:163–186
González-Guarda E, Petermann-Pichincura A, Tornero C et al (2018) Multiproxy evidence for leaf-browsing and closed habitats in extinct proboscideans (Mammalia, Proboscidea) from Central Chile. Proc Natl Acad Sci USA 115:9258–9263. https://doi.org/10.1073/pnas.1804642115
Grayson DK, Meltzer DJ (2015) Revisiting Paleoindian exploitation of extinct North American mammals. J Archaeol Sci 56:177–193
Gunter NL, Weir TA, Slipinksi A et al (2016) If dung beetles (Scarabaeidae: Scarabaeinae) arose in association with dinosaurs, did they also suffer a mass co-extinction at the K-Pg boundary? PLoS One 11:e0153570
Guthrie RD (2001) Origin and causes of the mammoth steppe: a story of cloud cover, woolly mammal tooth pits, buckles, and inside-out Beringia. Quat Sci Rev 20:549–574
Halligan JJ, Waters MR, Perrotti A, Owens IJ, Feinberg JM, Bourne MD, Fenerty B, Winsborough B, Carlson D, Fisher DC, Stafford TW, Dunbar JS (2016) Pre-Clovis occupation 14,550 years ago at the Page-Ladson site, Florida, and the peopling of the Americas. Sci Adv 2: e1600375
Helgen KM, Wells RT, Kear BP et al (2006) Ecological and evolutionary significance of sizes of giant extinct kangaroos. Aust J Zool 54:293–303
Hempson GP, Archibald S, Bond WJ (2015) A continent-wide assessment of the form and intensity of large mammal herbivory in Africa. Science 350:1056–1061
Holliday VT, Surovell T, Meltzer DJ et al (2014) The Younger Dryas impact hypothesis: a cosmic catastrophe. J Quat Sci 29:515–530
Janzen DH, Martin PS (1982) Neotropical anachronisms: the fruits the gomphotheres ate. Science 215:19–27
Johnson CN (2002) Determinants of loss of mammal species during the late Quaternary 'megafauna' extinctions: life history and ecology, but not body size. Proc R Soc B 269:2221–2227
Johnson CN (2009) Ecological consequences of Late Quaternary extinctions of megafauna. Proc R Soc B 276:2509–2519
Johnson CN, Rule S, Haberle SG et al (2015) Using dung fungi to interpret decline and extinction of megaherbivores: problems and solutions. Quat Sci Rev 110:107–113
Johnson CN, Rule S, Haberle SG et al (2016) Geographic variation in the ecological effects of extinction of Australia's Pleistocene megafauna. Ecography 39:109–116
Kershaw AP, McKenzie GM, Porch N et al (2007) A high-resolution record of vegetation and climate through the last glacial cycle from Caledonia Fen, southeastern highlands of Australia. J Quat Sci 22:481–500
Kistler L, Newsom LA, Ryan TM et al (2015) Gourds and squashes (*Cucurbita* spp.) adapted to megafaunal extinction and ecological anachronism through domestication. Proc Natl Acad Sci USA 112:15107–15112
Koch PL, Barnosky AD (2006) Late Quaternary extinctions: state of the debate. Annual Rev Ecol Evol Syst 37:215–250

Koch PL, Hoppe KA, Webb SD (1998) The isotopic ecology of late Pleistocene mammals in North America: part 1 Florida. Chem Geol 152:119–138
Kryger U, Cole KS, Tukker R et al (2006) Biology and ecology of *Circellium bacchus* (Fabricius 1781) (Coleoptera Scarabaeidae), a South African dung beetle of conservation concern. Trop Zool 19:185–207
Long JA, Archer M, Flannery T et al (2002) Prehistoric mammals of Australia and New Guinea: one hundred million years of evolution. Johns Hopkins University Press, Baltimore
Lundgren EJ, Ramp D, Ripple WJ et al (2018) Introduced megafauna are rewilding the Anthropocene. Ecography 41:857–866
Lyons SK, Smith FA, Brown JH (2004a) Of mice, mastodons and men: human-mediated extinctions on four continents. Evol Ecol Res 6:339–358
Lyons SK, Smith FA, Wagner PJ et al (2004b) Was a 'hyperdisease' responsible for the late Pleistocene megafaunal extinction? Ecol Lett 7:859–868
Malhi Y, Doughty CE, Galetti M et al (2016) Megafauna and ecosystem function from the Pleistocene to the Anthropocene. Proc Natl Acad Sci USA 113:838–846
Marean CW, Ehrhardt CL (1995) Paleoanthropological and paleoecological implications of the taphonomy of a sabertooth's den. J Hum Evol 29:515–547
Martin PS (1967) Prehistoric overkill. In: Martin PS, Wright HEJ (eds) Pleistocene extinctions: the search for a cause. Yale University Press, New Haven, pp 75–120
Martin PS (1984) Prehistoric overkill: the global model. In: Martin PS, Klein RG (eds) Quaternary extinctions: a prehistoric revolution. University of Arizona Press, Tucson, pp 354–403
Martin PS (2005) Twilight of the mammoths: ice age extinctions and the rewilding of North America. University of California Press, Berkeley
Martin PS, Steadman DW (1999) Prehistoric extinctions on islands and continents. In: MacPhee RDE, Sues HD (eds) Extinctions in near time. Springer, Boston, pp 17–52
Meltzer DJ (2015) Pleistocene overkill and North American mammalian extinctions. Ann Rev Anth 44:33–53
Menkhorst P, Knight F (2011) A field guide to the mammals of Australia. Oxford University Press, Oxford
Nagaoka L, Rick T, Wolverton S (2018) The overkill model and its impact on environmental research. Ecol Evol 8:9683–9696
Newsom LA, Mihlbachler MC (2006) Mastodons (*Mammut americanum*) diet foraging patterns based on analysis of dung deposits. In: First Floridians and Last Mastodons: the Page-Ladson site in the Aucilla River. Springer, Dordrecht, pp 263–331
Olff H, Ritchie ME, Prins HH (2002) Global environmental controls of diversity in large herbivores. Nature 415:901
Owen-Smith RN (1988) Megaherbivores: the influence of very large body size on ecology. Cambridge University Press, Cambridge
Pinter N, Scott AC, Daulton TL et al (2011) The Younger Dryas impact hypothesis: a requiem. Earth Sci Rev 106:247–264
Pires MM, Guimarães PR, Galetti M et al (2018) Pleistocene megafaunal extinctions and the functional loss of long-distance seed-dispersal services. Ecography 41:153–163
Price GJ, Ferguson KJ, Webb GE et al (2017) Seasonal migration of marsupial megafauna in Pleistocene Sahul (Australia–New Guinea). Proc R Soc B 284:20170785
Prideaux GJ, Ayliffe LK, DeSantis LR et al (2009) Extinction implications of a chenopod browse diet for a giant Pleistocene kangaroo. Proc Nat Acad Sci 106:11646–11650
Prins HHT (1998) The origins of grassland communities in northwestern Europe. In: Wallis de Vries MF, Bakker JP, van Wieren SE (eds) Grazing and conservation management. Kluwer Academic Publishers, Boston, pp 55–105
Ripple WJ, Estes JA, Beschta RL et al (2014) Status and ecological effects of the world's largest carnivores. Science 343:1241484
Ripple WJ, Newsome TM, Wolf C et al (2015) Collapse of the world's largest herbivores. Sci Adv 1:e1400103

Rivals F, Solounias N, Mihlbachler MC (2007) Evidence for geographic variation in the diets of late Pleistocene and early Holocene *Bison* in North America, and differences from the diets of recent Bison. Quat Res 68:338–346
Roberts RG, Flannery TF, Ayliffe LK et al (2001) New ages for the last Australian megafauna: continent-wide extinction about 46,000 years ago. Science 292:1888–1892
Robinson GS, Pigott Burney L, Burney DA (2005) Landscape paleoecology and megafaunal extinction in southeastern New York State. Ecol Monogr 75:295–315
Rule S, Brook BW, Haberle SG et al (2012) The aftermath of megafaunal extinction: ecosystem transformation in Pleistocene Australia. Science 335:1483–1486
Sandom C, Faurby S, Sandel B et al (2014) Global late Quaternary megafauna extinctions linked to humans, not climate change. Proc R Soc B 281:20133254
Sandom CJ, Ejrnæs R, Hansen MD et al (2015) High herbivore density associated with vegetation diversity in interglacial ecosystems. Proc Natl Acad Sci USA 111:4162–4167
Schweiger AH, Svenning JC (2018) Down-sizing of dung beetle assemblages over the last 53 000 years is consistent with a dominant effect of megafauna losses. Oikos 127:1–8
Sinclair ARE, Mduma S, Brashares JS (2003) Patterns of predation in a diverse predator–prey system. Nature 425:288
Smith FA, Lyons SK, Ernest SM et al (2003) Body mass of late Quaternary mammals. Ecology 84:3403–3403
Smith FA, Doughty CE, Malhi Y et al (2016) Megafauna in the earth system. Ecography 39:99–108
Smith FA, Smith REE, Lyons SK et al (2018) Body size downgrading of mammals over the late Quaternary. Science 360:310–313
Stuart AJ (1982) The occurrence of *Hippopotamus* in the British Pleistocene. Quartärpaläontologie 6:209–218
Stuart AJ (2015) Late quaternary megafaunal extinctions on the continents: a short review. Geol J 50:338–363
Stuart AJ, Lister AM (2011) Extinction chronology of the cave lion *Panthera spelaea*. Quat Sci Rev 30:2329–2340
Svenning JC, Pedersen PB, Donlan CJ et al (2016) Science for a wilder Anthropocene: synthesis and future directions for trophic rewilding research. Proc Natl Acad Sci USA 113:898–906
Tomlinson KW, van Langevelde F, Ward D, Prins HH, de Bie S, Vosman B, Sampaio EVSB, Sterck FJ (2016) Defence against vertebrate herbivores trades off into architectural and low nutrient strategies amongst savanna Fabaceae species. Oikos 125:126–136
Van Der Kaars S, Miller GH, Turney CS et al (2017) Humans rather than climate the primary cause of Pleistocene megafaunal extinction in Australia. Nat Commun 8:14142
Van Valkenburgh B, Hayward MW, Ripple WJ et al (2016) The impact of large terrestrial carnivores on Pleistocene ecosystems. Proc Natl Acad Sci USA 113:862–867
Waldram MS, Bond WJ, Stock WD (2008) Ecological engineering by a mega-grazer: white rhino impacts on a South African savanna. Ecosystems 11:101–112
Wallach AD, Lundgren EJ, Ripple WJ et al (2018) Invisible megafauna. Conserv Biol 32:1–4
Weber L (2013) Plants that miss the megafauna. Wildl Aust 50:22
Wilson DE, Reeder DM (eds) (2005) Mammal species of the world: a taxonomic and geographic reference. Johns Hopkins University Press, Baltimore
Woodman N, Athfield NB (2009) Post-Clovis survival of American mastodon in the southern Great Lakes region of North America. Quat Res 72:359–363
Wroe S, Crowther M, Dortch J et al (2004) The size of the largest marsupial and why it matters. Proc R Soc B 271:S34–S36
Wroe S, Field JH, Archer M et al (2013) Climate change frames debate over the extinction of megafauna in Sahul (Pleistocene Australia-New Guinea). Proc Natl Acad Sci USA 110:8777–8781
Yeakel JD, Guimarães PR, Bocherens H et al (2013) The impact of climate change on the structure of Pleistocene food webs across the mammoth steppe. Proc R Soc B 280:20130239
Zimov SA (2005) Pleistocene park: return of the mammoth's ecosystem. Science 308:796–798

Chapter 4
Morphological and Physiological Adaptations for Browsing and Grazing

Daryl Codron, Reinhold R. Hofmann, and Marcus Clauss

4.1 Introduction

Woody plants and grasses are two functionally, and ecologically distinct, components of terrestrial vegetation. For herbivores, these represent two distinct food groups—browse and grass, which differ in spatial distribution, architecture, height above ground, physico-mechanical, biochemical and fermentative properties, presenting different challenges and constraints to the animals that feed on them. Browse includes forbs and other non-woody dicots like herbs because, in many ways, these are structurally and biochemically similar to corresponding parts of woody plants. That large mammal herbivores, restricting our discussion to members of the 'ungulate' orders Proboscidea, Hyracoidea, Perissodactyla, and Artiodactyla, differ in feeding styles, with respect to whether species consume primarily browse or grass, was recognized early on, based on field studies of species' diet compositions in free-ranging environments (Van Zyl 1965; Gwynne and Bell 1968). But it was the work of Hofmann (Hofmann and Stewart 1972; Hofmann 1973, 1989) that truly formalized concepts that browser and grazer species have unique morphophysiological traits representing evolutionary adaptations to foraging differentially on browse or grass.

The evolutionary significance of browsing and grazing is generally understood, and debated, within the context of a morphophysiological adaptive landscape. Traits

D. Codron (✉)
Department of Zoology and Entomology, University of the Free State, Bloemfontein, South Africa
e-mail: CodronD@ufs.ac.za

R. R. Hofmann
Trompeterhaus, Baruth/Mark, Germany

M. Clauss
Clinic for Zoo Animals, Exotic Pets and Wildlife Vetsuisse Faculty, University of Zürich, Zürich, Switzerland
e-mail: mclauss@vetclinics.uzh.ch

I. J. Gordon, H. H. T. Prins (eds.), *The Ecology of Browsing and Grazing II*, Ecological Studies 239, https://doi.org/10.1007/978-3-030-25865-8_4

including body size, tooth and skull anatomy, morphophysiology of the gastrointestinal tract, and even various behavioural and ecological characteristics of large mammal herbivores, have been linked, both functionally and statistically, with browsing and grazing diet niches (Fortelius 1985; Gordon and Illius 1988; Janis 1988; Janis and Ehrhardt 1988; Hofmann 1989; Spencer 1995; Reed 1996; Owen-Smith 1997; Brashares et al. 2000; Clauss and Lechner-Doll 2001; Clauss et al. 2002; Mendoza et al. 2002; Clauss et al. 2003b; Mendoza and Palmqvist 2006; Clauss et al. 2008a; Mendoza and Palmqvist 2008; Clauss et al. 2009c, 2010a; Codron and Clauss 2010; Fraser and Theodor 2011; Hummel et al. 2011; Kaiser et al. 2013; Dittmann et al. 2015; Lazagabaster et al. 2016; Meier et al. 2016). Traits are often found to be convergent, across unrelated taxa or clades, making the ungulates one of the most conspicuous examples of adaptive radiation in the animal kingdom.

Some views have favoured an alternative model for herbivore differentiation, a diet quality-based niche structure driven by differences in feeding selectivity linked to body size (Gordon and Illius 1994; Robbins et al. 1995; Gordon and Illius 1996; Pérez-Barbería and Gordon 2001; Pérez-Barbería et al. 2001; Codron et al. 2007; Clauss et al. 2013). However, such a concept is mostly reconcilable with a browser-grazer based differentiation if one accepts that feeding selectivity (selective/unselective), and botanical composition of the diet do not covary, but that there may be unselective browsers and selective grazers (Demment and Longhurst 1987). Allowing for variability in selectivity within the botanical niche largely adds to our understanding of browser-grazer differences (Codron et al. 2007).

The process of large mammals acquiring nutrients from plant foods starts with ingestion (locating and biting food), followed by extensive oral processing, and finally achieving a level of digestive fermentation which is more intense than occurs in any other animal group (Karasov and Martínez del Rio 2007). The number of traits that have been discussed in this context is vast (in compiling data for this Chapter alone we collected data for a total of 155 variables!). However, the functional and/or statistical relevance (in terms of diet niche) of many of these characteristics is not always clear, and in fact dubious in many instances. A comprehensive review of the topic was presented more than a decade ago (Clauss et al. 2008b), which recognized multiple mismatches, in that several concepts linking traits to dietary function, while contributing substantially to the overall narrative, were not supported by empirical/statistical evidence. In this Chapter, we present a large set of statistical analyses of an up-to-date collection of datasets, allowing us to revisit old hypotheses, and to address recent developments over the past decade. This Chapter, therefore, complements the discussion presented in Clauss et al. (2008b), and we encourage readers to consult both to gain a complete perspective on herbivore adaptation. We reflect on specific anatomical and physiological features of browsing and grazing herbivores that allow them to overcome foraging challenges associated with each of the three stages, in succession, of foraging—ingestion, oral processing, and digestion.

4.2 Linking Form and Function

Any predicted link between an anatomical (or physiological) trait and diet niche, i.e., an association derived from the expected functional relevance of the trait, can be interrogated through one of two approaches: experimental or comparative. Experimental approaches seek qualitative changes in traits following manipulation, from feeding experiments and, potentially, defaunation, up to surgical removal of organs, to determine their function by observing the effect of their diminution or absence (Trautmann and Schmitt 1935; Sakaguchi et al. 1981; Tahas et al. 2017, 2018). Such approaches offer the most definitive support for functional significance of traits, for example, demonstrating that tooth wear in non-ruminants is more strongly influenced by external (grit) rather than internal (silica in forages) abrasives (Müller et al. 2014, 2015; and see **Saarinen** Chap. 2). But, in this example, the experiments do not tell us whether wear-resistant features of herbivore teeth evolved, in response to a shift in diet (to include foods with a higher grit load), or to prevailing environmental conditions. In any case, for mostly logistical reasons, only a few traits have, so far, been subject to such experimental manipulation. A more convenient hypothetico-deductive approach is the comparative method, a widely-used approach for studying adaptive trends in evolutionary biology (Harvey and Pagel 1991). Here, the relationship between traits and diet niches, across species, are evaluated statistically, and rejection of the null hypothesis allows us to conclude the value of the trait is an adaptive response to diet, or to something for which diet is a proxy.

4.2.1 Definition of Herbivore Diet Niches

Statistical evaluations of niche-trait associations, in herbivores, have relied on one of two ways to categorize diet niches. On the one hand, a categorical distinction can be made between browser and grazer species. This categorical usage is typical of most earlier studies (reviewed in Clauss et al. 2008b). In this scheme, mixed- or intermediate-feeders, species that regularly feed on both browse and grass, or switch diets across habitats and/or seasons, can be treated as a third category of feeding style. A categorical approach is not only heuristically valuable in reducing complex diets to simple rules, but has the advantage of being relatively non-sensitive to dietary variations within any one species. On the other hand, diet niches can be viewed as a continuum along the browser-grazer axis, typically depicted by the average percentage grass in the natural diet of each species (Janis 1995; Van Wieren 1996a; Clauss et al. 2003b; Pérez-Barbería et al. 2004). A continuous classification of diets has been favoured by most recent studies, especially following the introduction of new methods, such as stable carbon isotope analysis, that provide rapid estimates of proportions of browse:grass consumption, at least in subtropical savanna environments dominated by C_4 grasses (Cerling et al. 2003; Sponheimer et al. 2003; Codron et al. 2007, 2008b; Lazagabaster et al. 2016). The advantage of such an approach is that arbitrary boundaries between niche categories do not need to be established *a priori*; in a categorical scheme, the question is always at which level of browse/grass intake can a species be classified as a browser or grazer (75%?; 90%?).

In this Chapter we adopt the latter approach, representing the current state of the field, but also because the 'continuous diet niche' relaxes the assumption that species' niches (sensu Hutchinson 1959) are fixed. That is, we recognize that adoption of a specific trait does not prevent a species from feeding in a different niche space, depending upon circumstance. It simply means the species should be more efficient, i.e., more competitive, in the primary niche space to which it is adapted (Codron and Clauss 2010; Damuth and Janis 2011). At the same time, it should be noted that any morpho-physiological trait that may have evolved as an adaptation to a particular diet niche, does not necessarily reflect the average of that niche. In many instances, the trait may simply represent a threshold, above or below which a herbivore may become more efficient in a particular niche. While testing predictions about such nonlinear relationships requires the development of trait-specific models, outside the scope of a large-scale set of analysis such as we present here, treating diet as a continuous variable is a step closer towards achieving this ultimate goal.

4.2.2 Adaptive Value of Morphophysiological Traits

When dealing with a large number of traits, some authors have employed multivariate statistics to determine whether these can distinguish between herbivore feeding styles (Spencer 1995; Mendoza et al. 2002; Mendoza and Palmqvist 2006; Fraser and Theodor 2011), and indeed between diet niches, of other taxonomic groups (Stayton 2006; Martin et al. 2016). These methods offer cursory insights of relative species positions over the adaptive 'morphospace' landscape, but cannot link any single trait with a specific function, and so cannot explicitly resolve the relevance of particular traits. Univariate analyses, on the other hand, may be misleading in this regard, in cases where traits have developed as compensatory characteristics that evolved to accommodate features that do have a direct link to feeding behaviour (Raia et al. 2010). Actually, such compensation is also likely to raise, spuriously, the 'goodness-of-fit' of multivariate models as well. We have tried here to limit our discussion to only those traits for which there is a specific hypothesis about their functional relevance as a feeding tool, with some exceptions, which we note below. Additionally, we provide some examples of correlations between individual traits that we consider meaningful.

Comparative analyses, with species as the biological unit of interest, are performed on datasets that are inherently non-independent because species' traits are inherited from common ancestors (Garland et al. 2005). A common way to overcome this is to estimate the correlation between a trait/s and the study group's phylogenetic tree (Pagel 1999; Garland et al. 2005; Lajeunesse 2009). We follow the method employed by most recent studies on browser-grazer adaptation, using Phylogenetic Least Squares Regression (PGLS) of the caper package for R (Orme et al. 2013). Note, though, that from a deductive perspective, PGLS differs from non-phylogenetically controlled analysis, e.g., Generalized Least Squares (GLS), primarily in that only the former can be used to infer evolutionary convergence. By contrast, lack of a significant relationship between mean percentage grass in the diet with a specific trait does not necessarily mean lack of an adaptive response: the trait may well have evolved as a response to diet, but only within a specific clade that dominates the dataset. Therefore, we also provide results from GLS when these differ from PGLS.

4.2.3 Body Size

Body size is one of the most fundamental biological traits of species, influencing not only variables that reflect size, but also many characteristics related to physiology, shape and even ratios of two variables. In some approaches, particularly those expecting an effect of diet quality, body size has been treated as an alternative to diet niche as an explanatory factor for trait differentiation (Gordon and Illius 1994; Pérez-Barbería and Gordon 2001). Others take body size as a factor determining diet niche itself, with browsers being small and grazers being large, although recognition that browsers are represented across the herbivore size spectrum means that such an approach is not always substantiated. Figure 4.1 shows the typical pattern—while among very small species, there are no strict grazers, and among larger species, there are less browsers, there is no clear constraint put on feeding type by body mass. Here we treat body size as a covariate in all analyses, using body mass ($\log_{10}$-transformed) as a proxy for size. This approach not only allows for patterns to be inferred while controlling for body size variations across species, but—for traits which are also log-transformed—provides estimates of allometric scaling exponents (slopes in log-log regressions). Detailed discussion of differences in allometric scaling is outside our scope, but presentation of these results should be useful for stimulating further discussion about herbivore adaptations.

4.2.4 Data Compilation and Analysis

We compiled a database, from the published literature, of anatomical and physiological measures, as well as mean percentage grass in the natural diet. Additionally, we collated previously unpublished ruminant data on measures of the palate, the cranial and caudal rumen pillar thickness, the area of the intraruminal, ruminoreticular and reticuloomasal orifices (IRO, RRO, and ROO, respectively; calculated as ovals from two length measurements), the height of the *Papillae unguiculiformes*, the larger curvature of the abomasum, and liver mass. In total,

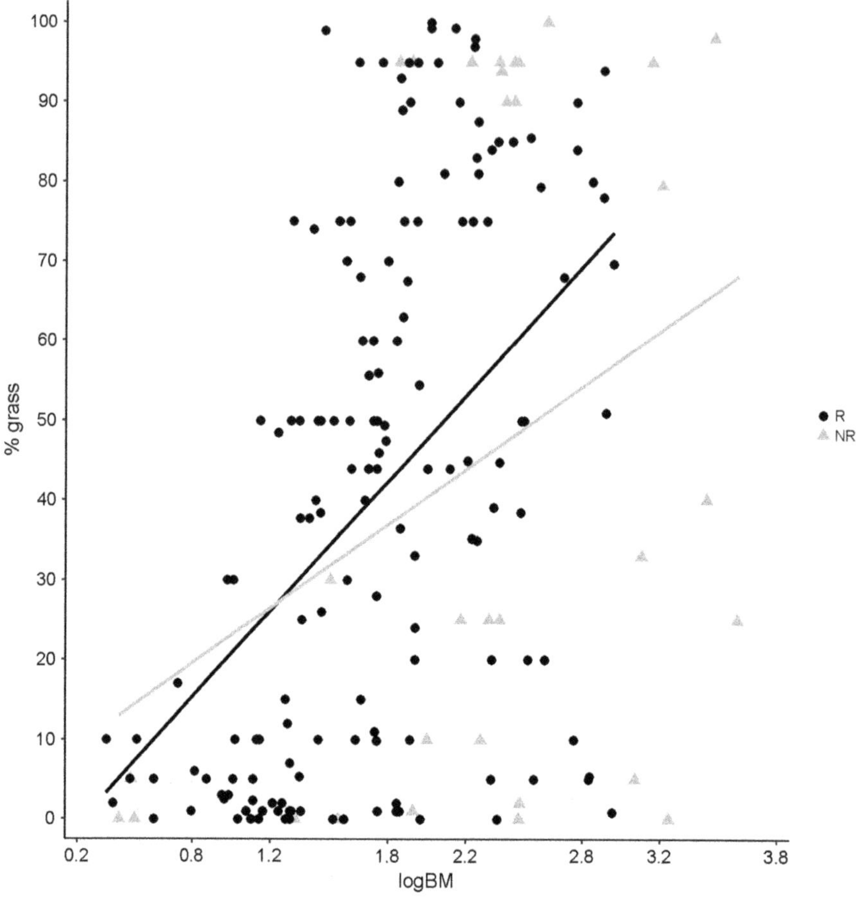

Fig. 4.1 Although there is a general linear relationship between body size and grazing amongst large mammal herbivores, browser and grazer species exist at all points along the body mass range (R = ruminants; NR = non-ruminants)

data for 188 large mammal herbivore species and 95 craniodental, skeletal, and soft tissue characteristics that have been (or can be) hypothesized to differ across diet niches were analyzed. The number of species included varied between datasets, ranging from 10 to 135. The complete dataset, including traits not included as part of our statistical analyses, is included as an electronic supplement with the online version of this Chapter. For PGLS, a single mammalian 'supertree' was used for phylogenetic correlation (Fritz et al. 2009b), pruned to incorporate the species included in each data set. Lambda (λ), depicting the strength of the phylogenetic signal (0 = no signal, values approaching 1 = strong phylogenetic correlation), is estimated using maximum likelihood estimates. We use a species' mean body mass (BM, in kg) as a proxy for body size, taking values reported in each study (where available) or species' means as reported in a global dataset (Smith et al. 2003).

Herbivore digestive strategies are dichotomously distributed between ruminants and non-ruminants (Clauss et al. 2015). Among large herbivores, non-ruminants are primarily hindgut fermenters, barring the Hippopotamidae. Traits related, not only to digestive physiology, but even biting and chewing of foods, are expected to respond differently between ruminants and non-ruminants. In many instances, therefore, studies have either been restricted to analysis of one group only (typically ruminants, or even single families within the Ruminantia, e.g., Bovidae), or analyses have been repeated on data subsets comprising species only from one group (Pérez-Barbería et al. 2004; Codron et al. 2008b; Lazagabaster et al. 2016). Here, we adopt a nesting approach, with digestive strategy treated as a binary variable within which BM and percentage of grass are nested. Thus, we test explicitly for differences in adaptive responses (differences in slopes) to percentage of grass between ruminants and non-ruminants, and reduce Type I error rates that would otherwise be inflated by conducting multiple analyses of overlapping datasets.

Ultimately, three models (four if both ruminants and non-ruminants are included in the dataset) are tested for each trait: BM and percentage grass in the diet as single effects, and BM + percentage grass as covariates. Model fits are compared by an Information Theoretic approach, Akaike's Information Criterion (AIC), applying corrections for small sample sizes (denoted by the subscript *c*). Only models with $\Delta AIC_c \leq 2$, where ΔAIC_c is the difference in AIC_c of a candidate model from the lowest AIC_c in the set, are reported when assembling tables of results (Burnham and Anderson 2001; Burnham and Anderson 2002). Thus, if factors like percentage grass, or ruminant vs. non-ruminant, do not feature in the best-supported models, we exclude them as effects driving variation in the specific trait.

4.3 Ingestion

4.3.1 Searching: Perception and Posture

One of the few advantages of being an herbivore, as compared to a carnivore, is that food is relatively easy to find, and it does not run away. This does not mean that plants are defenceless against herbivory, just that herbivores do not require too much in terms of perception and mobility when it comes to finding food. Nevertheless, browse plants and grasses are distributed differently at landscape, patch, and bite scales, presenting different searching and biting constraints for herbivores (Gross et al. 1993; Shipley 2007). Whereas grasses grow more-or-less continuously within a landscape or patch matrix, woody plants are more patchily dispersed. Browsers should, therefore, spend more time searching and moving between foraging patches than grazers. Whether browsers and grazers differ in sensory perception, as it may be associated with locating different food types, has so far received very little attention in empirical studies (Gordon 2003). Intuitively, we would expect that visual, hearing, and olfactory senses of herbivores are more likely to represent predator-detection systems rather than peculiar adaptations for locating stationary food items. Accordingly, brain and eye size, and indeed maximum visual acuity (determined by the number of cones per degree of visual angle), is not related to the percentage grass in species' diets (Table 4.1).

When suggesting that grazers have a higher density of lingual taste buds, Hofmann (1988) also suggested that browsers should rely more on their sense of smell. This hypothesis matches the finding that the area of the ethmoid bone is negatively related to percentage grass intake (Table 4.1). An increased ethmoid area means a larger area for nerves to be conducted to the nose and, ultimately, a more acute sense of smell. The resulting interpretation that, for diet selection, grazers rely more on their taste and browsers more on their smell, matches the concept that grasses contain less anti-digestive or toxic substances, and hence may not exert a strong selective pressure to evolve a pre-ingestive detection system (Fowler 1983; Mlambo et al. 2015).

Given the difference in spatial distribution of their foods, browsers and grazers could also be expected to differ in mobility. While limbs, particular hind limbs, tend to be longer amongst grazers (Table 4.1), such results should be treated with caution because many elements of limb morphology, including length, likely reflect the habitats in which species live. That is, grazers typically live in more open habitats and thus should be expected to take flight more often than browsers, which may often avoid predation simply by hiding. On the other hand, grasses are a seasonal resource, dying back in dry seasons (Tainton 1999). For this reason, grazing species are more often migratory, moving several hundred, or thousand, km to alternate foraging areas during limiting periods (Avgar et al. 2014); a phenomenon which could, in part, explain their longer limbs.

4.3.2 Biting: Face, Mouth, Lips, Tongue, and Palate

Browse and grass occur at different heights above ground. With the evident exception of bamboo, grasses are generally at, or near, ground-level, whereas browse foods are more heterogeneously distributed in vertical space, from trees of several metres in height to forbs located at or even below the grass layer (Tainton 1999). A vertical feeding height stratification amongst browser species accounts for, in part, the massive variation in body size of these animals (du Toit 1990). Grasses, however, are somewhat bimodally distributed in this regard, between tall, and short or 'lawn' grasses, and a feeding height stratification of grazers has been hypothesized from field studies (Bell 1971; Prins and Olff 1998; Murray and Illius 2000), and from investigations of craniodental morphology (Codron et al. 2008b). The various spatial arrangements of leaves, and fruit, of woody plants also means a more heterogeneous architecture than in grasses, in terms of individual bites presented to herbivores. Hence, grazers are typically expected to be less selective foragers, taking larger—and probably more nutritionally homogeneous—bites than browsers. A possible exception is amongst grazers feeding on low quality grasses—

Table 4.1 Models for traits related to searching

	n		Best Model/s ($\Delta AIC_c \leq 2$)					R v NR		
Trait	R	NR	Factors	r^2	λ	BM	%grass	Main	x BM	x %grass
Perception										
Brain mass (kg)	29	15	BM	0.873	0.972	0.55 (0.033)****	0	n.s	R < NR	n.s.
Eye mass (g)	15	8	Ruminant/(BM + %grass)	0.821	0.977	0.30 (0.063)***	0	R > NR	R > NR	R < NR
Visual acuity (cyc $^{\circ -1}$)	28	4	BM; %grass; BM + %grass	0.110	0.582	0.11 (0.056)	0	n.s	n.s.	n.s.
Ethmoid area (mm^2)	23	7	BM + %grass; BM	0.781	0.000	0.33 (0.052)****	–	R < NR	R > NR	n.s.
Postcranial anatomy										
Femur length (mm)	98	0	BM; BM+%grass	0.678	0.000	0.27 (0.019)****	0			
Tibia length (mm)	98	0	BM + %grass	0.643	0.620	0.19 (0.015)****	+			
Metatarsal length (mm)	98	0	BM + %grass	0.385	1.000	0.15 (0.021)****	+			
Humerus length (mm)	98	0	BM; BM+%grass	0.758	0.374	0.24 (0.014)****	0			
Ulna length (mm)	98	0	BM; BM+%grass	0.377	1.000	0.23 (0.031)****	0			
Radius length (mm)	98	0	BM	0.329	0.170	0.25 (0.036)****	0			
Metacarpal length (mm)	98	0	BM; BM + %grass	0.197	0.379	0.20 (0.042)****	0			

n = number of species; R = ruminants; NR = non-ruminants; λ = phylogenetic signal in PGLS; BM = body mass scaling exponent (s.e.); %grass = effect of mean % grass in species' natural diets on trait (0 = no effect, + = positive relationship, – = negative relationship [alternative result from GLS in parenthesis if different from PGLS]); Main = comparison of R v NR intercepts; x BM and x %grass = comparison of slopes (**** = $p < 0.0001$; *** = $p < 0.001$; ** = $p < 0.01$; * = $p < 0.05$; n.s. = interaction term not significant, i.e., $p > 0.05$)

often the taller, more fibrous grass taxa. In these cases, leaf:stem ratios are lower than amongst lawn grasses, meaning a lower nutritional value overall (because grass stems are generally tougher, more fibrous, and less proteinaceous than grass leaves) (Macandza et al. 2004; Benvenutti et al. 2006), and a possible requirement for smaller, more selective bite sizes in grazers feeding on this resource.

A more regular ground-level feeding behaviour of grazers is best reflected in the braincase angle—the angle between the basioccipital bone and the palate (Lazagabaster et al. 2016). A more acute angle translates into a steeper orientation of the jaw relative to the skull, and hence a strong negative relationship exists between the braincase angle and percentage grass in the diet (Table 4.2).

Feeding height differences between grazers and browsers are reflected, not only in head posture, but also the shape and size of the face. Grazers are expected to have a longer face than browsers, allowing the former to crop bites from short grasses, even during die-back in the dry season (Spencer 1995; Schuette et al. 1998). Even after accounting for species differences in body size, face depth, represented by the region in front of the orbit (measured as the distance between the orbit and the premolar-molar transition), increases across species with higher percentage grass in the diet, but less so in ruminants than in non-ruminants (Table 4.2). Also, amongst ruminants, browsers have a wider distance between the last molars than do grazers (see max palate width, Table 4.2), suggesting a more pointed face shape in browsers. A classic explanation for this finding is a typical *bauplan* constraint (e.g., Janis 1995): in order to accommodate the larger teeth of more hypsodont species (mostly grazers, see below), the orbita has to be moved posterior to the tooth row, leading, for example, to a negative relationship between the hypsodonty index and the masseteric fossa:face length ratio (Fig. 4.2a). Another measure that could be considered related to face depth, the length of the palate (either as total length, or as the length of the rugated portion), available for ruminants only, did not show a relationship with percentage grass intake (Table 4.2), which is in line with these measurements being independent from the orbita's position. This can be interpreted as an indication that it is the mentioned *bauplan* constraint, and not a general requirement for a snout, that leads to longer skulls in grazers.

The length of the rugated portion of the palate (anterior to the maxillary tooth row) corresponds somewhat to the length of the diastema (the distance between the base of the third incisor and the most anterior premolar present; Fig. 4.2b). The diastema has been hypothesized to be longer in grazers (Mendoza et al. 2002), although the functional relevance of this is unclear. We hypothesize that the diastema is linked to a functional dichotomy of the dental apparatus (Hemae 1967)—cropping by the incisors on the one hand, which requires that incisors can be brought into occlusion without impediment from the cheek teeth, and grinding movements of the cheek teeth on the other hand, which should not lead to concurrent incisor attrition. In species in which interlocking canines prevent a wide lateral grinding chewing stroke, such as suids, peccaries, hippos or tapirs (Kiltie 1981; Fortelius 1985; Herring 1985), this functional dilemma does not apply. The evolutionary loss of upper incisors in ruminants (including camelids), and the extremely loose fit of overhanging 'hinged' upper canines in those ruminant species that have them

Table 4.2 Models for traits related to food cropping

Trait	*n*		Best Model/s ($\Delta AIC_c \leq 2$)					R v NR		
	R	NR	Factors	r^2	λ	BM	% grass	Main	x BM	x % grass
Face and mouth										
Basicranial angle (°)	105	26	%grass	0.059	0.918	0.00 (0.003)	–	n.s	n.s.	R > NR
Face depth (cm)	107	28	Ruminant/(BM + % grass)	0.816	0.974	0.27 (0.025)****	+	R < NR	R > NR	R < NR
Muzzle width (cm)	107	28	Ruminant/(BM + % grass)	0.831	0.986	0.25 (0.023)****	+	R < NR	R > NR	R < NR
Mandible length, anterior diastema (cm)	107	28	Ruminant/(BM + % grass)	0.849	0.971	0.24 (0.021)****	0 (+)	R < NR	R > NR	R < NR
Diastema:lower tooth row length ratio	107	28	logBM	0.119	1.000	0.04 (0.009)****	0	n.s	R > NR	n.s.
Mandible length, posterior (cm)	107	28	Ruminant/(BM + % grass)	0.840	0.969	0.22 (0.019)****	0 (+)	n.s	R > NR	R < NR
Dentary shape index (length:width ratio)	34	0	%grass; BM + % grass	0.218	0.000	0.02 (0.029)	+			
Muzzle:palate width ratio	107	28	Ruminant/(BM + % grass)	0.481	0.988	0.08 (0.013)****	+	R < NR	R > NR	R < NR
I1 width (mm)	98	15	BM; BM + %grass	0.623	0.984	0.26 (0.019)****	0 (+)	n.s	R < NR	n.s.
Protrusion of incisor arcade	25	0	%grass; BM + % grass	0.788	1.000	0.33 (0.143)*	+			
I1:I3 width ratio	66	0	BM; BM + %grass	0.143	0.956	−0.14 (0.042)**	0			
Palate and tongue										
Palate width (dentary, cm)	107	28	Ruminant/(BM + % grass)	0.819	0.969	0.17 (0.016)****	+	R < NR	R > NR	R < NR
Palate dental pad length (cm)	52	0	BM; BM + %grass	0.761	0.000	0.28 (0.022)****	0			
Palate max width (rugae portion, cm)	46	0	BM + %grass; BM	0.820	0.593	0.25 (0.019)****	–			
Palate min width (rugae portion, cm)	44	0	BM	0.806	0.288	0.29 (0.022)****	0			

(continued)

Table 4.2 (continued)

Trait	n		Best Model/s ($\Delta AIC_c \leq 2$)					R v NR		
	R	NR	Factors	r^2	λ	BM	% grass	Main	x BM	x % grass
Palate max:min width ratio (rugae portion)	51	0	%grass; BM + % grass	0.071	1.000	0.00 (0.021)	+			
Palate length (rugae portion, cm)	52	0	BM; BM + %grass	0.781	0.733	0.33 (0.025)****	0 (−)			
Palate max width (between M3, cm)	51	0	BM + %grass	0.792	0.834	0.30 (0.023)****	+ (0)			
Palate min width (between P2, cm)	52	0	BM	0.824	0.000	0.31 (0.020)****	0			
Palate cheek tooth frame length (cm)	45	0	BM + %grass	0.905	0.000	0.29 (0.017)****	−			
Palate rugae (left)	52	0	BM; BM + %grass	0.293	0.000	0.09 (0.019)****	0			
Palate rugae (right)	52	0	BM	0.253	0.000	0.07 (0.018)***	0			
Palate length (total, cm)	45	0	BM	0.900	0.000	0.31 (0.016)****	0			
Tongue free portion (% of total)	63	0	BM + %grass	0.159	0.372	−0.02 (0.018)	−			
Tongue corpus max:min width ratio	56	0	BM	0.136	0.782	0.04 (0.012)**	0 (+)			

n = number of species; R = ruminants; NR = non-ruminants; λ = phylogenetic signal in PGLS; BM = body mass scaling exponent (s.e.); %grass = effect of mean % grass in species' natural diets on trait (0 = no effect, + = positive relationship, − = negative relationship [alternative result from GLS in parenthesis if different from PGLS]); Main = comparison of R v NR intercepts; x BM and x %grass = comparison of slopes (**** = $p < 0.0001$; *** = $p < 0.001$; ** = $p < 0.01$; * = $p < 0.05$; n.s. = interaction term not significant, i.e., $p > 0.05$)

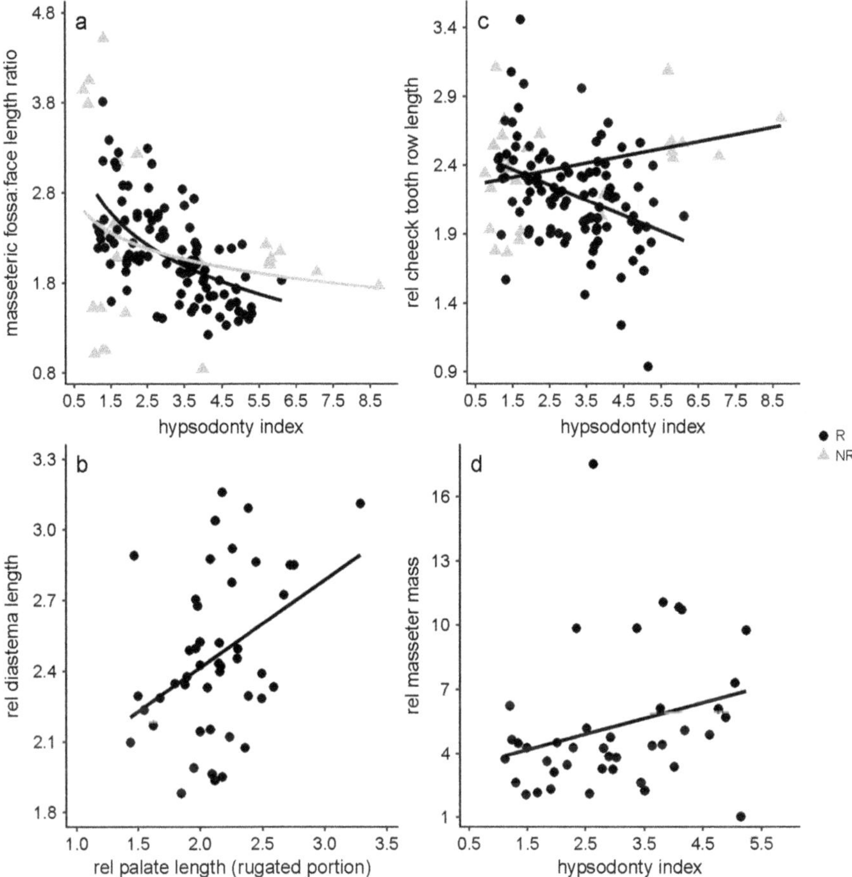

Fig. 4.2 Proxies for face depth (e.g., masseter:fossa face length ratio) are negatively related to M3 size (**a**—the hypsodonty index) and to each other (**b**—diastema length vs. palate length), and hypsodonty is also correlated with cheek tooth row length in ruminants, but not in non-ruminants (**c**) and with the size (mass) of the masseter (**d**). Diastema, palate and tooth row lengths are presented relative to $BM^{0.33}$, and masseter mass relative to $BM^{0.75}$

(Aitchison 1946), can be interpreted as adaptations to avoid disruption of the lateral chewing stroke by the front teeth. The cropping function of the anterior teeth is possible in species with a transverse chewing stroke because the mandible is generally less wide than the maxilla, and cheek teeth, therefore, are not in full occlusion when the incisors are. One important aspect of the transverse chewing stroke, in those herbivores that have it, is that it is not a strictly horizontal lateral movement, but, due to the inclination of the cheek teeth surface, a movement with both horizontal and vertical components. A longer diastema will increase the distance between front teeth due to the vertical component of the chewing stroke. If we accept that grazers generally have a lower occlusal relief in their cheek teeth (Fortelius and Solounias 2000), i.e., potentially a lesser vertical deflection during the chewing stroke, a longer diastema would, theoretically, compensate for this, ensuring the distance between the front teeth. If this was a major function of the diastema, we would also expect this effect to be larger in non-ruminants (with upper incisors) than in ruminants (without upper incisors). However, although the corresponding interaction terms indicated a shorter diastema in ruminants, diastema length was only significantly related to diet in GLS but not in PGLS, and neither was the diastema: cheek tooth row length ratio (Table 4.2), suggesting that it is not an important correlate of diet.

The shorter, more pointed face and mouth of browsers can be linked to a more selective feeding behaviour, aiming for leaves or fruits of relatively higher nutritional value, while attempting to avoid less nutritious stems, and defensive spines, within the available browse matrix. Consequently, browsers also have a relatively longer mobile portion of the tongue (Table 4.2), assisting with selective biting and with stripping leaves from spiny branches (Meier et al. 2016). Clearly, non-selective browsers such as elephants (although these animals also eat grass, and are more often classified as intermediate-feeders) lack this type of facial 'pointiness', and while their trunks do add to dexterity while foraging, they rely more on bulk intake of even low quality items like bark (Codron et al. 2006; Pretorius et al. 2016).

The biting apparatus of grazers reflects cropping of resources near the ground. Grazers have wider incisors, and a more protrusive incisor arcade (Table 4.2), that acts almost as a spade during cropping (Gordon and Illius 1988; Pérez-Barbería and Gordon 2001), and have a wider muzzle (represented by the absolute width of the premaxilla, the ratio of dentary length:breadth, or the ratio of muzzle:palate width, all showing the same relationship to diet; Table 4.2). Within ruminants, both the palate and the tongue showed a maximum:minimum width ratio (i.e., an hourglass shape) that increased with percentage grass intake (Table 4.2). However, the relationship only occurs in GLS for the tongue, and the two measures are not correlated with each other (Pearson's $r = -0.125$, $p = 0.398$), hence the relevance, if any, of the hourglass shape for the natural diet is not clear.

Amongst non-ruminants, the response of the face depth/muzzle width complex to percentage grass in the diet is stronger than amongst ruminants (steeper slopes for non-ruminants; Table 4.2). In an analysis that evaluated relationships between craniodental metrics with a presumed feeding height proxy amongst grazers, Codron et al. (2008b) showed that wide muzzles, and a steeper mandibular angle (a more direct measure of jaw position relative to the horizontal plane than the braincase angle), are most extreme amongst short/lawn grass grazers (principally the bovid tribe Alcelaphini—wildebeest (*Connochaetes* spp.), hartebeest and their kin). By contrast, members of the Hippotragini (specifically roan antelope *Hippotragus equiinus* and sable antelope *H. niger*) and Reduncini (waterbuck and reedbuck) have less-derived versions of these traits, almost resembling browsers. Subsequently, the evolution of at least two distinct evolutionary pathways to grazing (including between the Alcelaphini vs. Hippotragini/Reduncini clades) was detected in a phylogenetic analysis of African bovid diet niches (Louys and Faith 2015). If one extends this concept to include non-ruminants, the stronger response to percentage grass intake can be explained as a function of many non-ruminant grazers being large-bodied, ground-level 'lawnmowers', that use their wide mouths and prehensile lips to crop large mouthfuls, as especially evident in the square-lipped rhinoceros *Ceratotherium simum* or the common hippopotamus *Hippopotamus amphibius* (Owen-Smith 2013).

4.3.3 Intake Amount and Feeding Time

So far, we have mentioned intake quantity mainly as a function of variations in potential bite size. Such functions are not trivial, since bite size is the basic unit of intake for herbivores (Shipley 2007; for carnivores, ecological roles are determined more by the number of prey individuals taken). Nonetheless, for herbivores, for which the relatively low nutritional quality and digestibility of food is a limiting factor, achieving large amounts of total biomass intake is the primary foraging goal. In herbivores, intake scales higher than metabolic rate (i.e., at a scaling exponent higher than 0.75), which is possible because gut capacity also scales higher than metabolic rate (Müller et al. 2013; and see Table 4.3). Indeed, daily dry matter intake of mammalian herbivores is an order of magnitude greater than for similar-sized carnivores (Codron et al. 2016), the latter consuming diets of higher digestibility and digestible energy content. Consequently, herbivores require digestive tracts of larger volume (Chivers and Hladik 1980; Clauss et al. 2017). If one postulates a difference in digestible energy content between browse and grass, then one would also expect a corresponding difference in intake levels and gut capacity between browsers and grazers. So far, the available data for food intake from feeding experiments do not suggest such a difference (Table 4.3). This most likely reflects the fact that, although grass often contains higher digestible energy levels than browse (Hummel et al. 2006), regional, seasonal, plant species and plant part variability is so high (Paine et al. 2018) that food-specific intake levels would be senseless. Additionally, intake is often not aimed at meeting maintenance, but at maximizing energy gain, and, therefore, increases with forage diet quality in experimental settings (Van Soest 1965; Meyer et al. 2010). To judge whether intake really differs systematically between browser and grazer species, systematic feeding experiments aimed at specifically this question would probably be required.

By contrast, another characteristic of browse and grass, namely the speed at which the material can be fermented, has a more evident effect: browse typically reaches its maximum fermentation gain faster than grass (Hummel et al. 2006). Therefore, it makes sense that ruminant browsers have a higher frequency of feeding bouts, to faster replace the material that is digested quicker, whereas ruminant grazers have fewer, longer feeding bouts (Table 4.3). Differences in bout frequency between browsers and grazers also reflect differences in digestive strategies (Hummel et al. 2006; and see below), such as larger forestomach capacities in ruminant grazers.

As far as we are aware, differences in intake requirements of short- vs. tall-grass grazers have not been investigated. Theoretically, however, these two grazer types

Table 4.3 Models for traits related to intake amounts

Trait	n		Best Model/s ($\Delta AIC_c \leq 2$)					R v NR		
	R	NR	Factors	r^2	λ	BM	% grass	Main	x BM	x % grass
Dry matter intake (kg d^{-1})	24	12	BM; BM + %grass	0.939	0.781	0.80 (0.035)****	0	R > NR	R < NR	n.s.
GIT contents wet mass (kg)	25	8	BM; BM + %grass	0.975	0.000	1.06 (0.030)****	0	R > NR	R < NR	n.s.
Reticulo-rumen contents wet mass (kg)	39	0	BM + %grass	0.981	0.000	1.12 (0.028)****	+			
Feeding bouts d^{-1}	10	0	BM + %grass; % grass	0.737	0.000	−13.88 (9.610)	–			

n = number of species; R = ruminants; NR = non-ruminants; λ = phylogenetic signal in PGLS; BM = body mass scaling exponent (s.e.); %grass = effect of mean % grass in species' natural diets on trait (0 = no effect, + = positive relationship, – = negative relationship [alternative result from GLS in parenthesis if different from PGLS]); Main = comparison of R v NR intercepts; x BM and x %grass = comparison of slopes (**** = $p < 0.0001$; *** = $p < 0.001$; ** = $p < 0.01$; * = $p < 0.05$; n.s. = interaction term not significant, i.e., $p > 0.05$)

should achieve different instantaneous intakes at different feeding heights, with species like wildebeest achieving higher intake at lower sward levels than sympatric medium- or tall-grass grazers like hartebeest (*Alcelaphus* and *Damaliscus* spp.) (Murray 1993; Murray and Illius 2000), or even Reduncini and Hippotragini. Amongst non-ruminants, however, no difference in instantaneous intake was achieved by horses of three body size classes feeding at varying sward heights (Fleurance et al. 2009). Taking ever-increasing bite sizes during cropping might be a feature unique to ruminant grazers, and could explain the smaller effect of percentage grass in the diet on traits like muzzle width than in non-ruminants. Similarly, differences in cheek tooth anatomy—although these function mainly in chewing—reflect differences in bite sizes. In particular, ruminant grazers have reduced premolars compared to ruminant browsers (Table 4.4: premolar:molar row length ratio), possibly enabling larger bite sizes (Janis and Constable 1993; Reed 1996; Codron et al. 2008b; Copeland et al. 2009; Lazagabaster et al. 2016). The emphasis of mastication is then shifted to the back of the mouth (see below). Amongst tall-grass ruminant grazers, however, the premolar row is not reduced in this way, probably because of the smaller bite sizes these species achieve while cropping (Codron et al. 2008b). Similarly, non-ruminants do not have reduced premolars, which has been hypothesized as reflecting a requirement for higher food intake with instantaneous ingestive mastication than in ruminant grazers (Janis and Constable 1993). We view the relatively large premolars of non-ruminant grazers and of tall-grass ruminant grazers as convergent, both not necessarily reflecting a greater intake, but smaller, more consistent bite sizes, and an evenness of the distribution of chewing emphasis throughout the front and back of the mouth. This would also explain why intake rates of horses did not vary across sward heights.

4.4 Oral Processing

4.4.1 Chewing

Mammalian herbivores, more than any other animal group, rely on chewing as a necessary process in preparing food for digestion (Reilly et al. 2001). The main purpose of extensive mastication is to reduce ingesta to smaller particle sizes, enabling passage through the gastrointestinal tract and, more importantly, increasing the total surface area for fermentation by microbiota in the gut. Therefore, reducing digesta particle size is a means for herbivores to reduce the need for long digesta retention times (Clauss et al. 2009d), a trade-off particularly evident when comparing non-chewing herbivorous reptiles that have large digesta particles and long retention times, with herbivorous mammals that chew their ingesta and have smaller digesta particles and shorter retention times (Fritz et al. 2010; Franz et al. 2011). Among mammals, ruminants achieve distinctively finer particles in the lower digestive tract than non-ruminants (Fritz et al. 2009a; Clauss et al. 2015). Although a first

Table 4.4 Models for traits related to chewing

Trait	n		Best Model/s ($\Delta AIC_c \leq 2$)					R v NR		
	R	NR	Factors	r^2	λ	BM	% grass	Main	x BM	x % grass
Cheek Teeth										
Lower molar tooth row length (cm)	107	28	Ruminant/(BM + % grass)	0.831	0.971	0.19 (0.017)****	+	n.s	R > NR	R < NR
Lower premolar tooth row length (cm)	107	28	Ruminant/(BM + % grass)	0.649	0.983	0.17 (0.020)****	0	R < NR	R > NR	n.s.
Lower premolar:molar row length ratio	107	28	%grass; BM + % grass	0.043	1.000	−0.02 (0.013)	- (0)	n.s	n.s.	n.s.
Hypsodonty index (M3 enamel crown height:width ratio)	105	25	%grass; BM + % grass	0.266	0.912	0.01 (0.028)	+	n.s	n.s.	R < NR
Lower molar tooth volume (cm^3)	98	15	BM + %grass	0.901	0.644	0.91 (0.033)****	+	R < NR	R > NR	R < NR
Exposed enamel and dentine area (mm^2)	82	0	BM	0.822	0.242	0.53 (0.028)****	0			
Length of all internal enamel structures (mm^2)	82	0	BM + %grass	0.779	0.495	0.34 (0.023)****	+			
Enamel aligned 0 to 45 ° to chewing plane	37	0	BM + %grass; BM	0.328	0.000	0.21 (0.057)**	0			
Enamel aligned 10 to 40 ° to chewing plane	37	0	BM + %grass	0.523	0.000	0.28 (0.055)****	+			
Enamel aligned 0 to 40 ° to chewing plane	82	0	BM + %grass	0.329	0.978	0.08 (0.017)****	+ (0)			
Musculature										
Masseter mass (g)	47	0	BM; BM + %grass	0.987	0.000	0.96 (0.02)****	+			
Masseteric ridge length (cm)	107	28	Ruminant/(BM + % grass)	0.854	0.911	0.19 (0.018)****	+	R < NR	R > NR	R < NR
Masseteric fossa:face length ratio	107	28	Ruminant/(BM + % grass)	0.313	0.943	−0.06 (0.013)****	–	n.s	R < NR	R > NR

Mandibular muscle insertion area (mm^2)	51	0	BM + %grass	0.943	0.280	0.68 (0.037)****	+			
Length of mandibular 'lever'	51	0	BM + %grass; BM + %grass	0.853	1.000	0.36 (0.023)****	+			
Height of mandibular 'lever'	51	0	BM + %grass	0.926	0.000	0.39 (0.018)****	+			
Masseter moment arm (relative to zygomatic arch, °)	33	0	%grass	0.660	0.000	0.02 (0.021)	+			

n = number of species; R = ruminants; NR = non-ruminants; λ = phylogenetic signal in PGLS; BM = body mass scaling exponent (s.e.); %grass = effect of mean % grass in species' natural diets on trait (0 = no effect, + = positive relationship, − = negative relationship [alternative result from GLS in parenthesis if different from PGLS]); Main = comparison of R v NR intercepts; x BM and x %grass = comparison of slopes (**** = $p < 0.0001$; *** = $p < 0.001$; ** = $p < 0.01$; * = $p < 0.05$; n.s. = interaction term not significant, i.e., $p > 0.05$)

study with zoo animals indicated a more distinct particle size reduction in large ruminant grazers compared to browsers (Clauss et al. 2002), no similar difference was found between grazing and browsing rhinos (Steuer et al. 2010), and further studies with naturally feeding ruminant species did not reveal any difference in particle size between feeding types (Hummel et al. 2008; Lechner et al. 2010), indicating that ruminant browsers are only less efficient chewers on diets fed to them in captivity.

4.4.2 Cheek Teeth

Herbivores have evolved large, robust cheek teeth (premolars and molars) for chewing, and an extraordinary level of morphological diversity, across species and clades, is seen, even in early Eocene forms (Jernvall et al. 1996; and see **Saarinen** Chap. 2). Cheek tooth morphology has received perhaps the most research attention, in terms of traits that represent adaptations to various diet niches (Fortelius 1985; Janis 1988; Archer and Sanson 2002; Heywood 2010b; Kaiser et al. 2010).

One of the most common classifications of herbivore cheek tooth morphology is to separate species with low-crowned ‘brachydont’ from those with high-crowned ‘hypsodont’ teeth. Hypsodonty, usually measured as the ratio of the enamel crown height:occlusal width ratio of the M3 (the hypsodonty index), is associated with wear-resistance—an adaptation to ensure that sufficient enamel remains in occlusion throughout the life of the animal (reviewed in Damuth and Janis 2011). Therefore, hypsodonty is expected to be higher in species that consume more abrasive foods, as well as species in which attrition (wear that occurs from tooth-on-tooth contact during occlusion) is higher, although the latter has not been measured empirically. A positive relationship between the hypsodonty index and percentage grass in the diet (Table 4.4), based on the presumption that grasses are more abrasive foods than browse, has been shown repeatedly (Cerling et al. 2003; Sponheimer et al. 2003; Codron et al. 2007; Clauss et al. 2008b; Hummel et al. 2011; Kaiser et al. 2013; Lazagabaster et al. 2016). The hypsodonty index is probably the most widely-used proxy for interpreting diet niches of fossil mammals (Janis 1995; Palmqvist et al. 2003; Cerling et al. 2005; Janis 2008; Damuth and Janis 2011; and see **Saarinen** Chap. 2); a correlation between changes in percentage grass in the diet (or inferred from grass availability) and hypsodonty, over geological time scales, has even been found within lineages (Feranec 2003; Strömberg 2006). The development of a high-crowned M3 itself need not be interpreted as a factor restricting the diet niche; in most datasets, many intermediate-feeding species have as high a hypsodonty index, or even higher, than many grazers (Janis 1988; Copeland et al. 2009). Thus, even if the evolution of hypsodonty evolved at faster rates amongst grazing species, the trait does not preclude browsing, and so a high molar crown is probably an attractive option for all herbivores, to facilitate a broader niche (Feranec 2007; Rivals et al. 2010; Damuth and Janis 2011).

One factor driving the association between the hypsodonty index and percentage grass intake is that plant silica bodies (phytoliths), which are more abundant in grasses (Hodson et al. 2005), wear down tooth enamel. There has been discussion on whether phytoliths are harder than tooth enamel or not, and whether they can lead to tooth wear (for this reason, or irrespective of a hardness difference), or whether external abrasives (dust, grit) are more likely causative agents for cheek tooth wear (Baker et al. 1959; Mainland 2003; Sanson et al. 2007; Damuth and Janis 2011; Lucas et al. 2013; Erickson 2014; Rabenold and Pearson 2014; Xia et al. 2015). On the one hand, dental wear was caused in feeding experiments, in non-ruminants, by both diets with high phytolith and high external abrasives content (Müller et al. 2014, 2015). A similar result was produced in an in vitro study—using a chewing machine with horse teeth (Karme et al. 2016), and in an experiment with sheep—sand added to a hay diet led to changes in tooth microwear (Hoffman et al. 2015). On the other hand, other experimental findings suggest that external abrasives have less of an effect on the teeth of ruminants than expected, based on the findings in non-ruminants (Merceron et al. 2016; Ackermans et al. 2018). It has been suggested that ruminant digestive physiology reduces the effect of dust and grit on molar wear, because these external abrasives are probably washed off the plant material in the rumen before it is regurgitated for rumination. This hypothesis could explain why ruminants chew more cursorily during ingestion and more systematically during rumination (Dittmann et al. 2017), and why ruminant grazers are generally less hypsodont than non-ruminant grazers (Table 4.4), but this hypothesis awaits direct testing.

The higher hypsodonty index of grazers is interpreted as a response to i) ground-level feeding and ii) changing environmental conditions, with hypsodonty being greater in more arid and 'dusty' environments (Mendoza and Palmqvist 2008; Damuth and Janis 2011; but see results in Sanson and Read 2017 for lack of differences in buffalo *Syncerus caffer* tooth wear across substrates). As such, the hypsodonty index has been interpreted as a trait linked to species' habitats, rather than diet niches per se (Mendoza and Palmqvist 2008; Kaiser et al. 2013), and it has been promoted as a proxy for resolving palaeoclimatic conditions (Eronen et al. 2009, 2010).

Only one study has demonstrated an empirical link between hypsodonty, percentage grass in the diet, *and* total levels of silica consumed in free-ranging conditions (Hummel et al. 2011), but that study did not distinguish between silica of exogenous versus endogenous origin. In their study of feeding height stratification amongst grazers, Codron et al. (2008b) showed that the hypsodonty index is highest amongst short-grass grazers, which also points to an exogenous factor (there is no reason, at this stage, to believe levels of internal abrasives differ between short and tall grasses).

Regardless of the relative strength of diet versus environmental effects, a higher hypsodonty index, because this metric is measured on the third molar, reflects a shift in chewing emphasis from homogeneously distributed throughout the mouth (in browsers and in most non-ruminants) to the back of the mouth (in grazers, especially ruminant grazers). Such a shift, along with a reduced premolar row in many ruminant grazers (see above), may help reduce torsional forces during mastication (Greaves 1991). Actually, in ruminants there is a negative correlation between the lower cheek tooth row length and the hypsodonty index (Fig. 4.2c), adding weight to the functional interpretation of increased M3 hypsodonty shifting chewing towards the posterior part of the tooth row. However, in tall-grass ruminant grazers (e.g., Hippotragini and Reduncini), such a shift in chewing strategy has not occurred, as their premolars are far more pronounced, and their M3 hypsodonty index is much lower, than short-grass grazers (Codron et al. 2008b). Whether this difference reflects a difference in diet niche as well requires further investigation; measures of total enamel volume, rather than the hypsodonty of a single tooth, should answer this question. Although enamel volume is related positively with percentage grass intake, as is mandibular molar row length, and, also, the combined length of all internal enamel structures (Table 4.4), no quantitative comparisons between presumed short- and tall-grass grazers have yet been made based on these traits.

Regardless of whether the majority of chewing is located at the back, or throughout, the mouth, differences in masticatory traits between browsers and grazers should also reflect differences in the fracture properties of these food types. Fracture properties of leaves are associated with patterns of leaf venation, with polygonal particles emanating from browse and elongate particles from grass (Kelly and Sinclair 1989; Sanson 1989; Nultsch 2000). Thus, it is not only the volume of enamel, but variations in its structural distribution that differs between browsers and grazers (Archer and Sanson 2002). In line with the more heterogeneous fracture properties of browse, a more homogeneous cusp wear pattern was found in grazing rhinos compared with browsing species (Taylor et al. 2013). On a broader taxonomic scale, the development of fused cusps in Bovidae (antelope and buffalo), and the absence of this trait in Cervidae (deer), was proposed as an explanation for the fact that a strictly grazing diet niche is comparatively rare amongst cervids (Heywood 2010a).

The positioning of enamel ridges along the cheek teeth of herbivores also appears to reflect diet niche differences, in that a greater proportion of occlusal enamel is aligned at acute angles to the direction of the chewing plane in grazers (Heywood 2010b; Kaiser et al. 2010; and see Table 4.4). One interpretation is that, although chewing strokes are generally transverse in all herbivores (Fortelius 1985), grazers have more anterior-posterior jaw movements during chewing than do browsers. In the latter movement pattern, the alignment of occlusal enamel ridges in grazers could represent compensation for the straighter alignment of grass blades between occlusal surfaces (Kaiser et al. 2010).

4.4.3 Musculature, Teeth and Chewing Intensity

Adaptations for fracturing highly resistant plant foods include not only large, robust teeth, but also powerful chewing muscles, of which the masseter is the most pronounced (Hendrichs 1965). Presumably as a response to the more fracture-resistant properties of grass as compared with browse (see Paine et al. 2018), the size (mass) of the masseter is positively related to percentage grass intake (Table 4.4; Clauss et al. 2008a). Skeletal features of the mandible itself also reflect this trend, including the size of the masseteric ridge and insertion areas, and the area dimensions of the mandibular lever (Table 4.4). The latter is seen as a direct indicator that the masseteric action in grazers entails a higher workload than browsers, as does the fact that grazers have a longer moment arm, measured as the mean angle of the master relative to the zygomatic arch (Varela and Fariña 2015).

The mechanical advantages of larger masseters, with greater areas of action, implies that bite forces of grazers are greater than those of browsers. Such adaptations operate alongside the differences in cheek tooth morphology described above. Correspondingly, there are positive correlations between masseter and cheek tooth variables (Figs. 4.2a and d; see also Fraser and Rybczynski 2014). Notably, these traits would not preclude animals that have them (typically, grazers) from a browsing niche, but do appear to limit the niches of browsers.

Comparative investigations of chewing intensity in herbivores have been guided mostly by hypothesized differences between non-ruminants and ruminants. Indeed, early observations on the size of masseters showed that this muscle accounted for substantially less of the total jaw musculature of ruminants than non-ruminants (Hendrichs 1965). Overall oral chewing intensity is expected to be greater amongst non-ruminants (Turnbull 1970; Fortelius 1985), because, in ruminants, much of the mechanical breakdown of food, into smaller particles, is achieved through chewing by rumination (Trudell-Moore and White 1983; McLeod and Minson 1988). Experimental data provided some evidence for a more regular chewing pattern during ingestion in horses than in cattle and camels (Dittmann et al. 2017), and also for a greater chewing intensity amongst horses than cattle—although the latter result could not be supported statistically, because of a small sample size (Janis et al. 2010). The studies of Janis et al. (2010), and of Fletcher et al. (2010), did, however, forward a hypothesis that the more complex (and robust) chewing apparatus of non-ruminants, compared to that of ruminants, is a function of higher workloads amongst non-ruminants. This concept is corroborated by the finding of a lower strain measured in goats during rumination as compared to ingestive mastication (Williams et al. 2011). Again, the fact that the majority of particle size-reducing chewing occurs in ruminants during rumination, presumably on material that has likely been washed free of dust/grit, as well as being somewhat ‘softened’ by its residence in the rumen (Janis et al. 2010; Mihlbachler et al. 2016; Dittmann et al. 2017), is part of this hypothesis. A comparison of the properties of swallowed ingesta in ruminants with material regurgitated for rumination is required to address these hypotheses. We can nevertheless provide corroborative evidence from the comparative approach for a difference in chewing workloads of non-ruminants vs. ruminants: the responses of craniodental traits like the hypsodonty index, cheek tooth volume, and molar row length, as well as the masseteric ridge length and ratio of the masseteric fossa to face length (negative relationship) to percent grass intake are all stronger (with steeper slopes) amongst non-ruminants than ruminants (Table 4.4). Hence, the different digestive strategies of non-ruminants and ruminants differ not only in terms of how an herbivore adapts to make a living out of feeding on low quality diets, but also in terms of solutions to processing foods of increasing toughness. Under this scenario, the entire forestomach complex of ruminants can be seen as a feature relaxing the selective pressure on their chewing apparatus (and chewing intensity). This view differs from a traditional outlook that sees rumination as an advantageous digestive strategy, against which non-ruminants can only be competitive with if they achieve higher intake levels and thereby extract *more* nutrients per day than ruminants (Janis 1976; Duncan et al. 1990). Rather, we propose that lower demands on oral chewing apparatus is one of the key advantages to being a ruminant, and may even be the factor giving ruminants a competitive edge over non-ruminants that, ultimately, resulted in the former replacing the latter as the most speciose terrestrial herbivores since the Oligocene (Janis et al. 1994; Janis 2008). Indeed, changes in diversity patterns of non-ruminants and ruminants in the fossil record are broadly co-incident with the replacement of browsers by intermediate-feeders and grazers through the Miocene (Janis et al. 2000), and subsequently higher rates of speciation amongst ruminant grazers (Janis 2008; Codron In review).

4.5 Digestion

4.5.1 General Digestive Tract Capacity and Digesta Retention

With a past focus on comparative digestive anatomy of browsers and grazers among ruminants, few generalized expectations exist that apply to all large mammal herbivores, i.e., to include non-ruminants. Because browse ferments at a faster rate than grass, we would expect generally longer mean retention times (MRT) of digesta particles and fluids in the digestive tract of grazers, and a corresponding larger gut capacity (Hummel et al. 2006). The available data, however, does not corroborate this concept across herbivores (see results for gut contents wet mass in Table 4.3, and MRT in Table 4.5). However, across herbivores, the MRT difference between small particles and fluids (called the selectivity factor), in the whole digestive tract, increases significantly with percentage grass intake (Table 4.5; and see Steuer et al. 2010). For the ruminant forestomach, however, the expected effects can be demonstrated: a more capacious reticulorumen in grazers (results for reticulorumen wet contents in Table 4.3, and rumen height in Table 4.6) with longer MRT of small particles (but not of fluids) at this site (Table 4.5), and a resulting distinct increase in the MRT difference between small particles and fluids (Table 4.5). The effect of body mass on measures of MRT is much lower than previously suggested (reviewed in Clauss et al. 2013), and is even absent in the case of fluid MRT in the reticulorumen (Table 4.5). As a result of the larger rumen capacities and longer particle MRT, ruminant grazers have been reported to digest fibre better than ruminant browsers (Pérez-Barbería et al. 2004). Corresponding analyses that include both ruminants and non-ruminants are lacking, but the effect has been demonstrated across rhino species (Steuer et al. 2010). The shorter particle MRT in the

Table 4.5 Models for diet effects on mean retention time (MRT), and on rumination times in ruminants

	n		Best Model/s ($\Delta AIC_c \leq 2$)					R v NR		
Trait	R	NR	Factors	r^2	λ	BM	% grass	Main	x BM	x % grass
$MRT_{particles}$ (whole GIT, h)	41	14	BM	0.239	0.619	0.10 (0.025)***	0	n.s	R > NR	n.s.
$MRT_{particles}$ (reticulorumen, h)	37	0	BM + %grass	0.361	0.000	0.15 (0.040)***	+			
$MRT_{solutes}$ (whole GIT, h)	36	14	BM; BM + %grass; Ruminant/(BM + %grass)	0.147	0.000	0.05 (0.018)**	0	n.s	R > NR	n.s.
$MRT_{solutes}$ (reticulorumen, h)	38	0	BM; %grass	0.056	0.898	0.07 (0.047)	0			
Selectivity factor ($MRT_{particle}$: MRT_{solute} ratio, whole GIT)	36	14	BM + %grass; Ruminant/(BM + %grass)	0.336	0.882	0.06 (0.019)**	+	n.s	R > NR	R > NR
Selectivity factor ($MRT_{particle}$: MRT_{solute} ratio, reticulorumen)	36	0	BM + %grass	0.518	0.000	0.10 (0.032)**	+			
Rumination time, minimum % of day	20	0	%grass; BM + %grass	0.172	1.000	−0.03 (0.052)	+			
Rumination time, % of day	20	0	BM; %grass	0.156	0.000	0.10 (0.052)	0			
Rumination time in wet season, % of day	20	0	%grass; BM	0.135	0.000	0.05 (0.051)	0			

n = number of species; R = ruminants; NR = non-ruminants; λ = phylogenetic signal in PGLS; BM = body mass scaling exponent (s.e.); %grass = effect of mean % grass in species' natural diets on trait (0 = no effect, + = positive relationship, − = negative relationship [alternative result from GLS in parenthesis if different from PGLS]); Main = comparison of R v NR intercepts; x BM and x %grass = comparison of slopes (**** = $p < 0.0001$; *** = $p < 0.001$; ** = $p < 0.01$; * = $p < 0.05$; n.s. = interaction term not significant, i.e., $p > 0.05$)

Table 4.6 Models for diet effects on anatomy of the rumen, omasum, and abomasum, and on stomach contents

		Best Model/s ($\Delta AIC_c \leq 2$)				
Trait	n	Factors	r^2	λ	BM	% grass
Anatomy						
Rumen cranial pillar thickness (mm)	61	BM + %grass	0.708	0.000	0.34 (0.031)****	+
Rumen caudal pillar thickness (mm)	57	BM + %grass	0.675	0.000	0.32 (0.043)****	+
Rumen wall height (cm)	71	BM + %grass	0.856	0.282	0.35 (0.018)****	+
Reticulum height (cm)	70	BM	0.864	0.184	0.32 (0.015)****	0
Reticulum width (cm)	66	BM + %grass; BM	0.820	0.000	0.28 (0.019)****	0
Reticular crest height (mm)	59	BM + %grass	0.542	0.784	0.36 (0.055)****	+
Intra-ruminal orifice area (mm^2)	52	BM + %grass	0.909	0.000	0.75 (0.052)****	+
Reticulo-ruminal orifice area (mm^2)	41	BM; BM + %grass	0.715	0.386	0.48 (0.049)****	0
Reticulo-omasal orifice area (mm^2)	50	BM; BM + %grass	0.442	0.505	0.45 (0.073)****	0
Surface papillation in rumen (dorsal:atrium %)	57	BM + %grass	0.569	0.737	−0.34 (0.092)***	−
Max height of *papillae unguiculiformes* (mm)	29	BM + %grass; BM	0.323	0.561	0.24 (0.101)*	−
Omasal laminal surface area (cm^2)	33	BM + %grass	0.881	1.000	1.29 (0.096)****	+
Abomasum fundus mucosa thickness (mm)	41	BM; BM + %grass; %grass	0.043	1.000	0.05 (0.037)	0 (−)
Abomasum pylorus mucosa thickness (mm)	23	BM; BM + %grass	0.371	0.000	0.26 (0.075)**	0
Abomasum curvature (cm)	65	BM; BM + %grass	0.913	0.000	0.34 (0.013)****	0
Contents						
Dry matter difference (dorsal-ventral rumen)	13	%grass; BM + %grass	0.732	0.000	2.42 (1.466)	+
pH (rumen)	30	BM; %grass; BM + %grass	0.112	0.375	0.01 (0.008)	0
Crude fibre (reticulorumen, % of dry matter)	28	BM; BM + %grass	0.181	1.000	0.07 (0.030)*	0
Volatile fatty acid concentration (rumen, Mmol l^{-1})	29	BM	0.181	0.000	−0.07 (0.029)*	0
Volatile fatty acid concentration (caecum, Mmol l^{-1})	16	%grass	0.339	0.000	−0.06 (0.046)	−

n = number of species; λ = phylogenetic signal in PGLS; BM = body mass scaling exponent (s.e.); %grass = effect of mean % grass in species' natural diets on trait (0 = no effect, + = positive relationship, − = negative relationship [alternative result from GLS in parenthesis if different from PGLS]; **** = $p < 0.0001$; *** = $p < 0.001$; ** = $p < 0.01$; * = $p < 0.05$)

reticulorumen of browsing ruminants might explain why more material, with residual fermentability, reaches their large intestine, leading to higher concentrations of volatile fatty acids at this site (Table 4.6).

These results emphasize the difference between particle and fluid MRT for herbivores. It has been hypothesized that a differential movement of fluid and particles leads to a 'washing' of the digesta with a removal of a part of the microbiota (Müller et al. 2011). In foregut fermenters, including ruminants, this should theoretically lead to a higher inflow of microbes into the glandular stomach and small intestine (Dittmann et al. 2015; Hummel et al. 2015). This advantage has no relevance in hindgut fermenters, which might be the reason why the characteristic is generally more pronounced in ruminants (interaction effects for MRT in Table 4.5). For any herbivore (ruminant and non-ruminant), however, removal of microbes by 'washing' is bound to shift the metabolism of the microbiota from maintenance towards growth, putatively increasing their fermentative efficiency, and shifting fermentative processes from the production of CO_2 and methane towards microbial cell mass (reviewed in Clauss and Hummel 2017). Why such an option, which is in theory favourable for any kind of herbivore, may be not available to some browsers, is explained in the next section.

4.5.2 Ruminant Forestomach Morphophysiology

For a historical overview of the development of interpretations of observed differences in forestomach morphophysiology between ruminant browsers and grazers see Clauss and Hummel (2017). At first, differences were mainly linked to putative differences in fibre content between browse and grass (Hofmann 1989), but data on the crude fibre concentration of rumen contents do not support this interpretation (Table 4.6). Similarly, no difference in the pH, or the volatile fatty acid concentration, in the rumen contents are evident between browsers and grazers (Table 4.6). One of the components of the primary concept of rumen physiology had focused on the reticular groove—the anatomical structure facilitating the bypass of milk in suckling ruminants, and thus preventing it from entering the rumen. This structure was thought to remain functional in adult browsers (Rowell-Schäfer et al. 2001). Limited experimental data, however, did not corroborate this concept (Lechner et al. 2009).

In a second step, the focus shifted to another original observation by Hofmann (1989), namely a difference in the stratification of rumen contents between browsers and grazers (Clauss et al. 2003b), which was introduced by Hofmann et al. (2008), and in the predecessor to the current Chapter (Clauss et al. 2008b). A series of studies demonstrated this difference, using as proxies, either directly a difference in the dry matter concentration (the reciprocal of the moisture content) between dorsal (upper) and ventral (lower) rumen contents (with a higher difference in grazers; Table 4.6; and see Clauss et al. 2009b), a difference in the presence of a gas dome in the dorsal rumen (Tschuor and Clauss 2008), or indirectly the intraruminal papillation pattern

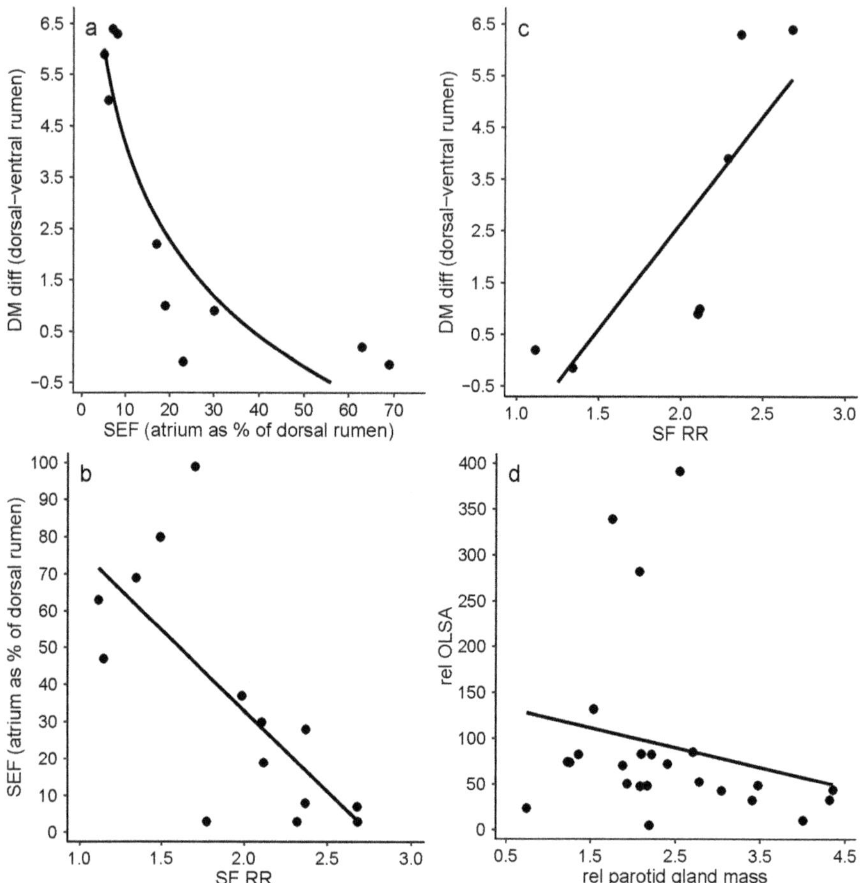

Fig. 4.3 Correlations between three proxies for rumen stratification (**a–c**)—the difference in dry matter (DM) of dorsal and ventral rumen contents, the proportion of surface papilla in the atrium relative to the dorsal rumen (SEF), and the selectivity factor (SF) of the reticulorumen contents (i.e., the ratio of particle:solute retention time). Coping with differences in fluid throughput also results in a correlation between parotid gland mass (expressed relative to $BM^{0.75}$) and omasum size (OLSA = omasal leaf surface area, expressed relative $BM^{0.67}$) (**d**)

(because stratified rumen contents should be linked to an inhomogenous papillation in the rumen; Clauss et al. 2009c, corresponding to a less homogenous papillation in grazers; Table 4.6). It was assumed that a higher fluid input in, and throughput through, the rumen leads to a more pronounced stratification in grazers. The correlations between the different stratification proxies (Fig. 4.3a–c) support this interpretation. Thus, a major difference between browsers and grazers is the ratio of small particle to fluid MRT in the reticulorumen in the latter (indicating a faster fluid turnover), and this ratio significantly increases with percentage grass in the diet (Table 4.5). In browsers, less fluid is put through the reticulorumen, and the rumen fluid is more viscous (Clauss et al. 2009a, b; Lechner et al. 2010), putatively trapping some of the fermentation gases, and resulting in less stratified rumen contents and an even intraruminal papillation. In grazers, the clear stratification comprises a dorsal gas dome, and a fibre 'raft' or 'mat' on top of a liquid layer (Hummel et al. 2009).

Focusing on the stratification creates a logically coherent interpretation of different observations. Significantly larger salivary glands in browsers than in grazers (Table 4.7; and see Hofmann et al. 2008), are no longer equated with a higher saliva output; larger glands are considered to produce saliva with a higher protein content (and hence viscosity), the proteins representing a defence against plant secondary metabolites, in particular against tannins (Austin et al. 1989; Hagermann and Robbins 1993; Fickel et al. 1998). The necessity to add these proteins to the saliva is considered a constraint on the amount of saliva that can be produced, and, therefore, smaller salivary glands—without the need to produce these proteins—are equated with higher amounts of saliva (Hofmann et al. 2008). The significantly thicker rumen muscles (the pillars) in grazers (Table 4.6) are considered an adaptation to the fibre 'mat' that putatively requires more force for mixing peristalsis than do homogenous rumen contents (Clauss et al. 2003b). Similarly, the larger intraruminal orifice in grazers (Table 4.6) may be better suited to allow a mixing of the fibre 'mat', which might tend to block a more narrow opening more easily. The significantly lower reticular crests in browsers (Table 4.6) are considered to lead to an incomplete emptying of the reticulum during contractions (Clauss et al. 2010a), which may be important because with the putatively more viscous rumen fluid and lower moisture saturation in the rumen of browsers, re-filling of the reticulum with fluid might be slower. Fluid reticulum contents are a prerequisite of the ruminant particle sorting mechanism, which operates on particle buoyancy and sedimentation (Lechner-Doll et al. 1991). Finally, grazers, with the higher fluid throughput through the reticulorumen, require larger omasa (Table 4.6)—with the main function of omasa being the resorption of fluid, to prevent too diluted digesta reaching the sites of auto-enzymatic digestion (Clauss et al. 2006). Therefore, across ruminants, we would expect a negative relationship between salivary gland size (with smaller glands producing higher amounts of salivary fluid) and omasum size (with larger omasa absorbing more of the fluid), as indicated in Fig. 4.3d. A thicker acid-producing fundic mucosa of the abomasum of browsers (Table 4.6; note that this is only the case in GLS but not in PGLS, and in contrast to the pyloric mucosa) is interpreted as an adaptation to putatively higher bicarbonate contents in the rumen fluid of browsers, due to the higher viscosity and CO_2 entrapment, which might require higher amounts of acid for acidification (Clauss et al. 2008b). No functional relevance, linked to the fluid throughput and viscosity, is attributed to the ruminoreticular orifice, or the reticuloomasal orifice, which both do not differ between browsers and grazers (Table 4.6); these openings do not have to accommodate the fibre 'mat' but pass on only a selection of rumen contents (ruminoreticular orifice), or those particles intended for passage into the lower digestive tract (reticuloomasal orifice). Neither the size of the reticulum nor that of the abomasum differ between the feeding types (Table 4.6), and are not linked to the concept of fluid throughput.

Table 4.7 Models for diet effects on salivary glands, liver, and other organs

Trait	*n*		Best Model/s ($\Delta AIC_c \leq 2$)					R v NR		
	R	NR	Factors	r^2	λ	BM	% grass	Main	x BM	x % grass
Salivary glands										
Parotid gland mass (g)	65	0	BM + %grass	0.889	0.517	0.68 (0.037)****	–			
Mandibular gland mass (g)	62	0	BM + %grass	0.908	0.541	0.72 (0.034)****	–			
Buccalis ventralis mass (g)	44	0	BM + %grass; BM	0.942	0.084	0.77 (0.032)****	0 (–)			
Sublingual gland mass (g)	28	0	BM + %grass	0.802	0.000	0.73 (0.080)****	–			
Liver										
Liver mass (kg)	57	10	Ruminant/(BM + %grass); BM + % grass; BM	0.963	0.572	0.83 (0.024)****	0 (–)	R < NR	R < NR	n.s.
Liver connective tissue (%)	33	0	BM + %grass	0.495	0.000	0.11 (0.024)****	+			
Other organs										
Heart mass (kg)	13	10	BM	0.957	0.986	0.87 (0.040)****	0	n.s	R < NR	n.s.
Kidney mss (kg)	13	10	BM; BM + %grass	0.965	0.272	0.87 (0.036)****	0	R < NR	R < NR	n.s.
Lung mass (kg)	12	10	Ruminant/(BM + %grass); BM	0.962	0.000	0.97 (0.061)****	0	n.s	R < NR	n.s.

n = number of species; R = ruminants; NR = non-ruminants; λ = phylogenetic signal in PGLS; BM = body mass scaling exponent (s.e.); %grass = effect of mean % grass in species' natural diets on trait (0 = no effect, + = positive relationship, – = negative relationship [alternative result from GLS in parenthesis if different from PGLS]); Main = comparison of R v NR intercepts; x BM and x %grass = comparison of slopes (**** = $p < 0.0001$; *** = $p < 0.001$; ** = $p < 0.01$; * = $p < 0.05$; n.s. = interaction term not significant, i.e., $p > 0.05$)

Thus, characteristics of grazers were interpreted as enhancing the stratification of the rumen contents (Clauss et al. 2008b), reinforcing a difference that was thought to exist between browse (with little propensity to stratify) and grass, based on in vitro experiments (Sutherland 1988; Wattiaux et al. 1992; Clauss et al. 2001). Note that this interpretation is still focused on a general difference between the diets, which is putatively amplified by the morphophysiological characteristics. The hypothesized adaptive value was supposed to lie in an enhanced 'filter bed effect', i.e., the entrapment of smaller particles in the fibre 'mat', with a corresponding longer particle retention and digestion, which appeared suitable for grass forage; additionally, it was thought that the greater amount of low-viscosity fluid, with a pre-sorting of particles already in the rumen, would facilitate a more efficient particle sorting mechanism (that should also lead to more intensive rumination in grazers; Clauss et al. 2003b). The, presumably, larger particles escaping the forestomach of browsers (Clauss et al. 2002) were considered an indication for a less efficient sorting mechanism, and the larger *papillae unguiculiformes*—papillae around the reticuloomasal orifice—in browsers (Table 4.6) were interpreted as a compensatory straining mechanism that partially prevents the escape of large particles (Nygren et al. 2001).

However, while the basis of fluid throughput as the main determinant of these morphophysiological interrelationships remains unchallenged, a revision of the interpretation of its adaptive value became necessary. During an experimental study, with fistulated non-domestic ruminants, in which differences in rumen contents stratification, rumen fluid viscosity and the ratio of particle to fluid MRT in the reticulorumen were demonstrated, no difference in the sorting mechanism (of large vs. small particles), or the size of particles escaping the forestomach, were evident, suggesting that these factors were not affected by adaptations related to fluid throughput and stratification (Lechner et al. 2010). The previously documented differences in particle size were related to the use of artificial diets in captivity (Hummel et al. 2008; Lechner et al. 2010), and the previous assumption that browse does not tend to stratify by particle size in in vitro systems was corrected (Clauss et al. 2009b; Lechner et al. 2010). Additionally, a 'filter-bed' effect could be detected on grass vs. browse diets, irrespective of the feeding type of the animals (Clauss et al. 2011a; Lauper et al. 2013), and no general difference in the time spent in rumination appears to exist between browsers and grazers (Table 4.5).

Therefore, the interpretation of the adaptive value of the various characteristics underwent another revision, and is currently focused on the relevance of the high fluid throughput itself. In order to avoid semantic circular reasoning, two digestive strategies in ruminants were defined (the 'moose-type' and the 'cattle-type'), that can then be compared to the feeding types of browsers and grazers (Clauss et al. 2010b). As stated above, a high fluid throughput most likely increases the harvest of microbes from the forestomach and, thus, makes this system more efficient, and should, therefore, be an adaptation in any ruminant, irrespective of feeding type (Clauss et al. 2010b; Clauss and Hummel 2017). Note that now, corresponding adaptations are not linked to a diet property, but to a putative optimization of the digestive tract itself. While any diet, that does not comprise extreme amounts of browse, appears to allow such adaptations, strict browsers, due to their salivary defence mechanisms, cannot use the advantages of a high fluid throughput (Codron and Clauss 2010). Correspondingly, while 'moose-type' ruminants appear to be constrained to pure browse diets (and are browsers), 'cattle-type' ruminants have a much broader diet spectrum. In particular, this new concept explains the lack of 'escalation' of morphophysiological adaptations from intermediate-feeding to grazing, i.e., many traits considered typical for grazers, including the selectivity factor, do not become more pronounced from intermediate-feeders to grazers (see also discussion on hypsodonty index above). Indeed, many of those species that show particularly extreme 'cattle-type' characteristics, such as the Bovini or the muskox (*Ovibos moschatus*), with a very high fluid throughput, a distinct rumen contents stratification, and intraruminal papillation pattern, and extremely large omasa, are not the strictest grazers, but often intermediate-feeders (Clauss and Hofmann 2014). The reasons for the differences within 'cattle-type' ruminants, as in the various characteristics linked to fluid throughput, are not considered direct reflections of a diet niche in this scenario, but expressions of different degrees of 'escalation' (*sensu* Vermeij 1994, 2013), during the evolutionary optimization of their digestive physiology.

4.5.3 Organ Size

The original suggestion that liver size is related to diet, with browsers putatively requiring larger livers for the detoxification of secondary plant compounds (Hofmann 1989; Duncan et al. 1998), was detectable in GLS, but not confirmed in PGLS in the current dataset, also not for ruminants (Table 4.7, interaction). However, the finding that among ruminants, the proportion of connective (i.e., non-functional) liver tissue increased significantly with percentage grass in the diet (Table 4.7), supports the concept that browsers require more functional liver tissue. The masses of the heart, kidney or lung show no relationship with diet (Table 4.7; even though it could be expected that some reduction in organ size compensates for larger rumens of grazers; Clauss et al. 2003a). The general tendencies for a lower organ size increase in ruminants compared to non-ruminants (Table 4.7, interaction) might correspond to such a concept of organ reduction in ruminants, in compensation for large forestomachs (Mortolaa and Lanthier 2005). However, more detailed data would be required to corroborate these findings.

4.6 Conclusions: Where to from Here?

The morphological and physiological traits of herbivores reflect different constraints imposed by browse- versus grass-based diet niches. The most prominent complex is that of putatively diet-specific adaptations in orocranial (and especially dental) anatomy. Morphophysiological characteristics of the digestive tract also show significant correlations with diet, even after accounting for the phylogenetic structure of datasets. Clearly, large mammalian herbivores remain an important model for our understanding of key macroevolutionary concepts such as adaptive radiation, and convergence.

An aspect of digestive physiology that has received comparatively little attention so far, in spite of an enormous increase in the availability of methodological approaches in recent years, is the bacterial, archaeal, fungal and protozoan microbiome of herbivores. Available data (e.g., wild ruminants; Henderson et al. 2015) has not yet been evaluated, either with respect to feeding type (browser vs grazer) or digestion type ('moose-type' vs 'cattle-type'); our own lack of methodological expertise prevented us from exploring those data here. Given that the microbiota may be involved in detoxification or defence processes, more studies would be highly interesting. To our knowledge, the only systematic comparison, with respect to the microbiome, is restricted to literature data (gained from traditional microscopic analyses) of the composition of the protozoal microbiome (Clauss et al. 2011b) in ruminants of different dietary niches. The case of rumen protozoa can serve as an illustration of why a detailed look at the data, and at the conceptual design of studies, is necessary to prevent fallacies and meaningless findings in the following paragraphs.

A subjective, put potentially widely shared impression, when reading literature about browsers and grazers, is that the categorisation of a species into a feeding type is often presented as an aim in itself, as if it had relevant heuristic value. Therefore, the impression is that studies involving morphophysiological measurements are often justified by their use of proxies to facilitate that categorisation. To single out—admittedly unfairly—a particular study as an example, it appears conceptually awkward to attempt to categorize a ruminant species, such as the mouflon (*Ovis ammon musimon*), in the feeding type spectrum by analysing its rumen protozoa as a proxy (Obidziński et al. 2017). When dealing with extant species, this approach has the evident flaw that the real proxy for a feeding type is the botanical investigation of the diet consumed; in the case of mouflon, stating that they are intermediate-feeders based on a review of their natural diet (Marchand et al. 2013) is the superior conceptual approach. The additional heuristic value of investigating their protozoal microbiome lies in revealing effects of differences in the actually consumed diet on that microbiome (as done very elegantly by Obidziński et al. 2017).

When dealing with extinct species, proxy-based approaches are the only ones available and very common, particularly for dental anatomical characters and dental wear proxies, often supplemented by stable isotope measurements (we refrain from reviewing that literature, but see **Saarinen** Chap. 2 for more detail). For dental anatomy and isotope values, such reconstructions often under-emphasize the degree of uncertainty linked to these proxies (for example, as mentioned above, that many extant ruminant intermediate-feeders have a higher hypsodonty index than strict grazers, and that the degree of wear experienced by herbivores differs between ruminants and non-ruminants), and the use of wear-related proxies is mostly not based on experimentally derived results.

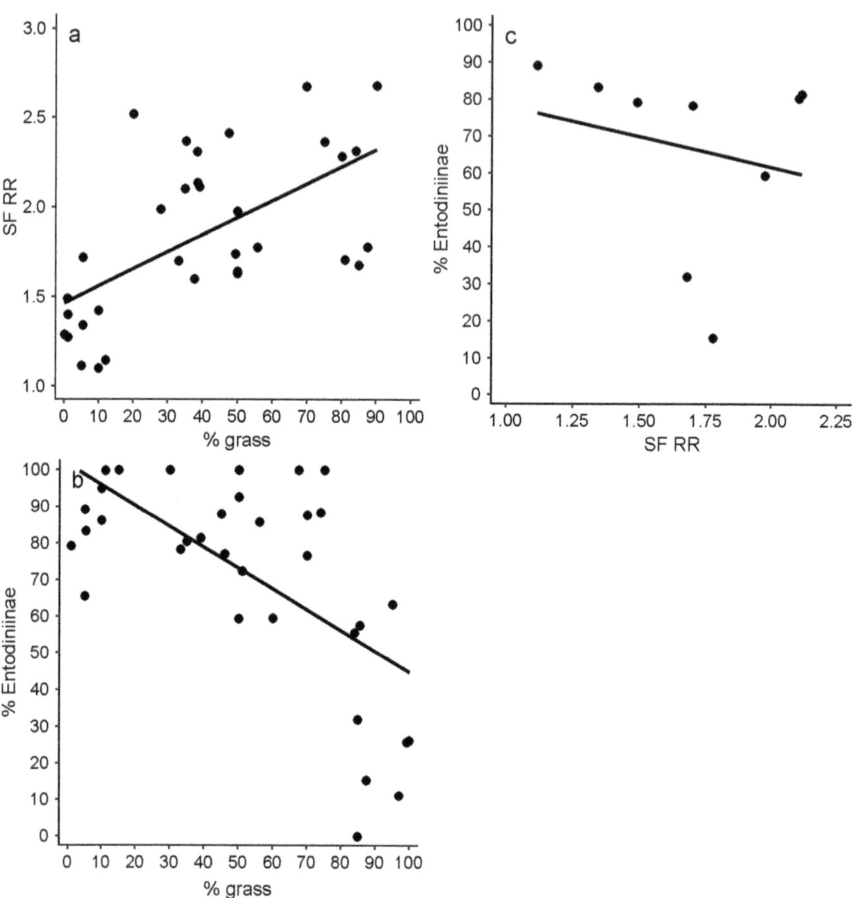

Fig. 4.4 Discordance in how morphophysiological traits respond to diet. The ratio of particle: solute mean retention time in the reticulorumen (called SF RR) increases dramatically in species with >15–20% grass in the natural diet (**a**), whereas the proportion of gut protozoa represented by the Entodiniinae decreases dramatically at around 80% grass in the natural diet (**b**; note that the proportion of Diplodiniinae increases at more or less the same rate at this diet). Consequently, the two traits are not correlated (**c**)

The actually observed pattern of the distribution of a character, across feeding types, is highly relevant for a functional understanding of digestive morphology and physiology. For example, differences in the ratio of small particle to fluid MRT in the reticulorumen (selectivity factor, Table 4.5; Dittmann et al. 2015), and in the composition of rumen protozoa (Clauss et al. 2011b), could be correctly summarized by the statement "in ruminants, browsers and grazers differ in characteristics of rumen fluid throughput and rumen protozoa". However, such a superficial view omits the fact that the seeming thresholds of change for the respective characteristic differs dramatically between the two datasets (Fig. 4.4a and b): while the selectivity factor increases dramatically at a threshold of 15–20% grass in the natural diet (as does the intraruminal papillation pattern; Codron and Clauss 2010), the proportion of Entodiniinae to Diplodiniinae protozoa changes at a threshold of about 80% grass in the natural diet. Consequently, the two measurements—the selectivity factor and the relative proportion of Entodiniinae—do not correlate well with each other (Fig. 4.4c). This closer look informs us that different mechanisms may be at play in these two complexes; at the moment we do not have a detailed understanding of these processes.

Another important aspect of comparative analyses is that statistical support for a link between a species' traits and their dietary niches should not lead to rigid views of exclusive solutions. Nor, on the other hand, should such links be interpreted to exclude species from certain niches (Codron and Clauss 2010). Rather, taxon-specific peculiarities need to be acknowledged that may deviate from general trends and might, depending on the number of corresponding species present in a dataset, allow, or prevent, the detection of these trends. For example, differences exist in the method of cropping plant material, whereby some species (like goats) use their lips, while others (like cattle) use their tongue more extensively (Meier et al. 2016). This does not invalidate the statistical finding that, across a certain dataset, ruminant browsers differ in tongue morphology from grazers in having higher proportions of a "freely mobile" part of their tongue. However, the outlier of cattle in this respect precludes the opposite conclusion, that species with a highly mobile tongue must be browsers, and also precludes the interpretation that grazers cannot have mobile tongues. Additionally, it highlights that conclusions might change, depending on the nature of the dataset: in a hypothetical dataset with mainly bovine grazers and only goat-type browsers, the analysis might have found a more mobile tongue in grazers than in browsers.

Other important examples are that, amongst grazers, some species have extremely wide muzzles (grazing rhinos, hippos, Alcelaphini), whereas others (Equidae, Hippotragini) have more pointed faces. Determining whether these differences reflect different dietary niches, for instance foraging at different sward heights, or merely different solutions to the same problem, requires taxon-specific hypotheses and, ideally, experimentation. The existence of outliers, for instance blackbuck (*Antilope cervicapra*), which have a high particle to fluid MRT ratio in the reticulorumen (i.e., a high fluid thoughput; Hummel et al. 2015), but a very small omasum (Sauer et al. 2016), emphasizes that general trends are not obligatory. Amongst ruminant browsers, cervids and giraffids are typically better represented in datasets than are bovids, and hence might unduly influence current interpretations. In particular, comparative physiological data for Tragelaphini are lacking in many datasets, but may represent an alternate solution to browsing from giraffids and cervids, as indicated by their deviation from the typical salivary gland size-diet relationship observed across ruminants (Robbins et al. 1995; Hofmann et al. 2008).

Apart from the self-evident request for more data on more species, and for additional corroboration of once-only measurements that are then analysed again and again with new statistical approaches, we believe that, in order to interrogate the question of exclusivity in trait-niche relationships, our description of diet niches themselves needs to shift from one based on species' means to a more inclusive approach depicting niche breadths. One such approach considers whether morphophysiological characteristics are more closely aligned with minimum or maximum levels of grass in species' natural diets (Codron and Clauss 2010), potentially allowing us to define thresholds around which species may or may not be excluded from certain niches. For example, there appears to be no continuous relationship between the amount of browse/grass in the diet and the digestion type ('moose-type' vs 'cattle-type') amongst ruminants, but rather a dichotomy between species that do, and do not, eat grass. Similarly, the hypsodonty index is not continuously distributed along a percentage grass in the diet axis; rather, species that lack sufficiently large tooth crowns typically do not eat grass, whereas species that have large tooth crowns may occupy almost any diet niche (Damuth and Janis 2011). These statements echo the one by Van Wieren (1996b) that "browsers are non-grazers". Broadening our representation of niches, to include 'niche breadths', would allow investigations not only of exclusiveness in adaptation, but also potentially raises important questions that could lead to a more complete functional interpretation. A potentially interesting addition to this niche breadth approach is to include the niche history of species, i.e., a measure of how long the species has existed in a given dietary niche space. Examples do exist of diet niche shifts within lineages, sometimes coupled with evidence for morphological shifts (Strömberg 2006; Feranec 2007; Codron et al. 2008a; Cerling et al. 2015; Ecker et al. 2018). These studies differ with respect to whether dietary niches were interpreted to have become more specialized or generalized over evolutionary time, further emphasizing a lack of dietary niche exclusivity.

What is also lacking from many discussions, and a topic we have only sporadically addressed here, is the covariance amongst traits. For the most part, investigations of multiple traits have attempted only to define species' distributions in multivariate morphospace. We expect, however, that covariance emanates because certain traits/organ systems have evolved as functional units, or as anatomical trade-offs. The relative size of salivary glands is an important example of this; only when considered together with the size of the omasum does the functional relevance of variations in both organ systems become clear. Thus, correlations between traits should be studied, and interpreted, in terms of both the overall *bauplan* of species, and the way in which components interact to make the organism more efficient at exploiting a particular niche. In doing so, we must be open to the outcome that what we considered an adaptation to a particular diet, or a particular characteristic of a diet, might ultimately be an adaptation to a different characteristic of the same diet, or an adaptation to a nondietary selective pressure that only correlated with the dietary niche.

References

Ackermans NL et al (2018) Controlled feeding experiments with diets of different abrasiveness reveal slow development of mesowear signal in goats (*Capra aegagrus hircus*). J Exp Biol 221:186411

Aitchison J (1946) Hinged teeth in mammals: a study of the tusks of muntjacs (*Muntiacus*) and Chinese water deer (*Hydropotes inermis*). Proc Zool Soc Lond 116:329–338
Archer D, Sanson G (2002) Form and function of the selenodont molar in southern African ruminants in relation to their feeding habits. J Zool 257:13–26. https://doi.org/10.1017/S0952836902000614
Austin PJ, Suchar LA, Robbins CT, Hagermann AE (1989) Tannin-binding proteins in saliva of deer and their absence in saliva of sheep and cattle. J Chem Ecol 15:1335–1347
Avgar T, Street G, Fryxell JM (2014) On the adaptive benefits of mammal migration. Can J Zool 92:481–490. https://doi.org/10.1139/cjz-2013-0076
Baker G, Jones LHP, Wardrop ID (1959) Cause of wear in sheep's teeth. Nature 184:1583–1584
Bell RHV (1971) A grazing ecosystem in the Serengeti. Sci Am 225:86–93
Benvenutti MA, Gordon IJ, Poppi DP (2006) The effect of the density and physical properties of grass stems on the foraging behaviour and instantaneous intake rate by cattle grazing an artificial reproductive tropical sward. Grass Forage Sci 61:272–281. https://doi.org/10.1111/j.1365-2494.2006.00531.x
Brashares JS, Garland T, Arcese P (2000) Phylogenetic analysis of coadaptation in behavior, diet, and body size in the African antelope. Behav Ecol 11:452–463
Burnham KP, Anderson DR (2001) Kullback-Leibler information as a basis for strong inference in ecological studies. Wildl Res 28:111–119
Burnham KP, Anderson DR (2002) Model selection and multimodel inference: a practical information-theoretic approach. Springer, New York
Cerling TE, Harris JM, Passey BH (2003) Diets of east African Bovidae based on stable isotope analysis. J Mammal 84:456–470
Cerling TE, Harris JM, Leakey MG (2005) Environmentally driven dietary adaptations in African mammals. In: Ehleringer JR, Cerling TE, Dearing MD (eds) A history of atmospheric CO_2 and its effects on plants, animals and ecosystems. Springer, New York, pp 258–272
Cerling TE et al (2015) Dietary changes of large herbivores in the Turkana Basin, Kenya from 4 to 1 ma. Proc Natl Acad Sci U S A 112:11467–11472. https://doi.org/10.1073/pnas.1513075112
Chivers DJ, Hladik CM (1980) Morphology of the gastrointestinal tract in primates: comparisons with other mammals in relation to diet. J Morphol 166:337–386
Clauss M, Hofmann RR (2014) The digestive system of ruminants, and peculiarities of (wild) cattle. In: Melletti M, Burton J (eds) Ecology, evolution and behaviour of wild cattle: implications for conservation. Cambridge University Press, Cambridge, pp 57–62
Clauss M, Hummel J (2017) Physiological adaptations of ruminants and their potential relevance for production systems. Rev Bras Zootec 46:606–613
Clauss M, Lechner-Doll M (2001) Differences in selective reticulo-ruminal particle retention as a key factor in ruminant diversification. Oecologia 129:321–327
Clauss M, Lechner-Doll M, Behrend A, Lason K, Lang D, Streich WJ (2001) Particle retention in the forestomach of a browsing ruminant, the roe deer (*Capreolus capreolus*). Acta Theriol 46:103–107
Clauss M, Lechner-Doll M, Streich WJ (2002) Faecal particle size distribution in captive wild ruminants: an approach to the browser/grazer-dichotomy from the other end. Oecologia 131:343–349
Clauss M et al (2003a) The maximum attainable body size of herbivorous mammals: morphophysiological constraints on foregut, and adaptations of hindgut fermenters. Oecologia 136:14–27
Clauss M, Lechner-Doll M, Streich WJ (2003b) Ruminant diversification as an adaptation to the physicomechanical characteristics of forage. A reevaluation of an old debate and a new hypothesis. Oikos 102:253–262
Clauss M et al (2006) The macroscopic anatomy of the omasum of free-ranging moose (*Alces alces*) and muskoxen (*Ovibos moschatus*) and a comparison of the omasal laminal surface area in 34 ruminant species. J Zool 270:346–358

Clauss M, Hofmann RR, Streich WJ, Fickel J, Hummel J (2008a) Higher masseter muscle mass in grazing than in browsing ruminants. Oecologia 157:377–385
Clauss M, Kaiser TM, Hummel J (2008b) The morphophysiological adaptations of browsing and grazing mammals. In: Gordon IJ, Prins HHT (eds) The ecology of browsing and grazing. Springer, Heidelberg, pp 149–178
Clauss M et al (2009a) Physical characteristics of rumen contents in two small ruminants of different feeding type, the mouflon (*Ovis ammon musimon*) and the roe deer (*Capreolus capreolus*). Zoology 112:195–205
Clauss M et al (2009b) Physical characteristics of rumen contents in four large ruminants of different feeding type, the addax (*Addax nasomaculatus*), bison (*Bison bison*), red deer (*Cervus elaphus*) and moose (*Alces alces*). Comp Biochem Physiol A Mol Integr Physiol 152:398–406
Clauss M, Hofmann RR, Fickel J, Streich WJ, Hummel J (2009c) The intraruminal papillation gradient in wild ruminants of different feeding types: implications for rumen physiology. J Morphol 270:929–942
Clauss M, Nunn C, Fritz J, Hummel J (2009d) Evidence for a tradeoff between retention time and chewing efficiency in large mammalian herbivores. Comp Biochem Physiol A Mol Integr Physiol 154:376–382
Clauss M, Hofmann RR, Streich WJ, Fickel J, Hummel J (2010a) Convergence in the macroscopic anatomy of the reticulum in wild ruminant species of different feeding types and a new resulting hypothesis on reticular function. J Zool 281:26–38
Clauss M, Hume ID, Hummel J (2010b) Evolutionary adaptations of ruminants and their potential relevance for modern production systems. Animal 4:979–992
Clauss M et al (2011a) The effect of size and density on the mean retention time of particles in the reticulorumen of cattle (*Bos primigenius* f. *taurus*), muskoxen (*Ovibos moschatus*) and moose (*Alces alces*). Br J Nutr 105:634–644
Clauss M, Müller K, Fickel J, Streich WJ, Hatt J-M, Südekum K-H (2011b) Macroecology of the host determines microecology of endobionts: protozoal faunas vary with wild ruminant feeding type and body mass. J Zool 283:169–185
Clauss M, Steuer P, Müller DWH, Codron D, Hummel J (2013) Herbivory and body size: allometries of diet quality and gastrointestinal physiology, and implications for herbivore ecology and dinosaur gigantism. PLoS One 8:e68714
Clauss M et al (2015) Faecal particle size: digestive physiology meets herbivore diversity. Comp Biochem Physiol A Mol Integr Physiol 179:182–191
Clauss M et al (2017) Reconstruction of body cavity volume in terrestrial tetrapods. J Anat (Lond) 230:325–336
Codron D (In review) Evolution of large mammal herbivores. In: Scogings P (ed) Herbivores and woody plants in savannas. Wiley
Codron D, Clauss M (2010) Rumen physiology constrains diet niche: linking digestive physiology and food selection across wild ruminant species. Can J Zool 88:1129–1138
Codron J, Lee-Thorp JA, Sponheimer M, Codron D, Grant RC, De Ruiter DJ (2006) Elephant (*Loxodonta africana*) diets in Kruger National Park, South Africa: spatial and landscape differences. J Mammal 87:27–34
Codron D, Lee-Thorp JA, Sponheimer M, Codron J, de Ruiter D, Brink JS (2007) Significance of diet type and diet quality for ecological diversity of African ungulates. J Anim Ecol 76:526–537
Codron D, Brink JS, Rossouw L, Clauss M (2008a) The evolution of ecological specialization in southern African ungulates: competition or physical environmental turnover? Oikos 117:344–353
Codron D et al (2008b) Functional differentiation of African grazing ruminants: an example of specialized adaptations to very small changes in diet. Biol J Linn Soc 94:755–764
Codron D, Codron J, Sponheimer M, Clauss M (2016) Within-population isotopic niche variability in savanna mammals: disparity between carnivores and herbivores. Front Ecol Evol 4:15. https://doi.org/10.3389/fevo.2016.00015

Copeland SR, Sponheimer M, Spinage CA, Lee-Thorp JA, Codron D, Reed KE (2009) Stable isotope evidence for impala *Aepyceros melampus* diets at Akagera National Park. Rwanda Afr J Ecol 47:490–501
Damuth J, Janis CM (2011) On the relationship between hypsodonty and feeding ecology in ungulate mammals, and its utility in palaeoecology. Biol Rev (Camb) 86:733–758
Demment MW, Longhurst WH (1987) Browsers and grazers: constraints on feeding ecology imposd by gut morphology and body size. Proceedings of the IVth international conference on goats. Brazilia, Brazil, pp 989–1004
Dittmann MT et al (2015) Digesta kinetics in gazelles in comparison to other ruminants: evidence for taxon-specific rumen fluid throughput to adjust digesta washing to the natural diet. Comp Biochem Physiol A Mol Integr Physiol 185:58–68. https://doi.org/10.1016/j.cbpa.2015.01.013
Dittmann MT, Kreuzer M, Runge U, Clauss M (2017) Ingestive mastication in horses resembles rumination but not ingestive mastication in cattle and camels. J Exp Zool A Ecol Integr Physiol 327:98–109
du Toit JT (1990) Feeding-height stratification among African browsing ruminants. Afr J Ecol 28:55–61
Duncan P, Foose TJ, Gordon IJ, Gakahu CG, Lloyd M (1990) Comparative nutrient extraction from forages by grazing bovids and equids: a test of the nutritional model of equid/bovid competition and coexistence. Oecologia 84:411–418
Duncan P, Tixier H, Hofmann RR, Lechner-Doll M (1998) Feeding strategies and the physiology of digestion in roe deer. In: Andersen R, Duncan P, Linell JDC (eds) The European roe deer: the biology of success. Scandinavian University Press, Oslo, pp 91–116
Ecker M, Brink JS, Rossouw L, Chazan M, Horwitz LK, Lee-Thorp JA (2018) The palaeoecological context of the Oldowan-Acheulean in southern Africa. Nat Ecol Evol. https://doi.org/10.1038/s41559-018-0560-0
Erickson KL (2014) Prairie grass phytolith hardness and the evolution of ungulate hypsodonty. Hist Biol 26:737–744
Eronen JT, Evans AR, Fortelius M, Jernvall J (2009) The impact of regional climate on the evolution of mammals: a case study using fossil horses. Evolution 64:398–408
Eronen JT et al (2010) Precipitation and large herbivorous mammals II: application to fossil data. Evol Ecol Res 12:235–248
Feranec RS (2003) Stable isotopes, hypsodonty, and the paleodiet of *Hemiauchenia* (Mammalia: Camelidae): a morphological specialization creating ecological generalization. Paleobiol 29:230–242
Feranec RS (2007) Ecological generalization during adaptive radiation: evidence from Neogene mammals. Evol Ecol Res 9:555–577
Fickel J, Göritz F, Joest BA, Hildebrandt T, Hofmann RR, Breves G (1998) Analysis of parotid and mixed saliva in roe deer (*Capreolus capreolus* L.). J Comp Physiol B 168:257–264
Fletcher TM, Janis CM, Rayfield EJ (2010) Finite element analysis of ungulate jaws: can mode of digestive physiology be determined? Palaeontol Electron 13:21A
Fleurance G, Fritz H, Duncan P, Gordon IJ, Edouard N, Vial C (2009) Instantaneous intake rate in horses of different body sizes: influence of sward biomass and fibrousness. Appl Anim Behav Sci 117:84–92
Fortelius M (1985) Ungulate cheek teeth: developmental, functional and evolutionary interrelations. Acta Zool Fenn 180:1–76
Fortelius M, Solounias N (2000) Functional characterization of ungulate molars using the abrasion-attrition wear gradient: a new method for reconstructing paleodiets. Am Mus Novit 3301:1–36
Fowler ME (1983) Plant poisoning in free-living wild animals: a review. J Wildl Dis 19:34–43
Franz R, Hummel J, Müller DWH, Bauert M, Hatt J-M, Clauss M (2011) Herbivorous reptiles and body mass: effects on food intake, digesta retention, digestibility and gut capacity, and a comparison with mammals. Comp Biochem Physiol A Mol Integr Physiol 158:94–101
Fraser D, Rybczynski N (2014) Complexity of ruminant masticatory evolution. J Morphol 275:1093–1102

Fraser D, Theodor JM (2011) Comparing ungulate dietary proxies using discriminant function analysis. J Morphol 272:1513–1526. https://doi.org/10.1002/jmor.11001
Fritz J, Hummel J, Kienzle E, Arnold C, Nunn C, Clauss M (2009a) Comparative chewing efficiency in mammalian herbivores. Oikos 118:1623–1632
Fritz SA, Bininda-Emonds ORP, Purvis A (2009b) Geographical variation in predictors of mammalian extinction risk: big is bad, but only in the tropics. Ecol Lett 12:538–549
Fritz J, Hummel J, Keinzle E, Streich WJ, Clauss M (2010) To chew or not to chew: faecal particle size in herbivorous reptiles and mammals. J Exp Zool A Ecol Integr Physiol 313:579–586
Garland TJ, Bennett AF, Rezende EL (2005) Phylogenetic approaches in comparative physiology. J Exp Biol 218:3015–3035
Gordon IJ (2003) Browsing and grazing ruminants: are they different beasts? For Ecol Manag 181:13–21
Gordon IJ, Illius AW (1988) Incisor arcade structure and diet selection in ruminants. Funct Ecol 2:15–22
Gordon IJ, Illius AW (1994) The functional significance of the browser-grazer dichotomy in African ruminants. Oecologia 98:167–175
Gordon IJ, Illius AW (1996) The nutritional ecology of African ruminants: a reinterpretation. J Anim Ecol 65:18–28
Greaves W (1991) A relationship between premolar loss and jaw elongation in selenodont artiodactyls. Zool J Linnean Soc 101:121–129
Gross JE, Hobbs NT, Wunder BA (1993) Independent variables for predicting intake rate of mammalian herbivores: biomass density, plant density, or bite size? Oikos 68:75–81
Gwynne MD, Bell RHV (1968) Selection of vegetation components by grazing ungulates in the Serengeti National Park. Nature 220:390. https://doi.org/10.1038/220390a0
Hagermann AE, Robbins CT (1993) Specifity of tannin-binding salivary proteins relative to diet selection by mammals. Can J Zool 71:628–633
Harvey PH, Pagel MD (1991) The comparative method in evolutionary biology. Oxford University Press, Oxford
Hemae KM (1967) Masticatory function in the mammals. J Dent Res 46:883–893
Henderson G et al (2015) Rumen microbial community composition varies with diet and host, but a core microbiome is found across a wide geographical range. Sci Rep 5:14567
Hendrichs H (1965) Vergleichende Untersuchung des Wiederkauverhaltens. Biol Zentralbl 84:681–751
Herring SW (1985) Morphological correlates of masticatory patterns in peccaries and pigs. J Mammal 66:603–617
Heywood JJN (2010a) Explaining patterns in modern ruminant diversity: contingency or constraint? Biol J Linn Soc 99:657–672
Heywood JJN (2010b) Functional anatomy of bovid upper molar occlusal surfaces with respect to diet. J Zool 281:1–11
Hodson MJ, White PJ, MEad A, Broadley MR (2005) Phylogenetic variation in the silicon composition of plants. Ann Bot 96:1027–1046
Hoffman JM, Fraser D, Clementz MT (2015) Controlled feeding trials with ungulates: a new application of in vivo dental molding to assess the abrasive factors of microwear. J Exp Biol 218:1538–1547
Hofmann RR (1973) The ruminant stomach. East African Literature Bureau, Nairobi
Hofmann RR (1988) Morphological evolutionary adaptations of the ruminant digestive system. In: Dobson A, Dobson MJ (eds) Aspects of digestive physiology in ruminants. Comstock Publishing Associates, Cornell University Press, Ithaca, NY
Hofmann RR (1989) Evolutionary steps of ecophysiological adaptation and diversification of ruminants: a comparative view of their digestive system. Oecologia 78:443–457
Hofmann RR, Stewart DRM (1972) Grazer or browser: a classification based on the stomach-structure and feeding habit of east African ruminants. Mammalia 36:226–240

Hofmann RR, Streich WJ, Fickel J, Hummel J, Clauss M (2008) Convergent evolution in feeding types: salivary gland mass differences in wild ruminant species. J Morphol 269:240–257
Hummel J, Südekum K-H, Streich WJ, Clauss M (2006) Forage fermentation patterns and their implications for herbivore ingesta retention times. Funct Ecol 20:989–1002
Hummel J et al (2008) Differences in fecal particle size between free-ranging and captive individuals of two browser species. Zoo Biol 27:70–77
Hummel J et al (2009) Physical characteristics of reticuloruminal contents of cattle in relation to forage type and time after feeding. J Anim Physiol Anim Nutr 93:209–220
Hummel J et al (2011) Another one bites the dust: faecal silica levels in large herbivores correlate with high-crowned teeth. Proc R Soc Lond B Biol Sci 278:1742–1747. https://doi.org/10.1098/rspb.2010.1939
Hummel J, Hammer S, Hammer C, Ruf J, Lechenne M, Clauss M (2015) Solute and particle retention in a small grazing antelope, the blackbuck (*Antilope cervicapra*). Comp Biochem Physiol A Mol Integr Physiol 182:22–26
Hutchinson GE (1959) Homage to Santa Rosalia, or why are there so many kinds of animals? Am Nat XCIII(870):137–145
Janis CM (1976) The evolutionary strategy of the Equidae and the origins of rumen and caecal digestion. Evolution 30:757–774
Janis CM (1988) An estimation of tooth volume and hypsodonty indices in ungulate mammals and the correlation of these factors with dietary preferences. In: Russell DE, Santoro J-P, Signogneau-Russell D (eds) Teeth revisited. Proceedings of the VIIth international symposium on dental morphology. Mémoires du Muséum national d'Histoire Naturelle, Paris (serie C), vol 53, pp 367–387
Janis CM (1995) Correlations between craniodental morphology and feeding behavior in ungulates: reciprocal illumination between living and fossil taxa. In: Thomason JJ (ed) Functional morphology in vertebrate paleontology. Cambridge University Press, Cambridge, pp 76–98
Janis CM (2008) An evolutionary history of browsing and grazing ungulates. In: Gordon IJ, Prins HHT (eds) The ecology of browsing and grazing. Springer, Heidelberg, pp 21–45
Janis CM, Constable E (1993) Can ungulate craniodental features determine digestive physiology? J Vertebr Paleontol 13:43A
Janis CM, Ehrhardt D (1988) Correlation of relative muzzle width and relative incisor width with dietary preference in ungulates. Zool J Linnean Soc 92:267–284
Janis CM, Gordon IJ, Illius AW (1994) Modelling equid/ruminant competition in the fossil record. Hist Biol 8:15–29
Janis CM, Damuth J, Theodor JM (2000) Miocene ungulates and terrestrial primary productivity: where have all the browsers gone? Proc Natl Acad Sci U S A 97:7899–7904
Janis CM, Constable EC, Houpt KA, Streich WJ, Clauss M (2010) Comparative ingestive mastication in domestic horses and cattle: a pilot investigation. J Anim Physiol Anim Nutr 94:e402–e409. https://doi.org/10.1111/j.1439-0396.2010.01030.x
Jernvall J, Hunter JP, Fortelius M (1996) Molar tooth diversity, disparity, and ecology in Cenozoic ungulate radiations. Science 274:1489–1492
Kaiser TM, Fickel J, Streich WJ, Hummel J, Clauss M (2010) Enamel ridge alignment in upper molars of ruminants in relation to their natural diet. J Zool 281:12–25
Kaiser DP, Müller DWH, Fortelius M, Schulze E, Codron D, Clauss M (2013) Hypsodonty and tooth facet development in relation to diet and habitat in herbivorous ungulates: implications for understanding tooth wear. Mammal Rev 43:34–46. https://doi.org/10.1111/j.1365-2907.2011.00203.x
Karasov WH, Martínez del Rio C (2007) Physiological ecology: how animals process energy, nutrients, and toxins. Princeton University Press, Princeton, NJ
Karme A, Rannikko J, Kallonen A, Clauss M, Fortelius M (2016) Mechanical modelling of tooth wear. J R Soc Interface 13. https://doi.org/10.1098/rsif.2016.0399
Kelly KE, Sinclair BR (1989) Size and structure of leaf and stalk components of digesta regurgitated for rumination in sheep offered five forage diets. N Z J Agric Res 32:365–374

Kiltie RA (1981) The function of interlocking canines in rain forest peccaries (Tayassuidae). J Mammal 62:459–469

Lajeunesse MJ (2009) Meta-analysis and the comparative phylogenetic method. Am Nat 174:369–381

Lauper M et al (2013) Rumination of different-sized particles in muskoxen (*Ovibos moschatus*) and moose (*Alces alces*) on grass and browse diets, and implications for rumination in different ruminant feeding types. Mamm Biol 78:142–152

Lazagabaster IA, Rowan J, Kamilar JM, Reed KE (2016) Evolution of craniodental correlates of diet in African Bovidae. J Mamm Evol 23:385–396

Lechner I et al (2009) No 'bypass' in adult ruminants: passage of fluid ingested vs. fluid inserted into the rumen in fistulated muskoxen (*Ovibos moschatus*), reindeer (*Rangifer tarandus*) and moose (*Alces alces*). Comp Biochem Physiol A Mol Integr Physiol 154:151–156

Lechner I et al (2010) Differential passage of fluids and different-sized particles in fistulated oxen (*Bos primigenius* f. *taurus*), muskoxen (*Ovibos moschatus*), reindeer (*Rangifer tarandus*) and moose (*Alces alces*): rumen particle size discrimination is independent from contents stratification. Comp Biochem Physiol A Mol Integr Physiol 155:211–222

Lechner-Doll M, Kaske M, von Engelhardt W (1991) Factors affecting the mean retention time of particles in the forestomach of ruminants and camelids. In: Tsuda T, Sasaki Y, Kawashima R (eds) Physiological aspects of digestion and metabolism in ruminants. Academic, San Diego, CA, pp 455–482

Louys J, Faith JT (2015) Phylogenetic topology mapped onto dietary ecospace reveals multiple pathways in the evolution of the herbivorous niche in African Bovidae. J Zool Syst Evol Res 53:140–154

Lucas PW et al (2013) Mechanisms and causes of wear in tooth enamel: implications for hominin diets. J R Soc Interface 10:2012.0923

Macandza VA, Owen-Smith N, Cross PC (2004) Forage selection by African buffalo in the late dry season in two landscapes. S Afr J Wildl Res 34:113–121

Mainland IL (2003) Dental microwear in grazing and browsing Gotland sheep (*Ovis aries*) and it implications for dietary reconstruction. J Archaeol Sci 30:1513–1527

Marchand P, Redjadj C, Garel M, Cugnasse J-M, Maillard D, Loison A (2013) Are mouflon (*Ovis gmelini musimon*) really grazers? A review of variation in diet composition. Mamm Rev 43:275–291

Martin SA, Alhajeri BH, Steppan SJ (2016) Dietary adaptations in the teeth of murine rodents (Muridae): a test of biomechanical predictions. Biol J Linn Soc 119:766–784

McLeod MN, Minson DJ (1988) Large particle breakdown by cattle eating ryegrass and alfalfa. J Anim Sci 66:992–999

Meier AR, Schmuck U, Meloro C, Clauss M, Hofmann RR (2016) Convergence of macroscopic tongue anatomy in ruminants and scaling relationships with body mass or tongue length. J Morphol 277:351–362

Mendoza M, Palmqvist P (2006) Characterizing adaptive morphological patterns related to diet in Bovidae (Mammalia: Artiodactyla). Acta Zool Sin 52:988–1008

Mendoza M, Palmqvist P (2008) Hypsodonty in ungulates: an adaptation for grass consumption or for foraging in open habitat? J Zool 274:134–142

Mendoza M, Janis CM, Palmqvist P (2002) Characterizing complex craniodental patterns related to feeding behaviour in ungulates: a multivariate approach. J Zool 258:223–246

Merceron G et al (2016) Untangling the environmental from the dietary: dust does not matter. Proc R Soc Lond B Biol Sci 283:20161032

Meyer K, Hummel J, Clauss M (2010) The relationship between forage cell wall content and voluntary food intake in mammalian herbivores. Mammal Rev 40:221–245. https://doi.org/10.1111/j.1365-2907.2010.00161.x

Mihlbachler MC, Campbell D, Ayoub M, Chen C, Ghani I (2016) Comparative dental microwear of ruminant and perissodactyl molars: implications for paleodietary analysis of rare and extinct ungulate clades. Paleobiology 42:98–116

Mlambo V, Marume U, Gajana CS (2015) Utility of the browser's behavioural and physiological strategies in coping with dietary tannins: are exogenous tannin-inactivating treatments necessary? S Afr J Anim Sci 45:441–451
Mortolaa JP, Lanthier C (2005) Breathing frequency in ruminants: a comparative analysis with non-ruminant mammals. Respir Physiol Neurobiol 145:265–277
Müller DWH et al (2011) Phylogenetic constraints on digesta separation: variation in fluid throughput in the digestive tract in mammalian herbivores. Comp Biochem Physiol A Mol Integr Physiol 160:207–220. https://doi.org/10.1016/j.cbpa.2011.06.004
Müller DWH et al (2013) Assessing the Jarman–Bell principle: scaling of intake, digestibility, retention time and gut fill with body mass in mammalian herbivores. Comp Biochem Physiol A Mol Integr Physiol 164:129–140
Müller J et al (2014) Growth and wear of incisor and cheek teeth in domestic rabbits (*Oryctolagus cuniculus*) fed diets of different abrasiveness. J Exp Zool 321A:283–298
Müller J et al (2015) Tooth length and incisal wear and growth in Guinea pigs (*Cavia porcellus*) fed diets of different abrasiveness. J Anim Physiol Anim Nutr 99:591–604
Murray MG (1993) Comparative nutrition of wildebeest, hartebeest and topi in the Serengeti. Afr J Ecol 31:172–177
Murray MG, Illius AW (2000) Vegetation modification and resource competition in grazing ungulates. Oikos 89:501–508
Nultsch W (2000) Allgemeine Botanik (General Botany), 11th edn. Georg Thieme Verlag, Stuttgart
Nygren K, Lechner-Doll M, Hofmann RR (2001) Influence of papillae on post-ruminal regulation of ingesta passage in moose (*Alces alces*). J Zool 254:375–380
Obidziński A, Miltko R, Bolibok L, Wajdzik M, Skubis J, Nasiadka P (2017) Variation of natural diet of free ranging mouflon affects their ruminal protozoa composition. Small Rumin Res 157:57–64
Orme D et al. (2013) Caper: comparative analyses of phylogenetics and evolution in R. R package version 0.5.2. https://CRAN.R-project.org/package=caper
Owen-Smith N (1997) Distinctive features of the nutritional ecology of browsing versus grazing ruminants. Z Säugetierkd 62:176–191
Owen-Smith N (2013) Megaherbivores. In: Levin S (ed) Encyclopedia of biodiversity, 2nd edn. Academic
Pagel M (1999) Inferring the historical patterns of biological evolution. Nature 401:877–884
Paine OCC et al (2018) Grass leaves as potential hominin dietary resources. J Hum Evol 117:44–52
Palmqvist P, Gröcke DR, Arribas A, Fariña RA (2003) Paleoecological reconstruction of a lower Pleistocene large mammal community using biogeochemical ($\delta^{13}C$, $\delta^{15}N$, $\delta^{18}O$, Sr:Zn) and ecomorphological approaches. Paleobiology 29:205–229
Pérez-Barbería FJ, Gordon IJ (2001) Relationships between oral morphology and feeding style in the Ungulata: a phylogenetically controlled evaluation. Proc R Soc Lond B Biol Sci 268:1021–1030
Pérez-Barbería J, Gordon I, Illius A (2001) Phylogenetic analysis of stomach adaptation in digestive strategies in African ruminants. Oecologia 129:498–508
Pérez-Barbería FJ, Elston DA, Gordon IJ, Illius AW (2004) The evolution of phylogenetic differences in the efficiency of digestion in ruminants. Proc R Soc Lond B Biol Sci 271:1081–1090. https://doi.org/10.1098/rspb.2004.2714
Pretorius Y et al (2016) Why elephant have trunks and giraffe long tongues: how plants shape large herbivore mouth morphology. Acta Zool 97:246–254. https://doi.org/10.1111/azo.12121
Prins HHT, Olff H (1998) Species-richness of African grazer assemblages: towards a functional explanation. In: Newbery DM, Prins HHT, Brown N (eds) Dynamics of tropical communities. Blackwell Science, Oxford, pp 449–490
Rabenold D, Pearson OM (2014) Scratching the surface: a critique of Lucas et al. (2013)'s conclusion that phytoliths do not abrade enamel. J Hum Evol 74:13–133
Raia P, Carotenuto F, Meloro C, Piras P, Pushkina D (2010) The shape of contention: adaptation, history, and contingency in ungulate mandibles. Evolution 64:1489–1503

Reed KE (1996) The palaeoecology of Makapansgat and other African Plio-Pleistocene hominid localities. Unpublished PhD thesis, University of Michigan, Michigan

Reilly SM, McBrayer LD, White TD (2001) Prey processing in amniotes: biomechanical and behavioral patterns of food reduction. Comp Biochem Physiol A Mol Integr Physiol 128:397–415

Rivals F, Semprebon GM, Solounias N (2010) Advances in ungulate dental wear techniques reveal new patterns of niche breadth and expansion throughout the Cenozoic. J Vertebr Paleontol SVP Program and Abstracts Book:151A–152A

Robbins CT, Spalinger DE, Van Hoven W (1995) Adaptation of ruminants to browse and grass diets: are anatomical-based browser-grazer interpretations valid? Oecologia 103:208–213

Rowell-Schäfer A, Lechner-Doll M, Hofmann RR, Streich WJ, Güven B, Meyer HHD (2001) Metabolic evidence of a "rumen bypass" or a "ruminal escape" of nutrients in roe deer (*Capreolus capreolus*). Comp Biochem Physiol A Mol Integr Physiol 128:289–298

Sakaguchi E, Itoh J, Shinohara H, Matsumoto T (1981) Effects of removal of the forestomach and caecum on the utilization of dietary urea in golden hamsters (*Mesocricetus auratus*) given two different diets. Br J Nutr 46:503–512

Sanson GD (1989) Morphological adaptations of teeth to diets and feeding in the Macropodoidea. In: Grigg G, Jarman PJ, Hume I (eds) Kangaroos, Wallabies and Rat-kangaroos. Surrey-Beatty, Sydney, pp 151–168

Sanson GD, Read SKJ (2017) Dietary exogenous and endogenous abrasives and tooth wear in African buffalo. Biosurf Biotribol 3:211–223

Sanson GD, Kerr SA, Gross KA (2007) Do silica phytoliths really wear mammalian teeth? J Archaeol Sci 34:526–531

Sauer C, Bertelsen MF, Hammer S, Lund P, Weisbjerg MR, Clauss M (2016) Macroscopic digestive tract anatomy of two small antelopes, the blackbuck (*Antilope cervicapra*) and the Arabian sand gazelle (*Gazella subgutturosa marica*). Anat Histol Embryol 45:392–398

Schuette JR, Leslie DMJ, Lochmiller RL, Jenks JA (1998) Diets of hartebeest and roan antelope in Burkina Faso: support of the long-faced hypothesis. J Mammal 79:426–436

Shipley LA (2007) The influence of bite size on foraging at larger spatial and temporal scales by mammalian herbivores. Oikos 116:1964–1974

Smith FA et al (2003) Body mass of late quaternary mammals. Ecology 84:3403. (Ecological Archives: E3084-3094)

Spencer LM (1995) Morphological correlates of dietary resource partitioning in the African Bovidae. J Mammal 76:448–471

Sponheimer M et al (2003) Diets of southern African Bovidae: stable isotope evidence. J Mammal 84:471–479

Stayton CT (2006) Testing hypotheses of convergence with multivariate data: morphological and functional convergence among herbivorous lizards. Evolution 60:824–841

Steuer P et al (2010) Comparative investigations on digestion in grazing (*Ceratotherium simum*) and browsing (*Diceros bicornis*) rhinoceroses. Comp Biochem Physiol A Mol Integr Physiol 156:380–388

Strömberg CAE (2006) Evolution of hypsodonty in equids: testing a hypothesis of adaptation. Paleobiology 32:236–258. https://doi.org/10.1666/0094-8373(2006)32[236:eohiet]2.0.co;2

Sutherland TM (1988) Particle separation in the forestomach of sheep. In: Dobson A, Dobson MJ (eds) Aspects of digestive physiology in ruminants. Cornell University Press, Ithaca, NY, pp 43–73

Tahas SA et al (2017) Gross measurements of the digestive tract and visceral organs of addax antelope (Addax nasomaculatus) following a concentrate or forage feeding regime. Anat Histol Embryol 46:282–293

Tahas SA et al (2018) Microanatomy of the digestive tract, hooves and some visceral organs of addax antelope (Addax nasomaculatus) following a concentrate or forage feeding regime. Anat Histol Embryol 47:254–267. https://doi.org/10.1111/ahe.12351

Tainton NM (1999) The ecology of the main grazing lands of South Africa: the savanna biome. In: Tainton NM (ed) Veld management in South Africa. University of Natal Press, Pietermaritzburg, pp 23–53
Taylor LA et al (2013) Detecting inter-cusp and inter-tooth weat patterns in rhinocerotids. PLoS One 8:e80921
Trautmann A, Schmitt I (1935) Experimentelle Untersuchungen zur Frage der Psalterfunktion (Experimental investigations on omasum function). Dtsch Tierarztl Wochenschr 12:177–179
Trudell-Moore J, White RG (1983) Physical breakdown of food during eating and rumination in reindeer. Acta Zool Fenn 175:47–49
Tschuor A, Clauss M (2008) Investigations on the stratification of forestomach contents in ruminants: an ultrasonographic approach. Eur J Wildl Res 54:627–633
Turnbull WD (1970) Mammalian masticatory apparatus. Fieldiana Geol 18:147–356
Van Soest PJ (1965) Symposium on factors influencing the voluntary intake of herbage by ruminants: voluntary intake in relation to chemical composition and digestibility. J Anim Sci 24:834–843
Van Wieren SE (1996a) Browsers and grazers: foraging strategies in ruminants. In: Van Wieren SE (ed) Digestive strategies in ruminants and nonruminants. Thesis Landbouw, University of Wageningen, Wageningen, pp 119–146
Van Wieren SE (1996b) Digestive strategies in ruminants and non-ruminants. PhD thesis. University of Wageningen, Wageningen
Van Zyl JHM (1965) The vegetation of the S.A. Lombard nature reserve and its utilization by certain antelope. Zool Afr 1:55–71
Varela L, Fariña RA (2015) Masseter moment arm as a dietary proxy in herbivorous ungulates. J Zool 296:295–304
Vermeij GJ (1994) The evolutionary interaction among species: selection, escalation, and coevolution. Annu Rev Ecol Syst 25:219–236
Vermeij GJ (2013) On escalation. Annu Rev Earth Planet Sci 41:1–19
Wattiaux MA, Satter LD, Mertens DR (1992) Effect of microbial fermentation on functional specific gravity of small forage particles. J Anim Sci 70:1262–1270
Williams SH, Stover KK, Davis JS, Montuelle SJ (2011) Mandibular corpus bone strains during mastication in goats (*Capra hircus*): a comparison of ingestive and rumination chewing. Arch Oral Biol 56:960–971
Xia J et al (2015) New model to explain tooth wear with implications for microwear formation and diet reconstruction. Proc Natl Acad Sci U S A 112:10669–10672. https://doi.org/10.1073/pnas.1509491112

Chapter 5
Feeding Ecology of Large Browsing and Grazing Herbivores

Jan A. Venter, Mika M. Vermeulen, and Christopher F. Brooke

5.1 Introduction

Different ungulate species, from a variety of taxonomic groups, tend to focus on either the grass or browse component of vegetation as a food source (Shipley 1999). They interact with forage resources across several temporal and spatial scales (Bailey et al. 1996; Owen-Smith et al. 2010), and forage in the form of browse versus grass are presented to the herbivore in distinct ways at each of these scales (Fig. 5.1 and Plates 5.1 and 5.2: A–H). Grasses, forbs and woody plants differ in chemistry, architecture, and heterogeneity with resultant physiological and behavioural adaptations of herbivores to these differing forage types (Shipley 1999; Searle and Shipley 2008) (Table 5.1). One can, therefore, expect differences in how grazers and browsers engage with their environment and forage resources.

5.2 The Functional Response

An ungulate's feeding behaviour is based on its functional traits and the traits of the plants it forages on (Searle and Shipley 2008). The functional response is the change in an animal's intake rate in relation to the availability of a given food source (Spalinger and Hobbs 1992; Holling 1959), and is an important component of food web interactions and population and community ecology (Sarnelle and Wilson 2008) (Fig. 5.2).

J. A. Venter (✉) · M. M. Vermeulen · C. F. Brooke
School of Natural Resource Management, Faculty of Science, George Campus, Nelson Mandela University, George, South Africa
e-mail: Jan.Venter@mandela.ac.za; Christopher@brooklands.co.za

I. J. Gordon, H. H. T. Prins (eds.), *The Ecology of Browsing and Grazing II*, Ecological Studies 239, https://doi.org/10.1007/978-3-030-25865-8_5

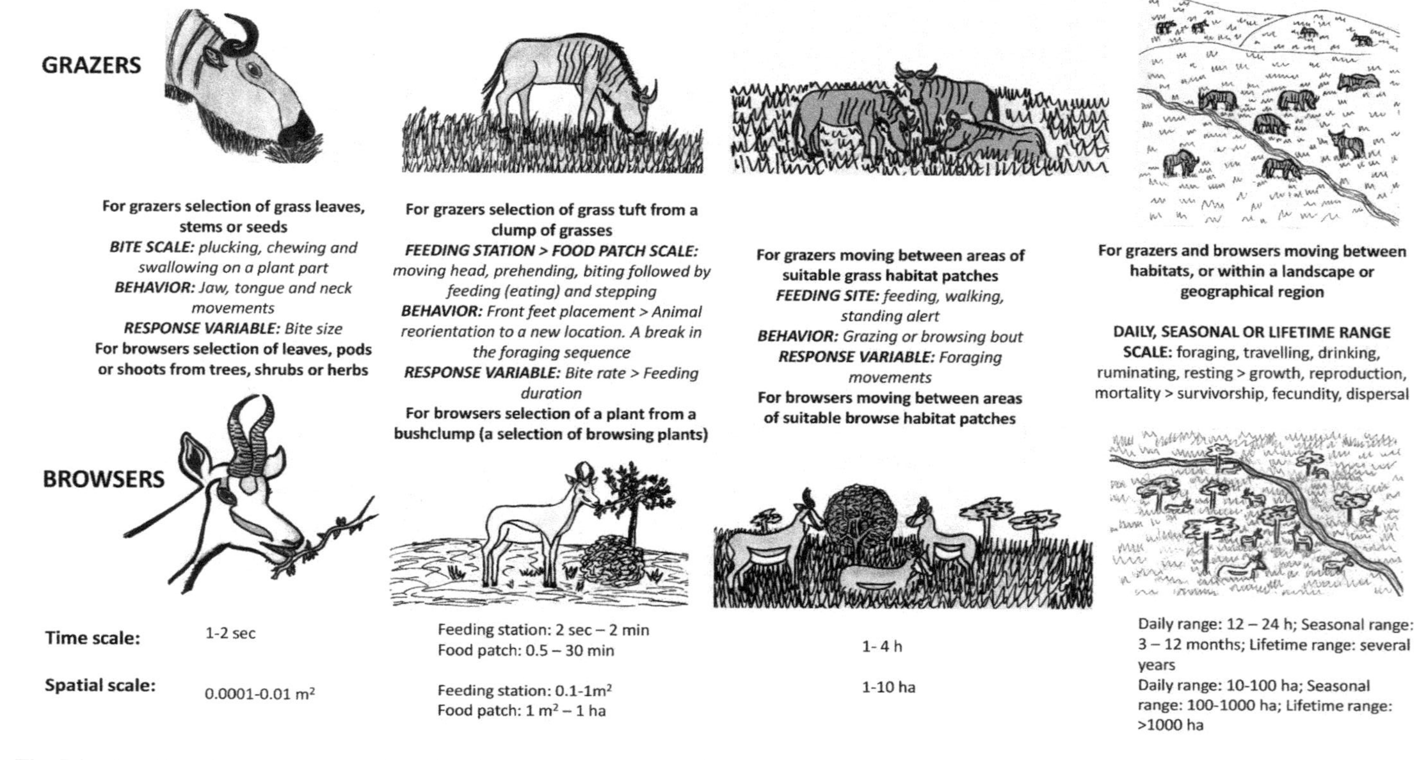

Fig. 5.1 The different temporal and spatial scales that grazers and browsers interact with forage resources adapted from Bailey and Provenza (2008). (Sketches by Sorina Venter)

Plate 5.1 (**A1–A3**) A series of grass bites taken by gemsbok (*Oryx gazelle*), within seconds from each other, captured by camera trap in Karoo National Park; (**B**) Blue wildebeest (*Connochaetes taurinus*) selecting short grass in the Kgalagadi Transfrontier Park; (**C**) Buffalo (*Syncerus caffer*) foraging on grass seeds in Madikwe Game Reserve; (**D**) White rhino (*Ceratotherium simum*) selecting short grass on a grazing lawn in Pilanesberg National Park

Herbivore foraging can be separated into three main components, i.e., searching for forage, cropping it and chewing it (cropping and chewing constitutes handling forage) (Spalinger and Hobbs 1992). Cropping competes with searching (reducing foraging velocity) and chewing (the mouth can only do one or the other) (Spalinger and Hobbs 1992). Chewing does not interfere with searching (an animal can move and search while chewing) and only slows down cropping (because an animal can move while cropping) (Spalinger and Hobbs 1992).

Plate 5.2 (**E1–E3**) A series of browse bites taken by springbok (*Antidorcas marsupialis*), within seconds from each other, captured by camera trap in Karoo National Park; (**F**) Giraffe (*Giraffa giraffa*) foraging on tall trees in the Kruger National Park; (**G**) African elephant (*Loxodonta africana*) foraging on medium sized trees in Addo Elephant National Park; (**H**) Nyala (*Tragelaphus angasii*) foraging on herbs and dwarf shrubs in the Karoo

Holling (1959) described three main types of functional responses linking the amount of food that is consumed to the amount of food that is on offer (Fig. 5.3 and Table 5.2). These functional response types were—Type 1: the mathematically simplest model shows a predator's searching rate remains constant at all prey densities (i.e., the number of kills would be linearly related to prey density); Type 2: is slightly more complex where the number of prey attacked by a predator is high with the initial increase in prey density, but begins slowing immediately until it slowed leading up to a fixed level of predation is neared (approaches, but does not

Table 5.1 A comparison of general chemical and structural characteristics of grasses (monocots) and browses (woody and herbaceous plants), adapted from Shipley (1999) and Clauss et al. (2008)

Subject groups	Characteristic	Browse	Grass
Growth pattern/ location	Landscape	Forests and woodlands/ spatially more structured	Open
	Growth pattern	At different heights	Mostly close to ground
	Nutritional homogeneity of a "bite"	Less	More
Chemical composition	Protein content	Higher (including nitrogenous secondary compounds)	Lower
	Fibre content	Lower but more lignified	Higher but less lignified
	Pectin content	Higher	Lower
	Secondary compounds	More	Less
Physical characteristics	Abrasive silica, adhering grit, resistance to chewing	Less	More
	Fracture pattern	Polygonal	Longish, fibre-like
	Change in specific gravity during fermentation	Less	More
Digestion/fermentation	Overall digestibility	Lower	Higher
	Speed of digestion	Fast	Slow

Many changes occur, but this table serves to highlight general patterns

necessarily reach a plateau); and Type 3: depicts a sigmoidal curve whereby the function is similar to a Type II functional response, but there is an initial slow increase in predation rate as a result of searching, followed by a rapid increase in predation rate before finally slowing again or decreasing as a result of increasing prey density (Holling 1959) (Fig. 5.3 and Table 5.2). Since Holling (1959) first published the three functional response types a fourth has been described in grazing studies. Heuermann et al. (2011) described a decrease in intake rate for small herbivores in high biomass grasslands where it will be harder to move through tall grass (Jarman 1974; Bell 1971), but high biomass grass may also be more difficult to handle for all types of herbivores (Fig. 5.3 and Table 5.2). The shape of the type 4 functional response is different from the other functional response types described by Holling (1959) as intake decreases at higher biomasses, and it fitted significantly better for small and large bodied herbivores. This is most likely because in consuming vegetation matter herbivores face a decrease of food quality with increasing food quantity (Olff et al. 2002; Prins and Olff 1998).

Spalinger and Hobbs (1992) defined three functional response processes relating foraging in mammalian herbivores to plant characteristics (Table 5.3), i.e., process 1, where plants are both dispersed and hidden within the landscape; process 2, where

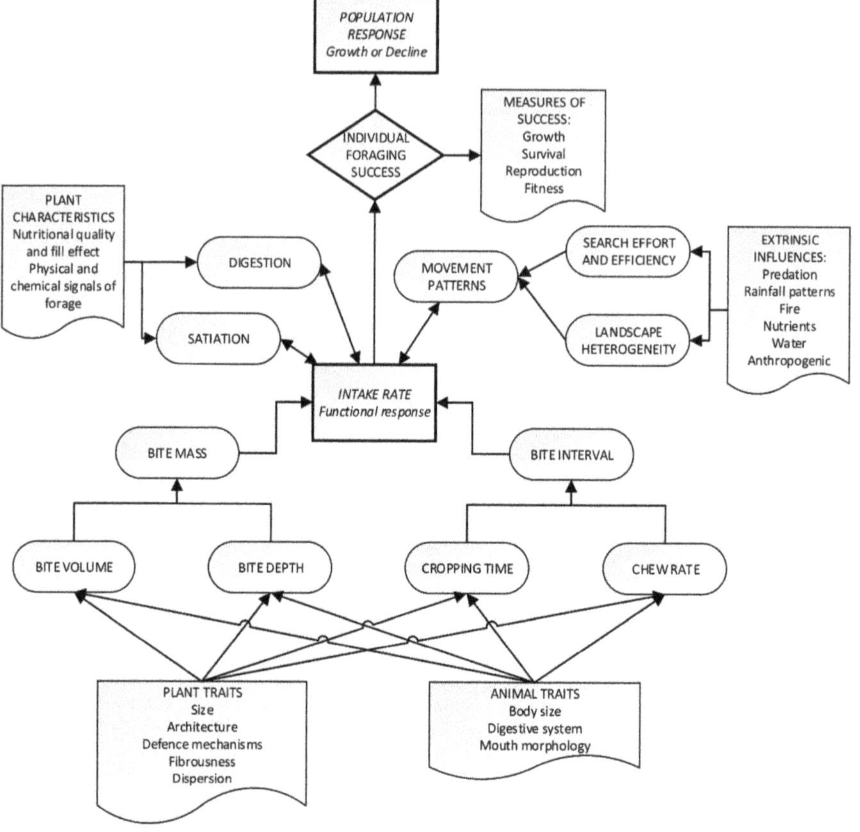

Fig. 5.2 A conceptual diagram demonstrating linkages between forage resources, herbivores, intake rate, behaviour and population dynamics adapted from Searle and Shipley (2008) and Baumont et al. (2000)

plants are dispersed but apparent within the environment; and process 3, where plants are both concentrated and apparent within the environment.

Of course these models often oversimplify intake rate into only two components; namely searching rate and handling time (Baker et al. 2010), that comes from the interactions between herbivore morphology and vegetation structure (Owen-Smith and Novellie 1982; Spalinger and Hobbs 1992). Fundamentally, energy and nutrient intake of herbivores influence their functional response and is determined by the time they spend feeding (including handling time) and the forage they consume, and as a result, influences the time that is spent on non-feeding activities (Spalinger and Hobbs 1992). Animals exploit resources in different ways in order to fulfil their individual energy requirements such as growth, survival and reproduction (Bjørneraas et al. 2012), which are the measures of individual foraging success (Fig. 5.2). The concept of functional response is useful in grasping the behaviour of herbivores and its application leads to further explaining the

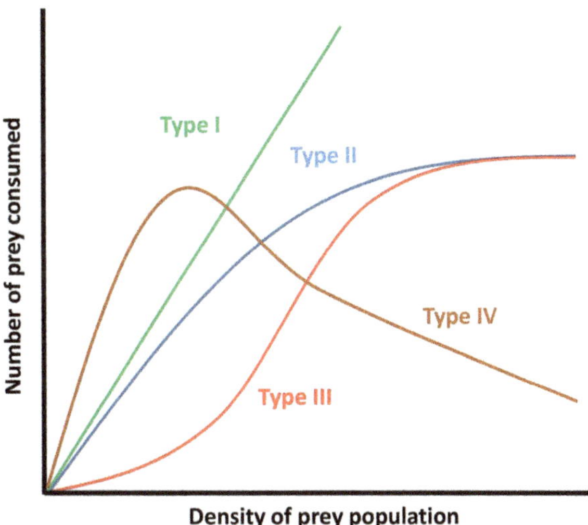

Fig. 5.3 Holling (1959) functional response curves, labelled with type of response adapted from Pettorelli et al. (2015) and the fourth functional response from Heuermann et al. (2011)

Table 5.2 Hollings three functional responses adapted from Pettorelli et al. (2015) and the fourth functional response from Heuermann et al. (2011)

Model	Formula	Relationship
Type I	$Kill\ rate = attack\ rate * prey\ density$	Linear
Type II	$Kill\ rate = \frac{attack\ rate * prey\ density}{1 + attack\ rate * prey\ density * handling\ time}$	Asymptotic
Type III	$Kill\ rate = \frac{b * prey\ density^2}{1 + c * prey\ density + b * handling\ time * prey\ density^2}$	Sigmoid
Type IV	$Kill\ rate = \frac{attack\ rate * prey\ density}{b + prey\ density + c * prey\ density^2}$	Lognormal

Note that in these equations b and c are constants

influential role in patch dynamics and nutrient cycling within an area (Spalinger and Hobbs 1992).

It is important to understand the shape of this functional response as this aids predictions of how an ungulate will respond to environmental change (Smart et al. 2008). Furthermore, the shape of the functional response is important for predicting and understanding population dynamics within species (Heuermann et al. 2011), and how animals cope with both the costs and benefits provided by forage resources (Bjørneraas et al. 2012).

Grass tends to form a more continuous vegetation layer with one bite linked to the next, whilst browse is characterised by a discontinuous, patchy multi-layered vegetation mosaic (Bond and Parr 2010); therefore, the functional response of browsers can only consider the availability of browse within reach of the herbivore rather than the browse available within the environment (de Knegt et al. 2008; Owen-Smith 2002).

Table 5.3 The three functional response processes developed by Spalinger and Hobbs (1992)

Process	Plant density	Plant apparency	Formula
1	Dispersed	Hidden	$\mathit{Intake\ rate} = \frac{\mathit{bite\ density} * \mathit{maximum\ travel\ velocity} * \mathit{forage\ path\ width}}{1 + \mathit{velocity\ decrease\ due\ to\ cropping} * \mathit{forage\ path\ width} * \mathit{bite\ density}} * \text{bite size}$
2	Dispersed	Apparent	$\mathit{Intake\ rate} = \frac{\mathit{maximum\ travel\ velocity} * \sqrt{\mathit{bite\ density}}}{1 + \mathit{forage\ path\ width} * \sqrt{\mathit{bite\ density}}} * \mathit{bite\ size}$
3	Concentrated	Apparent	$\mathit{Intake\ rate} = \frac{\mathit{rate\ of\ food\ processing} * \mathit{bite\ size}}{\mathit{rate\ of\ food\ processing} * \mathit{average\ bite\ crop\ time} + \mathit{bite\ size}}$

5.3 Components of the Functional Response

Herbivores have evolved in response to plants to be able to obtain food efficiently (Stephens and Krebs 1986). These co-evolutionary processes have shaped herbivore traits differently for grazing and browsing herbivores (Hofmann and Stewart 1972; Jarman 1974; Gordon 2003) (Table 5.4). Plant and animal traits, therefore, influence each of the components of the functional response. The functional response in ungulates is governed by two key processes, (1) the rate at which an ungulate can consume vegetation (R_{max}, i.e., the maximum cropping rate, and h the time it takes an ungulate from one bite to the next) and (2) the availability and spatial distribution of vegetation within the environment (Fraser and Theodor 2011) (i.e., searching and handling time) (Fig. 5.2). The rate at which ungulates can consume vegetation is governed by a number of key factors including bite volume, bite depth, cropping time and chew rate (Searle and Shipley 2008). These factors are further influenced by the traits of the plants being consumed, the traits of the animal consuming the plants, as well as extrinsic factors which influence search effort and efficiency and landscape/habitat heterogeneity (Searle and Shipley 2008).

5.3.1 *Cropping*

The time needed to crop bites is influenced by fibre composition, dispersion of plant parts, and structural defences of plants (Searle and Shipley 2008). There are significant differences amongst these factors between grasses and browse (Table 5.1).

Table 5.4 A comparison of different herbivore traits, related to digestive anatomy, of browsers and grazers, adapted from Searle and Shipley (2008) and Shipley (1999)

Characteristic	Grazers	Browsers
Foregut	Large, subdivided, smaller opening between reticulum and omasum, sparser more uneven papillae	Small, simple, larger opening between reticulum and omasum, denser, more even papillae
True stomach (abomasum)	Smaller	Larger
Hindgut	Smaller cecum and intestines	Larger cecum and intestines
Salivary glands	Smaller parotid salivary glands	Larger parotid salivary glands
Liver	Smaller	Larger
Mouth	Wide muzzle, smaller mouth opening, stiffer lips	Narrow muzzle, larger mouth opening, flexible, often muscular lips, long tongue
Teeth	Lower incisors of similar size, wider contact area, wide incisor row, incisors projected forward, high crowned teeth, less surface area, more shearing action, more ridges	Central incisors broader than outside ones, less contact area for manipulation, narrow incisor row, incisors more uprights, low-crowned teeth, large surface area of contact between lower and upper teeth for crushing action

Grasses generally have thicker cell walls, consisting of greater quantities of low digestible fibres such as cellulose. The thicker and more fibrous cell walls in grasses means that herbivores need to spend more energy and time to effectively crop grasses, particularly when stems are present (Gordon and Benvenutti 2006). Browse consists mainly of leaves with thinner cell walls that are more completely digestible and woody stems containing high levels of lignin (Searle and Shipley 2008). But browse often have physical defences such as thorns or spines which limit intake and/or contain higher levels of lignin which inhibits digestion (Shipley 1999; Tomlinson et al. 2016). Plant parts of browse are often more dispersed (i.e., thorns, spines and branches slow cropping; Belovsky et al. 1991) compared to homogenous grasses and thus browsers generally spend more time cropping compared to grazers (Searle and Shipley 2008; Illius and Gordon 1987).

5.3.2 Bite Size

Bite size is determined by the bite area and the depth of the bite and this affects the volume of vegetation that can be included in a bite (Pretorius et al. 2016). The shape of an animal's muzzle are often adapted to the food the animal eats (Fraser and Theodor 2011). Development of different feeding strategies, based on forage preferences, have resulted in the development of a wide range of differing mouth morphologies (Table 5.4 and Plates 5.1 and 5.2: A–H). Characteristically, grazers have larger muzzles and wider incisor arcades than similar sized browsers (Table 5.4) (Gordon and Illius 1988). This allows grazers to take larger bites of more uniform vegetation (Plate 5.1: A1–A3) and crop grasses shorter (Fraser and Theodor 2011; Gordon and Illius 1988), resulting in the more vegetation being consumed. However, at the same time, these larger bites may sacrifice the intake of energy resulting from the consumption of lower quality of vegetation (Fraser and Theodor 2011). For example in a type 4 functional response Heuermann et al. (2011) found, in a study on geese, that larger geese had larger bites but lower intake rates when compared to smaller geese. Additionally most grazers have hypsodont (high crowned) molar teeth (Damuth and Janis 2011; see also **Saarinen** Chap. 2; **Rowan and Faith** Chap. 3), which helps them deal with the higher levels of silica in grasses. Browsers have a narrow incisor arcade and long muzzle, relative to grazers, with prehensile lips that allow them to effectively strip whole branches of vegetation (Pretorius et al. 2016) (Plate 5.2: E1–E3), and brachydont (low crowned) teeth (Hummel et al. 2010) (Table 5.4). These differences in tooth structures between browsers and grazers affect their longevity and reproductive success (Hummel et al. 2010).

Bite size is often not limited by the available biomass in a patch, but rather by the animal's mouth morphology and their ability to effectively crop vegetation (Bailey et al. 1996; Pretorius et al. 2016). Bite area refers to the effective surface area of a food source that can be cropped at one time, whereas bite size refers to the amount of vegetation that can be removed from a plant in a single bite (Pretorius et al. 2016). Bite size scales linearly with mouth volume and interdental morphology. During the process of taking bites chewing and cropping rates are negatively linearly related, and both are affected by the chemical and structural characteristics of the forage resource (Pretorius et al. 2016).

The relationship between intake rate and available vegetation biomass has often resulted in the supposition that bite size scales linearly with body mass (Wilmshurst et al. 2000; Pretorius et al. 2016). This assumption can also be derived from the relationship between body mass and gut capacity (Demment and Van Soest 1985). In contrast, relationships between bite rate and body mass are uncommon and studies that have investigated these relationships often find none (Shipley et al. 1994; Fortin 2006). Bite rate is thus influenced by bite mass, especially when grass swards are short (Murray and Baird 2008), although bite mass is believed to have a greater influence on intake rate (R_{max}) than is bite rate (Pretorius et al. 2016). Finally intake rate of forage can be split into two broad categories: (1) the time required to take a bite of vegetation, which is independent to bite mass, and (2) the time required to chew the harvested bite which is related to bite mass (Newman et al. 1994; Baumont et al. 2000).

5.3.3 Chewing

Chewing is an important process in which herbivores effectively break down food particles prior to digestion (Hummel et al. 2008). The thicker cell walls found in grasses make them more energy-expensive to break down (crop and chew) than the more delicate leaves of browse plants (Shipley 1999). The differences in tooth structure are one of the key adaptations in herbivores to deal with fibrous food (Hummel et al. 2008). Like bite size, hypsodonty in ungulates can give an indication of the level of chewing that takes place to process plant material. Grasses are generally considered to be more abrasive than browse and, therefore, hypsodonty is an adaptation of many grazers to slow tooth wear from the consumption of grass (Damuth and Janis 2011). However, abrasion of teeth during chewing is not only limited to the physical properties of grasses (Shipley 1999; Damuth and Janis 2011), but can also be associated with the intake of soil and grit that may adhere to the food surface, and again this is more pronounced in grazers than browsers (Damuth and Janis 2011). However, there are also other morphological adaptations to dealing with potentially abrasive food possible (see **Rowan and Faith** Chap. 3; **Codron et al.** Chap. 4). Increased chewing by grazers may also be one method to extract nutrients from fibrous grasses and could be thought to be a selective advantage driving hypsodonty among grazers (Damuth and Janis 2011), especially in longer grass swards (Edouard et al. 2010).

Chewing is as essential to digestion as are digestive processes in the gut (Fritz et al. 2009). Importantly, the smaller the food particles are, the faster they can be digested (Clauss et al. 2009; Fritz et al. 2009). Therefore, increased chewing may be effective to increase the intake of energy during digestion (Clauss et al. 2009), but chewing is also a major cause of energy expenditure and can increase the cost of feeding, especially on longer grass swards (Edouard et al. 2010). Plant material that is more completely chewed can pass through the gut faster, and the smaller particles result in higher intake of energy from food (Shipley 1999). In a study on grazing herbivores, Clauss et al. (2009) suggested that, in order to increase digestive efficiency, herbivores needed to increase digestion time of plant material, and break down food more effectively through chewing. Likewise Baumont et al. (2000) noted that rumen retention time depended mainly on the degradation rate of food within the rumen and this in increased if food particles smaller when they enter the rumen. In the case of ruminants, when plant particles are too large to pass through the opening between the reticulum and omasum, particles are regurgitated and re-chewed until they are small enough (Shipley 1999). Grazers tend to have a smaller opening between the reticulum and omasum, forcing food to be chewed more before they can pass through. This appears to be an adaptation to slow the passage of food allowing more complete fermentation of plant fibres (Shipley 1999). Because late wet season forage is of lower quality than early wet season grass, it needs more rumination time for ruminants. During the season when plant material is high in fibre, rumination can compete for time with grazing, especially when bite size becomes small and rumination time is high (Beekman and Prins 1989).

5.3.4 Encountering

A herbivore's encounter rate with forage is determined by how an animal makes use of its environment, and the scale at which they are making foraging decisions. Furthermore the encounter rate with preferred food patches is affected by abiotic factors such as climate, topography and distance to water (Bailey et al. 1996). At different foraging scales a negative relationship between the quality and quantity of vegetation may force herbivores to make trade-offs when selecting areas to forage (Heuermann et al. 2011).

The strength of the relationships between an animal and the vegetation upon which they feed governs an animal's functional response. Encounter rate with forage controls intake rate in herbivores whenever chewing time is shorter than the time needed to travel between plants (process 1 and 2, Spalinger and Hobbs 1992). Alternatively intake is controlled by the processing time of a herbivore when vegetation is concentrated within an area and the time taken to travel between plants is shorter than the processing time of a herbivore (process 3, Spalinger and Hobbs 1992). In this second scenario the spatial arrangement of plants has no effect on the functional response in herbivores (Fortin 2006).

Within habitats there are a number of trade-offs that ungulates need to consider when making foraging decision, these include the quality and quantity of food (e.g., Prins and Van Langevelde 2008a), finding shelter and avoiding predation (Godvik et al. 2009).

5.3.5 Satiation and Digestion

During a foraging bout the rate of intake is the highest at the beginning and then decreases continuously as satiation proceeds towards satiety (Baumont et al. 2000). Intake is controlled by stretch- and mechano-receptors in the rumen wall (Leek 1977), and increasing rumen fill with indigestible material reduces intake rate (Faverdin et al. 1995). The main determinant of the vegetation digestibility in ungulates is the fibre content of forage which affects both the processing and passage time in the gut. As vegetation increases in biomass so too does its fibre content (Heuermann et al. 2011). Due to the high lignin content in the woody stems of browse, browsers have adopted a diet selectively feeding selectively on younger leaves with reduced lignin content (Fraser and Theodor 2011). Browsing herbivores have a greater degree of behavioural flexibility when it comes to trade-offs between selecting bites and bite size than do grazers. Moose (*Alces alces*), for example, have been shown to have an increase in the size of twigs that they consume as resource quality decreases (Searle et al. 2005). Generally browsers favour a larger number of smaller bites and select for high quality vegetation as opposed to larger, more indiscriminate, bites (Searle et al. 2005). Grazers too are known to make changes in their bite size and rate in response to changing forage characteristics; for example cattle (*Bos taurus*) have been shown to favour larger bites when vegetation is homogenous (Laca et al. 1994). Grazing herbivores tend to select bite size based on the maximisation of energy intake rate and commonly favour the top of the grass sward as this where leafy material is most prevalent (Searle et al. 2005).

The challenges of balancing trade-offs between quality and quantity of vegetation provide evidence that herbivores may select areas of higher biomass and lower quality of resources when there is increased competition among individuals for similar resources (Herfindal et al. 2009; Prins 1989). Similarly herbivores may avoid vegetation with high levels of anti-herbivore defences, such as tannins, lignin, silica, thorns and spines (Murray and Baird 2008) or that grows in areas with high levels of predation (White et al. 2003; Ford et al. 2014). Under these circumstances plant defences may be considered more limiting to herbivores than the quality of the vegetation (Murray and Baird 2008).

Aggregation of vegetation, and the defences that many plants have against herbivory, make it apparent that small herbivores and browsers are more limited by the spatial array and anti-herbivory defences of plants (Murray and Baird 2008; de Knegt et al. 2008). Illius and Gordon (1987) and Murray and Baird (2008) suggest that more selective species (i.e., smaller species) may outcompete less selective (i.e., larger) species by selectively removing high quality vegetation. However, a herbivores mouth morphology may be more important in determining selectivity in a species. Herbivores with narrow mouths can selectively feed on high quality vegetation, irrelevant of size. For example, red hartebeest (*Alcelaphus buselaphus caama*) and topi (*Damaliscus lunatus topi*) are both relatively large grazers, but have narrow muzzles and incisor arcades and are highly selective relative to their body mass (Venter et al. 2014; Murray 1993). Both species are adapted to feed selectively in high biomass grasslands, and because of their narrower mouths and longer muzzles are able to crop green shoots between senescent shoots. They can thus competitively exclude other similar size species, such as blue wildebeest (*Connochaetes taurinus*), from forage patches because these species are unable to be selective due to their broad flat dentition and muzzles (Murray and Baird 2008). However the reverse can also be true where larger, less selective herbivores indiscriminately remove biomass, thus cropping the available high quality vegetation before more selective species are able to utilize it.

5.3.6 Vigilance

Herbivores exposed to predators need to be vigilant whilst feeding (Fortin et al. 2004). Vigilance, the process where herbivores lift their head during feeding to scan for predators, is a factor that influences the foraging of animals (Baker et al. 2011). But mammalian herbivores can also use other senses (auditive and olfactory) for which they do not necessarily have to lift their heads to be vigilant (Nersesian et al. 2012).

Vigilance is assumed to play a role in the reduction of intake of food, by taking time away from searching and processing (Fortin et al. 2004). Herbivores need to manage trade-offs of feeding in areas less exposed to predators while consuming adequate high quality vegetation. Recently there have been several functional response models that have explicitly made reference to vigilance as an influencing factor, and have allowed for the interactions between vigilance, handling time and searching (Baker et al. 2011; Baker et al. 2010; Smart et al. 2008; Fortin et al. 2004). Unlike many other mammals, herbivores are able to continue processing food while simultaneously searching for predators and their next bite, and, as a result, their short-term intake rate may not be directly influenced by being vigilant (Fortin et al. 2004; Illius and FitzGibbon 1994). The processing and handling time of forage may be considered 'spare time' where an animal could search for predators or their next bite (Fortin et al. 2004). The influence of vigilance on feeding time may be further negated by behavioural adjustments in ungulates such as feeding in larger groups or feeding in patches where the likelihood of predation is less.

5.4 Searching for Forage

Finding suitable forage in a heterogeneous landscape, where patches change dynamically, both spatially and temporally, can be challenging to large ungulates, especially if they have no prior knowledge of the location of the forage patches (Venter et al. 2017; Bailey et al. 1996; Senft et al. 1987). Large herbivores may gain *a priori* knowledge using memory (from a previous visit to the patch) (Brooks and Harris 2008; Dumont and Petit 1998; Edwards et al. 1996; Fortin 2003; Hewitson et al. 2005) or through visual cues (Edwards et al. 1997; Howery et al. 2000; Renken et al. 2008) or prediction (Prins 1996). There is little evidence that browsers and grazers differ, fundamentally, in their spatial perception and memory when searching for forage (Searle and Shipley 2008). In the case of browsers the use of memory could be a potentially useful strategy due to the spatial and seasonal predictability of the forage resource; for example, trees are more "permanent" in a landscape (compared to individual grass plants) and a browser could use memory to revisited these over many years, at appropriate times when trees are flushing or bearing fruits (Janmaat et al. 2006). For browse, visual cues could potentially also be more prominent (Searle and Shipley 2008). For grazers memory could be important when grazing patches are clearly defined in a landscape of unsuitable forage areas, such as meadows in boreal forests (Merkle et al. 2014) and riverine deltas in an otherwise dry savanna (Prins 1996). But in some environments grasses are less predictable due to its dynamic short-term responses to factors such as rainfall, temperature and fire and the use of memory or visual cues may be of less value (Venter et al. 2017).

The use of visual cues by herbivores is scale dependent. A number of foraging studies have linked movement patterns to the use of memory (Brooks and Harris 2008; Ramos-Fernandez et al. 2003) or visual cues, at finer scales (e.g., bite, feeding station, and food-patch scales) (Howery et al. 2000; Laca 1998). At the broader habitat scale it seems that large browsing and grazing herbivores do not rely on visual cues but rather adapt their movement (accelerating movement when they move to patches that are not visible) to increase forage efficiency (Venter et al. 2017).

5.5 Patch Selection and Movement Between Patches

Large herbivores react to spatial patterns in topography and forage distribution (Bailey et al. 1996), mainly because resource heterogeneity occurs at different spatial and temporal scales (Senft et al. 1987; Bailey et al. 1996). Patch selection is scale-dependent, and although herbivores can often afford to be selective on a fine scale (plant part or species), this may not be the case at coarse scales (habitat scale) because of energetic constraints (van Beest et al. 2010). Foraging ungulate's food occurs in discrete patches (Prins 1996; Bailey and Provenza 2008; Prins and Van Langevelde 2008a) that are reasonably homogeneous with respect to some environmental features (Bailey et al. 1996; Bailey and Provenza 2008; Owen-Smith et al. 2010). Large grazing herbivores feed within forage patches and then move through areas where no or little acceptable food is encountered (Bailey et al. 1996; Prins 1996; Owen-Smith 2002); for example, blue wildebeest seeking out short grass patches and moving through tall grass habitats to get to them (Owen-Smith and Traill 2017).

The forage resources for grazers are spread out horizontally over the landscape. For browsers, forage resources are more complex because the food resource provides a less contiguous assemblage of plant parts of differing nutritional value, spaced both horizontally as well as vertically. Grazing and browsing herbivores utilize high-value food, in terms of quantity, quality or both, by adjusting their movements to habitat structure (Fortin 2003; de Knegt et al. 2007; Venter et al. 2015). They accelerate when moving between forage patches (Shipley et al. 1996; Venter et al. 2017) and spend more time in better forage patches (Distel et al. 1995; Courant and Fortin 2012). At different scales, plants may offer different amounts or concentrations of nutrients and, to reach a complete and balanced diet, animals may have to assemble a diet from different patches (Prins and Van Langevelde 2008b; Prins and Beekman 1989). Furthermore, large herbivores adjust their use of forage patches to the inter-patch distance (Shipley and Spalinger 1995), patch size (Venter et al. 2014), patch density (de Knegt et al. 2007), patch structure (Pietrzykowski et al. 2003), forage quality (Courant and Fortin 2012) and predation risk (Venter et al. 2014). Habitat patches can occur at a coarse scale where large patches are separated by long distances, or as a fine-scale mosaic separated by short distances (Cromsigt and Olff 2006). When patch heterogeneity is coarse, herbivores move long distances between suitable patches but with more localized heterogeneity, herbivores move more frequently over short distances between smaller patches (Hopcraft et al. 2010).

Animal movement is a core mechanism that influences a number of ecological processes at individual (e.g., home ranges, foraging), population (e.g., metapopulation connectivity, invasion dispersal), community (e.g., assemblages, species coexistence), and ecosystem levels (e.g., nutrient cycling, spread of disease, seed dispersal, trampling) (Turchin 1996; Fryxell et al. 2008; Nathan 2008; Delgado et al. 2009). Animal search movements consist of a discrete series of displacements (i.e., step lengths) separated by successive re-orientation events (i.e., turning angles) (Bartumeus et al. 2005). These movement parameters can be used to determine if an animal is foraging (encamped movement) or moving purposefully (exploratory movement) through the landscape (Fryxell et al. 2008). Encountering a patch of high-quality food plants can trigger a switch from exploratory to an encamped movement mode (Fryxell et al. 2008). The question as to whether these parameters differ between grazers and browsers has not been adequately investigated yet. The proportion of time spent feeding during foraging spells, lasting an hour or more, is typically 80–90% for grazers, and 65–80% for browsers (Owen-Smith 2002), indicating that browsers should have a higher movement rate, while foraging, compared to grazers. However, in a study by Venter et al. (2017), comparing eland (*Taurotragus oryx*), a browser, with zebra (*Equus quagga*) and red hartebeest, both grazers, the movement rate seemed to be more related to their digestive system (e.g., ruminant versus non-ruminant) than feeding preference.

Movement rate is not only determined by forage resources but also by water dependency, fear of predators and factors such as snow depth (Owen-Smith et al. 2010; Fryxell et al. 2008; Fortin 2003). Animals have to trade-off food quality and quantity with being close enough to water or being in habitats where they are less vulnerable to predators (de Boer et al. 2010).

5.6 Migration

Migration is the a predictable, round-trip, seasonal movement of an animal between two or more non-overlapping locations (Gnanadesikan et al. 2017). Migration is considered a behavioural strategy that provides explicit benefits to migratory species in the form of access to better forage, environmental conditions or breeding opportunities (Gnanadesikan et al. 2017; White et al. 2014; Fryxell et al. 1988). There are a number of dominant factors that drive mass migrations of ungulates, including, seasonal forage availability, snow depth, surface water availability and body-heat-balance (Harris et al. 2009; Zeng et al. 2010).

World-wide twenty four ungulate species migrate, or have migrated recently, in large aggregations (Harris et al. 2009). Fourteen of these species are from Africa, seven from Eurasia and four from North America (with caribou *Rangifer tarandus* occurs in both northern hemisphere continents) (Harris et al. 2009). The majority of these migrations involves species that are grazers, with only a few mixed feeders and browsers migrating (Fig. 5.1). Most grazers that migrate are from warm temperate and equatorial areas in Africa (Fig. 5.1). Possibly the most well-known grazer-based migratory system is the Serengeti/Maasai Mara where three grazing species namely; blue wildebeest (*Connochaetes taurinus*), plains zebra (*Equus quagga*) and Thomsons gazelle (*Gazella thomsonii*), migrate in succession, one species facilitating for another, according to phenological changes in the grass layer (McNaughton 1976; Bell 1971), or through a "parasite removal system" (Odadi et al. 2011).

Of the hundreds of ungulate species (see **Mishra et al.** Chap. 7), only 24 species are considered migratory. A number of other species show movements in reaction to seasonal differences in mountainous areas e.g., (Zeng et al. 2010) but perhaps the majority of ungulate species are not migratory, and many even quite sedentary within small home ranges. Ungulate ecologists should perhaps be aware that "the Serengeti Discourse" has been very dominant in recent decades. Most browser and mixed feeder migratory systems are in more extreme climatic regions, namely, arid, nival and polar (Fig. 5.4). In these systems the browse layer is affected by extreme temperatures, water stress or snow cover causing temporal and spatial changes in browse availability (for instance, leave drop by trees and shrubs). But most ungulates do not live in extreme climatic regions (see **Mishra et al.** Chap. 7).

5.7 Habitat Selection and Distribution Patterns

Large herbivores occur in heterogeneous environments and, consequently, multiple complex interactions, both biotic and abiotic, are involved in determining their habitat selection (Redfern et al. 2003; Kraaij and Novellie 2010). Large herbivore habitat selection is influenced not only by their dietary requirements, that is, actively seeking habitats as dictated by the availability of their preferred food source (Watson et al. 2011; Prins and Van Langevelde 2008b), but also their functional response (Shipley 2007), water availability (Ryan et al. 2006), escape cover and the associated

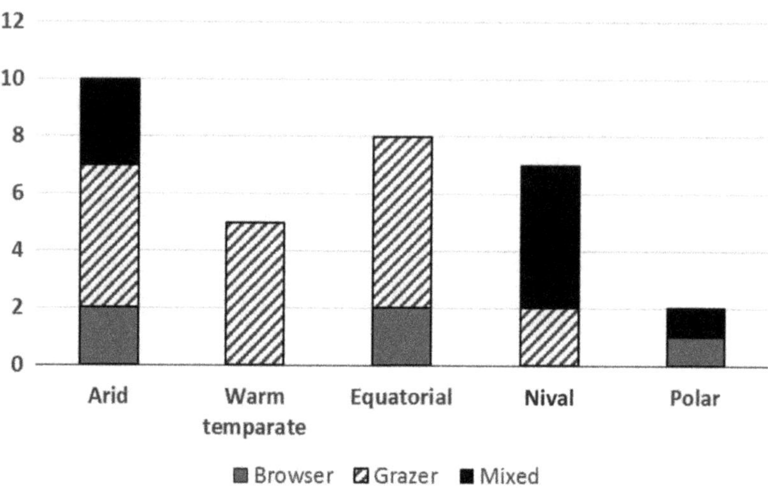

Fig. 5.4 The number of large scale migratory systems of browsing and grazing ungulates in the different world climatic regions. The global climatic regions that are used were derived from Kottek et al. (2006) and migratory systems from Harris et al. (2009)

predator avoidance strategies (Abu Baker and Brown 2014; Martin and Owen-Smith 2016). Generally browsers are largely water independent whilst grazers are considered water dependent (Western 1975). The differences in how water availability influences grazer and browser habitat selection is well illustrated in Kruger National Park, South Africa. Here the habitat selection of browsers appeared to be weakly associated with the availability and distribution of water, whilst grazers showed a stronger relationship (Redfern et al. 2003; Ryan et al. 2006). Grass quantity (and hence grazers) is more strongly positively associated with rainfall than are browse plants and browsers, because the storage organs and the depth of root penetration for browse plant forms allow trees a greater level of independence to rainfall events (Scholes and Archer 1997). This suggests that rainfall patterns are more influential on the distribution and habitat selection of grazers rather than browsers, because of the stronger immediate positive association of grass rather than browse production to rainfall.

Both browser and grazer selection of habitats is sensitive to the risks of predation. Species such as the common duiker (*Sylvicapra grimmia*), elk (*Cervus elaphus*), Nubian ibex (*Capra nubiana*) and white-tailed deer (*Odocoileus virginianus*) avoid open habitat because of its associated predation risk (Creel and Winnie 2005; Rieucau et al. 2007; Iribarren and Kotler 2012; Abu Baker and Brown 2014). African buffalo are most vulnerable to lion (*Panthera leo*) predation in the transition between grassland and closed vegetation (Prins and Iason 1989), whilst species such as the blue wildebeest remain in open, short grassed habitat when predators are present (Martin and Owen-Smith 2016). Large herbivore response, in terms of habitat selection (related to the threat of predation), is thus largely species specific.

Fire influences habitat selection of browsing and grazing ungulates as both show a preference for recently burnt vegetation due to improved plant quality in vegetation regrowth post-fire (Kraaij and Novellie 2010; Grossman et al. 1999; Klop et al. 2007). However, the preference for recently burnt vegetation does differ amongst ungulates with different feeding types (Venter et al. 2014; Klop et al. 2007; Klop and Prins 2008). Short grass grazers have a strong preference for burnt vegetation of up to 2 years of age, whilst bulk grazers will utilise veld up to 4 years old and browsers for up to 5 years of veld age (Kraaij and Novellie 2010). Many browsers also prefer recently burnt areas where they select new fire-induced spouts from forbs and trees (Klop et al. 2007) but more research could be done into understanding the preferences of different species in relation to fire history.

5.8 Variability in Ungulate Diets

A multitude of large herbivore feeding type categories have been developed over the years, often in relation to the relative proportion of plant forage types, or stomach structure of the species in question (Hofmann and Stewart 1972; Gagnon and Chew 2000). However, traditionally feeding guilds typically refer to browsers, grazers and mixed feeders along a continuum with varying distinctions between the guilds (Gagnon and Chew 2000). The use of a simple classification scheme referring simply to browsers, grazers and mixed feeders is useful, but it is important to understand that the same ungulate species occupy various different biomes and habitat types (Gagnon and Chew 2000; Cerling et al. 2003), and that a single classification scheme may not reveal local differences in ungulate diet, both regionally and seasonally (Gagnon and Chew 2000; Codron et al. 2007; cf. **Codron et al.** Chap. 4). Ungulate diets are influenced by habitats and associated forage availability (Djagoun et al. 2016), and substantial amounts of research has focussed on African ungulate feeding ecology and their associated evolutionary adaptations and diversification as well as their influence on ecosystem processes (du Toit 2003; Djagoun et al. 2016). Research on the relative consumption of graze and browse plant forms in the diets of both grazing and browsing African ungulates, of the same species, indicates variability across temporal and spatial scales. Grazers have been shown to be more responsive than browsers to changes in habitat type across biomes (e.g., adapting their diet to availability of graze versus browse), in terms of the relative availability of browse and graze, as reflected in the composition of their diets (i.e., grazers incorporating more browse in their diet than browsers would incorporate grass). For example, the diets of blue wildebeest, gemsbok (*Oryx gazella),* Cape mountain zebra (*Equus zebra)* and red hartebeest, generally considered to be grazers, were all shown to differ, in total contribution of grass versus browse, across multiple biomes (Cerling et al. 2003; Sponheimer et al. 2003; Vermeulen 2018). The greater adaptability in the diet of wild grazers than browsers is supported in the literature on the diets of domestic grazers and browsers, where it has been shown that the digestive system of the grazer is more efficiently suited to digest browse than the browser to digest graze (Clauss et al. 2010). The ability of grazers to handle greater variability in their diet, as compared to browsers, could, therefore, imply that the distinction between a grazer and a mixed feeder is not as distinct as the traditional browser, grazer and mixed feeder classification approach suggests (cf. **Codron et al.** Chap. 4). The variability in ungulate diets, between biomes, appears to be largely related to rainfall seasonality and amount, and the resulting abundance of available grass and tree cover for ungulate consumption. For example, in habitats where browse shrubs predominant, such as the Nama Karoo of South Africa, grazing ungulates consume a greater proportion of browse than what they do in biomes such as the savannas of Africa where grass is more widely available (Vermeulen 2018). The differences in ungulate diets, that occur both seasonally and at a biome scale, emphasise how important it is to have an understanding of ungulate feeding ecology at a local scale, especially given the threats (and opportunities) posed by climate change and the resultant change in vegetation structure and fire regimes (Kgope et al. 2010; Wigley et al. 2010; cf. **Van Langevelde** Chap. 10).

5.9 Conclusion

Human alteration of the global environment has triggered the Anthropocene, a major extinction event, in the history of life, which is causing widespread changes in the global distribution of biodiversity (Dirzo et al. 2014; WWF 2018). This wealth of biodiversity plays a significant role in a large proportion of the economy and livelihoods of many urban and rural people all over the world (Dirzo et al. 2014; Hempson et al. 2017). Human interference alter ecosystem processes and change the resilience of ecosystems to environmental change (Chapin et al. 2000). Many of the great ecosystems which house a diversity of large mammalian herbivores are exposed to alarming defaunation rates (Dirzo et al. 2014) driven by a variety of factors, some of which are well understood e.g., the effect of poaching (Chase et al. 2016) and overutilization (Chapin et al. 2000), and others not, e.g., change in ecosystem function (Hempson et al. 2017). Grazers and browsers interact with their environment and forage resources in different ways. An understanding of their behaviour is imperative for their long term persistence and the development of effective measures for their conservation; this is especially important when grasslands become less dominant and woody cover increases, as a consequence of climate change, leading to changes in large herbivore assemblages (Smit and Prins 2015).

References

Abu Baker MA, Brown JS (2014) Foraging and habitat use of common duikers, *Sylvicapra grimmia*, in a heterogeneous environment within the Soutpansberg, South Africa. Afr J Ecol 52:318–327

Bailey DW, Provenza FD (2008) Mechanisms determining large herbivore distribution. In: Prins HHT, Van Langevelde F (eds) Resource ecology: spatial and temporal dynamics of foraging. Springer, New York

Bailey DW, Gross JE, Laca EA, Rittenhouse LR, Coughenour MB, Swift DM, Sims PL (1996) Mechanisms that result in large herbivore grazing distribution patterns. J Range Manag 49:386–400

Baker DJ, Stillman RA, Smith BM, Bullock JM, Norris KJ (2010) Vigilance and the functional response of granivorous foragers. Funct Ecol 24:1281–1290

Baker DJ, Stillman RA, Smart SL, Bullock JM, Norris KJ (2011) Are the costs of routine vigilance avoided by granivorous foragers? Funct Ecol 25:617–627

Bartumeus FF, da Luz MGE, Viswanathan GM, Catalan J (2005) Animal search strategies: a quantitative random-walk analysis. Ecology 86:3078–3087

Baumont R, Prache S, Meuret M, Morand-Fehr P (2000) How forage characteristics influence behaviour and intake in small ruminants: a review. Livest Prod Sci 64:15–28

Beekman JH, Prins HHT (1989) Feeding strategies of sedentary large herbivores in East Africa, with emphasis on the African buffalo, *Syncerus caffer*. Afr J Ecol 27:129–147. https://doi.org/10.1111/j.1365-2028.1989.tb00937.x

Bell RHV (1971) A grazing ecosystem in the Serengeti. Sci Am 225:86–93. https://doi.org/10.1038/scientificamerican0771-86

Belovsky GE, Schmitz OJ, Slade J, Dawson T (1991) Effects of spines and thorns on Australian arid zone herbivores of different body masses. Oecologia 88:521–528

Bjørneraas K, Herfindal I, Solberg EJ, Sæther B-E, van Moorter B, Rolandsen CM (2012) Habitat quality influences population distribution, individual space use and functional responses in habitat selection by a large herbivore. Oecologia 168:231–243

Bond WJ, Parr CL (2010) Beyond the forest edge: ecology, diversity and conservation of the grassy biomes. Biol Conserv 143:2395–2404

Brooks CJ, Harris S (2008) Directed movement and orientation across a large natural landscape by zebras, *Equus burchelli antiquorum*. Anim Behav 76:277–285. https://doi.org/10.1016/j.anbehav.2008.02.005

Cerling TE, Harris JM, Passey BH (2003) Diets of East African Bovidae based on stable isotope analysis. J Mammal 84:456–470. https://doi.org/10.1644/1545-1542(2003)084<0456:DOEABB>2.0.CO;2

Chapin FS, Zavaleta ES, Eviner VT, Naylor RL, Vitousek PM, Reynolds HL, Hooper DU, Lavorel S, Sala OE, Hobbie SE, Mack MC, Diaz S (2000) Consequences of changing biodiversity. Nature 405:234–242

Chase MJ, Schlossberg S, Griffin CR, Bouché PJC, Djene SW, Elkan PW, Ferreira S, Grossman F, Kohi EM, Landen K, Omondi P, Peltier A, Selier SAJ, Sutcliffe R (2016) Continent-wide survey reveals massive decline in African savannah elephants. PeerJ 4:e2354. https://doi.org/10.7717/peerj.2354

Clauss M, Kaiser TM, Hummel J (2008) The morphophysiological adaptations of browsing and grazing mammals. In: Gordon IJ, Prins HHT (eds) The ecology of browsing and grazing. Springer, Heidelberg, pp 47–88

Clauss M, Nunn C, Fritz J, Hummel J (2009) Evidence for a tradeoff between retention time and chewing efficiency in large mammalian herbivores. Comp Biochem Physiol A Mol Integr Physiol 154:376–382

Clauss M, Hume I, Hummel J (2010) Evolutionary adaptations of ruminants and their potential relevance for modern production systems. Animal 4:979–992

Codron D, Codron J, Lee-Thorp JA, Sponheimer M, de Ruiter D, Sealy J, Grant R, Fourie N (2007) Diets of savanna ungulates from stable carbon isotope composition of faeces. J Zool 273:21–29. https://doi.org/10.1111/j.1469-7998.2007.00292.x

Courant S, Fortin D (2012) Time allocation of bison in meadow patches driven by potential energy gains and group size dynamics. Oikos 121:1163–1173. https://doi.org/10.1111/j.1600-0706.2011.19994.x

Creel S, Winnie JA Jr (2005) Responses of elk herd size to fine-scale spatial and temporal variation in the risk of predation by wolves. Anim Behav 69:9

Cromsigt JPGM, Olff H (2006) Resource partitioning among savanna grazers mediated by local heterogeneity: an experimental approach. Ecology 87:1532–1541

Damuth J, Janis CM (2011) On the relationship between hypsodonty and feeding ecology in ungulate mammals, and its utility in palaeoecology. Biol Rev 86:733–758

de Boer WF, Vis MJ, de Knegt HJ, Rowles C, Kohi EM, van Langevelde F, Peel M, Pretorius Y, Skidmore AK, Slotow R (2010) Spatial distribution of lion kills determined by the water dependency of prey species. J Mammal 91:1280–1286

de Knegt HJ, Hengeveld GM, van Langevelde F, de Boer WF, Kirkman KP (2007) Patch density determines movement patterns and foraging efficiency of large herbivores. Behav Ecol 18:1065–1072. https://doi.org/10.1093/beheco/arm080

de Knegt HJ, Groen TA, van de Vijver CADM, HHT P, van Langevelde F (2008) Herbivores as architects of savannas: inducing and modifying spatial vegetation patterning. Oikos 117:543–554

Delgado M d M, Penteriani V, Nams VO, Campioni L (2009) Changes of movement patterns from early dispersal to settlement. Behav Ecol Sociobiol 64:35–43

Demment MW, Van Soest PJ (1985) A nutritional explanation for body-size patterns of ruminant and nonruminant herbivores. Am Nat 125:641–672

Dirzo R, Young HS, Galetti M, Ceballos G, Isaac NJ, Collen B (2014) Defaunation in the Anthropocene. Science 345:401–406

Distel RA, Laca EA, Griggs TC, Demment MW (1995) Patch selection by cattle: maximization of intake rate in horizontally heterogeneous pastures. Appl Anim Behav Sci 45:11–21. https://doi.org/10.1016/0168-1591(95)00593-h

Djagoun CAMS, Codron D, Sealy J, Mensah GA, Sinsin B (2016) Isotopic niche structure of a mammalian herbivore assemblage from a West African savanna: body mass and seasonality effect. Mamm Biol – Zeitschrift für Säugetierkunde 81:644–650. https://doi.org/10.1016/j.mambio.2016.09.001

du Toit JT (2003) Large herbivores and savanna heterogeneity. In: Du Toit JT, Rodgers KH, Biggs HC (eds) The Kruger experience: ecology and management of savanna heterogeneity. Island Press, Washington, DC, pp 292–309

Dumont B, Petit M (1998) Spatial memory of sheep at pasture. Appl Anim Behav Sci 60:43–53

Edouard N, Duncan P, Dumont B, Baumont R, Fleurance G (2010) Foraging in a heterogeneous environment: an experimental study of the trade-off between intake rate and diet quality. Appl Anim Behav Sci 126:27–36. https://doi.org/10.1016/j.applanim.2010.05.008

Edwards GR, Newman JA, Parsons AJ, Krebs JR (1996) The use of spatial memory by grazing animals to locate food patches in spatially heterogeneous environments: an example with sheep. Appl Anim Behav Sci 50:147–160. https://doi.org/10.1016/0168-1591(96)01077-5

Edwards GR, Newman JA, Parsons AJ, Krebs JR (1997) Use of cues by grazing animals to locate food patches: an example with sheep. Appl Anim Behav Sci 51:59–68. https://doi.org/10.1016/S0168-1591(96)01095-7

Faverdin P, Baumont R, Ingvartsen KL (1995) Control and prediction of feed intake in ruminants. In: 4th International Symposium on the Nutrition of Herbivores, Clermont Ferrand (France), 11–15 Sep 1995. INRA, Paris

Ford AT, Goheen JR, Otieno TO, Bidner L, Isbell LA, Palmer TM, Ward D, Woodroffe R, Pringle RM (2014) Large carnivores make savanna tree communities less thorny. Science 346:346–349. https://doi.org/10.1126/science.1252753

Fortin D (2003) Searching behavior and use of sampling information by free-ranging bison (*Bos bison*). Behav Ecol Sociobiol 54:194–203

Fortin D (2006) The allometry of plant spacing that regulates food intake rate in mammalian herbivores. Ecology 87:1861–1866

Fortin D, Boyce MS, Merrill EH, Fryxell JM (2004) Foraging costs of vigilance in large mammalian herbivores. Oikos 107:172–180

Fraser D, Theodor JM (2011) Anterior dentary shape as an indicator of diet in ruminant artiodactyls. J Vertebr Paleontol 31:1366–1375

Fritz J, Hummel J, Kienzle E, Arnold C, Nunn C, Clauss M (2009) Comparative chewing efficiency in mammalian herbivores. Oikos 118:1623–1632

Fryxell JM, Greever J, Sinclair ARE (1988) Why are migratory ungulates so abundant? Am Nat 131:781–798. https://doi.org/10.2307/2461813

Fryxell JM, Hazell M, Borger L, Dalziel BD, Haydon DT, Morales JM, McIntosh T, Rosatte RC (2008) Multiple movement modes by large herbivores at multiple spatiotemporal scales. Proc Natl Acad Sci 105:6

Gagnon M, Chew AE (2000) Dietary preferences in extant African Bovidae. J Mammal 81:490–511. https://doi.org/10.1644/1545-1542(2000)081<0490:DPIEAB>2.0.CO;2

Gnanadesikan GE, Pearse WD, Shaw AK (2017) Evolution of mammalian migrations for refuge, breeding, and food. Ecol Evol 7:5891–5900. https://doi.org/10.1002/ece3.3120

Godvik IMR, Loe LE, Vik JO, Veiberg V, Langvatn R, Mysterud A (2009) Temporal scales, trade-offs, and functional responses in red deer habitat selection. Ecology 90:699–710

Gordon IJ (2003) Browsing and grazing ruminants: are they different beasts? For Ecol Manag 181:13–21. https://doi.org/10.1016/S0378-1127(03)00124-5

Gordon IJ, Benvenutti M (2006) How ruminant livestock interact with sown sward architecture at the bite scale. In: Bell V (ed) Feeding in domestic vertebrates: from structure to behaviour. CABI Publishing, Wallingford, pp 263–277

Gordon IJ, Illius AW (1988) Incisor arcade structure and diet selection in ruminants. Funct Ecol 2:15–22

Grossman D, Holden P, Collinson R (1999) Veld management on the game ranch. In: Veld management in South Africa. University of Natal Press, Pietermaritzburg, pp 261–279

Harris G, Thirgood S, Hopcraft JGC, Cromsigt JP, Berger J (2009) Global decline in aggregated migrations of large terrestrial mammals. Endanger Species Res 7:55–76

Hempson GP, Archibald S, Bond WJ (2017) The consequences of replacing wildlife with livestock in Africa. Sci Rep 7:17196. https://doi.org/10.1038/s41598-017-17348-4

Herfindal I, Tremblay JP, Hansen BB, Solberg EJ, Heim M, Sæther BE (2009) Scale dependency and functional response in moose habitat selection. Ecography 32:849–859

Heuermann N, van Langevelde F, van Wieren S, Prins H (2011) Increased searching and handling effort in tall swards lead to a type IV functional response in small grazing herbivores. Oecologia 166:659–669. https://doi.org/10.1007/s00442-010-1894-8

Hewitson L, Dumont B, Gordon IJ (2005) Response of foraging sheep to variability in the spatial distribution of resources. Anim Behav 69:8

Hofmann RR, Stewart DRM (1972) Grazer or browser: a classification based on the stomach-structure and feeding habitats of East African ruminants. Mammalia 36:226–240. https://doi.org/10.1515/mamm.1972.36.2.226

Holling CS (1959) Some characteristics of simple types of predation and parasitism. Can Entomol 91:385–398. https://doi.org/10.4039/Ent91385-7

Hopcraft JGC, Olff H, Sinclair ARE (2010) Herbivores, resources and risks: alternating regulation along primary environmental gradients in savannas. Trends Ecol Evol 25:119–128. https://doi.org/10.1016/j.tree.2009.08.001

Howery LD, Bailey DW, Ruyle GB, Renken WJ (2000) Cattle use visual cues to track food locations. Appl Anim Behav Sci 67:1–14. https://doi.org/10.1016/S0168-1591(99)00118-5

Hummel J, Fritz J, Kienzle E, Medici EP, Lang S, Zimmermann W, Streich WJ, Clauss M (2008) Differences in fecal particle size between free-ranging and captive individuals of two browser species. Zoo Biol 27:70–77. Published in affiliation with the American Zoo and Aquarium Association

Hummel J, Findeisen E, Südekum K-H, Ruf I, Kaiser TM, Bucher M, Clauss M, Codron D (2010) Another one bites the dust: faecal silica levels in large herbivores correlate with high-crowned teeth. Proc R Soc Lond B Biol Sci 278:1742–1747

Illius A, FitzGibbon C (1994) Costs of vigilance in foraging ungulates. Anim Behav 47:481–484

Illius AW, Gordon IJ (1987) The allometry of food intake in grazing ruminants. J Anim Ecol 56:989–999. https://doi.org/10.2307/4961

Iribarren C, Kotler BP (2012) Foraging patterns of habitat use reveal landscape of fear of Nubian ibex Capra nubiana. Wildl Biol 18:194–201
Janmaat KRL, Byrne RW, Zuberbühler K (2006) Evidence for a spatial memory of fruiting states of rainforest trees in wild mangabeys. Anim Behav 72:797–807. https://doi.org/10.1016/j.anbehav.2005.12.009
Jarman PJ (1974) The social organisation of antelope in relation to their ecology. Behavior 48:215–267. https://doi.org/10.1163/156853974X00345
Kgope BS, Bond WJ, Midgley GF (2010) Growth responses of African savanna trees implicate atmospheric [CO2] as a driver of past and current changes in savanna tree cover. Austral Ecol 35:451–463
Klop E, Prins HHT (2008) Diversity and species composition of West African ungulate assemblages: effects of fire, climate and soil. Glob Ecol Biogeogr 17:778–787. https://doi.org/10.1111/j.1466-8238.2008.00416.x
Klop E, van Goethem J, de Iongh HH (2007) Resource selection by grazing herbivores on post-fire regrowth in a West African woodland savanna. Wildl Res 34:77–83
Kottek M, Grieser J, Beck C, Rudolf B, Rubel F (2006) World map of the Köppen-Geiger climate classification updated. Meteorol Z 15:259–263
Kraaij T, Novellie PA (2010) Habitat selection by large herbivores in relation to fire at the Bontebok National Park (1974–2009): the effects of management changes. Afr J Range Forage Sci 27:21–27. https://doi.org/10.2989/10220111003703450
Laca EA (1998) Spatial memory and food searching mechanisms of cattle. J Range Manag 51:370–378
Laca E, Ungar E, Demment M (1994) Mechanisms of handling time and intake rate of a large mammalian grazer. Appl Anim Behav Sci 39:3–19
Leek B (1977) Abdominal and pelvic visceral receptors. Br Med Bull 33:163–168
Martin J, Owen-Smith N (2016) Habitat selectivity influences the reactive responses of African ungulates to encounters with lions. Anim Behav 116:163–170
McNaughton SJ (1976) Serengeti migratory wildebeest: facilitation of energy flow by grazing. Science 191:3
Merkle JA, Fortin D, Morales JM (2014) A memory-based foraging tactic reveals an adaptive mechanism for restricted space use. Ecol Lett 17:924–931. https://doi.org/10.1111/ele.12294
Murray MG (1993) Comparative nutrition of wildebeest, hartebeest and topi in the Serengeti. Afr J Ecol 31:172–177. https://doi.org/10.1111/j.1365-2028.1993.tb00530.x
Murray MG, Baird DR (2008) Resource-ratio theory applied to large herbivores. Ecology 89:1445–1456
Nathan R (2008) An emerging movement ecology paradigm. Proc Natl Acad Sci 105:2
Nersesian CL, Banks PB, McArthur C (2012) Behavioural responses to indirect and direct predator cues by a mammalian herbivore, the common brushtail possum. Behav Ecol Sociobiol 66:47–55
Newman J, Parsons A, Penning P (1994) A note on the behavioural strategies used by grazing animals to alter their intake rates. Grass Forage Sci 49:502–505
Odadi WO, Jain M, Van Wieren SE, Prins HH, Rubenstein DI (2011) Facilitation between bovids and equids on an African savanna. Evol Ecol Res 13:237–252
Olff H, Ritchie ME, Prins HHT (2002) Global environmental controls of diversity in large herbivores. Nature 415:901–904
Owen-Smith N (2002) Adaptive herbivore ecology: from resources to populations in variable environments. Wits University Press, Johannesburg
Owen-Smith N, Novellie P (1982) What should a clever ungulate eat? Am Nat 119:151–178
Owen-Smith N, Traill LW (2017) Space use patterns of a large mammalian herbivore distinguished by activity state: fear versus food? J Zool 303:281–290. https://doi.org/10.1111/jzo.12490
Owen-Smith N, Fryxell JM, Merrill EH (2010) Foraging theory upscaled: the behavioural ecology of herbivore movement. Philos Trans R Soc 365:2267–2278

Pettorelli N, Hilborn A, Duncan C, Durant SM (2015) Chapter two – individual variability: the missing component to our understanding of predator–prey interactions. In: Pawar S, Woodward G, Dell AI (eds) Advances in ecological research, vol 52. Academic, Waltham, MA, pp 19–44. https://doi.org/10.1016/bs.aecr.2015.01.001

Pietrzykowski E, McArthur C, Fitzgerald H, Goodwin AN (2003) Influence of patch characteristics on browsing of tree seedlings by mammalian herbivores. J Appl Ecol 40:458–469. https://doi.org/10.1046/j.1365-2664.2003.00809.x

Pretorius Y, de Boer WF, Kortekaas K, van Wijngaarden M, Grant RC, Kohi EM, Mwakiwa E, Slotow R, Prins HH (2016) Why elephant have trunks and giraffe long tongues: how plants shape large herbivore mouth morphology. Acta Zool 97:246–254

Prins H (1989) Buffalo herd structure and its repercussions for condition of individual African buffalo cows. Ethology 81:47–71

Prins HHT (1996) Behavior and ecology of the African buffalo: social inequality and decision making. Chapman & Hall, London

Prins H, Beekman J (1989) A balanced diet as a goal for grazing: the food of the Manyara buffalo. Afr J Ecol 27:241–259

Prins HHT, Iason GR (1989) Dangerous lions and nonchalant buffalo. Behaviour 108:262–296

Prins HHT, Olff H (1998) Species richness of African grazer assemblages: towards a functional explanation. In: Newbery DM, HHT P, Brown ND (eds) Dynamics of tropical communities, British ecological society symposium. Blackwell Science, Oxford, pp 449–490

Prins HHT, Van Langevelde F (2008a) Assembling a diet from different places. In: Prins HHT, Van Langevelde F (eds) Resource ecology: spatial and temporal dynamics of foraging. Springer, Dordrecht, pp 129–158

Prins HHT, Van Langevelde F (2008b) Resource ecology: spatial and temporal dynamics of foraging. Springer, New York

Ramos-Fernandez G, Mateos JL, Miramobtes O, Cocho G, Larralde H, Ayala-Orozco B (2003) Levy walk patterns in the foraging movements of spider monkeys (*Ateles geoffroyi*). Behav Ecol Sociobiol 55:223–230

Redfern JV, Grant R, Biggs H, Getz WM (2003) Surface-water constraints on herbivore foraging in the Kruger National Park, South Africa. Ecol Soc Am 84:2092–2107

Renken WJ, Howery LD, Ruyle GB, Enns RM (2008) Cattle generalise visual cues from the pen to the field to select initial feeding patches. Appl Anim Behav Sci 109:128–140. https://doi.org/10.1016/j.applanim.2007.03.014

Rieucau G, Vickery W, Doucet G, Laquerre B (2007) An innovative use of white-tailed deer (Odocoileus virginianus) foraging behaviour in impact studies. Can J Zool 85:839–846

Ryan SJ, Knechtel CU, Getz WM (2006) Range and habitat selection of African buffalo in South Africa. J Wildl Manag 70:764–776. https://doi.org/10.2193/0022-541x(2006)70[764:rahsoa]2.0.co;2

Sarnelle O, Wilson AE (2008) Type III functional response in Daphnia. Ecology 89:1723–1732

Scholes R, Archer S (1997) Tree-grass interactions in savannas. Annu Rev Ecol Syst 28:517–544

Searle KR, Shipley LA (2008) The comparative feeding bahaviour of large browsing and grazing herbivores. In: HHT P, Van Langevelde F (eds) The ecology of browsing and grazing. Springer, Berlin, pp 117–148

Searle KR, Thompson Hobbs N, Shipley LA (2005) Should I stay or should I go? Patch departure decisions by herbivores at multiple scales. Oikos 111:417–424. https://doi.org/10.1111/j.0030-1299.2005.13918.x

Senft RL, Coughenour MB, Bailey DW, Rittenhouse LR, Sala OE, Swift DM (1987) Large herbivore foraging and ecological hierarchies. Bioscience 37:789–799. https://doi.org/10.2307/1310545

Shipley LA (1999) Grazers and browsers: how digestive morphology affects diet selection. In: Launchbaugh KL, Sanders KD, Mosley JC (eds) Grazing behavior of livestock and wildlife. University of Idaho, Moscow, ID

Shipley LA (2007) The influence of bite size on foraging at larger spatial and temporal scales by mammalian herbivores. Oikos 116:1964–1974. https://doi.org/10.1111/j.2007.0030-1299.15974.x

Shipley LA, Spalinger DE (1995) Influence of size and density of browse patches on intake rates and foraging decisions of young moose and white-tailed deer. Oecologia 104:112–121

Shipley LA, Gross JE, Spalinger DE, Hobbs NT, Wunder BA (1994) The scaling of intake rate in mammalian herbivores. Am Nat 143:1055–1082

Shipley LA, Spalinger DE, Gross JE, Thompson Hobbs N, Wunder BA (1996) The dynamics and scaling of foraging velocity and encounter rate in mammalian herbivores. Funct Ecol 10:234–244

Smart SL, Stillman RA, Norris KJ (2008) Measuring the functional responses of farmland birds: an example for a declining seed-feeding bunting. J Anim Ecol 77:687–695

Smit IP, Prins HH (2015) Predicting the effects of woody encroachment on mammal communities, grazing biomass and fire frequency in African savannas. PLoS One 10:e0137857

Spalinger DE, Hobbs NT (1992) Mechanisms of foraging in mammalian herbivores: new models of functional response. Am Nat 140:325–348

Sponheimer M, Lee-Thorp JA, DeRuiter DJ, Smith JM, van der Merwe NJ, Reed K, Grant CC, Ayliffe LK, Robinson TF, Heidelberger C, Marcus W (2003) Diets of Southern African bovidae stable isotope evidence. J Mammal 84:9. https://doi.org/10.1644/1545-1542(2003)084<0471:DOSABS>2.0.CO;2

Stephens DW, Krebs JR (1986) Foraging theory. Princeton University Press, Princeton, NJ

Tomlinson KW, van Langevelde F, Ward D, Prins HH, de Bie S, Vosman B, Sampaio EV, Sterck FJ (2016) Defence against vertebrate herbivores trades off into architectural and low nutrient strategies amongst savanna Fabaceae species. Oikos 125:126–136

Turchin P (1996) Fractal analyses of animal movement: a critique. Ecology 77:5

van Beest FM, Mysterud A, Loe LE, Milner JM (2010) Forage quantity, quality and depletion as scale-dependent mechanisms driving habitat selection of a large browsing herbivore. J Anim Ecol 79:910–922. https://doi.org/10.1111/j.1365-2656.2010.01701.x

Venter JA, Nabe-Nielsen J, Prins HT, Slotow R (2014) Forage patch use by grazing herbivores in a South African grazing ecosystem. Acta Theriol 59:457–466. https://doi.org/10.1007/s13364-014-0184-y

Venter JA, Prins HHT, Mashanova A, de Boer WF, Slotow R (2015) Intrinsic and extrinsic factors influencing large African herbivore movements. Eco Inform 30:257–262. https://doi.org/10.1016/j.ecoinf.2015.05.006

Venter JA, Prins HHT, Mashanova A, Slotow R (2017) Ungulates rely less on visual cues, but more on adapting movement behaviour, when searching for forage. PeerJ 5:e3178. https://doi.org/10.7717/peerj.3178

Vermeulen MM (2018) Exploring feeding ecology and population growth rate responses of ungulates in southern African arid biomes. Nelson Mandela University, Port Elizabeth

Watson LH, Kraaij T, Novellie P (2011) Management of rare ungulates in a small park: habitat use of bontebok and cape mountain zebra in Bontebok National Park assessed by counts of dung groups. S Afr J Wildl Res 41:158–166

Western D (1975) Water availability and its influence on the structure and dynamics of a savannah large mammal community. Afr J Ecol 13:265–286

White CA, Feller MC, Bayley S (2003) Predation risk and the functional response of elk–aspen herbivory. For Ecol Manag 181:77–97. https://doi.org/10.1016/S0378-1127(03)00119-1

White KS, Barten NL, Crouse S, Crouse J (2014) Benefits of migration in relation to nutritional condition and predation risk in a partially migratory moose population. Ecology 95:225–237

Wigley BJ, Bond WJ, Hoffman M (2010) Thicket expansion in a South African savanna under divergent land use: local vs. global drivers? Glob Chang Biol 16:964–976

Wilmshurst JF, Fryxell JM, Bergman CM (2000) The allometry of patch selection in ruminants. Proc R Soc Lond B Biol Sci 267:345–349

WWF (2018) Living planet report – 2018: aiming higher. WWF, Galnd

Zeng Z-G, Beck PS, Wang T-J, Skidmore AK, Song Y-L, Gong H-S, Prins HH (2010) Effects of plant phenology and solar radiation on seasonal movement of golden takin in the Qinling Mountains, China. J Mammal 91:92–100

Chapter 6
Population Dynamics of Browsing and Grazing Ungulates in the Anthropocene

Christian Kiffner and Derek E. Lee

6.1 Introduction

Ungulates (here defined as terrestrial artiodactyls and perissodactyls; hoofed animals within an average female body mass of 1–1000 kg) are a highly diverse group of grazing and browsing animals. Despite their cultural, economic, and ecological importance across their nearly worldwide range, the status of many ungulate populations is worrisome (Ripple et al. 2015). Therefore, we address the topic of ungulate population dynamics from the perspective of conservation biology. Our framework is the "declining population paradigm", which aims at identifying demographic causes and mechanisms that underlie observed changes in population growth rates (Caughley 1994). We mainly focus on ungulate assemblages in temperate zones of Europe and North America, subtropical deciduous forests of South Asia, and savannas and woodlands of Africa, since most relevant research on ungulate population ecology has been carried out in these systems. Due to the high diversity of ungulates in African savannas (Olff et al. 2002), and our own experience, several examples in this Chapter were drawn from this region. Ungulates comprise a huge diversity of species that occur on most continents (Olff et al. 2002), and can functionally be grouped according to their feeding strategy as grazer (eating grass) or browser (eating woody and non-woody dicots) (Hofmann and Stewart 1972). This classification is generally not dichotomous because many species are intermediate (or mixed) feeders (Gagnon and Chew 2000; **Codron et al.** Chap. 4), and even archetype browsers such as giraffes (*Giraffa camelopardalis*) may occasionally feed

C. Kiffner (✉)
Center for Wildlife Management Studies, The School for Field Studies, Karatu, Tanzania
e-mail: ckiffne@gwdg.de

D. E. Lee
Wild Nature Institute, Concord, NH, USA

Department of Biology, Pennsylvania State University, University Park, PA, USA

I. J. Gordon, H. H. T. Prins (eds.), *The Ecology of Browsing and Grazing II*,
Ecological Studies 239, https://doi.org/10.1007/978-3-030-25865-8_6

on grasses (Seeber et al. 2012), just as typical grazers may occasionally browse (Owen-Smith 2008). We used this dichotomy even though we are aware that this is an oversimplification (**Codron et al.** Chap. 4; **Gordon and Prins** Chap. 16). For analyses, in this Chapter we characterize ungulates as either predominantly grazers (≥50% grass in diet) or browsers (≥50% dicots in diet). With this coarse dichotomy, and its inherent limitations in mind, we will focus on the following specific questions:

1. How does spatial and temporal variability of vegetation productivity affect grazers and browsers?
2. How do densities of grazers and browsers relate to body mass?
3. Which demographic rate contributes most to population growth of browsers and grazers?
4. What are the causal factors of population growth in ungulate populations?
5. Are browsers or grazers more susceptible to anthropogenic changes?

6.2 Spatial and Temporal Variability in Grass and Browse Availability

Forage abundance is the principal driver of second- and third-order habitat selection of animals (i.e., the distribution of home ranges and space utilization within the home range, respectively), and thus largely controls the distribution and density of ungulates (Johnson 1980; Pettorelli et al. 2009; Waltert et al. 2009); although scale-dependent trade-offs between forage quantity and quality exist (Van Beest et al. 2010).

The global distribution of tree cover is mainly affected by climate, but at intermediate precipitation and mild seasonality, fire is the main force differentiating savannas from forests (Langevelde et al. 2003; Staver et al. 2011). Without fire, closed forests could double in their extent (Bond et al. 2004). In arid and semi-arid regions, woody cover is limited by precipitation, fire, and herbivory, which interact to limit woody cover. Competition with grasses also limits recruitment of woody vegetation (de Waal et al. 2011; Morrison et al. 2018). In areas with precipitation exceeding 650 mm, savannas may transform to forests (and vice versa) following perturbations (Sankaran et al. 2005; Murphy and Bowman 2012). The presence or absence of ungulates can also affect vegetation structure and quantity both directly and indirectly. For example, browsing may limit woody species expansion, and thus indirectly stimulate grass growth, which increases fuel load and fire intensity, which further reduces woody cover (Langevelde et al. 2003). Alternatively, high densities of grazers can remove ground fuel to the point where fire prevalence is reduced and woody plant cover increases (Roques et al. 2001). Indeed, fire and herbivory strongly interact. For example during times of culling programs (and thus reduced herbivore densities) the lowered grazing pressure substantially led to increases in the extent of fires whereas the opposite was true during times after the culling programs when herbivores doubled in biomass density (Smit and Archibald 2019; cf. **Smit and Coetsee** Chap. 13). Particularly for grazers, these two alternative stable biome states are important because closed-canopy forests have almost no grasses (Ratnam et al. 2011) and reduced productivity in the ground vegetation layer (Melis et al. 2009).

Temporally, browse availability is relatively constant across years, but is influenced by recent precipitation (Rutherford 1984). Browse availability often varies seasonally, as during the dry season in sub-tropical deciduous forests, and during winter in temperate broad-leaved forests, when most of the woody vegetation sheds leaves. Grass availability is more strongly influenced by seasonal and inter-annual differences in rainfall (O'Connor et al. 2001; Ogutu and Owen-Smith 2003). In African savannas, the protein content of both browse and grasses (and thus the nutritional quality) is usually highest during the early rainy season and lowest during the dry season (Pellew 1983; Prins 1988; Robbins 1993).

Ungulates have adopted two main strategies to cope with the spatiotemporal variability in food resources. Mixed feeders, such as red deer (*Cervus elaphus*) or impala (*Aepyceros melampus*), can adjust their feeding strategies and mainly feed on grasses during the grass growing season and increase intake of woody vegetation during winter or dry seasons (Meissner et al. 1996; Verheyden-Tixier et al. 2008). Other ungulate species track the spatiotemporal variation in plant phenology by migrating to areas of higher forage quantity and quality (Merkle et al. 2016). Seasonal migrations have been documented for browsers such as roe deer (*Capreolus capreolus*) and moose (*Alces alces*), and grazers such as wildebeest (*Connochaetes taurinus*) and saiga antelope (*Saiga tatarica*), but most of the farthest long-distance migrations are undertaken by grazers (Teitelbaum et al. 2015).

6.3 Population Densities of Grazers and Browsers

To describe patterns and assess correlates of population densities of grazers and browsers, we compiled a database of density estimates of ungulate populations (n = 964) across the globe (available at http://www.wildnatureinstitute.org/uploads/5/5/7/7/5577192/kiffner___lee_ungulate_densities.xlsx). We are aware that the broad distinction into grazers and browsers (and even a trichotomy of browsers, mixed feeders, and grazers) is too simplistic from evolutionary and morphological perspectives (**Codron et al.** Chap. 4). Yet, in order to find broad patterns in population densities, a simplification into two categories (and thus sufficient sample sizes for each "feeding" category) was necessary to allow for our quantitative comparisons.

Population densities of browsers and grazers are highly variable (Fig. 6.1). Although most populations range around a few individuals km^{-2}, both grazing and browsing species can reach very high population densities (with 267 ind.km^{-2}, chital *Axis axis* had the highest density in our dataset; Wegge and Storaas 2009). In

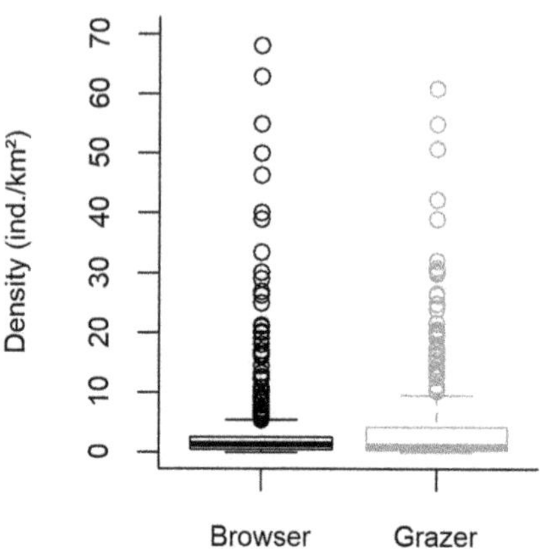

Fig. 6.1 Boxplot showing the range of population density estimates for browsing and grazing ungulates. The highest density in our dataset (267 chital.km^{-2}) was removed

this and subsequent sections, we outline how this variation in population densities of ungulates can be explained.

Body mass is one of the most fundamental traits of organisms (Peters 1983), and its relationship with population density has received substantial attention from a macro-ecological perspective (Damuth 1981; Blackburn and Gaston 1999; Olff et al. 2002; White et al. 2007). Negative relationships between the $\log_{10}$ body mass and $\log_{10}$ density of animals often have been reported, but the relationships are often non-linear and explain relatively little of the observed variation in animal population densities (Blackburn et al. 1994; Silva et al. 2001). The frequently observed explanation provided for this pattern is that animal abundances are limited by energy availability, but this explanation has been substantially challenged (Blackburn and Gaston 1999; White et al. 2007). To assess whether body mass is a strong predictor of ungulate densities, we plotted trend lines using best fitting (based on sample-size corrected AIC_c scores) general additive models in R (R Core Team 2016; Wood et al. 2016) to non-transformed data.

Our data indicate that relationships between ungulate density and average female body mass are—if at all—rather weak when analysed separately for grazers and browsers, or combined for all ungulates, and that body mass explains very little of the observed variation (Fig. 6.2). Particularly among grazers, biomass density (density x average female body mass) seems to have a bimodal distribution, with highest biomass densities in species of about 200 kg body mass. Rather than a linear body mass–density relationship, our data indicate that high ungulate densities are realized in specific body mass ranges. Globally, highest population densities (10% of highest densities in our dataset) occur in relatively small-bodied browsers (range: 20–233 kg; median: 45 kg) whereas highest densities in grazers are realized across a wider body mass range and typically in larger species (range: 17–325 kg; median: 137.5 kg). Extending this selection to the 20% highest population densities yielded similar body mass ranges for both browsers (range: 5.5–233 kg, median 45 kg) and grazers (range: 17–325 kg, median 50 kg), lending further support for upper and

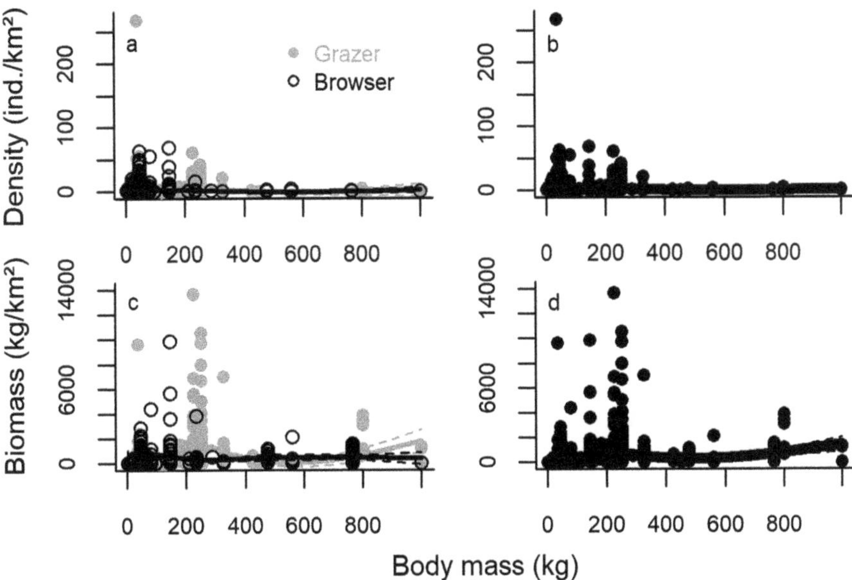

Fig. 6.2 Global patterns of grazer and browser (**a**) densities and (**c**) biomass densities in relation to average female body mass as well as the overall relationship between (**b**) ungulate density and (**d**) biomass and body mass

lower body size thresholds, beyond which smaller- or larger-sized species cannot obtain densities that are occasionally realized by medium-sized ungulates (Fig. 6.2).

Regional body mass–population density relationships (Fig. 6.3) occasionally support a negative, nonlinear but rather weak trend, particularly among browsers in temperate ecosystems of North America (Fig. 6.3a) and Europe (Fig. 6.3b), and grazers in South Asia (Fig. 6.3c). In African savannas, medium-sized grazers have the highest population densities (Fig. 6.3d)—a pattern found also in savannas of the Tarangire-Manyara ecosystem in northern Tanzania (Fig. 6.3f, g). However, in Miombo woodlands of East Africa, no body mass–population density relationships are apparent (Fig. 6.3e). In temperate zones (North America, Europe), browsers reach higher densities than grazers, whereas in tropical and subtropical regions, grazers tend to reach highest densities. This pattern is even more pronounced in body mass–biomass density relationships. In the northern hemisphere, browsers usually contribute more to overall ungulate biomass (Fig. 6.4a, b), whereas in tropical and subtropical regions, grazers tend to reach higher biomass densities (Fig. 6.4c, g). In South Asia, large-bodied, predominantly grazing gaurs (*Bos gaurus*: 800 kg), contribute substantially to overall ungulate biomass. In African savannas, grazers between ~200–400 kg of body mass contribute most to biomass densities. Although some wild ungulates can reach exceptionally high densities, livestock species frequently surpass densities of wild species in areas where wildlife and livestock coexist (Fig. 6.3h), and their contribution to overall herbivore biomass (Fig. 6.4h)

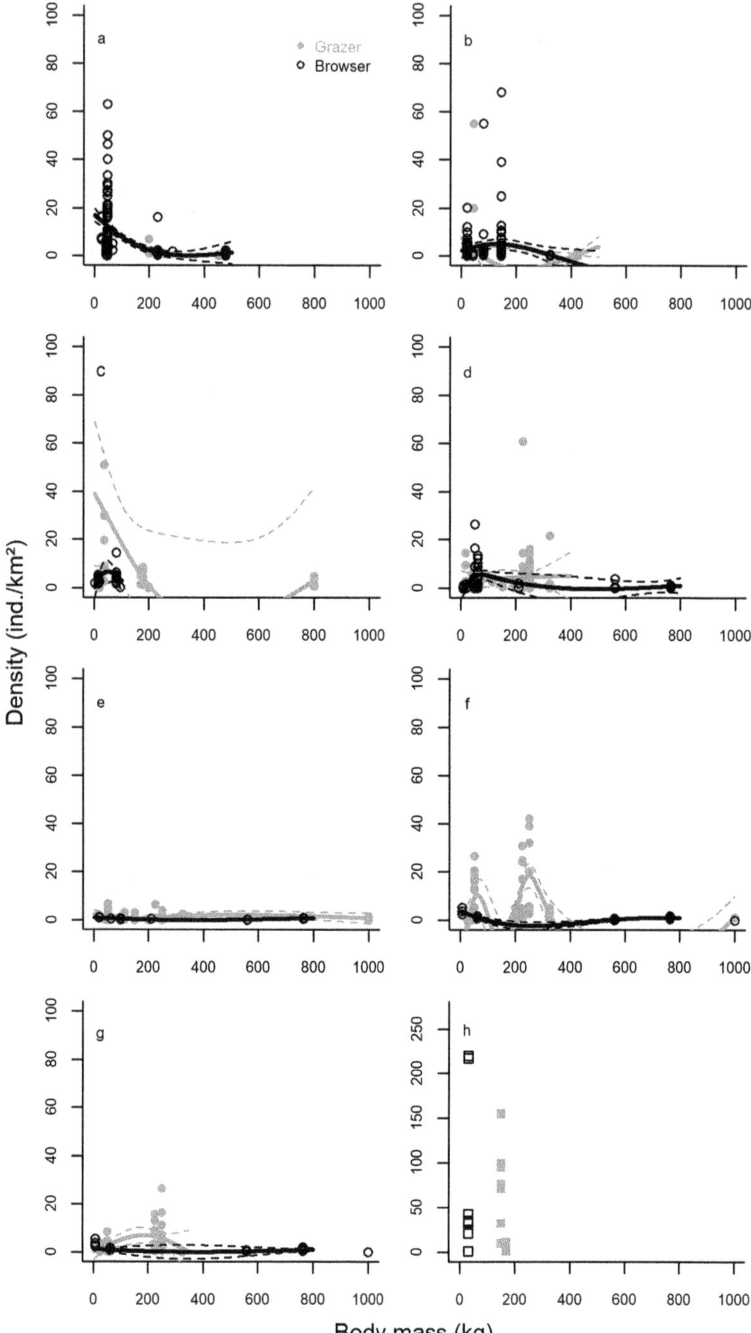

Fig. 6.3 Densities (ind.km^{-2}) of grazing and browsing wild ungulates in relation to the average female body mass (kg) in (**a**) Europe, (**b**) North America, (**c**) Southeast Asia, (**d**) and Africa. For the

usually exceeds those of wild ungulates (Prins 1992; Kiffner et al. 2016; Hempson et al. 2017; **Mishra et al.** Chap. 7).

These global-, continent-, and ecosystem-wide analyses indicate that the body mass–density relationships are not very strong, and suggest that medium-sized ungulates usually realize the highest densities. A possible explanation for this pattern could be found in the physiological constraints associated with body mass (see also **Codron et al.** Chap. 4). Small-sized species typically require high forage quality to sustain their proportionally high energy demands, and high-quality forage is usually rare in the environment. At the other body-size extreme, very large herbivores usually require proportionally greater amounts of forage (Müller et al. 2013). These considerations imply that small ungulates are mainly constrained by availability of high-quality forage; very large ungulates are mainly constrained by forage quantity in the environment; whereas medium-sized ungulates are less severely constrained by forage quality and quantity. These physiological considerations may partly explain observed patterns of highly abundant medium-sized ungulates, and low abundances of very small and very large ungulates.

Beyond among-species density differences, within-species variation in densities can be substantial. Indeed, within-species coefficients of variation (CV = standard deviation of density estimates / mean density) were clustered between 0 and 2 (Fig. 6.5). CVs in both grazers and bowsers were not significantly correlated with sample size (grazer: tau = 0.12, p = 0.37, n = 27; browser: tau = 0.30, p = 0.07, n = 20). In grazers, variability in density is generally negatively correlated with body mass, and above a body mass of ~100 kg tends to be lower than for browsers (Fig. 6.5a). In browsers, the body mass–coefficient of variation relationship is hump shaped with highest variability in densities among browsing species of 200–600 kg body mass. Among grazers, our dataset indicates particular high variability in densities of fallow deer (*Dama dama*), chital, southern reedbuck (*Redunca arundinum*), and African buffalo (*Syncerus caffer*). Browsers with highly variable densities were red deer, wild boar (*Sus scrofa*), bushbuck (*Tragelaphus scriptus*), and eland (*Taurotragus oryx*). Across all ungulates, we found variability in densities was negatively associated with body mass (Fig. 6.5b). In the following sections, we will outline how abiotic and biotic factors cause variation in demographic processes leading to variation in population growth.

Fig. 6.3 (continued) African continent, wild ungulate densities are presented separately for (**e**) Miombo ecosystems, (**f**) national parks and (**g**) community-based conservation areas in the Tarangire-Manyara ecosystem (Tanzania). (**h**) Depicts densities of livestock species (note the different y-axis scale) in community-based conservation projects of the Tarangire-Manyara ecosystem (Tanzania)

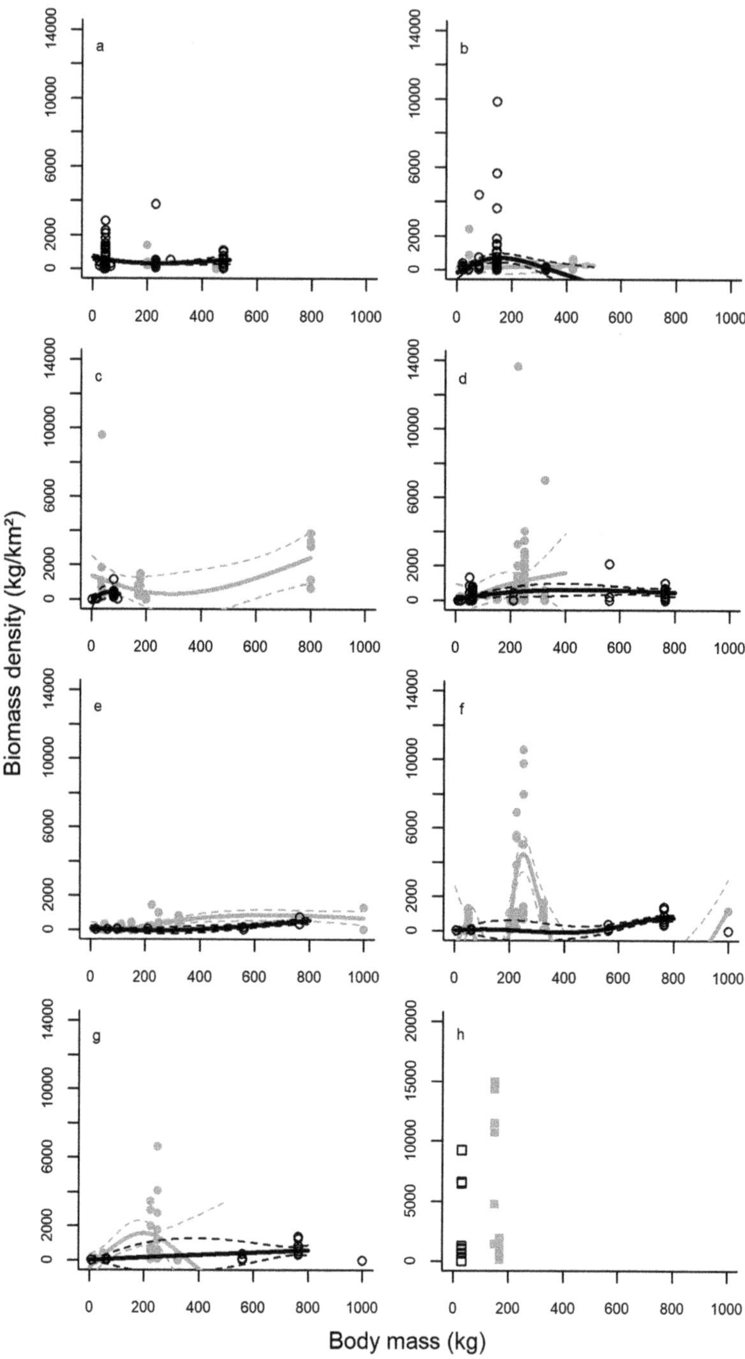

Fig. 6.4 Biomass densities (kg.km^{-2}) of grazing and browsing wild ungulates in relation to the average female body mass (kg) in (**a**) Europe, (**b**) North America, (**c**) Southeast Asia, (**d**) and

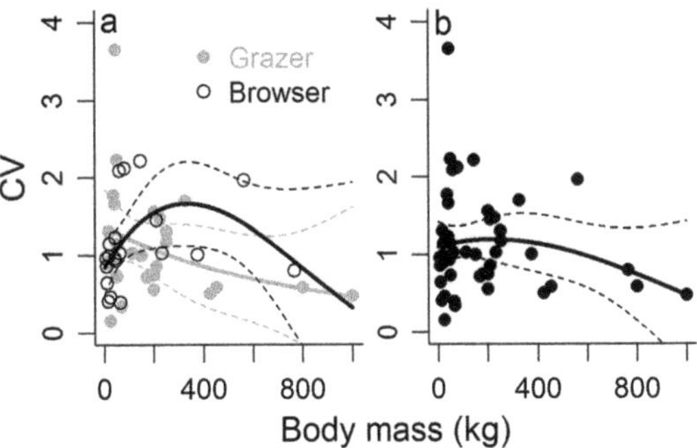

Fig. 6.5 Coefficient of variation (CV; 1 = 100%) of population densities (**a**) in grazing and browsing ungulates and (**b**) all grazers and browsers combined in relation to body mass. We only computed CVs for species with ≥3 density estimates

6.4 Demographic Patterns and Processes Underlying Population Dynamics

The numbers of individuals in animal populations fluctuate over time and across space (Lack 1966; Levins 1969; Sinclair 1977). Changes in population sizes are ultimately due to the demographic processes of births, deaths, and movements (immigration and emigration). Identifying which specific demographic rate (birth rate, juvenile survival, adult survival, age of first reproduction, immigration, or emigration) contributes most to changes in population growth (i.e., the "key demographic rate") can be considered the "holy grail" of population ecology (Gaillard et al. 1998, 2000; Morris and Doak 2002; Coulson et al. 2005), and is of great interest to studies of life history evolution and conservation biology. Our current synthesis of population ecology studies suggests that no specific demographic rate is central in governing all or most changes in growth rates, and that environmental variation in resource availability and predation directly and indirectly affect vital rates. In addition, indirect effects of perturbations can create cohort effects or alter age structures. This leads to transient population dynamics because different cohorts and age classes have different demographic rates. The relative contribution of direct vs. indirect effects can be dependent upon the life history strategy (slow versus fast) of the species (Gamelon et al. 2016). This suggests that all age groups contribute to changes in population growth (Gamelon et al. 2016). Furthermore, there is increasing evidence that ungulates can flexibly adjust reproductive tactics (and thus influence population growth) in response to environmental variation such as pulsed increases in food quantity and quality (Gamelon et al. 2017). This understanding

Fig. 6.4 (continued) Africa. For the African continent, wild ungulate biomass in (**e**) Miombo ecosystems, (**f**) national parks and (**g**) community-based conservation projects in the Tarangire-Manyara ecosystem (Tanzania) are presented as well. (**h**) Depicts biomass densities of livestock species (note the different y-axis scale) in community-based conservation projects of the Tarangire-Manyara ecosystem (Tanzania)

of ungulate population ecology has developed over time, which we will now briefly summarize.

Early syntheses of empirical studies of ungulate population dynamics examined temporal variation in demographic rates in relation to population growth rates (Gaillard et al. 1998, 2000). Their main findings were that adult female survival showed low inter-annual variation, fecundity of prime-aged females was moderately variable, and juvenile survival and young female fecundity showed the greatest inter-annual variation. Interestingly, although matrix population models indicated adult survival theoretically makes the greatest contribution to population growth, it had very low observed inter-annual variability, leaving little room for adult survival to have an appreciable effect on population growth. Conversely, juvenile survival theoretically had a low contribution to population growth rate, but exhibited large temporal variation that was primarily responsible for observed changes in population size, and thus was identified as the key demographic rate (Wisdom et al. 2000; Lehman et al. 2018).

Albon et al. (2000) sought the key demographic rate for red deer on Isle of Rum, Scotland and made a slightly more nuanced conclusion, finding that birth rate was the dominant component of relative population growth rate when the population was growing rapidly, but during a period when population size fluctuated near carrying capacity, variation in adult female survival (along with covariation of adult survival and calf survival) contributed most to relative variation in population growth rate. Clutton-Brock and Coulson (2002) also found that variation in the survival of mature animals contributed more to changes in population size than juvenile survival. Subsequent work (Coulson et al. 2005) indicated that the most influential demographic rates varied among populations of red deer and bighorn sheep (*Ovis canadensis*) depending on whether the population was growing or fluctuating near carrying capacity, and according to site-specific differences in ecological processes such as disease, predation, and density dependence. Recent studies have found that in declining populations, variation in adult survival, due to natural or anthropogenic predation, can be the most important factor affecting variation in population growth rates (Johnson et al. 2010; Lee et al. 2016a). Importantly, covariation among demographic rates within a population is a critical feature that should be considered when seeking the demographic causes of variation in population growth rate (Coulson et al. 2005).

Environmental variation among years such as temperature- or precipitation-dependent timing of plant phenology relative to timing of birth can affect all the newborns in an area similarly, creating cohort effects (Clutton-Brock and Coulson 2002). In years when food is scarce for all pregnant females in an area, offspring birth weights can be low and bodily growth of juveniles can be slower, and this can lead to lower demographic rates throughout the lives of all individuals born in a "bad" year cohort (Post and Stenseth 1999). Indeed, up to 50% of variation in individual performance within a population can be explained by early life environment in ungulates (Hamel et al. 2009).

Stochastic variation in population age structure (the distribution of different-aged animals in a population) is important because different ages have different demographic rates. Environmental variation can alter the age-structure distribution which causes transient population dynamics that are mediated by life history (Owen-Smith and Mason 2005; Haridas et al. 2009; Coulson et al. 2010). Fast-paced species (with a short generation time) usually increase population growth rates after disturbance, whereas slow-paced species (with a long generation time) frequently decrease growth rates after disturbance (Gamelon et al. 2014).

The demographic mechanisms underlying observed population dynamics are clearly complex, and suggest strong context dependencies (Clutton-Brock and Coulson 2002). Conservation and species recovery programs are most effective when system-specific contributions of demographic rates to population growth rates are known. In identifying demographic rates driving the dynamics of populations, analyses should incorporate transient dynamics, and actual variation in demographic rates. This requires data on demographic rate means, variances and covariances, and population sizes divided into age or stage distributions (Johnson et al. 2010). Integrated population models (Kery and Schaub 2012) and transient life table response experiments (Koons et al. 2016, 2017) that incorporate environmental stochasticity (Tuljapurkar 1982), correlations among demographic rates (Coulson et al. 2005), and non-stationarity (Jenouvrier et al. 2014) are useful tools for analysing demographic mechanisms underlying population fluctuations (Maldonado-Chaparro et al. 2018).

Metapopulation analyses are rarely conducted for large herbivores (Lee and Bolger 2017), but the theory of metapopulation dynamics that has arisen from studies of other species should be tested for applicability to grazers and browsers. Particularly important from a conservation perspective in increasingly anthropogenically fragmented habitats, is the idea that metapopulations can buffer subpopulation oscillations and reduce subpopulation extinction probabilities (Goodman 1987; Gilpin and Hanski 1991; Hess 1996).

6.5 Global and Local Causal Factors Underlying Ungulate Population Dynamics

In recent years, ecologists have moved away from mono-causal hypotheses explaining animal population dynamics, and developed more complex models, which propose that primary production, predation, droughts, fire, and land conversion (and possible other factors) all interact synergistically in their regulation of herbivore populations to create indirect-, additive-, reciprocal-, and interaction-modifying relationships (Hopcraft et al. 2010). There is increasing quantitative evidence that abiotic factors determine the relative importance of predation, forage quantity, and forage quality in regulating herbivores of different body sizes, and this alters the relative strength of the connections between biotic and abiotic components in ecosystems. Species with smaller body masses are often subject to greater levels of top-down control (mainly owing to their susceptibility to a more diverse set of predators), whereas body mass thresholds for escaping predation regulation, appear context dependent (Hopcraft et al. 2010). In sum, larger-sized species are mainly limited by food supply whereas the effect of predation may be most influential in relatively small species (Hopcraft et al. 2010), and in less productive environments (Melis et al. 2009).

At the core, theory of ungulate population dynamics needs to explicitly incorporate temporal and spatial aspects of environmental variation (Boyce et al. 2006; Hempson et al. 2015). Forage availability determines individual body condition and, therefore, survival and reproduction in ungulates (Parker et al. 2009), so resource availability ("bottom up regulation") is the ultimate causal factor determining population size and trajectory (Sinclair and Krebs 2002; Sinclair 2003; Pettorelli et al. 2009). Indeed, the observed variation in population densities of herbivores is primarily driven by primary production, which itself is mainly governed by soil fertility and precipitation. Thus, primary productivity is the main determinant of maximum density for a population (Coe et al. 1976; East 1984; Fritz and Duncan 1994; Pettorelli et al. 2009).

Internal feedbacks of population density (i.e., density dependence) may affect population growth of large ungulates to some degree as well (Bonenfant et al. 2009). High population density, that approaches or exceeds local carrying capacity, generally results in body mass decreases, increases in age of first breeding, and decreases in all aspects of reproduction from ovule production to weaning success, thereby generally reducing recruitment (Bonenfant et al. 2009). Survival during the first year is the demographic rate most frequently reported to be density dependent, and it also shows the greatest variation with density, but prime-aged adult survival and costs of reproduction are also density dependent (Bonenfant et al. 2009). Dispersal may also be affected by density (Matthysen 2005). There is substantial evidence that the relative importance of density dependence for regulating large herbivore populations is itself dependent on spatiotemporal variation in resources and predation. Temporal environmental variability has been associated with density dependence caused by forage deficits (Wang et al. 2006), and—among ungulates in the northern hemisphere—predation and spatial resource heterogeneity may weaken the density-dependent effects (Wang et al. 2009). At least some larger ungulate species may flexibly adjust their reproductive allocation in response to resource availability, such as pulsed resource availability caused by mast seeding (Gamelon et al. 2017). These examples show that spatiotemporal variation in resource availability and predation interact with density dependence and reinforce the notion that resource availability is the ultimate factor affecting population growth and density of ungulates.

The concept of the key resource for ungulate populations is defined as that resource which determines the demographic rate that exerts the most influence on the population trajectory (Illius and O'Connor 1999). Thus identifying the key resource for a population is useful for determining the specific pathway to population regulation. Resource availability often varies over time and annual variation in primary productivity is largely determined by atmospheric oscillations (El Niño Southern and North Atlantic Oscillations) affecting precipitation and temperature patterns, with direct impacts on vegetation phenology and primary productivity and, therefore, populations of ungulates (Post and Stenseth 1999; Ogutu et al. 2008; Hagen et al. 2017). Theoretically, spatial heterogeneity can buffer populations against temporal variability by allowing herbivores to access forage resources in the most nutritious state (Wang et al. 2006; Hobbs and Gordon 2010), via dispersal before forage is depleted (Owen-Smith 2004). However, localized weather extremes (e.g., reduced dry season rainfall), and increasingly restricted animal movement caused by fencing or other forms of habitat fragmentation, may amplify negative effects of large-scale climatic variation on ungulate populations (Ogutu and Owen-Smith 2003).

Stochastic disturbances, such as natural- and human-induced variation in climate extremes (droughts, cold, flooding), can directly (via increased mortality) and indirectly (via changes in available food resources) affect ungulate population growth rates. Sudden shifts in local primary productivity, due to perturbations such as fire, flood, or land conversion, can rearrange the dynamics of an ecosystem briefly or semi-permanently into a new state (van de Koppel et al. 2002). In temperate latitudes, harsh winters can strongly affect mortality rates, particularly among younger age classes, through a combination of greater thermoregulatory costs and decreased forage availability because of deep snow (Post and Stenseth 1999; Jacobson et al. 2004). In tropical or subtropical systems, droughts can directly affect mortality rates (Owen-Smith 1990), with more sedentary, grazing, and mixed-feeding species at highest risk from increasing drought intensity (Duncan et al. 2012). Human-caused increases in the atmospheric CO_2 concentration can act as fertilizer for plants in general, but woody vegetation appears to benefit most from CO_2 enrichment (Bond and Midgley 2000). Changes in herbivory, precipitation, and fire frequency may also affect woody versus herb–grass plant community composition locally (Morrison et al. 2016a), but CO_2 is considered to be the key underlying causal factor of shrub encroachment in savanna ecosystems (Devine et al. 2017).

Predation and diseases ("top down regulation") can reduce populations below their resource-determined potential carrying capacity. A famous example of top-down regulation is the six-fold increase in the Serengeti wildebeest population (Fig. 6.6a) after the population was released from the rinderpest virus (Sinclair 1979; Holdo et al. 2009). A more pessimistic example is the trajectory of the black rhinoceros (*Diceros bicornis*) population in Lake Manyara National Park, Tanzania (Fig. 6.6b) that was extirpated within few years due to poaching (Kiffner et al. 2017). Albeit classic top-down theory involves direct (mortality) effects, there is growing evidence that the mere presence of predators can affect demography and reproduction of ungulate species through behavioural and physiological effects of fear (Creel et al. 2007; LaManna and Martin 2016), but non-lethal effects of predation have yet to be fully integrated into models of population regulation (Peers et al. 2018). Although population growth is usually negatively correlated with population density, population growth rate and density can be positively associated at low abundances (Courchamp et al. 1999). This phenomenon (often named *Allee effect*) can be caused by predation or reduced reproduction, and can lead to increased local extinction risk of ungulates that occur at low densities (Wittmer et al. 2005; Bourbeau-Lemieux et al. 2011).

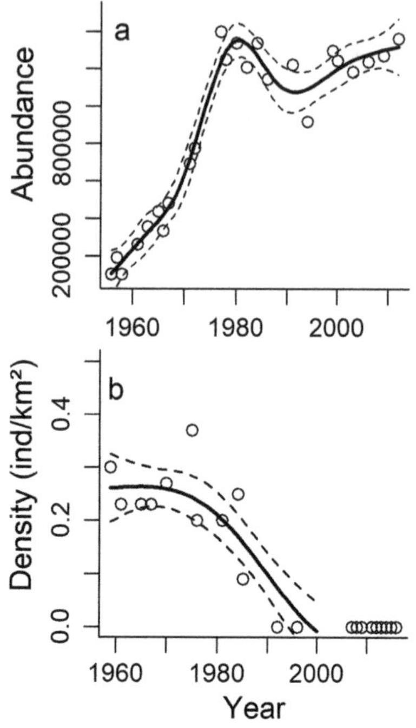

Fig. 6.6 Time series of wildebeest (*Connochaetus taurinus*) population sizes in the Serengeti ecosystem (data from Hopcraft et al. 2015) and densities of black rhinoceros (*Diceros bicornis*) in Lake Manyara National Park (data from Kiffner et al. 2017)

In most ecosystems, multiple ungulate species co-exist, which may cause competition over commonly used food resources. Indeed, correlative studies suggest that grazing ungulates in East Africa can be limited by competition with buffalo (de Boer and Prins 1990; Kiffner et al. 2017). In Europe, high red deer densities have a negative effect on body masses of roe deer fawns (Richard et al. 2010), and time series of herbivore assemblages suggest that interspecific competition affects ungulate population dynamics in temperate forests (Jędrzejewska et al. 1997). Facilitation within herbivore assemblages has been documented in tropical and subtropical ungulate communities as an important process governing coexistence (Olff et al. 2002). While competition over resources usually occurs during times of resource scarcity (when vegetation is dormant), facilitation mainly occurs during the growing season when species such as zebras (*Equus quagga*) stimulate grass growth (Sinclair and Norton-Griffiths 1982; Arsenault and Owen-Smith 2002; Wegge et al. 2006). The relative importance of competition versus facilitation is particularly relevant in areas where livestock species coexist with wildlife (Spear and Chown 2009). Exclusion experiments in Kenya's Laikipia landscape suggest that facilitation mainly occurs during the growing (wet) season, whereas wildlife and livestock compete for grasses during the dry season (Odadi et al. 2011a, b). Indirect effects, such as apparent competition and apparent mutualism among species, mediated by a shared predator, are also possible (Estes et al. 2013), but rarely quantified (Chaneton

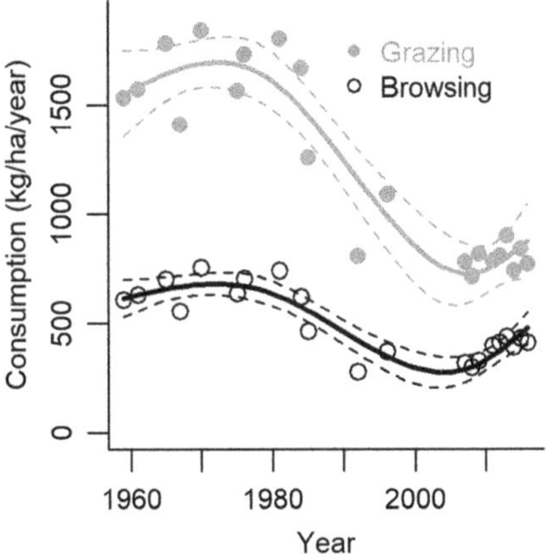

Fig. 6.7 Time series of estimated grass and browse consumption (kg·ha^{-1}·year^{-1}) as a function of population fluctuations of thirteen herbivore species and estimated daily grass and browse intake in Lake Manyara National Park, Tanzania (data from Kiffner et al. 2017)

and Bonsall 2000; Lee et al. 2016b). Beyond competition and facilitation, evidence from temperate and tropical biomes indicates that wild and domestic ungulates can have substantial cascading effects on plant regeneration, structure, and functioning (Goheen et al. 2018; Ramirez et al. 2018). In turn, this can shape the relative contribution of grazing and browsing (Fig. 6.7), as well as ecosystem structure and functioning (Dirzo et al. 2014; Hempson et al. 2017).

Advanced models to adequately depict ungulate population dynamics, therefore, need to (1) include biotic interactions (which could be additive, reciprocal, indirect, and interaction modifying) between resource availability, competition and facilitation, diseases, and predation; (2) incorporate spatiotemporal variation in abiotic factors, which determine resource availability, the relative strength of competition and facilitation, predation, and diseases in regulating herbivores of different body sizes; and (3) explicitly address possible feedback loops between abiotic factors, biotic interactions, and ungulate populations.

6.6 Predicted Effects of Anthropogenic Perturbations

Developing such models will be particularly important to assess the viability of ungulate populations in increasingly human-dominated landscapes. The most influential anthropogenic perturbations that affect ungulate populations are likely to be (1) land use change; (2) climate change; (3) invasive species (e.g., livestock); (4) increase in atmospheric CO_2; and (5) direct, unsustainable exploitation (Sala et al. 2000; Ripple et al. 2015). Some of these upheavals act on global scales (CO_2 enrichment, climate change), whereas others occur more localized (land-use change, invasive species, and direct exploitation). Considering these multiple upheavals, the large diversity of ungulates, strong context dependence, and lack of long time series for most ungulate populations, we used qualitative threat assessment methodology

(Burgman et al. 1993) to provide a general indication of the level of threat for grazers and browsers in temperate and tropical and sub-tropical biomes (Table 6.1). Given the difficulties in predicting indirect effects, we mainly focused on likely direct effects on ungulate populations.

We excluded climate change from the table because direct and indirect climate effects on distribution ranges and physiology of ungulates are likely to be case specific (Bellard et al. 2012; see also **Boone** Chap. 8). For example, in temperate zones and the arctic, milder winters are projected to reduce winter mortality (Loison et al. 1999; Post and Stenseth 1999), for both grazers and browsers. However, altered thawing and refreezing of surface snow in the arctic may substantially affect ungulate movement and possibly mortality (Bartsch et al. 2010). In general, climate change may particularly affect grazers, since variation in climatic conditions will lead to variable grass growth (Ogutu et al. 2008).

Despite the coarse nature of our assessment, Table 6.1 provides a narrative that suggests grazing ungulates are likely to be more negatively affected by human activities compared to browsers—a conclusion that is in line with the prediction that most biodiversity changes will occur in grassland biomes (Sala et al. 2000; Smit and Prins 2015; **Mishra et al.** Chap. 7). Moreover, current mainstream conservation efforts, such as REDD+, focus mainly on woodland conservation or afforestation (Collins et al. 2011), and globally elevated CO_2 concentrations favour woody vegetation to a greater extent than grasses (Devine et al. 2017). Yet, several species which we broadly classified as "grazers" are indeed mixed feeders, and may thus cope relatively well if grasslands transform to woodlands or shrublands as exemplified by sustained and even increasing densities of impalas in changing environments (Kiffner et al. 2016, 2017). However, obligate grazers typically require unrestricted access to large areas of grasslands (Fryxell et al. 2005)—a scenario that is scarce in a world of sustained human population growth (Gerland et al. 2014)—and may thus be particularly impacted by structural landscape alterations such as shrub encroachment, and agricultural and settlement expansions.

Environmental and anthropogenic perturbations rarely act independently from each other on ungulate populations (Dirzo et al. 2014). Indeed, negative effects of single perturbations may be amplified by changes in additional environmental conditions. For example, die-offs of saiga antelope due to bacterial infections were likely facilitated by temperature and humidity anomalies (Kock et al. 2018). Similarly, ungulates in continuous landscapes may be able to cope with seasonal or climate-induced shifts in plant phenology (Cleland et al. 2007), but populations in fragmented landscapes may be substantially affected (Jackson and Sax 2010; Morrison et al. 2016b). Importantly, the human-caused loss of large herbivorous mammals is not only a symptom of the Anthropocene but is now a major causal factor of ecological change (Dirzo et al. 2014). For instance, replacing large, wild ungulates with livestock can reduce fire frequencies, which usually increases woody cover (Hempson et al. 2017). Similarly, loss of mega-herbivores may release woody vegetation from strong herbivore pressure, which may, in return, have cascading effects on vegetation structure, other animal taxa, ecosystem functioning, and ecosystem services (Dirzo et al. 2014; cf. **Sabo** Chap. 11; **Katona and Coetsee** Chap. 12).

Table 6.1 Hypothesized effects of specific anthropogenic perturbations (**+** positive effect on populations; **−** negative effect on populations; **0** no strong effect) on population growth of browsing and grazing ungulate populations in temperate (-Temp) and tropical and subtropical (-Trop) regions

	Land use change	Invasive (livestock) species	CO_2	Direct exploitation
Browser-Temp	+: Rewilding, i.e., abandonment of agricultural areas increases habitat area (Perreira and Navaro 2015) **0**: Relative stable forest cover in many parts (Hansen et al. 2013) −: Loss of boreal forests (Hansen et al. 2013)	**0**: Unlikely to be of high relevance due to little overlap in food resources and husbandry practices in most temperate regions	+: Dicots likely to show increased growth due to CO_2 enrichment in the atmosphere (Norby et al. 2005)	**0**: Predominately sustainably managed hunting in North America and Europe (Milner et al. 2006) −: Several ungulate species threatened by harvesting in North Asia (Milner et al. 2006)
Browser-Trop	−: Deforestation highest in tropics and subtropics (Hansen et al. 2013) **0**: REDD+ projects may partially buffer against deforestation (Collins et al. 2011)	**0**: Unlikely to be of high relevance due to little overlap in food resources	+: Dicots likely to show increased growth due to CO_2 enrichment in the atmosphere (Norby et al. 2005)	−: Large species most vulnerable, mainly due to bushmeat hunting (Wilkie et al. 2011; Ripple et al. 2017)
Grazer-Temp	+: Targeted conservation and reintroduction projects for mega grazers in place (e.g., Kuemmerle et al. 2011) −: Range contraction of grasslands in North Asia (Olson et al. 2011; Buuveibaatar et al. 2016)	**0**: Unlikely to be of high relevance due to husbandry practices in most temperate regions	−: Potential for shrub encroachment due to CO_2 enrichment in the atmosphere (Devine et al. 2017)	**0**: Predominately sustainably managed hunting in North America and Europe and several large grazers strictly protected (Milner et al. 2006) −: Several ungulate species threatened by harvesting in North Asia (Milner et al. 2006; Olson et al. 2014)
Grazer-Trop	−: Conversion to agricultural lands highest in grassland biomes (Sala et al. 2000) −: Fragmentation of migratory ranges (Bolger et al. 2008) **0**: Annual range of only few populations entirely protected (Hopcraft et al. 2015)	−: Possibility for apparent competition with grazing livestock, particularly in areas with high stocking rates of cattle during the dry season (Odadi et al. 2011a, b)	−: Evidence for shrub encroachment due to CO_2 enrichment in the atmosphere (Devine et al. 2017)	−: Large species most vulnerable, mainly due to bushmeat hunting (Lindsey et al. 2013; Ripple et al. 2017)

6.7 Conclusions

The status of many large herbivores is a cause for concern (Ripple et al. 2015), and our qualitative analysis indicates that grazing ungulate species may be particularly threatened due to multiple anthropogenic perturbations hypothesized to negatively affect their populations (Prins and Gordon 2008; Gordon and Prins 2008). On a more optimistic note, there is ample evidence that large herbivores can thrive outside fully protected areas (e.g., Kiffner et al. 2016; Lee 2018), and that integrating livestock with wildlife can be beneficial for the environment and human well-being (Gordon 2018; Keesing et al. 2018). As a case in point, roe deer and wild boar populations in central Europe, and deer (*Odocoileus* spp.) in North America seem extraordinarily resilient and thrive in human-dominated landscapes to the point that they are considered "overabundant" (Côté et al. 2004; Burbaitė and Csányi 2009; Massei et al. 2015). Although there are some generalities how animals adjust their behaviour in human-dominated landscapes, such as shifting activity to nighttimes (Gaynor et al. 2018) and reducing movement (Tucker et al. 2018), there is a lack of quantitative, integrated, and systematic analyses that investigate how ungulates respond to anthropogenic change with respect to phenology, space use, and physiology (Bolger et al. 2008; Bellard et al. 2012), and how these responses affect population growth. A first (but rarely implemented) step in this direction would be systematic and ecosystem-wide population monitoring to describe the often substantial spatial and temporal variation of ungulate densities. Ideally, such monitoring would be coupled with large-scale metapopulation studies that allow estimation of site-specific demographic rates, to link variation in population growth rates with demographic processes and environmental and anthropogenic perturbations. Such process-oriented understanding is needed to guide effective conservation measures (including restoration attempts) of ungulates and ecosystems (Sinclair et al. 2018).

Acknowledgments We thank Herbert Prins and Iain Gordon for the invitation to write this Chapter and for very constructive feedback on this Chapter. We thank Monica Bond for constructive discussions and comments.

References

Albon SD, Coulson TN, Brown D, Guinness FE, Pemberton JM, Clutton-Brock TH (2000) Temporal changes in key factors and key age groups influencing the population dynamics of female red deer. J Anim Ecol 69(6):1099–1110

Arsenault R, Owen-Smith N (2002) Facilitation versus competition in grazing herbivore assemblages. Oikos 97(3):313–318

Bartsch A, Kumpula T, Forbes BC, Stammler F (2010) Detection of snow surface thawing and refreezing in the Eurasian Arctic with QuikSCAT: implications for reindeer herding. Ecol Appl 20(8):2346–2358

Bellard C, Bertelsmeier C, Leadley P, Thuiller W, Courchamp F (2012) Impacts of climate change on the future of biodiversity. Ecol Lett 15(4):365–377

Blackburn TM, Gaston KJ (1999) The relationship between animal abundance and body size: a review of the mechanisms. Adv Ecol Res 28:181–210
Blackburn TM, Brown VK, Doube BM, Greenwood JD, Lawton JH, Stork N (1994) The relationship between abundance and body size in natural animal assemblages. J Anim Ecol 62:519–528
Bolger DT, Newmark WD, Morrison TA, Doak DF (2008) The need for integrative approaches to understand and conserve migratory ungulates. Ecol Lett 11(1):63–77
Bond WJ, Midgley GF (2000) A proposed CO_2-controlled mechanism of woody plant invasion in grasslands and savannas. Glob Change Biol 6:865–869
Bond WJ, Woodward FI, Midgley GF (2004) The global distribution of ecosystems in a world without fire. New Phytol 165:525–538
Bonenfant C et al (2009) Emprirical evidence of density-dependence in populations of large herbivores. Adv Ecol Res 41:313–357
Bourbeau-Lemieux A, Festa-Bianchet M, Gaillard J-M, Pelletier F (2011) Predator-driven component Allee effects in a wild ungulate. Ecol Lett 14:358–363
Boyce MS, Haridas CV, Lee CT, The NCEAS Stochastic Demography Working Group (2006) Demography in an increasingly variable world. Trends Ecol Evol 21:141–148
Burbaitė L, Csányi S (2009) Roe deer population and harvest changes in Europe. Est J Ecol 58 (3):169–180
Burgman M, Ferson S, Akçakaya HR (1993) Risk assessment in conservation biology. Chapman and Hall, London
Buuveibaatar B et al (2016) Human activities negatively impact distribution of ungulates in the Mongolian Gobi. Biol Conserv 203:168–175
Caughley G (1994) Directions in conservation biology. J Anim Ecol 63:215–244
Chaneton EJ, Bonsall M (2000) Enemy-mediated apparent competition: empirical patterns and the evidence. Oikos 88:380–394
Cleland EE, Chuine I, Menzel A, Mooney HA, Schwartz MD (2007) Shifting plant phenology in response to global change. Trends Ecol Evol 22(7):357–365
Clutton-Brock TH, Coulson T (2002) Comparative ungulate dynamics: the devil is in the detail. Philos Trans R Soc B 35:1285–1298
Coe MJ, Cumming DH, Phillipson J (1976) Biomass and production of large African herbivores in relation to rainfall and primary production. Oecologia 22(4):341–354
Collins MM, Milner-Gulland EJJ, Macdonald EAA, Macdonald DWW (2011) Pleiotropy and charisma determine winners and losers in the REDD+ game: all biodiversity is not equal. Trop Conserv Sci 4(3):261–266
Côté SD, Rooney TP, Tremblay J-P, Dussault C, Waller DM (2004) Ecological impacts of deer overabundance. Annu Rev Ecol Evol Syst 35:113–147
Coulson T, Gaillard J-M, Festa-Bianchet M (2005) Decomposing the variation in population growth into contributions from multiple demographic rates. J Anim Ecol 74:789–801
Coulson T, Tuljapurkar S, Childs DZ (2010) Using evolutionary demography to link life-history, quantitative genetics and population ecology. J Anim Ecol 79(6):1226–1240
Courchamp F, Clutton-Brock T, Grenfell B (1999) Inverse density dependence and the Allee effect. Trends Ecol Evol 14(10):406–410
Creel S, Christianson D, Liley S, Winnie JA Jr (2007) Predation risk affects reproductive physiology and demography of elk. Science 315:960
Damuth J (1981) Population-density and body size in mammals. Nature 290:699–700
de Boer WF, Prins HHT (1990) Large herbivores that strive mightily but eat and drink as friends. Oecologia 82:264–274
de Waal C et al (2011) Scale of nutrient patchiness mediates resource partitioning between trees and grasses in a semi-arid savanna. J Ecol 99:1124–1133
Devine AP, McDonald RA, Quaife T, Maclean IMD (2017) Determinants of woody encroachment and cover in African savannas. Oecologia 183:939–951
Dirzo R, Young HS, Galetti M, Ceballos G, Isaac NJB, Collen B (2014) Defaunation in the Anthropocene. Science 345(6195):401–406

Duncan C, Chauvenet ALM, McRae LM, Pettorelli N (2012) Predicting the future impact of droughts on ungulate populations in arid and semi-arid environments. PLoS One 7(12):e51490
East R (1984) Rainfall, soil nutrient status and biomass of large African savanna mammals. Afr J Ecol 22(4):245–270
Estes JA, Brashares JS, Power ME (2013) Predicting and detecting reciprocity between indirect ecological interactions and evolution. Am Nat 181:S76–S99
Fritz H, Duncan P (1994) On the carrying capacity for large ungulates of African savanna ecosystems. Proc R Soc B 256:77–82
Fryxell JM, Wilmshurst JF, Sinclair ARE, Haydon DT, Holt RD, Abrams PA (2005) Landscape scale, heterogeneity, and the viability of Serengeti grazers. Ecol Lett 8(3):328–335
Gagnon M, Chew AE (2000) Dietary preferences in extant African Bovidae. J Mammal 81:490–511
Gaillard J-M, Festa-Bianchet M, Yoccoz NG (1998) Population dynamics of large herbivores: variable recruitment with constant adult survival. Trends Ecol Evol 13(2):58–63
Gaillard J-M, Festa-Bianchet M, Yoccoz NG, Loison A, Toïgo C (2000) Temporal variation in fitness components and population dynamics of large herbivores. Annu Rev Ecol Syst 31:367–393
Gamelon M, Gimenez O, Baubet E, Coulson T, Tuljapurkar S, Gaillard J-M (2014) Influence of life-history tactics on transient dynamics: a comparative analysis across mammalian populations. Am Nat 184:673–683
Gamelon M, Gaillard J-M, Gimenez O, Coulson T, Tuljapurkar S, Baubet E (2016) Linking demographic responses and life-history from longitudinal data in mammals. Oikos 125 (3):395–404
Gamelon M, Foccardi S, Baubet E, Brandt S, Franzetti B, Rochni F, Venner S, Sæther B-E, Gaillard J-M (2017) Reproductive allocation in pulsed resource environments: a comparative study in two populations of wild boar. Oecologia 183(4):1065–1076
Gaynor KM, Hojnowski CE, Carter NH, Brashares JS (2018) The influence of human disturbance on wildlife nocturnality. Science 360(6394):1232–1235
Gerland P et al (2014) World population stabilization unlikely this century. Science 346:234–237
Gilpin ME, Hanski I (1991) Metapopulation dynamics: empirical and theoretical investigations. Linnaean Society of London and Academic Press, London
Goheen JR et al (2018) Conservation lessons from large-mammal manipulations in East African savannas: the KLEE, UHURU, and GLADE experiments. Ann N Y Acad Sci 1429(1):31–49
Goodman D (1987) The demography of chance extinction. In: Soulé ME (ed) Viable populations for conservation. Cambridge University Press, Cambridge, pp 11–34
Gordon IJ (2018) Review: Livestock production increasingly influences wildlife across the globe. Animal 12(S2):s372–s382
Gordon IJ, Prins HHT (2008) Grazers and browsers in a changing world: conclusions. In: Gordon IJ, Prins HHT (eds) The ecology of browsing and grazing. Ecological studies, vol 195. Springer, Berlin, pp 309–321
Hagen R, Heurich M, Storch I, Hanewinkel M, Kramer-Schadt S (2017) Linking annual variations of roe deer bag records to large-scale winter conditions: spatio-temporal development in Europe between 1961 and 2013. Eur J Wildl Res 63:97
Hamel S, Gaillard J, Festa-Bianchet M, Côté S (2009) Individual quality, early-life conditions, and reproductive success in contrasted populations of large herbivores. Ecology 90:1981–1995
Hansen MC et al (2013) High resolution global maps of 21st-century forest cover change. Science 342(6160):850–853
Haridas CV, Tuljapurkar S, Coulson T (2009) Estimating stochastic elasticities directly from longitudinal data. Ecol Lett 12(8):806–812
Hempson GP, Illius AW, Hendricks HH, Bond WJ, Vetter S (2015) Herbivore population regulation and resource heterogeneity in a stochastic environment. Ecology 96(8):2170–2180
Hempson GP, Archibald S, Bond WJ (2017) The consequences of replacing wildlife with livestock in Africa. Sci Rep 7:17196

Hess GR (1996) Linking extinction to connectivity and habitat destruction in metapopulation models. Am Nat 148:226–236
Hobbs NT, Gordon IJ (2010) How does landscape heterogeneity shape dynamics of large herbivore populations? In: Owen-Smith N (ed) Dynamics of large herbivore populations in changing environments. Wiley-Blackwell, Hoboken, NJ, pp 141–164
Hofmann RR, Stewart DRM (1972) Grazer or browser: a classification based on the stomach-structure and feeding habits of East African ruminants. Mammalia 36(2):226–240
Holdo R et al (2009) A disease-mediated trophic cascade in the Serengeti and its implications for ecosystem C. PLoS Biol 7(9):e1000210
Hopcraft JGC, Olff H, Sinclair ARE (2010) Herbivores, resources and risks: alternating regulation along primary environmental gradients in savannas. Trends Ecol Evol 25(2):119–128
Hopcraft JGC, Holdo RM, Mwangomo E, Mduma SAR, Thirgood SJ, Borner M, Fryxell JM, Olff H, Sinclair ARE (2015) Why are wildebeest the most abundant herbivore in the Serengeti ecosystem? In: Sinclair ARE, Metzger KL, Mduma SAR, Fryxell JM (eds) Serengeti IV: sustaining biodiversity in a coupled human-natural system. The University of Chicago Press, Chicago, pp 125–174
Illius AW, O'Connor TG (1999) On the relevance of nonequilibrium concepts to arid and semiarid grazing systems. Ecol Appl 9:798–813
Jackson T, Sax DF (2010) Balancing biodiversity in a changing environment: extinction debt, immigration credit and species turnover. Trends Ecol Evol 25(3):153–160
Jacobson AR, Provenzale A, Hardenberg A, Bassano B, Festa-Bianchet M (2004) Climate forcing and density dependence in a mountain ungulate population. Ecology 85(6):1598–1610
Jędrzejewska B, Jędrzejewski W, Bunevich AN, Miikowski L, Krasiński ZA (1997) Factors shaping population densities and increase rates of ungulates in Bialowieża primeval forest (Poland and Belarus) in the 19th and 20th centuries. Acta Theriol 42(4):399–451
Jenouvrier S, Holland M, Stroeve J, Serreze M, Barbraud C, Weimerskirch H, Caswell H (2014) Projected continent wide declines of the emperor penguin under climate change. Nat Clim Chang 4:715–718
Johnson DH (1980) The comparison of usage and availability measurements for evaluating resource preference. Ecology 61:65–71
Johnson HE, Mills LS, Stephenson TR, Wehausen JD (2010) Population-specific vital rates contributions influence management of an endangered ungulate. Ecol Appl 20(6):1753–1765
Keesing F et al (2018) Consequences of integrating livestock and wildlife in an African savanna. Nat Sust 1:566–573
Kery M, Schaub M (2012) Bayesian population analysis using WinBUGS: a hierarchical perspective. Academic, Boston, MA
Kiffner C, Nagar S, Kollmar C, Kioko J (2016) Wildlife species richness and densities in wildlife corridors of Northern Tanzania. J Nat Conserv 34:82–92
Kiffner C, Rheault H, Miller E, Scheetz T, Enriquez V, Swafford R, Kioko J, Prins HHT (2017) Long-term population dynamics in a multi-species assemblage of large herbivores in East Africa. Ecosphere 8(12):e02027
Kock RA et al (2018) Saigas on the brink: multidisciplinary analysis of the factors influencing mass mortality events. Sci Adv 4:eaao2314
Koons DN, Iles DT, Schaub M, Caswell H (2016) A life history perspective on the demographic drivers of structured population dynamics in changing environments. Ecol Lett 19:1023–1031
Koons DN, Arnold TW, Schaub M (2017) Understanding the demographic drivers of realized population growth rates. Ecol Appl 27(7):2102–2115
Kuemmerle T et al (2011) Cost-effectiveness of strategies to establish a European bison metapopulation in the Carpathians. J Appl Ecol 48(2):317–329
Lack D (1966) Population studies of birds. Oxford University Press, Oxford
LaManna JA, Martin TE (2016) Costs of fear: behavioral and life-history responses to risk and their demographic consequences vary across species. Ecol Lett 19:403–413

Langevelde F et al (2003) Effects of fire and herbivory on the stability of savanna ecosystems. Ecology 84(2):337–350

Lee DE (2018) Evaluating conservation effectiveness in a Tanzanian community wildlife management area. J Wildl Manag 82(8):1767–1774

Lee DE, Bolger DT (2017) Movements and source-sink dynamics among subpopulations of giraffe. Popul Ecol 59:157–168

Lee DE, Bond ML, Kissui BM, Kiwango YA, Bolger DT (2016a) Spatial variation in giraffe demography: a test of 2 paradigms. J Mammal 97:1015–1025

Lee DE, Kissui BM, Kiwango YA, Bond ML (2016b) Migratory herds of wildebeests and zebras indirectly affect calf survival of giraffes. Ecol Evol 6(23):8402–8411

Lehman CP, Rota CT, Raithel JD, Millspaugh JJ (2018) Pumas affect elk dynamics in absence of other large carnivores. J Wildl Manag 82(2):344–353

Levins R (1969) Some demographic and genetic consequences of environmental heterogeneity for biological control. Bull Entomol Soc Am 15:237–240

Lindsey PA et al (2013) The bushmeat trade in African savannas: impacts, drivers, and possible solutions. Biol Conserv 160:80–96

Loison A, Langvatn R, Solberg EJ (1999) Body mass and winter mortality in red deer calves: disentangling sex and climate effects. Ecography 22(1):20–30

Maldonado-Chaparro AA, Blumstein DT, Armitage KB, Childs DZ (2018) Transient LTRE analysis reveals the demographic and trait-mediated processes that buffer population growth. Ecol Lett 21:1693–1703

Massei G et al (2015) Wild boar populations up, numbers of hunters down? A review of trends and implications for Europe. Pest Manag Sci 71(4):492–500

Matthysen E (2005) Density-dependent dispersal in birds and mammals. Ecography 28(3):403–416

Meissner HH, Pieterse E, Potgieter JHJ (1996) Seasonal food selection and intake by male impala *Aepyceros melampus* in two habitats. S Afr J Wildl Res 26(2):56–63

Melis C et al (2009) Predation has a greater impact in less productive environments: variation in roe deer, *Capreolus capreolus*, population density across Europe. Glob Ecol Biogeogr 18 (6):724–734

Merkle JA et al (2016) Large herbivores surf waves of green-up during spring. Proc R Soc B 283:20160456

Milner JM, Nilsen EB, Andreassen HP (2006) Demographic side effects of selective hunting in ungulates and carnivores. Conserv Biol 21(1):36–47

Morris WF, Doak DF (2002) Quantitative conservation biology. Sinauer Associates, Sunderland, MA

Morrison TA, Holdo RM, Anderson TM (2016a) Elephant damage, not fire or rainfall, explains mortality of overstorey trees in Serengeti. J Ecol 104(2):409–418

Morrison TA, Link WA, Newmark WD, Foley CAH, Bolger DT (2016b) Tarangire revisited: consequences of declining connectivity in a tropical ungulate population. Biol Conserv 197:53–60

Morrison TA, Holdo RM, Rugemalila DM, Nzunda M, Anderson TM (2018) Grass competition overwhelms effects of herbivores and precipitation on early tree establishment in Serengeti. J Ecol. https://doi.org/10.1111/1365-2745.13010

Müller DWH et al (2013) Assessing the Jarman – Bell principle: scaling of intake, digestibility, retention time and gut fill with body mass in mammalian herbivores. Comp Biochem Physiol A 164:129–140

Murphy BP, Bowman DMJS (2012) What controls the distribution of tropical forest and savanna? Ecol Lett 15:748–758

Norby RJ et al (2005) Forest response to elevated CO_2 is conserved across a broad range of productivity. Proc Natl Acad Sci U S A 102(50):18052–18056

O'Connor TG, Haines LM, Snyman HA (2001) Influence of precipitation and species composition on phytomass of a semi-arid grassland. J Ecol 89:850–860

Odadi WO, Karachi MK, Abdulrazak SA, Young TP (2011a) African wild ungulates compete with or facilitate cattle depending on season. Science 333:1753–1755

Odadi WO, Jain M, van Wieren SE, Prins HHT, Rubenstein DI (2011b) Facilitation between Bovids and Equids in an African Savanna. Evol Ecol Res 13:237–252

Ogutu JO, Owen-Smith N (2003) ENSO, rainfall and temperature influences on extreme population declines among African savanna ungulates. Ecol Lett 6(5):412–419

Ogutu JO, Piepho H-P, Dublin HT, Bhola N, Reid RS (2008) El Niño-Southern Oscillation, rainfall, temperature, and Normalized Difference Vegetation Index fluctuations in the Mara-Serengeti ecosystem. Afr J Ecol 46(2):132–143

Olff H, Ritchie ME, Prins HHT (2002) Global environmental controls of diversity in large herbivores. Nature 415:901–904

Olson KA et al (2011) Death by a thousand huts? Effects of household presence on density and distribution of Mongolian gazelles. Conserv Lett 4(4):304–312

Olson KA et al (2014) Survival probabilities of adult Mongolian gazelles. J Wildl Manag 78 (1):35–41

Owen-Smith N (1990) Demography of a large herbivore, the greater kudu *Tragelaphus strepsiceros*, in relation to rainfall. J Anim Ecol 59:893–913

Owen-Smith N (2004) Functional heterogeneity in resources within landscapes and herbivore population dynamics. Landsc Ecol 19:761–771

Owen-Smith N (2008) The comparative population dynamics of browsing and grazing ungulates. In: Gordon IJ, Prins HHT (eds) The ecology of browsing and grazing. Springer, Berlin, pp 149–177

Owen-Smith N, Mason DR (2005) Comparative changes in adult vs. juvenile affecting population trends of African ungulates. J Anim Ecol 74:762–773

Parker KL, Barboza PS, Gillingham MP (2009) Nutrition integrates environmental responses of ungulates. Funct Ecol 23:57–69

Peers MJ, Majchrzak YN, Neilson E, Lamb CT, Hämäläinen A, Haines JA, Garland L, Doran-Myers D, Broadley K, Boonstra R, Boutin S (2018) Quantifying fear effects on prey demography in nature. Ecology 99(8):1716–1723

Pellew RA (1983) The giraffe and its food resource in the Serengeti. I. Composition, biomass and production of available browse. Afr J Ecol 21:241–267

Perreira HM, Navaro LM (2015) Rewilding European landscapes. Springer, Heidelberg

Peters RH (1983) The ecological implications of body size. Cambridge University Press, Cambridge

Pettorelli N, Bro-Jørgensen J, Durant SM, Blackburn T, Carbone C (2009) Energy availability and density estimates in African ungulates. Am Nat 173(5):698–704

Post E, Stenseth NC (1999) Climatic variability, plant phenology, and northern ungulates. Ecology 80:1322–1339

Prins HHT (1988) Plant phenology patterns in Lake Manyara National Park, Tanzania. J Biogeogr 15:465–480

Prins HHT (1992) The pastoral road to extinction: competition between wildlife and traditional pastoralism in East Africa. Environ Conserv 19:117–123

Prins HHT, Gordon IJ (2008) Introduction: grazers and browsers in a changing world. In: Gordon I, Prins HHT (eds) The ecology of browsing and grazing. Ecological studies, vol 195. Springer, Berlin, pp 1–20

R Development Core Team (2016) R: a language and environment for statistical computing. R Foundation for Statistical Computing, Vienna. ISBN 3-900051-07-0. http://www.R-project.org

Ramirez JI, Jansen PA, Poorter L (2018) Effects of wild ungulates on the regeneration, structure and functioning of temperate forests: a semi-quantitative review. For Ecol Manag 424:406–419

Ratnam J, Bond WJ, Fensham RJ, Hoffmann WA, Archibald S, Lehman CER, Anderson MT, Higgins SI, Sankaran M (2011) When is a 'forest' a savanna, and why does it matter? Glob Ecol Biogeogr 20:653–660

Richard E, Gaillard JM, Saïd S, Hamann JL, Klein F (2010) High red deer density depresses body mass of roe deer fawns. Oecologia 163(1):91–97
Ripple WJ, Newsome TM, Wolf C, Dirzo R, Everatt KT, Galetti M, Hayward MW, Kerley GIH, Levi T, Lindsey PA, MacDonald DW, Malhi Y, Painter LE, Sandom CJ, Terborgh J, Van Valkenburgh B (2015) Collapse of the world's largest herbivores. Sci Adv 1:e1400103
Ripple W, Wolf C, Newsome TM, Hoffmann M, Wirsing AJ, McCauley DJ (2017) Extinction risk is most acute for the world's largest and smallest vertebrates. Proc Natl Acad Sci U S A 114 (40):10678–10683
Robbins CT (1993) Wildlife feeding and nutrition. Academic, New York
Roques KG, O'Connor TG, Watkinson AR (2001) Dynamics of shrub encroachment in an African savanna. Relative influences of fire, herbivory, rainfall and density dependence. J Appl Ecol 38 (2):268–180
Rutherford MC (1984) Relative allocation and seasonal phasing of growth of woody plant components in a South African savanna. Prog Biometerol 3:200–221
Sala OE et al (2000) Global biodiversity scenarios for the year 2100. Science 287:1770–1774
Sankaran M et al (2005) Determinants of woody cover in African savannas. Nature 438(8):846–849
Seeber P, Ndlovu HT, Duncan P, Ganswindt A (2012) Grazing behavior of the giraffe in Hwange National Park, Zimbabwe. Afr J Ecol 50(2):247–250
Silva M, Brimacombe M, Downing JA (2001) Effects of body mass, climate, geography, and census area on population density of terrestrial mammals. Glob Ecol Biogeogr 10:469–485
Sinclair ARE (1977) The African buffalo: a study of resource limitation of populations. University of Chicago Press, Chicago
Sinclair ARE (1979) The eruption of the ruminants. In: Sinclair ARE, Norton-Griffiths M (eds) Serengeti: dynamics of an ecosystem. Chicago University Press, Chicago, pp 82–103
Sinclair ARE (2003) Mammal population regulation, keystone processes and ecosystem dynamics. Philos Trans R Soc B 358:1729–1240
Sinclair ARE et al (2018) Predicting and assessing progress in the restoration of ecosystems. Conserv Lett 11(2):e12390
Sinclair ARE, Krebs CJ (2002) Complex numerical responses to top-down and bottom-up processes in vertebrate populations. Philos Trans R Soc B 357:1221–1231
Sinclair ARE, Norton-Griffiths M (1982) Does competition or facilitation regulate migrant ungulate populations in the Serengeti? A test of hypotheses. Oecologia 53(3):364–369
Smit IPJ, Archibald S (2019) Herbivore culling influences spatio-temporal patterns of fire in a semi-arid savanna. J Appl Ecol 56(3):711–721. https://doi.org/10.1111/1365-2664.13312
Smit IPJ, Prins HHT (2015) Predicting the effects of woody encroachment on mammal communities, grazing biomass and fire frequency in African savannas. PLoS One 10(9):e0137857
Spear D, Chown SL (2009) Non-indigenous ungulates as a threat to biodiversity. J Zool 279 (1):1–17
Staver AC, Archibald S, Levin SA (2011) The global extent and determinants of savanna and forest as alternative biome states. Science 334:230–232
Teitelbaum CS et al (2015) How far to go? Determinants of migration distance in land mammals. Ecol Lett 18(6):545–552
Tucker MA et al (2018) Moving in the Anthropocene: global reductions in terrestrial mammalian movements. Science 359(6375):466–469
Tuljapurkar S (1982) Population dynamics in variable environments. III evolutionary dynamics of r-selection. Theor Popul Biol 21:141–165
van Beest FM, Mysterud A, Loe LE, Milner JM (2010) Forage quantity, quality and depletion as scale-dependent mechanisms driving habitat selection of a large browsing herbivore. J Anim Ecol 79(4):910–922
van de Koppel J et al (2002) Spatial heterogeneity and irreversible vegetation change in semiarid grazing systems. Am Nat 159(2):209–218
Verheyden-Tixier H, Renaud P-C, Morellet N, Jamot J, Besle J-M, Dumont B (2008) Selection for nutrients by red deer hinds feeding on a mixed forest edge. Oecologia 156:715–726

Waltert M, Meyer B, Kiffner C (2009) Habitat availability, hunting or poaching: what affects distribution and density of large mammals in western Tanzanian woodlands? Afr J Ecol 47:737–746
Wang G et al (2006) Spatial and temporal variability modify density dependence in populations of large herbivores. Ecology 87(1):95–102
Wang G et al (2009) Density dependence in northern ungulates: interactions with predation and resources. Popul Ecol 51(1):123
Wegge P, Storaas T (2009) Sampling tiger ungulate prey by the distance method: lessons learned in Bardia National Park, Nepal. Anim Conserv 12:78–84
Wegge P, Shresta AK, Moe SR (2006) Dry season diets of sympatric ungulates in lowland Nepal: competition and facilitation in alluvial tall grasslands. Ecol Res 21:698–706
White EP, Ernest SKM, Kerkoff AJ, Enquist BJ (2007) Relationships between body size and abundance in ecology. Trends Ecol Evol 22:323–330
Wilkie DS, Bennett EL, Peres CA, Cunningham AA (2011) The empty forest revisited. Ann N Y Acad Sci 1223:120–128
Wisdom MJ, Mills LS, Doak DF (2000) Life stage simulation analysis: estimating vital rate effects on population growth for conservation. Ecology 81:628–641
Wittmer HU, Sinclair ARE, McLellan BN (2005) The role of predation in the decline and extirpation of woodland caribou. Oecologia 144:257–267
Wood SN, Pya N, Saefken B (2016) Smoothing parameter and model selection for general smooth models (with discussion). J Am Stat Assoc 111:1548–1575

Chapter 7
Community Dynamics of Browsing and Grazing Ungulates

Charudutt Mishra, Munib Khanyari, Herbert H. T. Prins, and Kulbhushansingh R. Suryawanshi

7.1 Introduction

Species do not exist in isolation. Multiple species typically coexist in time and space, giving rise to ecological communities or assemblages. These are, in turn, ultimately limited by the continental or regional species pools, and determined by the outcome of dynamic colonization and extinction processes operating locally or regionally (Andrewartha and Birch 1972; Krebs 2002).

The distribution and co-occurrence of ungulate species, and the factors that allow them to coexist, have long fascinated ecologists (Jarman and Sinclair 1979; Prins and Olff 1998; Mishra et al. 2002, 2016), considering that, on the face of it, these herbivorous species belong, largely, to the same feeding guild, namely, those that feed on terrestrial plants. In this Chapter, our aim is to better understand, and generate hypotheses, regarding the underlying mechanisms that influence the community structuring and co-occurrence of ungulate species worldwide. We do

C. Mishra · K. R. Suryawanshi (✉)
Nature Conservation Foundation, Mysore, India

Snow Leopard Trust, Seattle, WA, USA
e-mail: charu@snowleopard.org; kulbhushan@ncf-india.org

M. Khanyari
Nature Conservation Foundation, Mysore, India

Snow Leopard Trust, Seattle, WA, USA

Interdisciplinary Center for Conservation Sciences (ICCS), Oxford University, Oxford, UK

School of Biological Sciences, University of Bristol, Bristol, UK
e-mail: munib@ncf-india.org

H. H. T. Prins
Animal Sciences Group, Wageningen University, Wageningen, The Netherlands
e-mail: Herbert.prins@wur.nl

I. J. Gordon, H. H. T. Prins (eds.), *The Ecology of Browsing and Grazing II*, Ecological Studies 239, https://doi.org/10.1007/978-3-030-25865-8_7

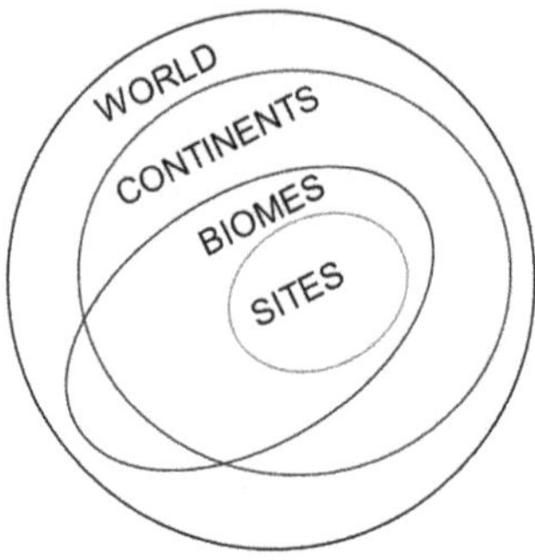

Fig. 7.1 Conceptual representation of the spatial structure of our dataset on the world's extant and extinct ungulates, their sub-guild (grazer, browser, mixed feeder), and body mass. The scales aren't completed nested as some biomes are found across more than one continent

this by examining the global patterns in species occurrences of ungulates (incl. Proboscoidea) with respect to their body mass and feeding guilds at continental, biome, and site scales (Fig. 7.1). Sites are defined here as locations, such as protected areas or contiguous habitats, with ungulate assemblages. The eight biomes we considered were tundra, taiga, temperate deciduous forests, rainforests, chaparral, grasslands, deserts and alpine tundra (Olson and Dinerstein 1998). We realise that there are many other biomes, but for the sake of analysis we kept the number small enough to allow for statistical exploration, and the biomes that we chose were contrasting enough to be able to find patterns. In our analyses, we included the African elephant (*Loxodonta africana)* and the Asian elephant (*Elephas maximus*), as these megaherbivorous species play an important role in shaping the herbivore assemblages where they occur (Kerley and Landman 2006; Landman et al. 2013). We extended our dataset to ungulates and proboscideans that went extinct during the transition from the late Pleistocene to the Holocene, in order to examine the influence of Pleistocene glaciations (see **Saarinen** Chap. 2; **Rowan and Faith** Chap. 3) on the structuring of today's ungulate assemblages. We also considered the influence of precipitation, which is presently the primary determinant of plant productivity (Sankaran et al. 2005; Yang et al. 2008). In all our analyses, our focal variable is (mean and median) ungulate body mass, which has a critical role in ungulate physiology and ecology (see other Chapters in this book).

7.2 Plants and Grazing and Browsing Ungulates

Approximately 300,000 species of plant species exist globally, forming the basis of terrestrial food webs (Kreft and Jetz 2007). On the face of it, this is a diverse set of food resources for the 240 extant ungulates occurring on the planet (Wilson and Reeder 2005). Yet, most plant material does not represent suitable forage for ungulates, either being of low digestibility, or being defended against herbivory by physical and chemical defences (Zangerl and Bazzaz 1992).

Plants typically occur as assemblages of a varying number of species, and, based on the dominance of particular life forms, form the major biomes of the world (Woodward et al. 2004). These include the various biomes, categorised on basis of leaf longevity and morphology of woody species, and the so called super-biome comprised by grasslands (Woodward et al. 2004). Based largely on latitude, and partly on temperature, precipitation, soil, and human influence, different areas of the world are typified by different biomes, and at smaller spatial scales, by plant assemblages (Mutke et al. 2011; Kerkhoff et al. 2014). Some areas are—depending on the water balance, nutrient status of the soil, and temperature—covered by tall woody vegetation, others by stunted woody vegetation or dwarf shrubs, and others by grasses.

From the perspective of ungulates, preferred plant species can be simplified into 'grass', comprising of monocots such as grasses and sedges, or 'browse', comprising of dicots including forbs, shrubs and trees. These two groups of plants have a number of basic differences in their physical and nutritional characteristics (e.g., grass species have more cell wall content and silica, while browse species have more nutritive cell content but also lignin which reduces the digestibility of the call wall) that influence their consumption by ungulates, and appear to have played a role in shaping the evolution and morphological adaptions of grazing and browsing large mammalian herbivores (Hofmann 1989; Gordon and Prins 2008; see also Chapters in this book). Based on their feeding habits, founded on morphological adaptions in mouthparts and the gut, ungulates can be broadly classified into three sub-guilds: 'grazers' (feeding predominantly on grasses and sedges), 'browsers' (feeding primarily on forbs, shrubs, and trees), and 'mixed feeders' (which consume a combination of graze and browse). Species may show some plasticity in their need and ability to shift their diet from being grass to browse dominated, as illustrated by the bharal (*Pseudois nayaur*) in Trans-Himalayan rangelands (Mishra et al. 2004; Suryawanshi et al. 2010) (see also **Venter et al.** Chap. 5). Evolutionarily, large herbivores are thought to have initially been omnivorous, evolving with changing climate and vegetation into species that browsed, and subsequently, with the spread of grasses, into grazers (see **Saarinen** Chap. 2).

Although individual biomes tend to be defined by predominant plant lifeforms (tree, herb, shrub, or monocots), typically, any given biome, or plant community, comprises plants belonging to more than one lifeform. Because grasses (monocots), and herbs or browse (dicots), are nearly always intermixed in varying proportions, at a basic level, this distinction amongst ungulate species, based on the extent of their consumption of monocots and dicots, offers opportunities for the co-existence of grazers and browsers through foraging niche partitioning. A nice illustration is the Black rhinoceroses (*Diceros bicornis*) apparently grazing the floor of the Ngorongoro Crater in Tanzania amidst Burchell's zebra (*Equus burchellii*): yet the rhinos browse the forbs that are 'hidden' between the grasses that the zebra feed on.

7.3 Physiological and Ecological Significance of Body Mass for Grazing and Browsing Species

Unlike the relatively high-energy food of carnivores or frugivores, processing of leafy forage by ungulates, that is relatively low in per unit nutritive value (e.g., Prins and Van Langevelde 2008), requires time, specialized gut anatomy and microbial symbionts to gain access to the energy and protein therein. Hence, in addition to the feeding sub-guilds of 'grazers', 'browsers' and 'mixed feeders', ungulates may be divided into two main types based on gut morphology and associated physiological adaptations—'ruminants' and 'hindgut fermenters'. In ruminants, fermentation of forage, and absorption of nutrients, takes place before the passage of plant through the lower gut, while in hindgut fermenters, energy is derived as food passes through the stomach, and additional energy is liberated through fermentation in the hind gut. Although hindgut fermenters are, to a certain extent, able to hasten the processing of forage in their guts, the majority of the extant ungulates are ruminants, which, being fore-gut fermenters, are relatively limited in their daily forage intake and have longer food retention times (Demment and Van Soest 1985). Of course, this is another simplification of the rich array of evolved life forms, and pigs, peccaries, hippos, etc., do not fit into this simple dichotomy. Yet, simplification into a few categories may assist in discovering patterns in assemblage structure, while in examining the emerging 'exceptions' one may test any emerging rule.

Gut capacity in ungulates scales linearly with body mass, while basal metabolic rate scales as a power function (kg body mass$^{0.75}$). The different shapes of these allometric relationships render a critical significance to ungulate body mass in terms of the kind of forage a species can effectively utilize (Peters and Peters 1986; du Toit and Owen-Smith 1989; Prins and Van Langevelde 2008; Franz et al. 2011). These allometric relationships imply that larger bodied ruminants are able to utilize and process greater volumes of relatively low nutritive value forage, while smaller bodied species must rely on relatively easily digestible forage, thus providing another mechanism for species differing in body masses to coexist. While this does not preclude competition between larger and smaller bodied species, it does create further resource diversification that could allow multiple species to co-exist. There are, however, limits to body mass beyond which the ruminant system is unable to process sufficient quantities of forage and still get enough energy out of the microbial digestion as compared to what the microbial mass gets, and, perhaps as a consequence, megaherbivores tend to be hindgut fermenters. (*cf.* Van Soest 2018).

It is useful here to link the allometric constraints of body size in ungulates to the forage characteristics of grasses and browse. As mentioned earlier, compared to grasses, browse generally tends to have a higher level of soluble cell content and nitrogen, albeit having constraints for browsing ungulates in the form of lignin and secondary metabolites (e.g., alkaloids, tannins, etc.). Unsurprisingly, small bodied ungulates that must rely on more nutritive, high energy forage due to their allometric constraints, tend to be selective browsers (Gordon and Illius 1996). Following a similar line of reasoning, we expect grazing species, in general, to have larger body masses.

We created a dataset of wild ungulates recorded from 78 sites across all the continents (except Antarctic and Australia where they do not occur naturally), and biomes of the world (Map 7.1), using an online literature review of peer-reviewed scientific evidence (using multiple online portals such as Google Scholar and Web of Science; we only interrogated literature published in English and searched while based in India [Bangalore] and the UK [Oxford]; we terminated the search

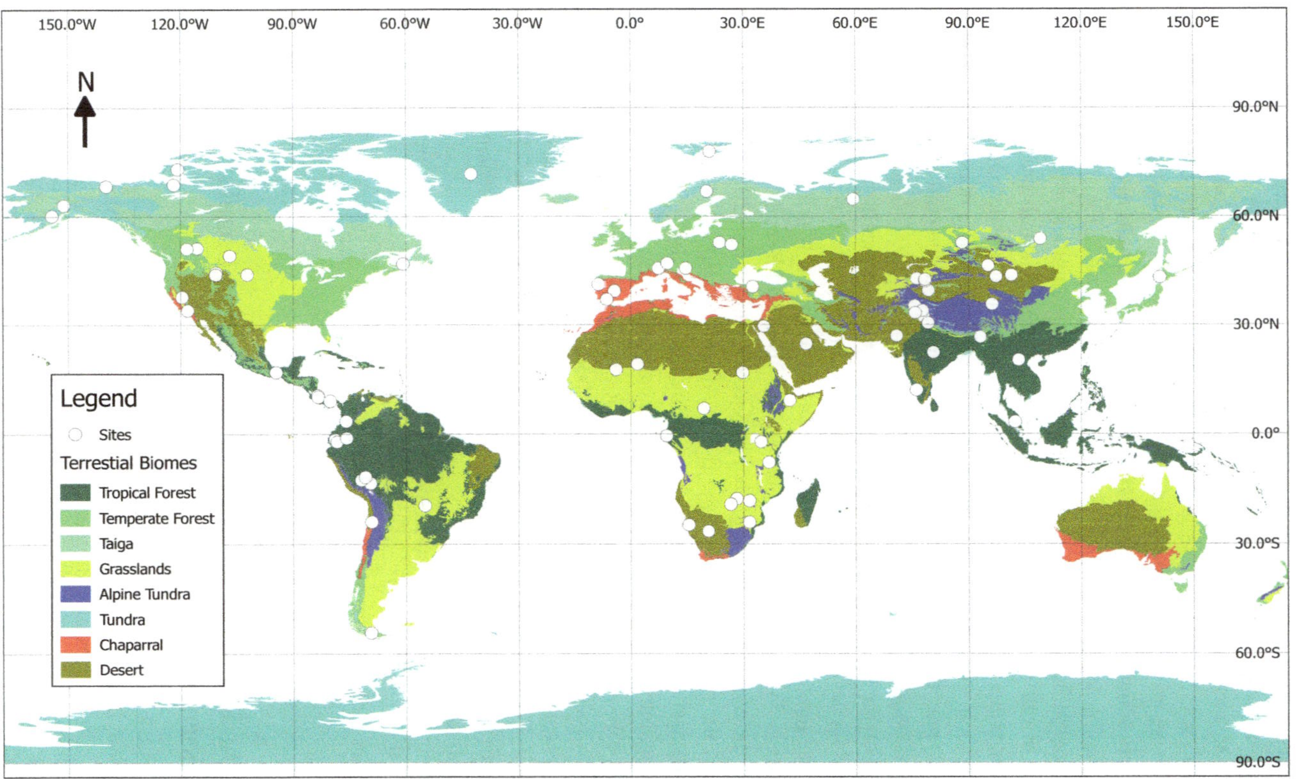

Map 7.1 Map of all (n = 78) sites for which data on extant species of herbivores was used in our analysis. The sites were spread across eight major terrestrial biomes of world (Olson et al. 2001)

at [Nov–Dec 2017 and Oct–Nov 2018 in Bangalore and April 2018 in Oxford]). It is important to note that not all ungulate species that exist in the world were represented due to epistemic uncertainties (i.e., taxonomic ambiguity, paucity of information on their distribution and body mass), and lack of representation of range-restricted species. As a side-remark, we observe that, from the point of view of conducting meta-analyses, the recent upsurge in species-splitting is of deep concern. The sites were chosen to contain an ungulate assemblage. Additionally, species that are exclusively found on islands were removed from the database as concepts of island biogeography are well studied (MacArthur and Wilson 2001), and would have played a significant role in structuring island communities, which was not the intention of our investigation with this dataset. Because islands are physically (and often even temporally) well-defined and of limited extent, the dominance of findings based on their studies in the recent development of ecological notions (often called 'theories') may be of much smaller use if one wants to understand the distribution of species on continents.

Based on the literature review, we categorized each species as 'grazer', 'browser', or 'mixed feeder'. For each ungulate we also recorded body mass (in kg). Where the body mass range were reported, they were first averaged within each sex and then across sexes. We followed the taxonomy of Wilson and Reeder (2005) and did not consider the hyper-splitting suggested by Groves and Grubb (2011). In that respect we follow the recent decision of the IUCN African Antelope Specialist Group.

Our dataset (available from corresponding author) of 142 (59%) of the world's extant ungulate species, occurring in five continents, showed that the median body mass of browsers (33 kg) was significantly lower than the global median body mass of 50 kg for all ungulates, while the median body mass of grazers was significantly higher (69 kg). The median body mass for mixed feeders (45 kg), on the other hand, was similar to the global median body mass of all ungulates (Fig. 7.2). The differences in the global body mass and foraging sub-guild medians were tested using 10,000 boot-straps.

In terms of abundance, smaller-bodied browsers appear to attain highest densities, while in the case of grazers, species attaining higher densities tend to be relatively larger ones (see **Kiffner and Lee** Chap. 6).

7.4 The Role of Body Mass in Structuring Ungulate Assemblages

The role of body mass in enabling various species to utilize forage of different nutritive values, creates the potential for facilitative or positive interactions among ungulates. This can be mediated through different mechanisms. For instance, larger bodied species can physically improve access to suitable forage for smaller bodied species by modifying or opening up the habitat (Gordon 1988; Farnswoth et al. 2002). Alternately, amongst grazing species, the relationship between the nutritive

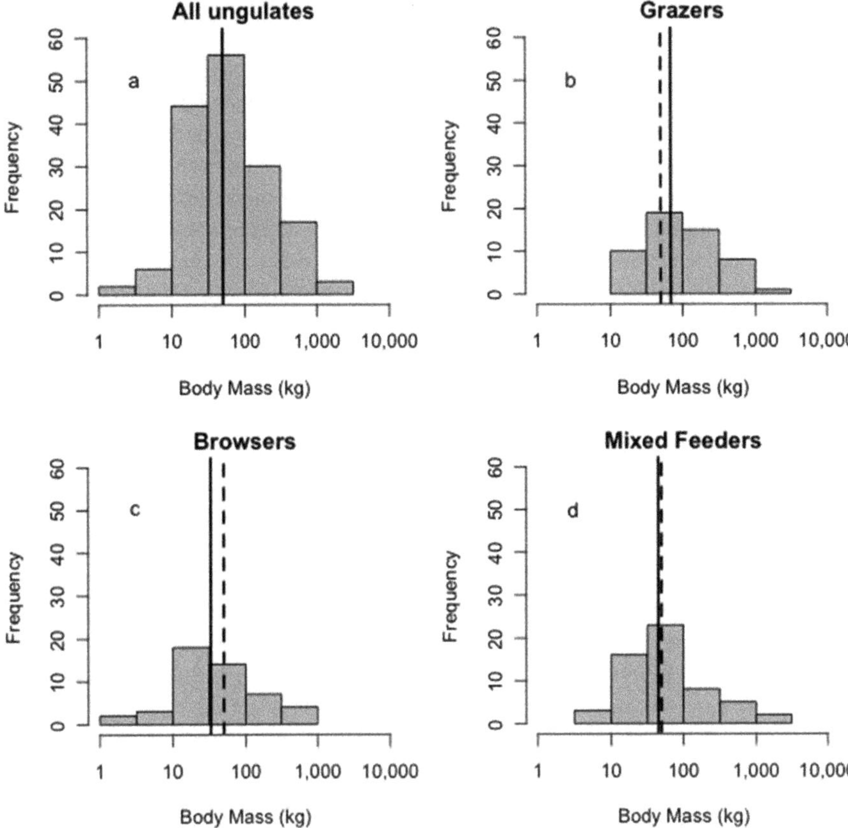

Fig. 7.2 Histogram of log-10 of Body mass of 142 ungulate species (including two proboscideans) from five continents. The solid line represents the global median (50 kg) of all ungulates and the dashed line represents median of grazers, browsers and mixed feeders in sub-figure b, c and d, respectively. Median for grazers (69 kg) was significantly higher ($p = 0.03$) than the global median for all ungulates while median for browsers (33 kg) was significantly lower ($p = 0.01$). Median for mixed feeders (45 kg) was similar to global median ($p = 0.31$). The smallest ungulate we considered weighed 2.75 kg, the Bate's Pygmy antelope (*Neotragus batesii*), while the largest two species in our dataset were the proboscideans the African elephant (*Loxodonta africana*) at 2925 kg, Asian elephant (*Elephas maximus*) at 2750 kg, followed by the ungulate white rhinoceros (*Ceratotherium simum*) 1500 kg. The log-10 transformed values for the Bate's Pygmy antelope and the African elephant were, respectively, 0.44 and 3.47, which may be of assistance to appreciate the scale

value of grass and its age, or standing biomass, can also lead to facilitative interactions among species. The quality of grass tends to decline with an increase in its standing biomass (Olff et al. 2002), and a larger bodied grazer, by reducing the standing biomass, through feeding and stimulating fresh growth, can thus enhance forage quality, thereby making it suitable for consumption by a smaller species.

It appears, therefore, that when two species within the same sub-guild, especially grazers, differ in their body mass, the larger species could facilitate the smaller one. If the species are very similar in body mass, they are likely to compete for resources, while if the size difference between them is very large, the smaller species might not be able to benefit from facilitation (Prins and Olff 1998 based on Hutchinson 1959; Hutchinson and MacArtur 1959). Such interplay of facilitative and competitive interactions can theoretically create specific niches that optimise body mass differences within ungulate assemblages, especially amongst grazers, where species body masses would tend to be neither too close to each other (or they would compete strongly), or too different (which would restrict the benefits of facilitation) (Prins and Olff 1998).

The structuring of body masses has received attention in the past in the grazer assemblages of Africa and Asia (Prins and Olff 1998; Mishra et al. 2002). In terms of the proximate mechanisms, structuring could result from species sorting over ecological timescales (Ricklefs 1987), or through character displacement (Brown and Wilson 1956). The potential for body mass structuring in browser assemblages has, to our knowledge, not been explored, even though competition amongst browsers can be intense, and there is the possibility of facilitation through habitat modification (Cameron and du Toit 2006).

We tested for body mass structuring amongst browsers and grazers in grassland (grazer dominated) and tropical forest biomes (browser dominated). If herbivore species interact strongly enough to structure an assemblage, then a strong linear relationship can be expected between the $\log_e$ body mass and the rank of the species (Prins and Olff 1998; Olff et al. 2002). Mixed feeders were considered together with grazers for this analysis following Prins and Olff (1998). We estimated the explained variance (r^2-values) for the relationship between $\log_e$ body mass and species rank number for grazers and browsers in the two biomes. We then tested whether the r^2 was higher than the 97.5 percentile of 10,000 randomly generated r^2 by resampling the species body mass range for each assemblage for equal number of hypothetical species (*cf.* Olff et al. 2002).

Our results point towards structuring of body masses amongst grazer species, in both grassland and tropical forest biomes (Fig. 7.3). The r^2s for grazers were 0.96 and 0.95 in grasslands and forests, respectively. In the case of browsers, although the relationship was not significant in grasslands, there seemed to be some evidence for body mass structuring, presumably resulting from competitive interactions, in tropical forests. The r^2s for browsers were 0.91 and 0.96 in grasslands and tropical forests, respectively.

Across all biomes, except grasslands, the median body masses of species converged towards the global median of 50 kg. In grasslands, the median body mass (65 kg) was significantly higher (Fig. 7.4). This is likely due to the preponderance of grazers in grasslands that, as seen earlier, tend to be heavier as compared to browsers and mixed feeders.

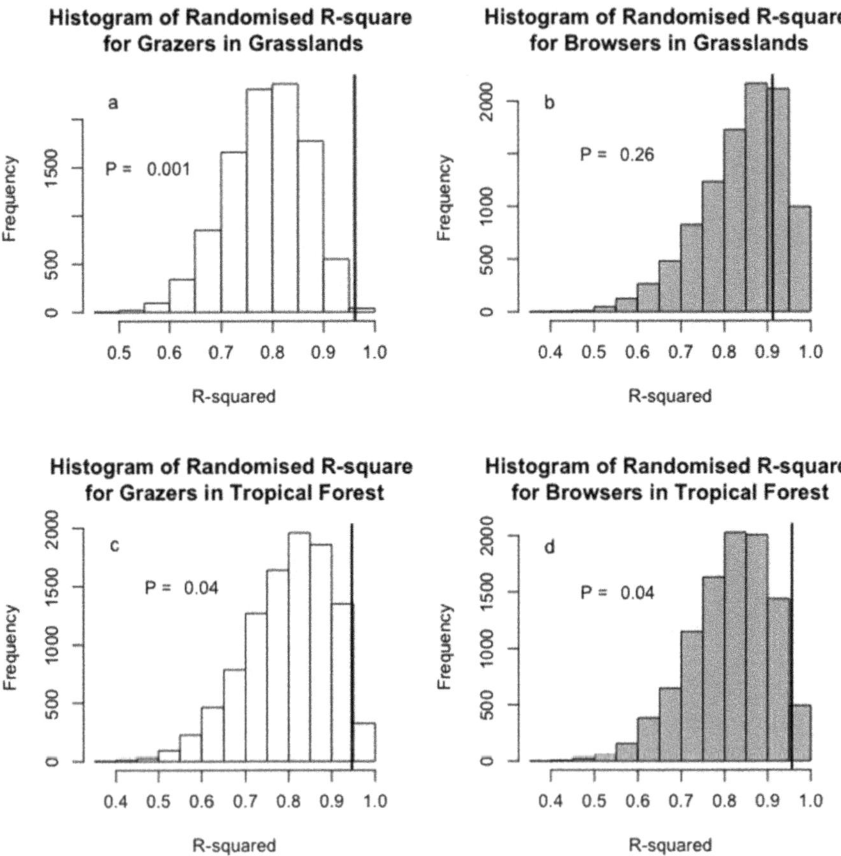

Fig. 7.3 Histogram of r^2 from 10,000 simulated data using the species richness and body size of (**a**) grazers in grassland biome, (**b**) browsers in grassland biome, (**c**) grazers in tropical forest biome, (**d**) browsers in tropical forest biome. Grazers and mixed feeders were clubbed together for this analysis. The vertical line indicates the actual r^2 from the real data. The p-value indicated the percentage of times the simulated data resulted in the same or higher r^2 than that from real data

7.5 Continental Structuring of Grazing and Browsing Ungulates

Across continents, as in the case of biomes, the median body mass of all ungulates was not significantly different from the global median of 50 kg of all ungulates (Fig. 7.5). However, the shapes of the frequency distributions suggest differences that are linked to continental identity. Compared to Africa and Asia, the assemblages of North America, South America, and Europe appear to have species missing at the lower and higher ends of the body mass gradient. What could explain this 'continental identity effect'?

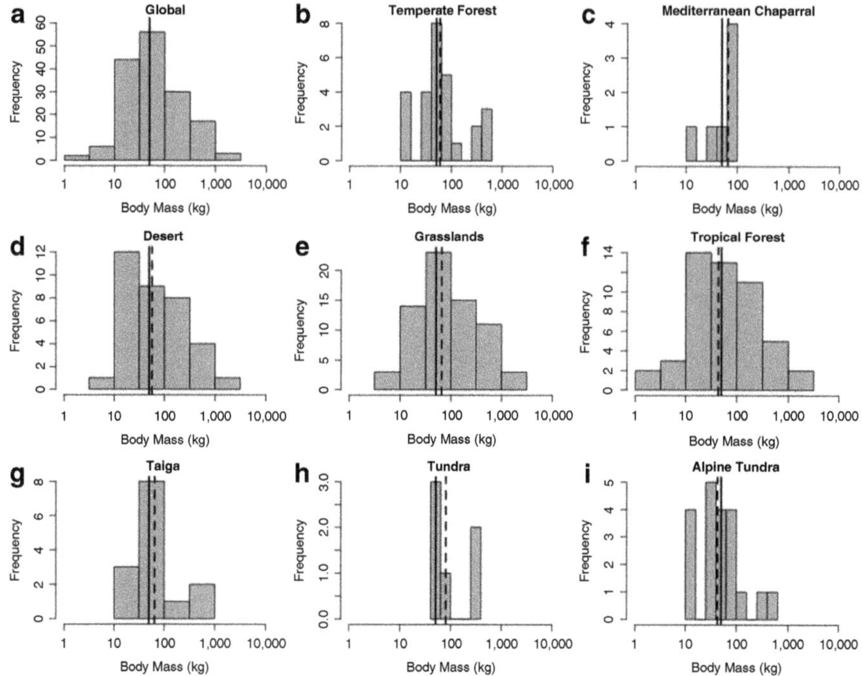

Fig. 7.4 Histogram of log-10 of Body mass of 142 ungulate species (including two proboscideans) from five continents by biome. The solid line represents the global median (50 kg) and the dashed line represents median for each biome. The median of grasslands (65 kg) was marginally higher than the global median ($p = 0.05$). Differences between median were tested using 10,000 bootstraps

Across vertebrate taxa, species extinctions during the late Pleistocene, are thought to have been an important determinant of today's extant (continental) species pools. It has been reported that South America lost 76% of all herbivore genera >5 kg during the late Pleistocene, mirrored closely by North America (75%) and to some extent by Europe (45%), in contrast to Africa that retained some 87% of the species (see Prins and Olff 1998); we have not estimated how many ungulate species were lost from South and SE Asia as it would require a more thorough review of ungulate extinctions which was beyond the scope of this Chapter. Among mammals, in general, medium and small bodied species are thought to have remained largely unaffected by late Pleistocene extinctions (Lambert and Holling 1998). It is, therefore, instructive to assess the extant continental assemblages of grazing and browsing species against the backdrop of Pleistocene extinctions.

We created a dataset of 101 ungulate species that are believed to have gone extinct during the Pleistocene. This dataset was established by first conducting an online keyword search for Pleistocene extinctions of ungulates as outlined above. Subsequently, we used a snowball sampling approach (Goodman 1961) to find

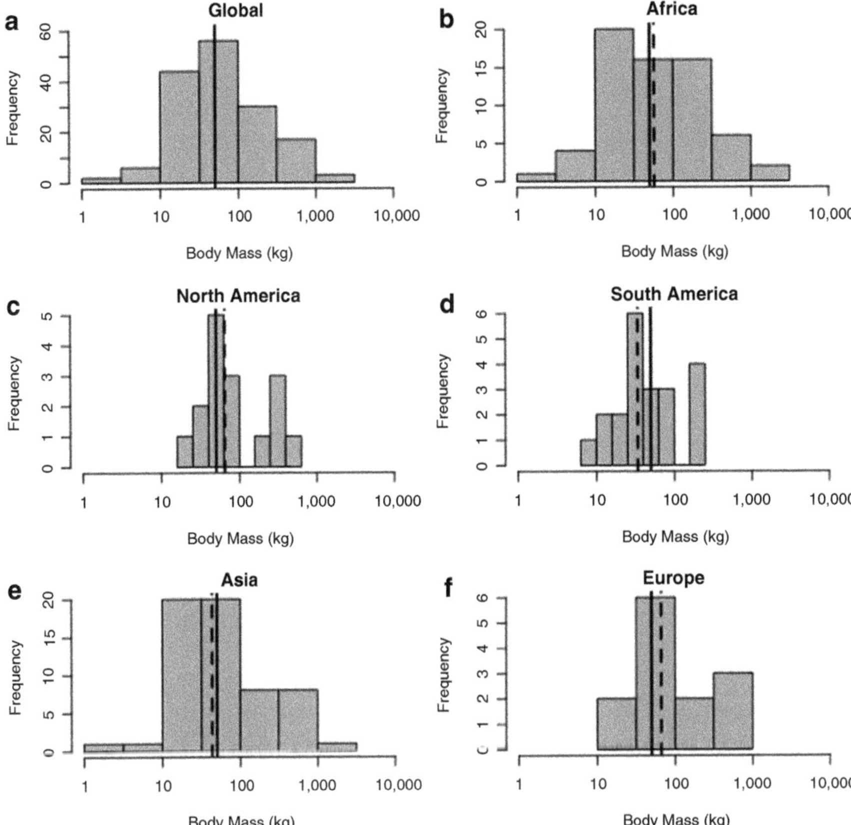

Fig. 7.5 Histogram of log-10 of Body mass (BM) of 142 ungulate extant species (including two proboscideans) from five continents. The solid line represents the global median (50 kg) of all ungulates in our sample (142) and the dashed line represents median values for each continent. None of the BM medians of the continents were statistically significantly different from the global median. Difference between medians was tested using 10,000 bootstraps

published literature from which we derived information on body mass and feeding sub-guilds of these extinct species.

Our data on body masses of extinct species seem to suggest a disproportionate extinction of larger bodied ungulates during the Pleistocene (Fig. 7.6), a pattern that has been reported earlier for mammals (e.g., Lambert and Holling 1998). However, frequency distributions of body masses of extant species seem to indicate missing, or extinct, species not just amongst the largest size categories, but also the smallest ones in the continents that saw the highest proportion of Pleistocene extinctions (see Fig. 7.5 and Table 7.1). Provided the fossil record, as well as our sampling of extinct species, is representative, one would expect a bi-modal distribution in the body mass frequency of extinct species, if both large bodied and small bodied species went

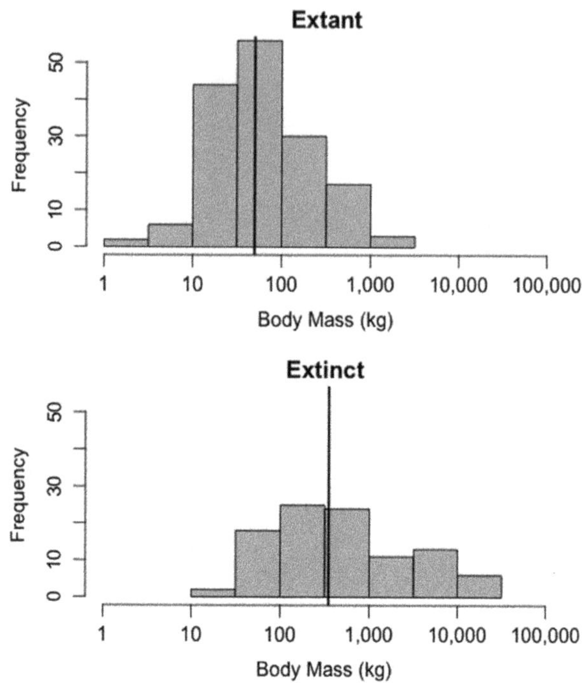

Fig. 7.6 Histogram of log-10 of body mass of 142 extant and 101 extinct ungulates and proboscideans globally that we had in our database. The vertical line indicates the median, which was 50 kg for the extant species, and 350 kg for the extinct ones. The smallest extinct species in our database weighed 20 kg (Gazella atlantica and Gazella tingitana), the largest extinct proboscidean weighed 22,000 kg (Paleoloxodon namadicus) and the largest extinct ungulate weighed 3900 kg (Hippopotamus gorgops). The log-10 transformed values for the smallest and the largest were, respectively, 1.30 and 4.34, which may be of assistance to appreciate the scale

Table 7.1 Median body masses of the world's living ungulate species, those that went extinct during the Pleistocene, and of the presumed ungulate assemblage of the Pleistocene (obtained by combining the extant and extinct species). The data indicate a disproportionate loss of larger bodied ungulates as a result of Pleistocene extinctions

	Median extinct (kg)	Median extant (kg)	Combined median (kg)
Global	350	50	115
Africa	300	57	90
Asia	950	43	95
North America	462	64.5	271
South America	265	34	131.5
Europe	1919	66.5	365

extinct. Our histogram of extinct species does not show this (Fig. 7.6). It, therefore, appears that the possible mass extinctions of smaller bodied ungulates may have either gone unrecorded, or that the smaller bodied species are less preserved in the fossil record.

To our knowledge, a possible disproportional 'mass extinction' of small ungulates in the Pleistocene has not been reported before. A potential explanation could be found in the decline of C_3 grasses, and C_3 trees, due to the extremely low CO_2 values during the hyper-arid period of the Last Glacial Maximum (some 40,000 BP) (Street-Perrott et al. 1997); C_3 plants have a higher digestibility than C_4 forage which could have strongly affected the small ungulates (Caswell et al. 1973). Therefore, the influence of size biased Pleistocene extinctions appears to be seen in today's continental assemblages, particularly in continents that are thought to have lost the most number of species. Indeed, there have been other ecological consequences of these extinctions, in the form of enhanced fire regimes, vegetation state shifts, and near-extinction of large fruiting plants, as seen in **Rowan and Faith** Chap. 3 (see also **van Langevelde et al.** Chap. 10; **Smit and Coetsee** Chap. 13).

7.6 Body Masses of Ungulates Along a Precipitation Gradient

Precipitation is one of the most important determinants of primary plant productivity and standing plant biomass (Sankaran et al. 2005), and also of forage quality (Olff et al. 2002). We were able to obtain data on ungulate and proboscidean assemblages and annual precipitation from 77 sites across five continents (Fig. 7.7). It appears that ungulates attained the highest median body mass in the precipitation range of 350–1000 mm. This rainfall rangc corrcsponds to thc grassland-savanna habitat (Sankaran et al. 2005), and is consistent with the finding that grasslands support larger bodied ungulates (Fig. 7.4). Areas with more than 1000 mm of precipitation tend to be dominated by forest habitats which typically support browsers, which are expected to be smaller in body size. Although sites with lower than 300 mm precipitation are grasslands and deserts capable of supporting grazers, the low plant productivity may be unable to support the daily forage intake requirement of large bodied grazers (Prins and Olff 1998).

7.7 Conclusions

Body mass plays an important role in foraging ecology and community dynamics of ungulates. Consistency in the median body mass of extant ungulates across biomes and continents suggests 50 kg to be some kind of an optimum body mass amongst extant ungulates towards which the species seem to converge. Globally, grazers tend to be significantly heavier than this global median, browsers significantly lighter, while mixed feeders converge around the global median. Our results suggest that positive and negative species interactions, amongst ungulates, appear to give rise to

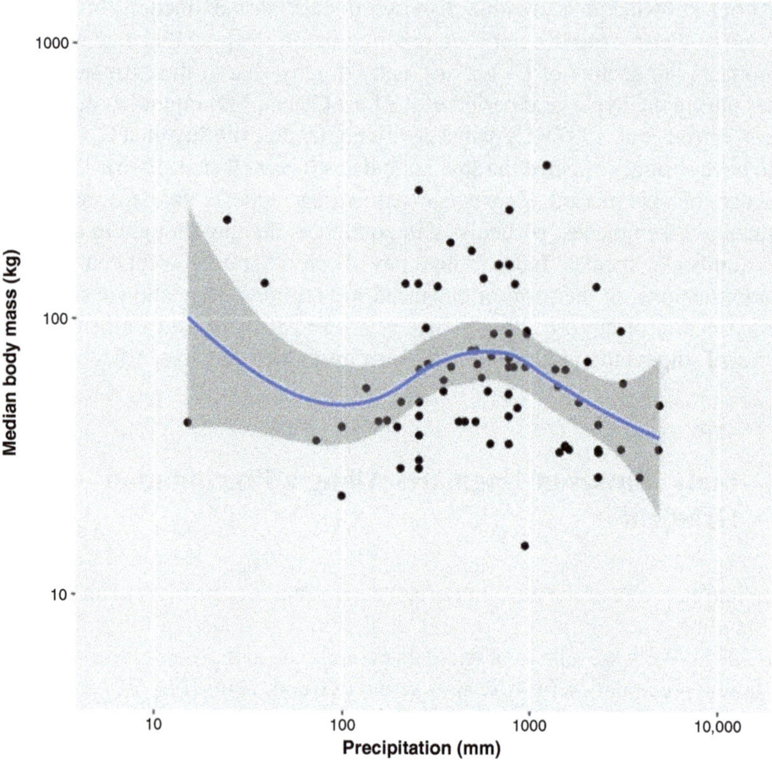

Fig. 7.7 Scatter plot of log-10 of median of ungulate and proboscidean body mass from 77 sites across five continents around the world and the log-10 of precipitation at the site. The line indicates a GAM function and the shaded area is the 95% CI. The left most point is Northern Atacama Desert. The low number of data below about 100 mm of annual precipitation precludes a very strong conclusion in that part of the precipitation gradient

body mass structuring in, not just grazer assemblages, but also amongst browsers, though the pattern appeared weaker in browsers as compared to grazers, with browsers showing the pattern in the forested tropical biome that has more browsing species. This merits further investigation. Our comparison of body mass structuring of Pleistocene assemblages, with the extant grazer and browser assemblages, shows a significant decline in overall body masses due to the extinction of larger bodied ungulates. However, it is intriguing that the extant assemblages of both North and South America and of Europe indicate that species may also be missing at the smaller end of the body mass spectrum. This could potentially imply disproportionate Pleistocene extinctions of smaller bodied ungulates that may have gone unrecorded so far, or that have not been captured adequately in the fossil record.

References

Andrewartha HG, Birch LC (1972) The distribution and abundance of animals. University of Chicago Press, Chicago
Brown WL, Wilson EO (1956) Character displacement. Syst Zool 5:49–64
Cameron EZ, du Toit JT (2006) Winning by a neck: tall giraffes avoid competing with shorter browsers. Am Nat 169:130–135
Caswell H, Reed F, Stephenson SN, Werner PA (1973) Photosynthetic pathways and selective herbivory: a hypothesis. Am Nat 107:465–480
Demment MW, Van Soest PJ (1985) A nutritional explanation for body-size patterns of ruminant and nonruminant herbivores. Am Nat 125:641–672
du Toit JT, Owen-Smith N (1989) Body size, population metabolism, and habitat specialization among large African herbivores. Am Nat 133:736–740
Farnsworth KD, Focardi S, Beecham JA (2002) Grassland-herbivore interactions: how do grazers coexist? Am Nat 159:24–39
Franz R, Hummel J, Müller DW, Bauert M, Hatt JM, Clauss M (2011) Herbivorous reptiles and body mass: effects on food intake, digesta retention, digestibility and gut capacity, and a comparison with mammals. Comp Biochem Physiol A Mol Integr Physiol 158:94–101
Goodman LA (1961) Snowball sampling. Ann Math Stat 32:148–170
Gordon IJ (1988) Facilitation of red deer grazing by cattle and its impact on red deer performance. J Appl Ecol 25:1–10
Gordon IJ, Illius AW (1996) The nutritional ecology of African ruminants: a reinterpretation. J Anim Ecol 65:18–28
Gordon IJ, Prins HH (2008) The ecology of browsing and grazing (No. 195). Springer, Berlin
Groves C, Grubb P (2011) Ungulate taxonomy. JHU Press, Baltimore
Hofmann RR (1989) Evolutionary steps of ecophysiological adaptation and diversification of ruminants: a comparative view of their digestive system. Oecologia 78:443–457
Hutchinson GE (1959) Homage to Santa Rosalia or why are there so many kinds of animals? Am Nat 93:145–159
Hutchinson GE, MacArthur RH (1959) A theoretical ecological model of size distributions among species of animals. Am Nat 93:117–125
Jarman PJ, Sinclair ARE (1979) Feeding strategy and the pattern of resource partitioning in ungulates. In: Sinclair ARE, Norton-Griffiths N (eds) Serengeti: dynamics of an ecosystem. Chicago University Press, Chicago, pp 130–163
Kerkhoff AJ, Moriarty PE, Weiser MD (2014) The latitudinal species richness gradient in new world woody angiosperms is consistent with the tropical conservatism hypothesis. Proc Natl Acad Sci 111:8125–8130
Kerley GI, Landman M (2006) The impacts of elephants on biodiversity in the Eastern Cape Subtropical Thickets: elephant conservation. S Afr J Sci 102:395–402
Krebs CJ (2002) Ecology: the experimental analysis of distribution and abundance. Prentice Hall, Upper Saddle River, NJ
Kreft H, Jetz W (2007) Global patterns and determinants of vascular plant diversity. Proc Natl Acad Sci 104:5925–5930
Lambert WD, Holling CS (1998) Causes of ecosystem transformation at the end of the Pleistocene: evidence from mammal body-mass distributions. Ecosystems 1998:157–175
Landman M, Schoeman DS, Kerley GI (2013) Shift in black rhinoceros diet in the presence of elephant: evidence for competition? PLoS One 8(7):e69771
MacArthur RH, Wilson EO (2001) The theory of island biogeography, vol 1. Princeton University Press, Princeton, NJ
Mishra C, Van Wieren SE, Heitkönig IM, Prins HHT (2002) A theoretical analysis of competitive exclusion in a Trans-Himalayan large-herbivore assemblage. Anim Conserv Forum 5:251–258

Mishra C, Van Wieren SE, Ketner P, Heitkönig IMA, Prins HHT (2004) Competition between domestic livestock and wild bharal *Pseudois nayaur* in the Indian Trans-Himalaya. J Appl Ecol 41:344–354
Mishra C, Bhatnagar YV, Suryawanshi KR (2016) Species richness and size distribution of large herbivores in the Himalaya. In: Ahrestani FS, Sankaran M (eds) The ecology of large herbivores in South and Southeast Asia. Ecological studies. Springer, Dordrecht, vol 225, pp 89–97
Mutke J, Sommer JH, Kreft H, Kier G, Barthlott W (2011) Vascular plant diversity in a changing world: global centres and biome-specific patterns. In: Zachos FE, Habel JC (eds) Biodiversity hotspots. Springer, Berlin, pp 83–96
Olff H, Ritchie MH, Prins HHT (2002) Global environmental determinants of diversity in large herbivores. Nature 415:901–904
Olson DM, Dinerstein E (1998) The Global 200: a representation approach to conserving the Earth's most biologically valuable ecoregions. Conserv Biol 12:502–515
Olson DM, Dinerstein E, Wikramanayake ED, Burgess ND, Powell GVN, Underwood EC, D'Amico JA, Itoua I, Strand HE, Morrison JC, Loucks CJ, Allnutt TF, Ricketts TH, Kura Y, Lamoreux JF, Wettengel WW, Hedao P, Kassem KR (2001) Terrestrial ecoregions of the world: a new map of life on earth. BioScience 51:933–938
Peters RH, Peters RH (1986) The ecological implications of body size, vol 2. Cambridge University Press, Cambridge
Prins HHT, Olff H (1998) Species-richness of African grazer assemblages: towards a functional explanation. In: Newberry DE, Brown ND, Prins HHT (eds) Dynamics of tropical communities: 37th symposium of the british ecological society. Cambridge University Press, Cambridge, pp 449–490
Prins HHT, van Langevelde F (2008) Assembling a diet from different places. In: Prins HHT, van Langevelde F (eds) Resource ecology: spatial and temporal dynamics of foraging. Springer, Dordrecht, pp 129–158
Ricklefs RE (1987) Community diversity: relative roles of local and regional processes. Science 235:167–171
Sankaran M, Hanan NP, Scholes RJ, Ratnam J, Augustine DJ, Cade BS, Gignoux J, Higgins SI, Le Roux X, Ludwig F, Ardo J (2005) Determinants of woody cover in African savannas. Nature 438:846–849
Street-Perrott FA, Huang Y, Perrott RA, Eglinton G, Barker P, Khelifa LB, Harkness DD, Olago DO (1997) Impact of lower atmospheric carbon dioxide on tropical mountain ecosystems. Science 278:1422–1426
Suryawanshi KR, Bhatnagar YV, Mishra C (2010) Why should a grazer browse? Livestock impact on winter resource use by bharal Pseudois nayaur. Oecologia 162:453–462
Van Soest PJ (2018) Nutritional ecology of the ruminant. Cornell University Press, Ithaca
Wilson DE, Reeder DM (eds) (2005) Mammal species of the world: a taxonomic and geographic reference, vol 1. JHU Press, Baltimore
Woodward FI, Lomas MR, Kelly CK (2004) Global climate and the distribution of plant biomes. Philos Trans R Soc Lond B Biol Sci 359:1465–1476
Yang Y, Fang J, Ma W, Wang W (2008) Relationship between variability in aboveground net primary production and precipitation in global grasslands. Geophys Res Lett 35:L23710. https://doi.org/10.1029/2008GL035408
Zangerl AR, Bazzaz FA (1992) Theory and pattern in plant defense allocation. In: Fritz RS, Simms EL (eds) Plant resistance to herbivores and pathogens: ecology evolution and genetics. Chicago University Press, Chicago, pp 363–391

Chapter 8
Weather and Climate Impacts on Browsing and Grazing Ungulates

Randall B. Boone

8.1 Introduction

Climate, the prevailing weather in an area over a long period, provides selection pressures that determine the taxonomic diversity, geographic ranges, physiology and behaviours of ungulates. Climatic constraints are evident in the biogeography of ungulates, such as there being more than 80 bovid species alone in Africa, but just two species of ungulates breeding in the far northern polar regions (muskox, *Ovibos moschatus*, and caribou or reindeer, *Rangifer tarandus*). Ancient shifts in climates promoted vicariance through favouring grasslands or forests, dividing ungulate populations and promoting community diversity (Lorenzen et al. 2012). Climate shifts, such as the Younger Dryas, caused extinctions in many mammals (Boulanger and Lyman 2014), such as Columbian mammoths (*Mammuthus columbi*), and extinctions of mammals from their continents of origin, such as Bactrian camels (*Camelus bactrianus*) in North America. More generally, climate has influenced the distribution of vegetation complexes, and those in turn influence the distribution of faunal groups, including ungulates (see **Saarinen** Chap. 2; **Rowan and Faith** Chap. 3). Ungulates have adapted to habitats as diverse as arid landscapes, such as those that are home to the African wild ass (*Equus africanus*), high alpine areas inhabited by wild yaks (*Bos mutus*), and the high arctic sites that are home to muskox.

Over life-spans or shorter time scales, environmental conditions for ungulates are integrated by the animals' nutritional state, and reflected in changes in body condition (Parker et al. 2009). Extreme weather can cause ungulate populations to decline, and the relatively slow reproductive rates of ungulates mean that years may pass while populations rebuild (e.g., Ellis and Swift 1988). Winter conditions that

R. B. Boone (✉)
Ecosystem Science and Sustainability, Colorado State University, Fort Collins, CO, USA
e-mail: Randall.Boone@ColoState.edu

I. J. Gordon, H. H. T. Prins (eds.), *The Ecology of Browsing and Grazing II*,
Ecological Studies 239, https://doi.org/10.1007/978-3-030-25865-8_8

continue into spring can stress ungulates that have used resources to survive the length of the normal winter. Body mass, relative to some standard, reflects body condition, and poor condition affects individual their survival and reproductive rates, which in aggregate influence population size (e.g., ungulates as the breeders that rely upon built capital of Festa-Bianchet et al. (1998) and others).

In this Chapter, I review some effects of climate and weather on the physiology of ungulates, and ways in which ungulate populations relate to changes in climate and weather. I then describe some effects of recent changes in weather and climate on ungulates, their populations, and distributions, and cite potential changes in the distributions of ungulates, as environmental conditions change. These sections are intended to introduce selected concepts; the ways in which ungulate distributions, abundances, and nutrition relate to climate and weather are richly represented in the literature. For example, the effects of climate change on livestock production systems (e.g., Jones and Thornton 2009; Thornton et al. 2009; Rojas-Downing et al. 2017), and the inverse (Goodland and Anhang 2009), have been reported on broadly, and are not expanded upon here.

8.2 Weather, Climate, and Individual Effects

Thermoregulation allows ungulates to maintain their body temperatures over a variety of ambient conditions. A range of temperatures correspond to a zone of thermal comfort for an animal. In those conditions, energy put towards thermoregulation is minimized, and the remainder can be put to other uses, such as fat storage or reproduction. Temperatures causing thermal stress can vary seasonally; for example, moose (*Alces alces*) respired more quickly when temperatures rose above −5 °C in winter, and in summer that threshold was crossed at 14 °C (Renecker and Hudson 1986); wind can narrow that temperature range. Across mammals, core body temperatures vary little, with some exceptions (e.g., monotremes, sloths and anteaters, Order Pilosa, have low body temperatures, and their body temperatures can vary through the day; Taylor 1970; Shrestha et al. 2012), and basal metabolism relates to body size (Benedict 1938), scaling to the 2/3rd power (White and Seymour 2003). Thus, differences in insulation are a major determinate of the thermal comfort zones in mammals, and insulate animals in both cold and hot conditions (Scholander et al. 1950; Jessen 2001). That insulation can be remarkably effective—Parker and Gillingham (1990) cite mule deer (*Odocoileus hemionus*), having skin temperatures like those on a warm summer day, in the early morning sun, on a day with −30 °C air temperature. The differences in insulation between species, and even within an individual, can be striking, as some animals grow thick coats in winter and shed these in preparation for summer. As I have already mentioned, ungulates exist from tropical to polar biomes, and the fundamental niches of species can be broad. Subspecies of elk (*Cervus elaphus*), for example, range in North America from central Canada (e.g., an average January temperature of −30 °C in the southwest Northern Territories) to the southern United States (e.g., an average June temperature of 32 °C in western Texas).

Animals will seek out means to remain within a thermal comfort zone through movement or other behaviours (Mysterud and Østbye 1999; Terrien et al. 2011). Thermal cover captures the concept of ungulates seeking landscape features that help maintain a positive thermal energy balance (e.g., Cook et al. 2004). Vegetation cover reduces wind speed, can reduce ambient temperature through evapotranspiration, blocks solar radiation through shading during the day, and traps long-wave radiation at night. Ground-cover availability, and its albedo, are relevant to microclimates; e.g., ungulates will seek shade during the heat of the day to rest and ruminate. Other behaviours include piloerection of fur, and vasoconstriction to limit blood flow to the skin and extremities (Feist and White 1989). Animals, such as muskox, huddle together in low temperatures to decrease their collective surface-to-volume ratio and conserve heat (Parker and Robbins 1985). Ungulates will consume more food in cold conditions, to offset the energy costs of thermoregulation, and increased muscle activity (Ragsdale et al. 1950), whereas in hot conditions they may reduce food intake to limit energy expenditure and decrease heat-producing digestive activity (Parker et al. 2009; reviewed in Arnold 1985).

Intriguing trade-offs exists in thermoregulation, in that muscle activity is the main means of heat production in ungulates (e.g., through shivering or locomotion associated with acquiring more energy to compensate for the cold), and is highly effective, but can be energetically expensive (Parker and Robbins 1985). A less energetically expensive way to maintain body temperature is by absorbing heat radiation ('basking'), which explains winter movements of golden takin (*Budorcas taxicolor bedfordi*) in the Qinling Mountains of China (Zeng et al. 2010, with evolutionary background provided by Geiser et al. 2002). In contrast, dissipating excess heat is mostly passive, or has low energy costs, through vasodilation, panting and sweating. But, whereas heat production at low temperatures can be increased if sufficient energy remains, means by which excess heat can be dissipated cannot readily be increased, making ungulates more at risk from high rather than low temperatures (Terrien et al. 2011). In high temperatures, ungulates will reduce activity, and seek areas with higher wind speeds or cool rock ledges, for modest changes in heat dissipation, or cooler shaded forests (Bailey and Provenza 2008; van Beest and Milner 2013). An exception is submersion in water, which dissipates heat well.

Winter, or dry seasons, are typically bottleneck periods with respect to the nutritional needs for ungulates. A given land area can typically support more animals in the spring, summer and autumn/fall, or in the wet season than in the winter or dry seasons. Fat reserves, reduced metabolism, insulative coats, and changes in behaviour can reduce the need for energy acquisition in stressful seasons. Seasonal migration allows ungulates to escape the harshest regions (Hobbs et al. 2008). In the winter, or dry season, browse can be key resources for ungulates (Gordon and Illius 1989), which can be higher in nutrients than grasses. For example, elk will use browse when grasses are unavailable (Hobbs et al. 1981), and camels (*Camelus dromedarius*) that used shrubs in droughts in Turkana, Kenya, proved critical to pastoralists' wellbeing (Coughenour et al. 1985). Evergreen shrubs and trees can be particularly valuable for browsers, when the season or conditions cause deciduous plants to drop their leaves (Reid and Ellis 1995; Illius and O'Connor 1999). Annuals, and drought-tolerant perennials, can also be important in dry periods.

Precipitation can reduce mobility and space use, increase thermoregulation costs, and snow can increase travel costs and limit access to forage (Cooperrider 1986). Ungulates may easily paw through snow up to 10 cm deep, and some species can manage much deeper snow depths; e.g., moose may be unencumbered by snow depths of 40–50 cm (Kelsall 1969). Deep snow impedes ungulate travel (if the species does not have wide hooves to move over packed snow, such as the adaptation of caribou or reindeer), and pushing through snow can be extremely energetically expensive, if it exceeds the animal's brisket height (Parker et al. 1984). High insolation and high temperatures in the day, or winter rain, combined with low nighttime temperatures, can cause the crusting of snow, most often in the spring. If the crust can support an animal, it can decrease travel costs, but reduce access to underlying forage (Fancy and White 1985; Richard et al. 2014). If the crust breaks, it can increase travel costs, but provide easier access to forage.

Increases in soil moisture, leading to increased forage production, is an example of an indirect effect of weather on ungulate populations. Extended snow cover can improve soil moisture, and lengthen the period over which new vegetation growth is available to ungulates in the spring and summer (Mysterud et al. 2001).

8.3 Weather, Climate, and Population Effects

Prolonged periods of energy deficit, associated with harsh weather or sudden storms, or other events that limit food intake, or induce hypothermia, have direct effects on body mass, mortality rates, and birth rates (reviewed in Hudson and White 1985). Animals with relatively low body masses have a greater risk of dying in harsh winters than do those with normal body masses (Loison et al. 1999). Losses of individual animals within a group, taken in aggregate, can have population effects (e.g., Picton 1984; Georgiadis et al. 2007). The dynamics of many populations have been associated with changes in weather and extreme events, such as drought and heavy snow (Young 1994). Climate as well affects populations, such as the effects on elk populations (Sauer and Boyce 1983; Picton 1984), and the relative carrying capacities of juveniles of elk, mule deer, and bighorn sheep (*Ovis canadensis*) (Picton 1984). Analyses may relate weather events directly to sudden declines in populations, or include integrative or lagged weather and climate in assessments of reproduction or population dynamics. An example, at broader scales, is from Mongolia, where a summer drought in 2016 was followed by two large winter storms (*dzuds*), which caused a drop in temperatures, wetted animals, and blanketed pastures in snow that prevented animals from feeding. In the winter of 2016–2017, 1.1 million livestock died (Reuters 2017); in the severe dzud of 2009–2010, 9.7 million livestock died, 20% of the entire national herd (Fernandez-Gimenez et al. 2015).

Drought can cause large-scale die-offs of both wild and domestic ungulates, through low primary production and loss of water sources, and floods too can kill many animals, either through drowning, loss of access to forage, or through favouring disease vectors such as ticks (Young 1994; Boone et al. 2000). Populations of ungulates, in semi-arid and arid areas, are most at risk, with primary production linked to precipitation that can be highly variable (Ellis and Galvin 1994). At broader scales, the dynamics of some ungulate populations relate to regional atmospheric pressure patterns that cycle across years or even decades (reviewed in Steptoe et al. 2018), such as the North Atlantic Oscillation (Forchhammer et al. 1998; Post and Stenseth 1998; Mysterud et al. 2001), El Niño-Southern Oscillation (Ogutu and Owen-Smith 2003; Ogutu et al. 2009), and Indian Ocean Dipole (Ogutu et al. 2011).

The degree to which climate may influence population dynamics of ungulates in arid systems has been debated (Vetter 2005). As perspectives expanded to appreciate non-equilibrium dynamics of ecosystems (DeAngelis and Waterhouse 1987), ecologists applied those concepts to arid rangelands to consider the strength of linkages between grazers and vegetation (Ellis and Swift 1988). Systems with precipitation that is relatively consistent from year-to-year will see animal populations reach an equilibrium, with density-dependent population changes associated with excess or insufficient forage. Frequent droughts in arid systems (Young 1994) can kill large numbers of animals, and their reproductive capacity is such that populations may take years to rebound (Ellis and Swift 1988). During recovery, ungulate populations may not be closely associated with forage availability (see also **Kiffner and Lee** Chap. 6). If the frequency of droughts is shorter than the period required for populations to recover, ungulate population dynamics are essentially density-independent, and controlled by abiotic climatic effects, i.e., the linkage between ungulate population dynamics and vegetation is weak. A number of authors have argued that what may appear as density-independent responses, may be actually animal populations limited by key resources (Illius and O'Connor 1999). The spatial variability of population dynamics can be spatially fine-grained and linked to fragmented landscapes (Boone 2007; Derry and Boone 2010), and the implications for pastoral people, and the rangelands they inhabit, have been hotly debated (reviewed in Vetter 2005). I take from the literature two main messages, that: (1) the spatial extent and granularity of arid rangelands allows for a diversity of ungulate populations' responses in landscape patches; and (2) equilibrium and non-equilibrium dynamics vary in a continuum throughout arid rangelands. Both messages encourage tempering of points of view regarding arid rangeland dynamics.

8.4 Climate Change Effects and Responses

The Earth's climate is changing more rapidly than ever in recorded history (Hansen et al. 2006), due to burning of fossil fuels and other anthropogenic sources of air pollution. Mining of organic materials locked away millions of years ago, has released CO_2 and other greenhouse gasses into the atmosphere. Ruminant ungulates contribute to emissions through methane production from digestion. The increasing demand for livestock products, due to human population growth, urbanization, and changing incomes and tastes, will exacerbate the problem, although some mitigation is possible (Thornton and Herrero 2010). Today, global temperatures are, on average, 0.8 °C warmer than they otherwise would be (GISTEMP 2018). Analyses, using ensemble global circulation models, forecast a further 1.5 °C to 4.5 °C increase in temperatures by 2100 (IPCC 2018). The effects of warming are now common, such as thawing arctic sea ice, rapid snow melt, shrinking glaciers, and rising sea levels. More frequent hot days, more frequent and extreme droughts, and shifts in precipitation are underway, and are likely to be more prominent in the future (IPCC 2018). Temperatures in Europe and America have now recovered from the Little Ice Age, and are back to the levels, or have exceeded those, of the High Middle Ages (Crowley 2000; Cronin et al. 2003; McCormick et al. 2012), but have not yet reached the highs of the Antonine period of the Roman Empire (Büntgen et al. 2016).

The ecological effects of climate change are diverse (reviewed in Walther et al. 2002; Parmesan and Yohe 2003). Phenological changes are commonly reported, such as earlier annual green-up of vegetation (Boone et al. 2007; Ma et al. 2013), and flowering dates (Menzel et al. 2006). Growing seasons have lengthened, and vegetation productivity, in the northern hemisphere, has generally increased because of CO_2 fertilization (Körner 2006). Carbon dioxide is used by plants in photosynthesis to create organic compounds, and many experiments have demonstrated that increased ambient CO_2 increased photosynthetic rates and vegetation productivity (many reviews cited in Körner 2006). Large-scale field experiments, too, have shown increased vegetation production with increased ambient CO_2 (reviewed by Ainsworth and Long 2005).

8.5 Ungulate Responses to Climate Change

The responses of ungulate populations to a changing climate depend upon their exposure to changing conditions, sensitivity to those changes, resilience, and capacity to adapt. Populations may shift their geographic ranges, adapt locally, or go locally extinct. Large-scale genomic responses to climate change are unlikely in ungulates, in that rates of climate change are much faster than the rate of evolution of most species (Quintero and Wiens 2013), and generation times in large ungulates are too long to allow rapid evolutionary responses (Hetem et al. 2014). Regional changes in land cover may be expected to favour given ungulate groups, however, such as expanding woody vegetation under future climates (Boone et al. 2018) favouring mixed feeders and browsers.

Rising temperatures may cause organisms to shift their ranges to higher elevations or latitudes (e.g., Büntgen et al. 2017). Chen et al. (2011) used meta-analyses to identify an 11 m decadal shift to higher elevations, and 16.9 km shift per decade to higher latitudes across many species, with shifts in ranges correlated with levels of warming and highly variable within taxa. Large mammals are more likely to respond to a changing climate by migration than small mammals, with those at higher elevations or latitudes most likely to respond (McCain and King 2014); for example, alpine chamois (*Rupicapa rupicapra*) moved upslope in warmer periods (but were forced to move to avoid domestic sheep even more) (Mason et al. 2014). That said, shifts to higher elevations should not be envisioned as animal or plant communities simply shifting upslope. The palaeoecological and palaeoclimatical record shows that mammals have shifted their ranges in the past, in response to climate change (reviewed in Li et al. 2006). Ancient shifts, by individual species, can be idiosyncratic, rather than as communities shifting in unison.

Conservation boundaries are mostly static, and as such, conserving species in currently protected areas is an insufficient approach (Bull et al. 2013). Managers must consider opportunities for ungulates to move toward the poles and to higher elevations. Shifts in physiological niche dimensions are expected to be slow in ungulates, because of their lengthy generation times (Hetem et al. 2014), but as those niches shift in response to climate change, changes will be overshadowed by realized niches constrained by competing species, including domestic livestock. Ungulate diets can overlap more than in other taxa, and in mountainous areas, shifts to higher elevations will compress ranges due to the negative relationship between elevation and surface area, and this will increase the likelihood for dietary competition.

Large-scale shifts in modern large ungulate populations are hampered by fragmented landscapes that are semipermeable or impermeable to animal movements (Hobbs et al. 2008). Ecological generalists are better equipped to respond to shifting climate regimes than specialists, animals adapted to disturbance and semi-urban areas are likely to adapt well, and temperate species appear to be more adaptable than tropical species (Hetem et al. 2014). Janzen (1967) reminds us that "mountain passes are higher in the tropics," meaning that species with niches packed tightly, in stable climates, are physiologically more stressed to move through variable landscapes than are species adapted to harsh seasons. Some ungulates will require human intervention to persist in the face of shifting habitats, such as scimitar-horned oryx (*Oryx dammah*), which Thuiller et al. (2006) projected to have their suitable habitat shift by thousands of kilometres under future climate scenarios.

Local adaptations are the main means by which ungulates are likely to respond to climate change. Genetic changes in ungulates linked to altered climates, are poorly studied (Hetem et al. 2014; Boutin and Lane 2014), but differences in genetic diversity in caribou, under different climates have been identified, for example (Yannic et al. 2014). Anatomical changes in ungulates may result from climate change; for example, Maloney et al. (2009) hypothesized that the proportion of dark Soay sheep (*Ovis aries*), in the population on St. Kilda (Scotland), has declined because warmer temperatures increase the value of a pelt with higher albedo (Gebremedhin et al. 1997). Phenological traits are more plastic than anatomical ones (Boutin and Lane 2014), such as red deer (*Cervus elaphus*) on the Isle of Rum (Scotland), which have shifted their reproductive behaviours to earlier in the year, in association with earlier plant growth (Moyes et al. 2011). We are reminded, however, that phenological plasticity can lead to changes in phenology that are both mis-timed and maladaptive, termed a trophic mismatch (Boutin and Lane 2014).

Changes in behaviour are the most plastic form of adaptation in response to changing climates, involving only decision making by individuals. Ungulates that are diurnal or nocturnal, and in predator-free environments, are better equipped to modify their behaviours, in response to rising temperatures, than those that must be active during the day. Grazing at night, or in the cool hours of the day, reduces heat stress brought about by higher temperatures (Merrill 1991; McCain and King 2014). Moose provide an excellent example of behavioural changes, as a species sensitive to maximum temperatures in every season; for example, in Minnesota, the number of moose declined roughly 60% from historic numbers, until stabilizing in recent years, and some of that decline has been attributed to warming temperatures. Dussault et al. (2004) reported moose seeking conifer stands, with high tree cover, to avoid high air temperatures, and moose immersed themselves in water (Hetem et al. 2014). Doing so improved their thermoregulation, but reduced their foraging time (unless thermal cover is also a good food source, which was not the case in the study of Dussault et al. 2004), and their body mass and condition declined. Hoy et al. (2018) identified shrinking body sizes (inferred from skull sizes), in yearling moose at their southern range limit, and shorter life spans associated with malnutrition, and related that to warmer winters, linking behavioural and phenotypic plasticity. How much of the observed changes is associated with changing adaptations, and how much is due to changing resources, is difficult to know. Changes in moose populations in Minnesota are now thought to relate most closely with increases in white-tailed deer (*Odocoileus virginianus*) populations, and the parasites they carry that cause death in moose (Wünschmann et al. 2015). But regardless, decreases in nutritional state can increase the susceptibility of moose to parasites, other diseases, predators, and other stressors (Hetem et al. 2014), suggesting declines in southern populations may be associated with warming temperatures in complex ways (Rempel 2011).

Changes in precipitation, and in potential evapotranspiration, associated with a changing climate, poses risks for ungulates dependent upon limited water sources. As water sources dry, populations of water-dependent species will decline, such as cattle and Thomson's gazelle (*Gazella* [*Eudorcas*] *thomsonii*), and those that are better adapted to low water availability will be favoured, such as camels and Grant's gazelle (*Gazella* [*Nanger*] *granti*) (Georgiadis et al. 2007; Doreau et al. 2012). Moreover, more precipitation will fall as rain rather than snow, changing the timing of runoff from mountain streams and rivers (Gleick 1987), and, therefore, water availability for ungulates.

More frequent extreme events are of concern in the context of ungulate biology and rangeland dynamics. As cited above, populations of wild and domestic ungulates recover between droughts or winter storms, and as those storms are likely to be more frequent under climate change (Masih et al. 2014), the period for recovery is likely to be shorter. Carried through to the long-term, more frequent extreme events will lead to perennially lower ungulate populations. More frequent droughts have been verified in Africa (Masih et al. 2014), and N. Bolormaa, Secretary-General of the Mongolian Red Cross Society, said, "In the past, frequency of dzud disasters was

5 times in 60 years. The old frequency was once in 12 years. For the last 27 years, Mongolia has experienced 7 dzud disasters, or one in every 3.8 years," (Reuters 2017, but see, e.g., Tachiiri and Shinoda 2012). Our understanding of the drivers of mortality in dzuds is improving, demonstrating interactions across spatial and temporal scales. Fast drivers (*sensu* Carpenter and Turner 2001), such as dzuds, may cause dramatic change that capture peoples' interests, but slow drivers of change may cause subtle shifts in systems that make them less resilient to those fast drivers. Specifically, dry summers or droughts, and dzuds, may interact to cause livestock losses, with droughts in months before winter snows weakening animals and degrading pastures, increasing livestock mortality (Tachiiri et al. 2008; Joly et al. 2018). Those authors identified a decline in the number of young produced by livestock prior to dzud, associated with reduced body condition. Moreover, slow drivers, affecting the frequency of winter disaster, are captured in the term 'hoof dzud,' introduced to identify dzuds precipitated by overgrazing (Tachiiri and Shinoda 2012). An example of variation across spatial scales is provided by Bayasgalan et al. (2009), who portray the spatial variability in risk to livestock losses from future dzuds for regions of Mongolia (see also Tachiiri and Shinoda 2012).

Risks from more frequent extreme events are compounded by fragmented landscapes; for example, a drought in the mid-1980s caused dramatic losses among the African buffalo (*Syncerus caffer*) population in the Serengeti ecosystem, but the buffalo population recovered in less than a decade. A drought in the early 1990s also killed many buffalo, but their populations has not fully recovered due to habitat fragmentation, and also competition with livestock, predation, and more frequent droughts (Dublin and Ogutu 2015).

Shrub expansion is a global phenomenon linked with fire suppression, warmer temperatures and changes in soil moisture and snow cover associated with climate change, CO_2 fertilization, as well as changes in the abundances of wild and domestic ungulates (Myers-Smith et al. 2011; Boone et al. 2018; IPCC 2018). Invasive shrubs, such as mesquite (*Prosopis juliflora*) in East Africa, contribute to the expansion due, in part, to benefits that exotic species enjoy, such as release from pathogens and specialized herbivores (Maundu et al. 2009; Prins and Gordon 2014). These changes in forage availability can be expected to benefit browsing ungulates, with corresponding shifts in ungulate communities, and browsers, in turn, can slow shrub expansion (e.g., Myers-Smith et al. 2011; Ravolainen et al. 2014). Declines in herbaceous cover, and in the abundance of grazing domestic ungulates, are a concern for those people reliant upon animals for commercial sale, and those who are often food insecure and rely upon livestock for subsistence (e.g., Moleele et al. 2002).

In this brief review I have focused on changes in individual species associated with climate change, but we may expect cascading effects as ungulates shift their distributions, or are eliminated from communities (Beschta and Ripple 2009; Estes et al. 2011). As they are most often consumers in the middle of trophic pyramids, ungulates have controlling effects on vegetation, which in turn can influence a wealth of other ecosystem functions (e.g., Teichman et al. 2013; **Sabo** Chap. 11;

Katona and Coetsee Chap. 12). In certain conditions ungulates may be controlled by predators, with feedbacks on predator populations, other ungulates, and other species to which predators may switch (e.g., Sinclair et al. 2003). Palmer et al. (2015) provides an example—they demonstrate a trophic cascade, where population increases in three cervid species, over more than a decade, were closely associated with declines in bird species, judged *a priori* to be sensitive to loss of understory habitat.

Feral pigs (*Sus scrofa*) provide a reminder that increases in ungulate abundances, associated with a warming climate (Vetter et al. 2015), presents complex problems. In North America, feral pigs are a destructive introduced species—Anderson et al. (2016) estimate $189 million in annual crop damage across 11 USA States, and total costs of damage and control have been estimated to be $1.5 billion per year (Pimental 2007). But vocal hunting constituencies value the increased opportunity to hunt feral pigs (e.g., Bevins et al. 2014). "The conundrum is that you've got one of the world's hundred worst invasive animals, and at the same time you've got a highly desirable game species" (Dr. JJ Mayer, quoted in Goode 2013). So-called winners and losers are part of many climate change scenarios.

At broader spatial scales, a changing climate will have mixed benefits for food production for ungulates. As cited above, throughout the Northern Hemisphere, greenness and productivity, particularly in browse species, have increased through CO_2 fertilization. Debates are ongoing about the longevity of the effect, and its interplay with reduced soil moisture associated with warming, etc. (Norby et al. 2010). But, in general, ungulates are likely to have more forage biomass available to them in the future (Boone et al. 2007), but the nutrient density of that forage may be lower as more biomass is produced on soils of limited fertility (Cotrufo et al. 1998).

8.6 Conclusions

The management of limited energy available to individuals is an effective means of understanding the linkages between climate and weather, and the distributions, population dynamics, adaptations, and behaviours of ungulates. For example, range shifts, the use of thermal cover, and other behavioral responses, may be viewed as means to conserve energy, and increase the degrees of freedom in ways that energy can be allocated. Compared to wild ungulates, the physiological responses of domestic ungulates to extremes in weather are well known, but many questions remain for both groups. Forecasting the ways in which ungulates will respond to climate change, associated with rising greenhouse gas emissions, involves many more unknowns, a few of which I have touched upon, but many that are beyond the scope of this Chapter. For example, how might increasing evaporative cooling, under warmer temperatures, affect ungulate water balances (e.g., Hetem et al. 2014)? What will be the effects of more rapid green-up of vegetation (e.g., Pettorelli et al. 2007)? Will predators, and other species interacting with ungulates, be able to match their range shifts, or could we expect dramatic community changes? How may biomes shift (e.g., Bergengren et al. 2011; Luo et al. 2015)? As populations shift to higher latitudes or to higher elevations, will the dispersing populations be sustainable? If they are, will drift and bottleneck effects change the genetic makeup of the populations in undesirable ways? Are habitats, in a changed landscape, sufficient to support the populations that people wish to be maintained (e.g., Beschta et al. 2013)? Are barriers to dispersal, either natural or human made, going to limit the movements of ungulates to such a degree that shifts are not possible? N.B. we know already that many migratory pathways are no longer available (e.g., Berger 2004; Bolger et al. 2008; Harris et al. 2009). The pulsing and pressing drivers of change, to which ungulates are responding, are with us long-term—greenhouse gases contributing to radiative forcing will continue to warm the planet for centuries to come, for example (Solomon et al. 2010). The futures of many ungulate species will be dependent upon their abilities to adapt to these drivers, and the foresight, planning, and priorities of wildlife ecologists and other conservation professionals, politicians, and citizens.

References

Ainsworth EA, Long SP (2005) What have we learned from 15 years of free-air CO_2 enrichment (FACE)? A meta-analytic review of the responses of photosynthesis, canopy properties and plant production to rising CO_2. New Phylol 165:351–371. https://doi.org/10.1111/j.1469-8137.2004.01224.x

Anderson A, Slootmaker C, Harper E et al (2016) Economic estimates of feral swine damage and control in 11 US states. Crop Prot 89:89–94

Arnold GW (1985) Regulation of forage intake. In: Hudson RJ, White RG (eds) Bioenergetics of wild herbivores. CRC Press, Boca Raton, FL, pp 81–101

Bailey WB, Provenza FD (2008) Mechanisms determining large-herbivore distribution. In: Prins HHT, van Langevelde F (eds) Resource ecology: spatial and temporal dynamics of foraging. Springer, Dordrecht, pp 7–28

Bayasgalan B, Mijiddorj R, Gombluudev P, Oyunbaatar D, Bayasgalan M, Tas A et al (2009) Climate change and sustainable livelihood of rural people in Mongolia. In: Devissher T, O'Brien G, O'Keefe P, Tellam I (eds) The adaptation continuum: groundwork for the future. ETC Foundation, Leusden, pp 193–213

Benedict FG (1938) Vital energetics: a study in comparative basal metabolism. Carnegie Institution of Washington, Washington, DC. Publication no. 503

Bergengren JC, Waliser DE, Yung YL (2011) Ecological sensitivity: a biospheric view of climate change. Clim Chang 107:433. https://doi.org/10.1007/s10584-011-0065-1

Berger J (2004) The last mile: how to sustain long-distance migration in mammals. Conserv Biol 18:320–331

Beschta RL, Ripple WJ (2009) Large predators and trophic cascades in terrestrial ecosystems of the western United States. Biol Conserv 142:2401–2414

Beschta RL, Donahue DL, DellaSala DA, Rhodes JJ, Karr JR, O'Brien MH, Fleischner TL, Deacon Williams C (2013) Adapting to climate change on western public lands: addressing the ecological effects of domestic, wild, and feral ungulates. Environ Manage 51:474–491

Bevins SN, Pedersen K, Lutman MW, Gidlewski T, Deliberto TJ (2014) Consequencies associated with the recent range expansion of nonnative feral swine. Bioscience 64:291–299

Bolger DT, Newmark WD, Morrison TA, Doak DF (2008) The need for integrative approaches to understand and concerve migratory ungulates. Ecol Lett 11:63–77

Boone RB (2007) Effects of fragmentation on cattle in African savannas under variable precipitation. Landsc Ecol 22:1355–1369. https://doi.org/10.1007/s10980-007-9124-4
Boone RB, Galvln KA, Smith NM, Lynn SJ (2000) Generalizing El Niño effects upon Maasai livestock using hierarchical clusters of vegetation patterns. Photogramm Eng Remote Sensing 66:737–744
Boone RB, Lackett JM, Galvin KA et al (2007) Links and broken chains: evidence of human-caused changes in land cover in remotely sensed images. Environ Sci Policy 10:135–149. https://doi.org/10.1016/j.envsci.2006.09.006
Boone RB, Conant RT, Sircely J et al (2018) Climate change impacts on selected global rangeland ecosystem services. Glob Chang Biol 24:1382–1393. https://doi.org/10.1111/gcb.13995
Boulanger MT, Lyman RL (2014) Northeastern North American Pleistocene megafauna chronologically overlapped minimally with Paleoindians. Quat Sci Rev 85:35–46. https://doi.org/10.1016/j.quascirev.2013.11.024
Boutin S, Lane JE (2014) Climate change and mammals: evolutionary versus plastic responses. Evol Appl 7:29–41. https://doi.org/10.1111/eva.12121
Bull JW, Suttle KB, Singh NJ, Milner-Gulland EJ (2013) Conservation when nothing stands still: moving targets and biodiversity offsets. Front Ecol Environ 11:203–210
Büntgen U, Myglan VS, Ljungqvist FC, McCormick M, Di Cosmo N et al (2016) Cooling and societal change during the Late Antique Little Ice Age from 536 to around 660 AD. Nat Geosci 9:231–236
Büntgen U, Greuter L, Bollmann K et al (2017) Elevational range shifts in four mountain ungulate species from the Swiss Alps. Ecosphere 8:e01761. https://doi.org/10.1002/ecs2.1761
Carpenter SR, Turner MG (2001) Hares and tortoises: interactions of fast and slow variables in ecosystems. Ecosystems 3:495–497
Chen I-C, Hill JK, Ohlemüller R et al (2011) Rapid range shifts of species associated with high levels of climate warming. Science 333:1024–1026. https://doi.org/10.1126/science.1206432
Cook JG, Irwin LL, Bryant LD, Riggs RA, Thomas JW (2004) Thermal cover needs of large ungulates: a review of hypothesis tests. In: Transactions of the 69th North American wildlife and natural resources conference, pp 708–726
Cooperrider A (1986) Inventory and monitoring of wildlife habitat. US Dept. Interior Bureau of Land Management, Denver, CO
Cotrufo MF, Ineson P, Scott A (1998) Elevated CO_2 reduces the nitrogen concentration of plant tissues. Glob Chang Biol 4:43–54. https://doi.org/10.1046/j.1365-2486.1998.00101.x
Coughenour MB, Ellis JE, Swift DM et al (1985) Energy extraction and use in a nomadic pastoral ecosystem. Science 230:619–625. https://doi.org/10.1126/science.230.4726.619
Cronin TM, Dwyer GS, Kamiya T, Schwede S, Willard DA (2003) Medieval warm period, little ice age and 20th century temperature variability from Chesapeake Bay. Glob Planet Change 36:17–29
Crowley TJ (2000) Causes of climate change over the past 1000 years. Science 289:270–277
DeAngelis DL, Waterhouse JC (1987) Equilibrium and nonequilibrium concepts in ecological models. Ecol Monogr 57:1–21. https://doi.org/10.2307/1942636
Derry JF, Boone RB (2010) Grazing systems are a result of equilibrium and non-equilibrium dynamics. J Arid Environ 74:307–309. https://doi.org/10.1016/j.jaridenv.2009.07.010
Doreau M, Corson MS, Wiedemann SG (2012) Water use by livestock: a global perspective for a regional issue? Anim Front 2:9–16
Dublin HT, Ogutu JO (2015) Population regulation of African buffalo in the Mara–Serengeti ecosystem. Wildl Res 42:382. https://doi.org/10.1071/WR14205
Dussault C, Ouellet J-P, Courtois R et al (2004) Behavioural responses of moose to thermal conditions in the boreal forest. Ecoscience 11:321–328
Ellis J, Galvin KA (1994) Climate patterns and land-use practices in the dry zones of Africa. BioScience 44:340–349
Ellis JE, Swift DM (1988) Stability of African pastoral ecosystems: alternate paradigms and implications for development. J Range Manag 41:450–459

Estes JA, Terborgh J, Brashares JS et al (2011) Trophic downgrading of planet earth. Science 333:301–306

Fancy SG, White RG (1985) Incremental cost of activity. In: Hudson RJ, White RG (eds) Bioenergetics of wild herbivores. CRC Press, Boca Raton, FL, pp 143–159

Feist DD, White RG (1989) Terrestrial mammals in cold. In: Wang LCH (ed) Animal adaptations to cold. Springer, Berlin, pp 327–360

Fernandez-Gimenez ME, Batkhishig B, Batbuyan B, Ulambayar T (2015) Lessons from the dzud: community-based rangeland management increases the adaptive capacity of Mongolian herders to winter disasters. World Dev 68:48–65

Festa-Bianchet M, Gaillard J-M, Jorgenson JT (1998) Mass- and density-dependent reproductive success and reproductive costs in a capital breeder. Am Nat 152:367–379

Forchhammer MC, Stenseth NC, Post E, Langvatn R (1998) Population dynamics of Norwegian red deer: density-dependence and climatic variation. Proc Biol Sci 265:341–350

Gebremedhin KG, Ni H, Hillman PE (1997) Modeling temperature profile and heat flux through irradiated fur layer. Trans ASAE 40:1441–1447

Geiser F, Goodship N, Pavey CR (2002) Was basking important for the evolution of mammalian endothermy? Naturwissenschaften 89:412–414

Georgiadis NJ, Olwero JGN, Ojwang G, Romañach SS (2007) Savanna herbivore dynamics in a livestock-dominated landscape: I. Dependence on land use, rainfall, density, and time. Biol Conserv 137:461–472. https://doi.org/10.1016/j.biocon.2007.03.005

GISTEMP Team (2018) GISS surface temperature analysis (GISTEMP). NASA Goddard Institute for Space Studies. http://data.giss.nasa.gov/gistemp/. Accessed 1 Nov 2018

Gleick PH (1987) Regional hydrologic consequences of increases in atmospheric CO_2 and other trace gases. Clim Chang 10:137–160. https://doi.org/10.1007/BF00140252

Goode E (2013) When one man's game is also a marauding pest. New York Times, New York

Goodland R, Anhang J (2009) Livestock and climate change: what if the key actors in climate change are … cows, pigs, and chickens? Worldwatch Institute, Washington, DC

Gordon IJ, Illius AW (1989) Resource partitioning by ungulates on the Isle of Rhum. Oecologia 79:383–389. https://doi.org/10.1007/BF00384318

Hansen J, Sato M, Ruedy R et al (2006) Global temperature change. Proc Natl Acad Sci USA 103:14288–14293. https://doi.org/10.1073/pnas.0606291103

Harris G, Thirgood S, Hopcraft JGC, Cromsigt JPGM, Berger J (2009) Global declines in aggregated migrations of large terrestrial mammals. Endang Species Res 7:55–76

Hetem RS, Fuller A, Maloney SK, Mitchell D (2014) Responses of large mammals to climate change. Temp (Austin) 1:115–127. https://doi.org/10.4161/temp.29651

Hobbs NT, Baker DL, Ellis JE, Swift DM (1981) Composition and quality of elk winter diets in Colorado. J Wildl Manage 45:156–171

Hobbs NT, Galvin KA, Stokes CJ et al (2008) Fragmentation of rangelands: implications for humans, animals, and landscapes. Glob Environ Chang 18:776–785. https://doi.org/10.1016/j.gloenvcha.2008.07.011

Hoy SR, Peterson RO, Vucetich JA (2018) Climate warming is associated with smaller body size and shorter lifespans in moose near their southern range limit. Glob Chang Biol. https://doi.org/10.1111/gcb.14015

Hudson RJ, White RG (1985) Bioenergetics of wild herbivores. CRC Press, Boca Raton, FL

Illius AW, O'Connor TG (1999) On the relevance of nonequilibrium concepts to arid and semiarid grazing systems. Ecol Appl 9:798–813

IPCC (International Panel on Climate Change) (2018) Summary for policymakers. In: Masson-Delmotte V, Zhai P, Pörtner HO, Roberts D, Skea J et al (eds) Global warming of 1.5°C. an IPCC special report on the impacts of global warming of 1.5°C above pre-industrial levels and related global greenhouse gas emission pathways, in the context of strengthening the global response to the threat of climate change, sustainable development, and efforts to eradicate poverty. World Meteorological Organization, Geneva

Janzen DH (1967) Why mountain passes are higher in the tropics. Am Nat 101:233–249

Jessen C (2001) Temperature regulation in humans and other mammals. Springer, Heidelberg. https://doi.org/10.1007/978-3-642-59461-8

Joly F, Sabatier R, Hurbert B, Munkhtuya B (2018) Livestock productivity as indicator of vulnerability to climate hazards: a Mongolian case study. Nat Hazards 92:S95–S107. https://doi.org/10.1007/s11069-017-2963-7

Jones PG, Thornton PK (2009) Croppers to livestock keepers: livelihood transitions to 2050 in Africa due to climate change. Environ Sci Pol 12:427–437. doi.org. https://doi.org/10.1016/j.envsci.2008.08.006

Kelsall JP (1969) Structural adaptations of moose and deer in snow. J Mammal 50:302–310

Körner C (2006) Plant CO_2 responses: an issue of definition, time and resource supply. New Phytol 172:393–411. https://doi.org/10.1111/j.1469-8137.2006.01886.x

Li M-H, Krauchi N, Gao S-P (2006) Global warming: can existing reserves really preserve current levels of biological diversity? J Integr Plant Biol 48:255–259. https://doi.org/10.1111/j.1744-7909.2006.00232.x

Loison A, Langvatn R, Solberg EJ (1999) Body mass and winter mortality in red deer calves: disentangling sex and climate effects. Ecography 22:20–30. https://doi.org/10.1111/j.1600-0587.1999.tb00451.x

Lorenzen ED, Heller R, Siegismund HR (2012) Comparative phylogeography of African savannah ungulates 1. Mol Ecol 21:3656–3670. https://doi.org/10.1111/j.1365-294X.2012.05650.x

Luo Z, Jiang Z, Tang S (2015) Impacts of climate change on distributions and diversity of ungulates on the Tibetan Plateau. Ecol Appl 25:24–38

Ma T, Zhou C, Pei T, Xie Y (2013) A comparative analysis of changes in the phasing of temperature and satellite-derived greenness at northern latitudes. J Geogr Sci 23:57–66. https://doi.org/10.1007/s11442-013-0993-y

Maloney SK, Fuller A, Mitchell D (2009) Climate change: is the dark Soay sheep endangered? Biol Lett 5:826–829. https://doi.org/10.1098/rsbl.2009.0424

Masih I, Maskey S, Mussá FEF, Trambauer P (2014) A review of droughts on the African continent: a geospatial and long-term perspective. Hydrol Earth Syst Sci 18:3635–3649. https://doi.org/10.5194/hess-18-3635-2014

Mason THE, Stephens PA, Apollonio M, Willis SG (2014) Predicting potential responses to future climate in an alpine ungulate: interspecific interactions exceed climate effects. Glob Chang Biol 20:3872–3882. https://doi.org/10.1111/gcb.12641

Maundu P, Kibet S, Morimoto Y et al (2009) Impact of *Prosopis juliflora* on Kenya's semi-arid and arid ecosystems and local livelihoods. Biodiversity 10:33–50. https://doi.org/10.1080/14888386.2009.9712842

McCain CM, King SRB (2014) Body size and activity times mediate mammalian responses to climate change. Glob Chang Biol 20:1760–1769. https://doi.org/10.1111/gcb.12499

McCormick M, Büntgen U, Cane MA, Cook ER, Harper K et al (2012) Climate change during and after the Roman empire: reconstructing the past from scientific and historical evidence. J Interdiscip Hist 43:231–236

Menzel A, Sparks TH, Estrella N et al (2006) European phenology response to climate change matches the warming pattern. Glob Chang Biol 12:1969–1976

Merrill EH (1991) Thermal constraints on use of cover types and activity time of elk. Appl Anim Behav Sci 29:251–267. https://doi.org/10.1016/0168-1591(91)90252-S

Moleele NM, Ringrose S, Matheson W, Vanderpost C (2002) More woody plants? The status of bush encroachment in Botswana's grazing areas. J Environ Manag 64:3–11. https://doi.org/10.1006/jema.2001.0486

Moyes K, Nussey DH, Clements MN et al (2011) Advancing breeding phenology in response to environmental change in a wild red deer population. Glob Chang Biol 17:2455–2469. https://doi.org/10.1111/j.1365-2486.2010.02382.x

Myers-Smith IH, Forbes BC, Wilmking M et al (2011) Shrub expansion in tundra ecosystems: dynamics, impacts and research priorities. Environ Res Lett 6:45509. https://doi.org/10.1088/1748-9326/6/4/045509

Mysterud A, Østbye E (1999) Cover as a habitat element for temperate ungulates: effects on habitat selection and demography. Wildl Soc Bull 27:385–394

Mysterud A, Stenseth NC, Yoccoz NG et al (2001) Nonlinear effects of large-scale climatic variability on wild and domestic herbivores. Nature 410:1096–1099. https://doi.org/10.1038/35074099

Norby RJ, Warren JM, Iversen CM, Medlyn BE, McMurtrie RE (2010) CO2 enhancement of forest productivity constrained by limited nitrogen availability. PNAS 107:19368–19373. https://doi.org/10.1073/pnas.1006463107

Ogutu JO, Owen-Smith N (2003) ENSO, rainfall and temperature influences on extreme population declines among African savanna ungulates. Ecol Lett 6:412–419. https://doi.org/10.1046/j.1461-0248.2003.00447.x

Ogutu JO, Piepho H-P, Dublin HT et al (2009) Rainfall extremes explain interannual shifts in timing and synchrony of calving in topi and warthog. Popul Ecol 52:89–102. https://doi.org/10.1007/s10144-009-0163-3

Ogutu JO, Owen-Smtih N, Piepho H-P, Said MY (2011) Conitinuing wildlife population declines and range contraction in the Mara region of Kenya during 1977–2009. J Zool 285:99–109

Palmer G, Stephens PA, Ward AI, Willis SG (2015) Nationwide trophic cascades: changes in avian community structure driven by ungulates. Sci Rep 5:15601. https://doi.org/10.1038/srep15601

Parker KL, Gillingham MP (1990) Estimates of critical thermal environments for mule deer. J Range Manag 43:73–81

Parker KL, Robbins CT (1985) Thermoregulation in ungulates. In: Hudson RJ, White RG (eds) Bioenergetics in wild herbivores. CRC Press, Boca Raton, FL, pp 161–182

Parker KL, Robbins CT, Hanley TA (1984) Energy expenditures for locomotion by mule deer and elk. J Wildl Manag 48:474–488

Parker KL, Barboza PS, Gillingham MP (2009) Nutrition integrates environmental responses of ungulates. Funct Ecol 23:57–69. https://doi.org/10.1111/j.1365-2435.2009.01528.x

Parmesan C, Yohe G (2003) A globally coherent fingerprint of climate change impacts across natural systems. Nature 421:37–42. https://doi.org/10.1038/nature01286

Pettorelli N, Pelletier F, von Hardenberg A, Festa-Bianchet M, Côté SD (2007) Early onset of vegetation growth vs. rapid green-up: impacts on juvenile mountain ungulates. Ecology 88:381–390

Picton HD (1984) Climate and the prediction of reproduction of three ungulate species. J Appl Ecol 21:869–879

Pimental D (2007) Environmental and economic costs of vertebrate species invasions into the United States. In: Managing vertebrate invasive species, 38

Post E, Stenseth NC (1998) Large-scale climatic fluctuation and population dynamics of moose and white-tailed deer. J Anim Ecol 67:537–543. https://doi.org/10.1046/j.1365-2656.1998.00216.x

Prins HHT, Gordon IJ (eds) (2014) Invasion biology and ecological theory: insights from a continent in transformation. Cambridge University Press, Cambridge

Quintero I, Wiens JJ (2013) Rates of projected climate change dramatically exceed past rates of climatic niche evolution among vertebrate species. Ecol Lett 16:1095–1103. https://doi.org/10.1111/ele.12144

Ragsdale AC, Thompson HJ, Worstell DM, Brody S (1950) Environmental physiology, with special reference to domestic animals. IX. Milk production and feed and water consumption responses of Brahman, Jersey, and Holstein cows to changes in temperature, 50° to 105° F and 50° to 8° F. Univ Missouri Agric Exp Stat, Columbia, Missouri, Research Bulletin 460

Ravolainen VT, Bråthan KA, Yoccoz NG, Nguyen JK, Ims RA (2014) Complementary impacts of small rodents and semi-domesticated ungulates limit tall shrub expansion in the tundra. J Appl Ecol 51:234–241

Reid RS, Ellis JE (1995) Impacts of pastoralists on woodlands in South Turkana, Kenya: livestock-mediated tree recruitment. Ecol Appl 5:978–992

Rempel RS (2011) Effects of climate change on moose populations: exploring the response horizon through biometric and systems models. Ecol Model 222:3355–3365. https://doi.org/10.1016/j.ecolmodel.2011.07.012

Renecker LA, Hudson RJ (1986) Seasonal energy expenditures and thermoregulatory responses of moose. Can J Zool 64:322–327. https://doi.org/10.1139/z86-052

Reuters (2017) Widespread deaths among livestock in Mongolia after second big freeze. In: Glob News. https://globalnews.ca/news/3254902/widespread-deaths-among-livestock-in-mongolia-after-second-big-freeze/. Accessed 26 Jan 2018

Richard JH, Wilmshurst J, Côté SD (2014) The effect of snow on space use of an alpine ungulate: recently fallen snow tells more than cumulative snow depth. Can J Zool 92:1067–1074. https://doi.org/10.1139/cjz-2014-0118

Rojas-Downing MM, Nejadhashemi AP, Harrigan T, Woznicki SA (2017) Climate change and livestock: impacts, adaptations, and mitigation. Clim Risk Manag 16:145–163. https://doi.org/10.1016/j.crm.2017.02.001

Sauer JR, Boyce MS (1983) Density dependence and survival of elk in northwestern Wyoming. J Wildl Manag 47:31. https://doi.org/10.2307/3808049

Scholander PF, Hock R, Walters V, Irving L (1950) Adaptation to cold in artic and tropical mammals and birds in relation to body temperature, insulation, and basal metabolic rate. Biol Bull 99:259–271. https://doi.org/10.2307/1538742

Shrestha AK, van Wieren SE, van Langevelde F, Fuller A, Hetem RS, Meyer LCR, de Bie S, Prins HHT (2012) Body temperature variation of South African antelopes in two climatically contrasting environments. J Thermal Biol 37:171–178. https://doi.org/10.1016/j.jtherbio.2011.12.008

Sinclair ARE, Mduma S, Brashares JS (2003) Patterns of predation in a diverse predator-prey system. Nature 425:288–290

Solomon S, Daniel JS, Sanford TJ, Murphy DM, Plattner G-K, Knutti R, Friedlingstein P (2010) Persistence of climate change due to a range of greenhouse gases. Proc Natl Acad Sci USA 43:18354–18359

Steptoe H, Jones SEO, Fox H (2018) Correlations between extreme atmospheric hazards and global teleconnections: implications for multihazard resilience. Rev Geophys 56:50–78. https://doi.org/10.1002/2017RG000567

Tachiiri K, Shinoda M (2012) Quantitative risk assessment for future meteorological disasters: reduced livestock mortality in Mongolia. Clim Chang 113:867–882

Tachiiri K, Shinoda M, Klinkenberg B, Morinaga Y (2008) Assessing Mongolian snow disaster risk using livestock and satellite data. J Arid Env 72:2251–2263. https://doi.org/10.1016/j.jaridenv.2008.06.015

Taylor CR (1970) Strategies of temperature regulation: effect on evaporation in East African ungulates. Am J Phys 219:1131–1135

Teichman JK, Nielsen ES, Roland J (2013) Trophic cascades: linking ungulates to shrub-dependent birds and butterflies. J Anim Ecol 82:1288–1299. https://doi.org/10.1111/1365-2656.12094

Terrien J, Perret M, Aujard F (2011) Behavioral thermoregulation in mammals: a review. Front Biosci 16:1428–1444

Thornton PK, Herrero M (2010) Potential for reduced methane and carbon dioxide emissions from livestock and pasture management in the tropics. Proc Natl Acad Sci USA 107:19667–19672. https://doi.org/10.1073/pnas.0912890107

Thornton PK, van de Steeg J, Notenbaert A, Herrero M (2009) The impacts of climate change on livestock and livestock systems in developing countries: a review of what we know and what we need to know. Agric Syst 101:113–127. https://doi.org/10.1016/j.agsy.2009.05.002

Thuiller W, Broennimann O, Hughes G et al (2006) Vulnerability of African mammals to anthropogenic climate change under conservative land transformation assumptions. Glob Chang Biol 12:424–440. https://doi.org/10.1111/j.1365-2486.2006.01115.x

Van Beest FM, Milner JM (2013) Behavioural responses to thermal conditions affect seasonal mass changes in a heat-sensitive northern ungulate. PLoS One 8(6):e65972. https://doi.org/10.1371/journal.pone.0065972

Vetter S (2005) Rangelands at equilibrium and non-equilibrium: recent developments in the debate. J Arid Environ 62:321–341. https://doi.org/10.1016/j.jaridenv.2004.11.015

Vetter SG, Ruf T, Bieber C, Arnold W (2015) What is a mild winter? Regional differences in within-species responses to climate change. PLoS One 10:e0132178. https://doi.org/10.1371/journal.pone.0132178
Walther G-R, Post E, Convey P et al (2002) Ecological responses to recent climate change. Nature 416:389–395. https://doi.org/10.1038/416389a
White CR, Seymour RS (2003) Mammalian basal metabolic rate is proportional to body mass$^{2/3}$. PNAS 100:4046–4049. https://doi.org/10.1073/pnas.0436428100
Wünschmann A, Armien AG, Butler E et al (2015) Necropsy findings in 62 opportunistically collected free-ranging moose (*Alces alces*) from Minnesota, USA (2003–2013). J Wildl Dis 51:157–165. https://doi.org/10.7589/2014-02-037
Yannic G, Pellissier L, Ortego J, Lecomte N, Couturier S, Cuyler C et al (2014) Genetic diversity in caribou linked to past and future climate change. Nat Clim Chang 4:132–137. https://doi.org/10.1111/ele.12376
Young TP (1994) Natural die-offs of large mammals: implications for conservation. Conserv Biol 8:410–418. https://doi.org/10.1046/j.1523-1739.1994.08020410.x
Zeng Z, Skidmore AK, Song Y, Wang T, Gong H (2010) Seasonal altitudinal movements of golden takin in the Qinling Mountains of China. J Wildl Manag 72:611–617. https://doi.org/10.2193/2007-197

Chapter 9
Impacts of Browsing and Grazing Ungulates on Soil Biota and Nutrient Dynamics

Judith Sitters and Walter S. Andriuzzi

9.1 Introduction on Soil Biota: Their Importance for Ecosystem Functioning

Far from being merely the surface on which herbivores walk, or the medium in which plants grow, soils are a fundamental component of terrestrial ecosystems. Soil functioning depends on the organisms that inhabit it, including bacteria, archaea, fungi, protists, and animals, which show a tremendous taxonomic and ecological diversity (Bardgett and van der Putten 2014). Soil biota control processes upon which plants and their consumers depend, such as the decomposition of plant and animal remains, recycling and redistribution of carbon (C) and nutrients, maintenance of soil structure, and oxygen and water availability to plant roots (Brussaard et al. 1997; Rillig and Mummey 2006). Many soil microbes and invertebrates also directly interact with plants as mutualists (e.g., mycorrhizal fungi, N_2-fixing bacteria), pathogens, or root-feeders (Wardle et al. 2004). Through their overall effect on soil fertility, or by facilitating or suppressing certain plant species or functional groups, soil biota can alter plant diversity and community composition (Ruijven et al. 2004; Partsch et al. 2006; Bonkowski and Roy 2012), with ecosystem-wide consequences on the structure and successional trajectory of terrestrial ecosystems (Streitwolf-Engel et al. 1998; De Deyn et al. 2003). The diversity and functional attributes of soil communities shape their interactions with plants and their effects on

J. Sitters (✉)
Ecology and Biodiversity, Department of Biology, Vrije Universiteit Brussel, Brussels, Belgium
e-mail: judith.sitters@vub.be

W. S. Andriuzzi
Department of Biology, School of Global Environmental Sustainability, Colorado State University, Fort Collins, CO, USA

Present Address: Nature Communications, Nature Research, Berlin, Germany
e-mail: walter.andriuzzi@nature.com

I. J. Gordon, H. H. T. Prins (eds.), *The Ecology of Browsing and Grazing II*, Ecological Studies 239, https://doi.org/10.1007/978-3-030-25865-8_9

soil processes (Wardle et al. 2004; Hättenschwiler and Gasser 2005; Kardol et al. 2016), and determines C and nutrient cycling and retention (Johnston et al. 2004). Therefore, soil communities may affect herbivores through various ecological processes, generally with plants as intermediary. In turn, browsing and grazing ungulates can affect soil organisms and biological processes via their effects on plants, and also by returning C and nutrients (through excreta and carcasses) and modifying soil physical properties (e.g., through trampling).

9.2 The Impact of Ungulates on Soil Biota

Ungulates may affect soil biota through trophic mechanisms, namely return of plant-derived C and nutrients constituting basal resources in the soil food web, and non-trophic mechanisms, namely physical disturbance. These mechanisms can impact soil biota simultaneously—e.g., reduced plant cover due to grazing may also affect soil biota by changing temperature and moisture (Yates et al. 2000; Odriozola et al. 2014). Ungulates can change the amount, and quality, of resources fuelling the soil food web, by chemical and/or physical changes in plant litter, resulting from physiological responses of plants and/or shifts in plant species composition (reviewed in Bardgett and Wardle 2003), or by providing fast pools of C and nutrients via their excreta (Mikola et al. 2009; Mosbacher et al. 2016; Buscardo et al. 2017) and carcasses (Carter et al. 2006; Szelecz et al. 2016). Browsed and grazed plants may also alter the allocation of C belowground, including root exudates (Hamilton et al. 2008; Sun et al. 2017), which are now recognized as a no less important energy source than is leaf litter to many soil food webs (Pollierer et al. 2007; Fujii et al. 2016).

Physical disturbance by ungulates, although often less appreciated by ecologists than plant feeding and excretion, may have large effects on soil structure, processes, and biota. Trampling by medium- and large-sized ungulates may cause significant soil compaction, especially on fine-textured soils (Van Haveren 1983; Schrama et al. 2013). Soil compaction, caused by trampling, may be strongest in relatively small areas of the landscapes, for instance along preferred trails (Ostermann-Kelm et al. 2009), but nonetheless reduces the abundance of sensitive soil biota across the landscape (Cluzeau et al. 1992; Beever and Herrick 2006; Sorensen et al. 2009). There is growing evidence that the impact of large ungulates on soil physical properties, especially moisture, partly determine the patchiness of plant communities in many grazed ecosystems (Howison et al. 2017). In addition, some ungulates affect soil physically by digging—e.g., wild boar (*Sus scrofa*) (Mohr et al. 2005; Bueno et al. 2013), African elephant *(Loxodonta spp.*) (Haynes 2012)—although the most important herbivore bioturbators are fossorial mammals such as rodents and lagomorphs (Davidson et al. 2012).

In general, grazing and browsing ungulates may be expected to have distinct effects belowground due to the plant communities in which they dominate. In early successional systems, non-woody plants with their high-quality litter, and increased nutrient uptake and growth due to grazing, are expected to promote soil microbial decomposers and animal detritivores, whereas opposite effects are expected in late successional systems because of increased dominance of woody plants due to browsing, which have low-quality litter (Bardgett and Wardle 2003). Quantitative syntheses, however, showed that negative effects of mammalian herbivores on the abundance of soil biota are far more common than positive effects (Andriuzzi and Wall 2017; Zhao et al. 2017). A meta-analysis, by Andriuzzi and Wall (2017), pointed to both climate and herbivore body size as controlling the response of soil biota, with declines in abundance of soil organisms and mineralization rates more likely in harsh climates and/or with large herbivores. Nonetheless, it is possible that soil biota to respond differently to browsers *vs.* grazers in a given ecosystem, as well as to browsers or grazers with different feeding behaviours. For example, in Europe horses (*Equus caballus*) tend to select against tall vegetation patches as compared to the foraging choices of cattle (*Bos taurus*), maintaining a mosaic of plants and, therefore, litter which cattle tend instead to homogenise (Menard et al. 2002). Moreover, by feeding on grass closer to the ground than cattle (Gordon 1989), horses could expose the soil microclimate to larger fluctuations.

9.2.1 Shifts in Soil Microbial and Invertebrate Communities

Through their effects on basal resources and soil physical properties, browsers and grazers may induce shifts in soil microbial and invertebrate communities. Herbivory may lead plants to allocate less C to mycorrhizal partners, and in fact negative effects of browsers on ectomycorrhizae have been reported from different ecosystems (Rossow et al. 1997; Markkola et al. 2004), although a meta-analysis found that mycorrhizal responses to herbivory are often negligible and may even be positive (Barto and Rillig 2010). When it results in the enhanced availability and decomposability of resources entering the soil food web, grazing may promote bacteria over fungi (Bardgett et al. 2001; Andres et al. 2016), whereas it is unclear whether the reverse shift occurs when grazing has negative effects on resources such as soil organic matter content. Little is known on the responses of different taxa, or functional groups, of bacteria and fungi, but there is some evidence that among bacteria Gram-positive taxa such as Actinobacteria may benefit from grazing (Yang et al. 2013; Eldridge et al. 2017). In Australian drylands, rabbits (*Oryctolagus cuniculus*) negatively affected both ectomycorrhizal and arbuscular mycorrhizal fungi, whereas sheep (*Ovis aries*) tended to suppress fungal pathogens of plants (Eldridge and Delgado-Baquerizo 2018).

Groups of soil invertebrates may also diverge in their responses to herbivore activities. Microarthropods and macro-predators tend to decline under grazing, whereas macro-detritivores, such as earthworms and some insect larvae, are more resistant and may even increase (Bardgett et al. 1993; Kay et al. 1999; Schon et al. 2012). However, responses *within* a taxonomic group may vary considerably; of the two most abundant groups of microarthropods, both animal decomposers in a broad sense, oribatid mites tend to be more negatively affected by grazing than collembola

(Andriuzzi and Wall 2017). This could be due to differences in life strategies such as development rate and fecundity (Behan-Pelletier 1999). Zooming into these microarthropod taxa, response traits vary widely among species, for instance in relation to physical disturbance (Maraun et al. 2003) and, in fact, some collembola are also negatively affected by grazing while some oribatid mites are not (Francini et al. 2014). Similarly, some earthworm species may benefit from grazing while others may be negatively affected (Mikola et al. 2009). Broadly defined functional groups—cross-taxa aggregations of species based on presumed similarity in ecological attributes, such as diet and life history strategies (Brussaard 1998)—may also obscure ecological differences between species (Heemsbergen et al. 2004; Fujii et al. 2016). These findings show that responses (or lack thereof) observed at coarse taxonomic resolution, or broad functional groups, may conceal ecologically meaningful changes in relation to mammalian herbivore activity detectable at finer taxonomic scales. A way to detect such changes is to identify and measure functional traits that link biota to ecosystem processes (Heemsbergen et al. 2004; Bardgett and van der Putten 2014; Kardol et al. 2016). Despite a growing interest in functional traits in various areas of ecology, their use in herbivore-soil research has lagged behind. For example, body size—a trait of fundamental importance in soil communities (Brose et al. 2012)—is increasingly being investigated in the context of climate change (Bokhorst et al. 2012; Lindo et al. 2012; Knox et al. 2017), but rarely in response to grazing (Mulder et al. 2008; Mills and Adl 2011; Schon et al. 2012; Comor et al. 2014; Andriuzzi and Wall 2018), let alone browsing.

Depending on the context, ungulates may cause either increases in the average size of soil fauna due to the increased availability of basal resources, or decreases due to physical disturbance (e.g., by compacting soil and thus reducing pore space). Recently, grazing, in a semi-arid steppe, was found to skew soil nematode communities toward larger body sizes by increasing soil fertility where it was lowest (Andriuzzi and Wall 2018), consistent with previous fertilization studies (Verschoor et al. 2001; Liu et al. 2015). On the other hand, findings from wetter soils, rich in organic matter, suggest that grazing disturbance may promote smaller body sizes in soil fauna (Mulder et al. 2008; Comor et al. 2014). Current knowledge on how browsers and grazers may affect other functional traits than body size in soil microbes and fauna is even more limited.

9.3 Cascading Effects of Ungulates on Soil Communities

9.3.1 Soil Food Web Changes

Ungulate-induced shifts in the community composition of microbes and invertebrates has implications for soil food web structure and functioning. Grazers and browsers may favour the bacterial over the fungal channel or vice versa, but predicting what controls the direction of change has proved elusive (Wardle et al. 2001; Mikola et al. 2001; Andriuzzi et al. 2013). Responses at higher trophic levels are also difficult to predict. In a large-scale study in New Zealand, browsing had idiosyncratic effects on soil biota across sites, however, in a given location responses tended to be consistent across trophic levels, e.g., where detritivorous nematodes and microarthropods declined, so did their predators (Wardle et al. 2001). On the other hand, in subarctic ecosystems root-feeding nematodes are often negatively affected by large ungulates whereas predatory nematodes are not (Andriuzzi et al. 2013), and the same was found in a temperate marsh (Yang et al. 2017); conversely, in alpine pastures root-feeding nematodes did not respond to grazing, whereas the other nematode trophic groups tended to benefit from herbivore removal (Vandegehuchte et al. 2016).

In some cases, the lack of soil food web responses may be explained by a weak, or negligible, effect of ungulates on soil properties. In sub-alpine grasslands, Vandegehuchte et al. (2016) found no consistent effects of grazing on soil nutrients, and in fact no substantial change in soil food web structure was detected, as the different trophic levels responded in synchrony (i.e., bacterial- and fungal-feeding nematodes, as well as their putative predators, all became more abundant). In other cases, responses in soil food web dynamics to browsing and grazing appear decoupled from those of plants, soil properties, or even the abundance of major groups of soil biota. In New Zealand forests, microbial biomass either decreased, increased, or was unchanged by browsing, and the responses at the next trophic level (microbial-feeding nematodes) were often neutral or even opposite, e.g., in some sites browsing increased microbial biomass but decreased microbial-feeder abundance (Wardle et al. 2001). This could indicate that microbes benefitted from an ungulate-induced decline in their consumers, or increased due to distinct sensitivity to the habitat changes caused by the ungulates (e.g., changes in soil microclimate due to trampling and alterations in litter thickness). A grassland study found that soil organic matter turnover and bacterial biomass greatly increased under grazing, but the only consumers to show a similar response were omnivorous nematodes, and no change in the amount of C processed by the fungal *vs.* the bacterial channel was found (Andres et al. 2016). This may be partly explained by the fact that abundance or biomass of soil biota, both microbes and animals, is not necessarily a good predictor of their effects on process rates (Pausch et al. 2015; Rousk 2016). Presently not much is known about how ungulates may affect, not only the composition of soil food webs, but also their functional attributes. This limits the understanding of what herbivore-soil interactions entail for ecosystem functioning. Given the context-dependency of soil ecological patterns, the available knowledge does not currently suffice to make robust generalizations. For instance, whilst the aforementioned study by Andres et al. (2016) found that soil texture determines whether grazing strengthens or weakens soil food web stability in semi-arid grassland, it is premature to say whether this is a common pattern.

9.3.2 Feedbacks Between the Soil Food Web and Ungulates

Soil communities do not merely respond to ungulates, but influence them as well. By altering plant biomass, traits, diversity and community composition, soil biota clearly have the potential to modulate the availability and quality of food resources available to browsing and grazing ungulates (Ingham et al. 1985; Moore et al. 2003; Wardle et al. 2004). Soil microbial decomposers can increase nutrient availability for plants (Hamilton et al. 2008) or compete with them (Bardgett et al. 2003), thereby, either boosting or depressing food quality for ungulates. It is well-established that mycorrhizal fungi, root-associated bacteria, and earthworms can alter the nutritional content and defence compounds in plant leaves and shoots, and, therefore, affect their resistance to insect herbivores (Babikova et al. 2013; Badri et al. 2013; Xiao et al. 2017), but if, and how, this affects vertebrate browsers and grazers is not well-known. As for soil fauna, microbivores and detritivores, across a wide range of sizes, from nematodes to earthworms, often enhance plant growth (Wu et al. 2013; van Groenigen et al. 2014; Trap et al. 2016), and may, therefore, benefit ungulates. Recent studies also highlighted more complex interactions; in African savannas, the formation of nutrient hotspots, preferred by ungulates—e.g., grazing lawns (Pretorius et al. 2011; Veldhuis et al. 2014; Cromsigt and te Beest 2014)—may be either facilitated or hindered by soil fauna. More broadly, evidence has been accruing that interactions of ungulates with ecosystem engineers, such as earthworms, termites, ants, and dung beetles, are a major contributor to the patchiness of grazed ecosystems (Howison et al. 2017). There is also evidence of the positive effects on plant resistance or resilience against climate variability in the presence of certain soil microbes and macroinvertebrates (Lau and Lennon 2012; Andriuzzi et al. 2015; Bonachela et al. 2015; Johnson et al. 2015), and at least some of the latter are closely associated with ungulates, for instance coprophagous specialists such as many species of dung beetles (Nichols et al. 2009).

9.4 The Role of Ungulates in Nutrient Cycling

Ungulates strongly impact nutrient cycling and availability, thereby determining plant community composition, productivity and the functioning of these ecosystems (Milchunas and Lauenroth 1993; Hobbs 1996; Bardgett and Wardle 2003). The classically accepted view is that they do this mainly through changing the quantity and/or quality of resources entering the soil, affecting the rate of organic matter decomposition and soil nutrient availability (reviewed in Bardgett and Wardle 2003). Ungulates influence the quantity of resource inputs to the soil through changes in plant biomass production and resource allocation, fine root dynamics and root exudation (Augustine and McNaughton 1998; Bardgett and Wardle 2003; Markkola et al. 2004; Pastor et al. 2006; Hamilton et al. 2008; Klumpp et al. 2009). They can also influence the quality of resource inputs into the soil, either directly by returning nutrients in the form of dung and/or urine, which are more easily decomposed than plant litter (Hobbs 1996; Augustine et al. 2003; Mikola et al. 2009; Buscardo et al. 2017; Sitters et al. 2017a; Veldhuis et al. 2018), or indirectly by altering plant nutrient concentrations and/or secondary metabolites, which change litter quality (Bardgett et al. 1998; Ritchie et al. 1998; Wardle et al. 2002). Additionally, over longer time scales, litter quantity and quality can be changed through ungulate-induced shifts in primary productivity and/or plant species composition (Davidson 1993; Ritchie et al. 1998; Côté et al. 2004; Ripple and Beschta 2012). Empirical studies have found positive (Hobbs et al. 1991; Frank and Evans 1997; McNaughton et al. 1997; Liu et al. 2018), negative (Pastor et al. 1993; Ritchie et al. 1998; Harrison and Bardgett 2004; Gass and Binkley 2011), or mixed effects (Olofsson et al. 2001; Singer and Schoenecker 2003; Stark et al. 2003; Bakker et al. 2009; Schrama et al. 2013; Sitters et al. 2017b) of ungulates on nutrient (mostly nitrogen) cycling.

It is assumed that the net effect of ungulates on nutrient cycling depends on the dominant leaf traits and plant physiological responses to consumption, which in turn depend on ecosystem fertility and successional state (Ritchie et al. 1998; Bardgett and Wardle 2003; Pastor et al. 2006). In early successional, nutrient-rich ecosystems, ungulates can promote the growth of plants that are tolerant of herbivory with high litter quality, thereby increasing mineralization rates and nutrient availability. Additionally, ungulates excrete a higher proportion of N in urine when the N content of plants is higher (Hobbs 1996), while diet N and dung N are also positively correlated (Sitters et al. 2017a). Ungulate urine and/or dung has been shown to stimulate soil microbial activity and nutrient cycling (Frank and McNaughton 1992; van der Wal et al. 2004; Fornara and du Toit 2008). In contrast, ungulates can selectively remove high quality plants with low carbon to nitrogen (C:N) ratios from late successional nutrient-poor systems, thereby strengthening the dominance of plants that are intolerant of herbivory with low quality litter, and decreasing mineralization rates and nutrient availability. Although this framework improved our understanding of above-belowground interactions, evidence gathered since suggests that its generalities hold only in some contexts (Stark and Grellmann 2002; Bakker et al. 2009; Cherif and Loreau 2013; Schrama et al. 2013; Stark et al. 2015; Sitters et al. 2017b). Several alternative mechanisms have been proposed to explain the accelerating or decelerating effects of mammalian ungulates on nutrient cycling (synthesized in Sitters and Olde Venterink 2015 and in Table 9.1). When present, these mechanisms will, together, determine the net effect ungulates have on nutrient cycling.

9.4.1 Ungulate Impacts on Other Nutrients than Nitrogen

The aforementioned examples, and ideas, on the role of mammalian ungulates in nutrient cycling are based on the cycling of N, while the cycling of other nutrients, potentially important for plant growth, such as phosphorus (P) or potassium (K), have often been ignored. However, the biogeochemical cycles of N, P and K differ

Table 9.1 Alternative mechanisms by which ungulates accelerate or decelerate nutrient cycling

Ungulate-induced mechanism	Pathway by which nutrient cycling is accelerated	Pathway by which nutrient cycling is decelerated	Refs.
Carbon respiration of consumed vegetation	In areas with high plant C:N ratios, the decrease of organic C supply decreases microbial immobilization when microbes are C-limited	In areas with low plant C:N ratios, the decrease of organic C supply decreases microbial mineralization when microbes are C-limited	Sankaran and Augustine (2004), Cherif and Loreau (2013)
Removal of aboveground vegetation	By an increase in the area of bare soil, which increases soil temperature and the activity of soil organisms	By an increase in the area of bare soil, which increases soil salinity and decreases the activity of soil organisms	van der Wal et al. (2004), Schrama et al. (2013)
Lateral transport of nutrients	By the deposition of urine, dung or carcasses in an area where ungulates have not consumed these nutrients	By consuming vegetation in one area and not returning these nutrients through deposition of urine, dung or carcasses in the same area	Augustine et al. (2003), Sitters et al. (2015), Veldhuis et al. (2018)
Soil compaction by trampling		By limiting oxygen or water availability in wet and dry soils respectively, particularly those with a fine texture (i.e., clay). Soils with intermediate soil moisture are less sensitive to compaction	Schrama et al. (2013)

(e.g., N availability is largely determined by microbiological transformations, while P availability is strongly affected by physico-chemical processes in the soil; Chapin et al. 2002), therefore, it cannot be assumed that an ungulate-induced change in one element will be matched by change in other elements. This is likely to have consequences for plant growth and ecosystem dynamics; for example, an important way in which ungulates can have differential impacts on N and P cycling is through the differential release of these nutrients in their waste products (i.e., dung and urine). The ratio of N to P released strongly depends on the stoichiometric composition of the ungulate's food and the ungulate's demand for these nutrients (Sterner 1990; Urabe et al. 1995; Elser and Urabe 1999). Conceptual models envisage that, because of their high bone mass and P demand, ungulates recycle more N than P and are consequently able to shift plant nutrient limitation to more P-limited conditions (Sterner 1990; Sterner and Elser 2002). In practice little is known about the N:P requirements of wild ungulates (Sardans et al. 2012; Sitters et al. 2017a) but two recent studies have corroborated these model predictions with observations in the field (Sitters et al. 2017b; Veldhuis et al. 2018).

9.5 The Impacts of Browsing vs. Grazing Ungulates on Nutrient Cycling

Decelerating and accelerating effects of ungulates on nutrient cycling rates have been ascribed to browsing and grazing systems, respectively (Bardgett and Wardle 2003; Pastor et al. 2006). This is because these effects are best known from the boreal forest of Isle Royale in North America (browsing system) and the Serengeti savanna in East Africa (grazing system) (Fig. 9.1). The former is an example of an infertile, unproductive ecosystem where long-term selective browsing by moose has caused acceleration of succession, leading to the domination of woody plants with low litter quality (Pastor et al. 1993). This system carries a low number of browsers which, consume a low percentage of primary productivity, resulting in a low return of labile nutrients in dung and/or urine that is not large enough to offset the negative effects through ungulate-induced changes in plant composition and quality. In contrast, the Serengeti is an example of a fertile, productive system where non-selective grazing facilitates increased nutrient uptake and growth by the grazing-tolerant grasses, creating grazing lawns that produce high quality litter (McNaughton 1984; Veldhuis et al. 2014; Hempson et al. 2015). This system carries a large number of grazers, which consume a high percentage of primary productivity, resulting in a high return of labile nutrients in dung and/or urine. Indeed, the return of nutrients in dung and urine is presumed to be the main driving mechanism for ungulates stimulating N cycling and availability in these grassland ecosystems (Bardgett and Wardle 2003). The studies in these two distinct ecosystems have led to the generalisation that browsers decelerate nutrient cycling, while grazers accelerate it (Pastor et al. 2006). This dichotomy is maintained as literature on positive consumer-resource interactions in plant-ungulate systems is dominated by studies on grassland systems, while our understanding of browsing systems is much more limited and based on few studies performed in relatively nutrient poor ecosystems (Cromsigt and Kuijper 2011).

Studies have shown that browsers can accelerate nutrient cycling (Stark et al. 2007; Fornara and du Toit 2008), such as in savanna where intense browsing actually increases resource availability and nutrient cycling in browsing hotspots (du Toit et al. 1990; Fornara and du Toit 2007; Cromsigt and Kuijper 2011; Fig. 9.1). Several studies have also observed that grazers decelerate nutrient cycling, such as reindeer (*Rangifer tarandus*) in tundra heath (Stark and Grellmann 2002; Stark et al. 2003; Fig. 9.1), and elk (*Cervus elaphus*) and cattle in grasslands (Singer and Schoenecker 2003; Zhou et al. 2017). These observations have prompted du Toit and Olff (2014) to strongly question the generality of the grazer-browser dichotomy as a controller of nutrient cycling. They propose that the influence of large ungulates on the rate of nutrient cycling is controlled by the mix of herbivory tolerance and resistance traits in the plant community, and not based on the relative dominance of browsing or grazing. However, neither the classical framework of Bardgett and Wardle (2003), nor this proposal incorporate any of the alternative mechanisms (Table 9.1).

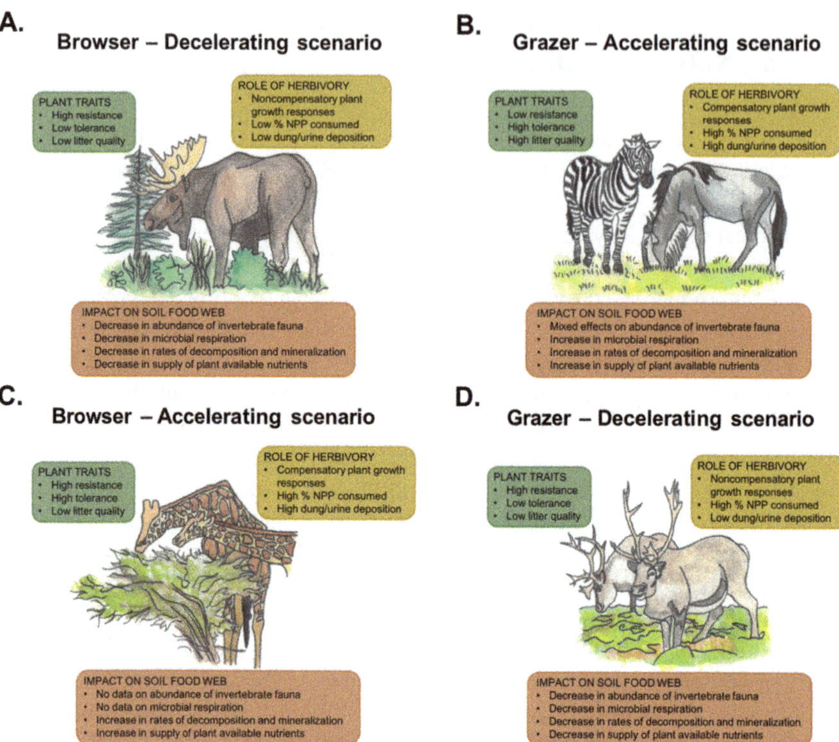

Fig. 9.1 Four possible scenarios of how browsing and grazing ungulates impact soil food webs, whereby they can change the abundance of soil fauna, and either decelerate or accelerate nutrient cycling. Scenario A and B represent the best-known examples of browsing by moose in boreal forest (Pastor et al. 1993) and grazing by zebra and wildebeest in savanna (McNaughton 1984), whereby the former decreases nutrient cycling while the latter increase it. The abundance of invertebrate fauna generally decreases in scenario A, likely due to a reduction in the amount of high quality leaf litter (Suominen et al. 1999), whereas in scenario B it may either increase or decrease (or be unaffected) depending on the combined effects of increase of provision of high quality resources (litter, root exudates, dung and/or urine) *vs.* strong soil compaction by trampling (Andriuzzi and Wall 2017). Scenario C shows that browsers, such as giraffe in savanna, can accelerate nutrient cycling by high deposition of dung and/or urine, even if litter quality is low due to strong nutrient resorption (Fornara and du Toit 2008). For this scenario we are unaware of any studies looking at the impact on soil fauna or microbial activity. Scenario D shows that grazers, such as reindeer in tundra heath, can decelerate nutrient cycling especially when reindeer migration causes a net resource output from the system (Stark and Grellmann 2002). Here the abundance of invertebrate fauna also decreases due to soil compaction by trampling (Francini et al. 2014). Note that reindeer can also be responsible for acceleration of nutrient cycling (Stark et al. 2002; Olofsson et al. 2004; Sitters et al. 2017b). The figure expands on the framework from Bardgett and Wardle (2003)

9.5.1 The Importance of Alternative Mechanisms in Explaining the Impact of Browsing vs. Grazing Ungulates on Nutrient Cycling

How important the proposed alternative mechanisms are in explaining the impact of browsing as compared to grazing ungulates on nutrient cycling is not known, but they likely do not play the same role in every ecosystem. For example, changes in soil properties, due to compaction through trampling, might play a more important role when grazers are the dominant ungulate, as they often congregate in large herds, while browsers are almost always solitary (Pastor et al. 2006). For example, the deceleration of nutrient cycling when the density of grazing livestock (cattle/sheep) is high (Zhou et al. 2017) might be related to the negative effects of trampling. Ungulates are also known to redistribute nutrients across the landscape (Singer and Schoenecker 2003; Augustine 2004; Moe and Wegge 2008; Sitters et al. 2015) and, thereby, strongly increase the spatial heterogeneity of, for example, N:P availability (Sitters et al. 2017a). Grazing and browsing ungulates are likely to have a differential impact on this spatial redistribution. In savannas, mega-grazers, such as white rhinoceros (*Ceratotherium simum*) and hippopotamus (*Hippopotamus amphibius*), do not feed and defecate in the same areas (Veldhuis et al. 2018). The latter in particular cause a major redistribution of nutrients from terrestrial grasslands to aquatic systems (Subalusky et al. 2015), while white rhinoceros are responsible for a negative nutrient balance in grasslands by the translocation of nutrients to their dung deposition points (Veldhuis et al. 2018).

Browsers seem to cause a net flow of nutrients from woodlands to grasslands (Sitters et al. 2014; Veldhuis et al. 2018), while meso-grazers tend to congregate and deposit dung in open sites (Augustine 2004; van der Waal et al. 2011). Both browsers and grazers likely do this to avoid predation (Riginos and Grace 2008). Variation in N:P ratios of the dung of ungulates can be linked to the stoichiometric composition of the ungulate's food, and thus to their feeding strategy. Dung from browsing ungulates tends to have higher N:P ratios than dung from grazing ungulates (Sitters et al. 2014; Valdés-Correcher et al. 2019), as N concentrations in dung increase with decreasing grass consumption (Codron et al. 2007). Rates of dung decomposition and nutrient release are in turn influenced by dung N:P ratios, with lower relative losses of the least abundant nutrient (Sitters et al. 2014). Browser dung is, therefore, likely to increase relative N availability more than P availability to the soil and plant components of the ecosystem. However, dung from browsers might contain higher amounts of secondary metabolites (e.g., tannin, lignin), which could decrease the release rate of N to the soil (Hobbs 1996). Hence, differences in nutrients and other compounds between grazer and browser dung are likely to have strong local effects upon N and P cycling.

9.6 Consequences of Altered Nutrients for Ecosystem Dynamics

When ungulates are responsible for the acceleration of nutrient cycling and availability, more nutrients are available to plants in a shorter amount of time and, hence, primary productivity and plant quality will increase. In contrast, when ungulates are responsible for the deceleration of nutrient cycling, less nutrients are available to plants in the same amount of time, reducing primary productivity and increasing the dominance of less nutritious species (Bardgett and Wardle 2003). Hence, ungulates have the capacity to force shifts in vegetation states, with either negative (Pastor et al. 1993; Rietkerk et al. 1997), or positive (McNaughton et al. 1997; van der Wal 2006), consequences for ecosystem functioning and the density of ungulates. For example, acceleration of nutrient cycling by reindeer caused transitions from moss- or shrub-dominated tundra to graminoid-dominated tundra, enhancing plant productivity and density of reindeer in the system (van der Wal 2006). By contrast, deceleration of nutrient cycling by moose in boreal forest reduced plant productivity and shifted dominance towards low-quality tree species, leading to a catastrophic collapse of the herbivore population (Pastor et al. 1993). Ungulates also have the potential to change the relative balance of nutrients, such as N and P, available to plants and are, therefore, able to shift plant nutrient limitation. Studies have observed ungulates to either increase N-limited (Carline et al. 2005), NP co-limited (Frank 2008) or P-limited conditions (Sitters et al. 2017b) for plant growth. It is, however, hard to make any general predictions on the stoichiometric impact of ungulates on plants (Sitters et al. 2017a), hence there is a pressing need to collect data from a broad range of sites across ecosystems around the world.

The stoichiometric impact of browsing versus grazing ungulates on plants is related to the stoichiometric variation (e.g., N:P ratio) in their dung, as differences in the nutrient release rates of dung will likely influence competitive interactions between plants and plant community composition. In savannas we, therefore, expect both grazing and browsing ungulates to play an important role in maintaining the tree-grass balance; i.e., the low N:P ratio of the dung of grazers benefits N_2-fixing tree seedlings, while the high N:P ratio of the dung of browsers favours the growth of grasses (Sitters and Olde Venterink 2018). In accordance with this concept, woody recruitment was constrained in abandoned livestock areas, where P levels were 25-fold higher than the surrounding savanna landscape (van der Waal et al. 2011). Moreover, a recent study (Valdés-Correcher et al. 2019) showed that differences in dung stoichiometry among ungulate species influences the diversity and composition of an experimental plant community. These studies highlight the importance of a diverse assemblage of ungulate species (both browsing and grazing) to maintain ecosystem productivity, richness and functioning.

9.7 Future Research Needs

To better understand the impacts of browsing and grazing ungulates on soil biota and nutrient dynamics, more studies on browsers are needed as so far most research has focused on grazers. To better distinguish between the impacts of browsing *vs.* grazing ungulates, studies are required where browsing and grazing effects are compared across ecosystem types. As changes in abundance and diversity of broad groups of soil biota (e.g., bacteria, soil mesofauna, etc.) do not necessarily indicate how ungulates affect soil functioning, research would benefit from studies focusing on the impacts of browsers and grazers (i) on functional traits of soil biota (e.g., body size) and functional attributes of soil food webs; and/or (ii) at finer taxonomic scales (e.g., with species or genus identification for fauna). Additionally, topical research on important, but little studied taxa (e.g., protists), and across a range of soil properties (e.g., soil texture, organic matter content), would be valuable. When it comes to the impact of browsing and grazing ungulates on nutrient cycling, there is a pressing need to develop and validate an integrated model that includes all the important and alternative feedbacks between herbivores, plants and soil nutrient cycling (Sitters and Olde Venterink 2015). Additionally, the inclusion of elements other than nitrogen (e.g., phosphorus, potassium) would help develop more general predictions on the stoichiometric impact of ungulates on soils and plants, and how this feedbacks on the ungulates themselves. For this, data needs to be collected across a broad range of sites among ecosystems around the world. Comparing the impact of co-occurring browsers and grazers will also contribute to better understanding under which conditions ungulates accelerate or decelerate nutrient cycling.

9.8 Conclusions

Not only do browsing and grazing ungulates impact aboveground vegetation, they also have a strong impact on soil biota and nutrient dynamics. Changes in belowground properties and processes may, in turn, impact aboveground vegetation through various mechanisms, among which availability of water or nutrients for plant growth, and eventually feedback on the ungulates themselves. Ungulates may affect soil biota through trophic mechanisms, by changing the return of basal resources to the soil, or through non-trophic mechanisms, such as changes in soil physical properties. These herbivore-induced changes can cause shifts in soil microbial and invertebrate communities, which will have consequences for soil food web structure and functioning. These changes, in turn, will feedback on to the ungulates themselves and have consequences for ecosystem functioning, as soil biota have the potential to modulate plant communities.

Ungulates strongly impact nutrient dynamics through changes in the quantity and/or quality of resources entering the soil, whereby it is assumed that the net effect of ungulates depends on the dominant leaf traits and plant physiological responses to consumption, which are in turn determined by ecosystem fertility and successional state. Differences between the effects of browsers and grazers can be interpreted mechanistically, for instance in terms of chemical content of plant litter and waste products. The classical theory of ungulate-induced changes in resource quantity and/or quality links browsing ungulates and grazing ungulates to deceleration and acceleration of nutrient cycling, respectively. However, recent evidence on alternative mechanisms (e.g., soil compaction by trampling and lateral transport of nutrients) questions this classical theory and should be incorporated to develop a more integrative framework. By changing nutrient dynamics, and balances, for plant growth, ungulates can shift plant nutrient limitation and vegetation states, which has consequences for ecosystem functioning and the density of ungulates.

References

Andres P, Moore JC, Simpson RT et al (2016) Soil food web stability in response to grazing in a semi-arid prairie: the importance of soil textural heterogeneity. Soil Biol Biochem 97:1–13. https://doi.org/10.1016/j.soilbio.2016.02.014

Andriuzzi WS, Wall DH (2017) Responses of belowground communities to large aboveground herbivores: meta-analysis reveals biome-dependent patterns and critical research gaps. Glob Chang Biol 23:3857–3868. https://doi.org/10.1111/gcb.13675

Andriuzzi WS, Wall DH (2018) Grazing and resource availability control soil nematode body size and abundance-mass relationship in semi-arid grassland. J Anim Ecol 87:1407–1417. https://doi.org/10.1111/1365-2656.12858

Andriuzzi WS, Keith AM, Bardgett RD, van der Wal R (2013) Soil nematode assemblage responds weakly to grazer exclusion on a nutrient-rich seabird island. Eur J Soil Biol 58:38–41. https://doi.org/10.1016/j.ejsobi.2013.05.009

Andriuzzi WS, Pulleman M, Schmidt O et al (2015) Anecic earthworms (*Lumbricus terrestris*) alleviate negative effects of extreme rainfall events on soil and plants in field mesocosms. Plant Soil 397:103–113. https://doi.org/10.1007/s11104-015-2604-4

Augustine DJ (2004) Influence of cattle management on habitat selection by impala on central Kenyan rangeland. J Wildl Manag 68:916–923. https://doi.org/10.2193/0022-541x(2004)068[0916,iocmoh]2.0.co;2

Augustine DJ, McNaughton SJ (1998) Ungulate effects on the functional species composition of plant communities: herbivore selectivity and plant tolerance. J Wildl Manag 62:1165–1183

Augustine DJ, McNaughton SJ, Frank DA (2003) Feedbacks between soil nutrients and large herbivores in a managed savanna ecosystem. Ecol Appl 13:1325–1337. https://doi.org/10.1890/02-5283

Babikova Z, Gilbert L, TJA B et al (2013) Underground signals carried through common mycelial networks warn neighbouring plants of aphid attack. Ecol Lett 16:835–843. https://doi.org/10.1111/ele.12115

Badri DV, Zolla G, Bakker MG et al (2013) Potential impact of soil microbiomes on the leaf metabolome and on herbivore feeding behavior. New Phytol 198:264–273. https://doi.org/10.1111/nph.12124

Bakker ES, Knops JMH, Milchunas DG et al (2009) Cross-site comparison of herbivore impact on nitrogen availability in grasslands: the role of plant nitrogen concentration. Oikos 118:1613–1622. https://doi.org/10.1111/j.1600-0706.2009.17199.x
Bardgett RD, van der Putten WH (2014) Belowground biodiversity and ecosystem functioning. Nature 515:505–511. https://doi.org/10.1038/nature13855
Bardgett RD, Wardle DA (2003) Herbivore-mediated linkages between aboveground and belowground communities. Ecology 84:2258–2268
Bardgett RD, Frankland JC, Whittaker JB (1993) The effects of agricultural management on the soil biota of some upland grasslands. Agric Ecosyst Environ 45:25–45. https://doi.org/10.1016/0167-8809(93)90057-V
Bardgett RD, Wardle DA, Yeates GW (1998) Linking above-ground and below-ground interactions: how plant responses to foliar herbivory influence soil organisms. Soil Biol Biochem 30:1867–1878
Bardgett RD, Jones AC, Jones DL et al (2001) Soil microbial community patterns related to the history and intensity of grazing in sub-montane ecosystems. Soil Biol Biochem 33:1653–1664. https://doi.org/10.1016/S0038-0717(01)00086-4
Bardgett RD, Streeter TC, Bol R (2003) Soil microbes compete effectively with plants for organic-nitrogen inputs to temperate grasslands. Ecology 84:1277–1287. https://doi.org/10.1890/0012-9658(2003)084
Barto EK, Rillig MC (2010) Does herbivory really suppress mycorrhiza? A meta-analysis. J Ecol 98:745–753. https://doi.org/10.1111/j.1365-2745.2010.01658.x
Beever EA, Herrick JE (2006) Effects of feral horses in Great Basin landscapes on soils and ants: direct and indirect mechanisms. J Arid Environ 66:96–112. https://doi.org/10.1016/j.jaridenv.2005.11.006
Behan-Pelletier VM (1999) Oribatid mite biodiversity in agroecosystems: role for bioindication. Agric Ecosyst Environ 74:411–423. https://doi.org/10.1016/S0167-8809(99)00046-8
Bokhorst S, Phoenix GK, Bjerke JW et al (2012) Extreme winter warming events more negatively impact small rather than large soil fauna: shift in community composition explained by traits not taxa. Glob Chang Biol 18:1152–1162. https://doi.org/10.1111/j.1365-2486.2011.02565.x
Bonachela JA, Pringle RM, Sheffer E et al (2015) Termites mounds can increase the robustness of dryland ecosystems to climatic change. Science 347:651–655
Bonkowski M, Roy J (2012) Decomposer community complexity affects plant competition in a model early successional grassland community. Soil Biol Biochem 46:41–48. https://doi.org/10.1016/j.soilbio.2011.11.013
Brose U, Dunne JA, Montoya JM et al (2012) Climate change in size-structured ecosystems. Philos Trans R Soc B Biol Sci 367:2903–2912. https://doi.org/10.1098/rstb.2012.0232
Brussaard L (1998) Soil fauna, guilds, functional groups and ecosystem processes. Appl Soil Ecol 9:123–135. https://doi.org/10.1016/S0929-1393(98)00066-3
Brussaard L, Behan-Pelletier VM, Bignell DE et al (1997) Biodiversity and ecosystem functioning in soil. Ambio 26:563–570
Bueno CG, Azorín J, Gómez-García D et al (2013) Occurrence and intensity of wild boar disturbances, effects on the physical and chemical soil properties of alpine grasslands. Plant Soil 373:243–256. https://doi.org/10.1007/s11104-013-1784-z
Buscardo E, Geml J, Schmidt SK et al (2017) Of mammals and bacteria in a rainforest: temporal dynamics of soil bacteria in response to simulated N pulse from mammalian urine. Funct Ecol. https://doi.org/10.1111/1365-2435.12998
Carline KA, Jones HE, Bardgett RD (2005) Large herbivores affect the stoichiometry of nutrients in a regenerating woodland ecosystem. Oikos 110:453–460
Carter DO, Yellowlees D, Tibbett M (2006) Cadaver decomposition in terrestrial ecosystems. Naturwissenschaften 94:12–24. https://doi.org/10.1007/s00114-006-0159-1
Chapin FS, Matson PA, Mooney HA (2002) Principles of terrestrial ecosystem ecology. Springer, New York

Cherif M, Loreau M (2013) Plant - herbivore - decomposer stoichiometric mismatches and nutrient cycling in ecosystems. Proc R Soc B-Biological Sci 280:20122453. https://doi.org/10.1098/rspb.2012.2453

Cluzeau D, Binet F, Vertes F et al (1992) Effects of intensive cattle trampling on soil-plant-earthworms system in two grassland types. Soil Biol Biochem 24:1661–1665. https://doi.org/10.1016/0038-0717(92)90166-U

Codron D, Lee-Thorp JA, Sponheimer M, Codron J (2007) Nutritional content of savanna plant foods: implications for browser/grazer models of ungulate diversification. Eur J Wildl Res 53:100–111. https://doi.org/10.1007/s10344-006-0071-1

Comor V, Thakur MP, Berg MP et al (2014) Productivity affects the density–body mass relationship of soil fauna communities. Soil Biol Biochem 72:203–211. https://doi.org/10.1016/j.soilbio.2014.02.003

Côté SD, Rooney TP, Tremblay J-P et al (2004) Ecological impacts of deer overabundance. Annu Rev Ecol Evol Syst 35:113–147. https://doi.org/10.1146/annurev.ecolsys.35.021103.105725

Cromsigt J, Kuijper DPJ (2011) Revisiting the browsing lawn concept: evolutionary interactions or pruning herbivores? Perspect Plant Ecol Evol Syst 13:207–215. https://doi.org/10.1016/j.ppees.2011.04.004

Cromsigt JPGM, te Beest M (2014) Restoration of a megaherbivore: landscape-level impacts of white rhinoceros in Kruger National Park, South Africa. J Ecol 102:566–575. https://doi.org/10.1111/1365-2745.12218

Davidson DW (1993) The effects of Herbivory and Granivory on terrestrial plant succession. Oikos 68:23–35

Davidson AD, Detling JK, Brown JH (2012) Ecological roles and conservation challenges of social, burrowing, herbivorous mammals in the world's grasslands. Front Ecol Environ 10:477–486. https://doi.org/10.1890/110054

De Deyn GB, Raaijmakers CE, Zoomer HR et al (2003) Soil invertebrate fauna enhances grassland succession and diversity. Nature 422:711–713. https://doi.org/10.1038/nature01548

du Toit JT, Olff H (2014) Generalities in grazing and browsing ecology: using across-guild comparisons to control contingencies. Oecologia 174:1075–1083. https://doi.org/10.1007/s00442-013-2864-8

du Toit JT, Bryant JP, Frisby K (1990) Regrowth and palatability of Acacia shoots following pruning by African savanna browsers. Ecology 71:149–154

Eldridge DJ, Delgado-Baquerizo M (2018) Functional groups of soil fungi decline under grazing. Plant Soil. https://doi.org/10.1007/s11104-018-3617-6

Eldridge DJ, Delgado-Baquerizo M, Travers SK et al (2017) Competition drives the response of soil microbial diversity to increased grazing by vertebrate herbivores. Ecology 98:1922–1931. https://doi.org/10.1002/ecy.1879

Elser JJ, Urabe J (1999) The stoichiometry of consumer-driven nutrient recycling: theory, observations, and consequences. Ecology 80:735–751. https://doi.org/10.1890/0012-9658(1999)080[0735,tsocdn]2.0.co;2

Fornara DA, du Toit JT (2007) Browsing lawns? Responses of *Acacia nigrescens* to ungulate browsing in an African savanna. Ecology 88:200–209

Fornara DA, du Toit JT (2008) Browsing-induced effects on leaf litter quality and decomposition in a southern african savanna. Ecosystems 11:238–249. https://doi.org/10.1007/s10021-007-9119-7

Francini G, Liiri M, Mannisto M et al (2014) Response to reindeer grazing removal depends on soil characteristics in low Arctic meadows. Appl Soil Ecol 76:14–25. https://doi.org/10.1016/j.apsoil.2013.12.003

Frank DA (2008) Ungulate and topographic control of nitrogen: phosphorus stoichiometry in a temperate grassland; soils, plants and mineralization rates. Oikos 117:591–601. https://doi.org/10.1111/j.2007.0030-1299.16220.x

Frank DA, Evans RD (1997) Effects of native grazers on grassland N cycling in Yellowstone National Park. Ecology 78:2238–2248. https://doi.org/10.1890/0012-9658(1997)078[2238:eongog]2.0.co;2

Frank DA, McNaughton SJ (1992) The ecology of plants, large mammalian herbivores, and drought in Yellowstone National Park. Ecology 73:2043–2058. https://doi.org/10.2307/1941454
Fujii S, Mori AS, Kominami Y et al (2016) Differential utilization of root-derived carbon among collembolan species. Pedobiologia 59:225–227. https://doi.org/10.1016/j.pedobi.2016.05.001
Gass TM, Binkley D (2011) Soil nutrient losses in an altered ecosystem are associated with native ungulate grazing. J Appl Ecol 48:952–960. https://doi.org/10.1111/j.1365-2664.2011.01996.x
Gordon IJ (1989) Vegetation community selection by ungulates on the isle of Rhum. II. vegetation community selection. J Appl Ecol 26:53. https://doi.org/10.2307/2403650
Hamilton E, Frank D, Hinchey P, Murray T (2008) Defoliation induces root exudation and triggers positive rhizospheric feedbacks in a temperate grassland. Soil Biol Biochem 40:2865–2873. https://doi.org/10.1016/j.soilbio.2008.08.007
Harrison KA, Bardgett RD (2004) Browsing by red deer negatively impacts on soil nitrogen availability in regenerating native forest. Soil Biol Biochem 36:115–126
Hättenschwiler S, Gasser P (2005) Soil animals alter plant litter diversity effects on decomposition. Proc Natl Acad Sci 102:1519–1524
Haynes G (2012) Elephants (and extinct relatives) as earth-movers and ecosystem engineers. Geomorphology 157–158:99–107. https://doi.org/10.1016/j.geomorph.2011.04.045
Heemsbergen DA, Berg MP, Loreau M et al (2004) Biodiversity effects on soil processes explained by interspecific functional dissimilarity. Science 306:1019–1020. https://doi.org/10.1126/science.1101865
Hempson GP, Archibald S, Bond WJ et al (2015) Ecology of grazing lawns in Africa. Biol Rev 90:979–994. https://doi.org/10.1111/brv.12145
Hobbs NT (1996) Modification of ecosystems by ungulates. J Wildl Manag 60:695–713. https://doi.org/10.2307/3802368
Hobbs NT, Schimel DS, Owensby CE, Ojima DS (1991) Fire and grazing in the tallgrass prairie - contingent effects on nitrogen budgets. Ecology 72:1374–1382. https://doi.org/10.2307/1941109
Howison RA, Olff H, van de Koppel J, Smit C (2017) Biotically driven vegetation mosaics in grazing ecosystems: the battle between bioturbation and biocompaction. Ecol Monogr 87:363–378. https://doi.org/10.1002/ecm.1259
Ingham RE, Trofymow JA, Ingham ER, Coleman DC (1985) Interactions of bacteria, fungi, and their nematode grazers: effects on nutrient cycling and plant growth. Ecol Monogr 55:119–140. https://doi.org/10.2307/1942528
Johnson SN, Lopaticki G, Barnett K et al (2015) An insect ecosystem engineer alleviates drought stress in plants without increasing plant susceptibility to an above-ground herbivore. Funct Ecol:894–902. https://doi.org/10.1111/1365-2435.12582
Johnston CA, Groffman P, Breshears DD et al (2004) Carbon cycling in soil. Front Ecol Environ 2:522–528. https://doi.org/10.1890/1540-9295(2004)002[0522:CCIS]2.0.CO;2
Kardol P, Throop HL, Adkins J, de Graaff MA (2016) A hierarchical framework for studying the role of biodiversity in soil food web processes and ecosystem services. Soil Biol Biochem 102:33–36. https://doi.org/10.1016/j.soilbio.2016.05.002
Kay FR, Sobhy HM, Whitford WG (1999) Soil microarthropods as indicators of exposure to environmental stress in Chihuahuan Desert rangelands. Biol Fertil Soils 28:121–128. https://doi.org/10.1007/s003740050472
Klumpp K, Fontaine S, Attard E et al (2009) Grazing triggers soil carbon loss by altering plant roots and their control on soil microbial community. J Ecol 97:876–885. https://doi.org/10.1111/j.1365-2745.2009.01549.x
Knox MA, Andriuzzi WS, Buelow HN et al (2017) Decoupled responses of soil bacteria and their invertebrate consumer to warming, but not freeze-thaw cycles, in the Antarctic dry valleys. Ecol Lett 20:1242–1249. https://doi.org/10.1111/ele.12819
Lau JA, Lennon JT (2012) Rapid responses of soil microorganisms improve plant fitness in novel environments. Proc Natl Acad Sci 109:14058–14062. https://doi.org/10.1073/pnas.1202319109

Lindo Z, Whiteley J, Gonzalez A (2012) Traits explain community disassembly and trophic contraction following experimental environmental change. Glob Chang Biol 18:2448–2457. https://doi.org/10.1111/j.1365-2486.2012.02725.x

Liu T, Guo R, Ran W et al (2015) Body size is a sensitive trait-based indicator of soil nematode community response to fertilization in rice and wheat agroecosystems. Soil Biol Biochem 88:275–281. https://doi.org/10.1016/j.soilbio.2015.05.027

Liu C, Wang L, Song X et al (2018) Towards a mechanistic understanding of the effect that different species of large grazers have on grassland soil N availability. J Ecol 106:357–366. https://doi.org/10.1111/1365-2745.12809

Maraun M, Salamon J-A, Schneider K et al (2003) Oribatid mite and collembolan diversity, density and community structure in a moder beech forest (*Fagus sylvatica*): effects of mechanical perturbations. Soil Biol Biochem 35:1387–1394. https://doi.org/10.1016/S0038-0717(03)00218-9

Markkola A, Kuikka K, Rautio P et al (2004) Defoliation increases carbon limitation in ectomycorrhizal symbiosis of *Betula pubescens*. Oecologia 140:234–240. https://doi.org/10.1007/s00442-004-1587-2

McNaughton SJ (1984) Grazing lawns - animals in herds, plant form, and coevolution. Am Nat 124:863–886. https://doi.org/10.1086/284321

McNaughton SJ, Banyikwa FF, McNaughton MM (1997) Promotion of the cycling of diet-enhancing nutrients by African grazers. Science 278:1798–1800. https://doi.org/10.1126/science.278.5344.1798

Menard C, Duncan P, Fleurance G et al (2002) Comparative foraging and nutrition of horses and cattle in European wetlands. J Appl Ecol 39:120–133. https://doi.org/10.1046/j.1365-2664.2002.00693.x

Mikola J, Yeates GW, Barker GM et al (2001) Effects of defoliation intensity on soil food-web properties in an experimental grassland community. Oikos 92:333–343. https://doi.org/10.1034/j.1600-0706.2001.920216.x

Mikola J, Setälä H, Virkajärvi P et al (2009) Defoliation and patchy nutrient return drive grazing effects on plant and soil properties in a dairy cow pasture. Ecol Monogr 79:221–244. https://doi.org/10.1890/08-1846.1

Milchunas DG, Lauenroth WK (1993) Quantitative effects of grazing on vegetation and soils over a global range of environments. Ecol Monogr 63:327–366

Mills AAS, Adl MS (2011) Changes in nematode abundances and body length in response to management intensive grazing in a low-input temperate pasture. Soil Biol Biochem 43:150–158. https://doi.org/10.1016/j.soilbio.2010.09.027

Moe SR, Wegge P (2008) Effects of deposition of deer dung on nutrient redistribution and on soil and plant nutrients on intensively grazed grasslands in lowland Nepal. Ecol Res 23:227–234. https://doi.org/10.1007/s11284-007-0367-y

Mohr D, Cohnstaedt LW, Topp W (2005) Wild boar and red deer affect soil nutrients and soil biota in steep oak stands of the Eifel. Soil Biol Biochem 37:693–700. https://doi.org/10.1016/J.SOILBIO.2004.10.002

Moore J, McCann K, Setälä H, de Ruiter P (2003) Top-down is bottom-up: does predation in the rhizosphere regulate aboveground dynamics? Ecology 84:846–857

Mosbacher JB, Kristensen DK, Michelsen A et al (2016) Quantifying muskox plant biomass removal and spatial relocation of nitrogen in a high arctic tundra ecosystem. Arctic Antarct Alp Res 48:229–240. https://doi.org/10.1657/AAAR0015-034

Mulder C, Den Hollander HA, Hendriks AJ (2008) Aboveground herbivory shapes the biomass distribution and flux of soil invertebrates. PLoS One 3:1–7. https://doi.org/10.1371/journal.pone.0003573

Nichols E, Gardner TA, Peres CA, Spector S (2009) Co-declining mammals and dung beetles: an impending ecological cascade. Oikos 118:481–487. https://doi.org/10.1111/j.1600-0706.2009.17268.x

Odriozola I, García-Baquero G, Laskurain NA, Aldezabal A (2014) Livestock grazing modifies the effect of environmental factors on soil temperature and water content in a temperate grassland. Geoderma 235–236:347–354. https://doi.org/10.1016/j.geoderma.2014.08.002

Olofsson J, Kitti H, Rautiainen P et al (2001) Effects of summer grazing by reindeer on composition of vegetation, productivity and nitrogen cycling. Ecography (Cop) 24:13–24. https://doi.org/10.1034/j.1600-0587.2001.240103.x
Olofsson J, Stark S, Oksanen L (2004) Reindeer influence on ecosystem processes in the tundra. Oikos 105:386–396. https://doi.org/10.1111/j.0030-1299.2004.13048.x
Ostermann-Kelm SD, Atwill EA, Rubin ES et al (2009) Impacts of feral horses on a desert environment. BMC Ecol 9:1–10. https://doi.org/10.1186/1472-6785-9-22
Partsch S, Milcu A, Scheu S (2006) Decomposers (Lumbricidae, Collembola) affect plant performance in model grasslands of different diversity. Ecology 87:2548–2558
Pastor J, Dewey B, Naiman RJ et al (1993) Moose browsing and soil fertility in the boreal forests of Isle Royale National Park. Ecology 74:467–480. https://doi.org/10.2307/1939308
Pastor J, Cohen Y, Hobbs NT (2006) The role of large herbivores in ecosystem nutrient cycles. In: Dannell R, Bergstrom R, Duncan P, Pastor J (eds) Large herbivore ecology, ecosystem dynamics and conservation. Cambridge University Press, Cambridge, pp 289–319
Pausch J, Kramer S, Scharroba A et al (2015) Small but active - pool size does not matter for carbon incorporation in below-ground food webs. Funct Ecol 30:479–489. https://doi.org/10.1111/1365-2435.12512
Pollierer MM, Langel R, Körner C et al (2007) The underestimated importance of belowground carbon input for forest soil animal food webs. Ecol Lett 10:729–736. https://doi.org/10.1111/j.1461-0248.2007.01064.x
Pretorius Y, de Boer FW, van der Waal C et al (2011) Soil nutrient status determines how elephant utilize trees and shape environments. J Anim Ecol 80:875–883. https://doi.org/10.1111/j.1365-2656.2011.01819.x
Rietkerk M, van den Bosch F, van de Koppel J (1997) Site-specific properties and irreversible vegetation changes in semi-arid grazing systems. Oikos 80:241–252. https://doi.org/10.2307/3546592
Riginos C, Grace JB (2008) Savanna tree density, herbivores, and the herbaceous community: bottom-up vs. top-down effects. Ecology 89:2228–2238
Rillig MC, Mummey DL (2006) Mycorrhizas and soil structure. New Phytol 171:41–53
Ripple WJ, Beschta RL (2012) Trophic cascades in Yellowstone: the first 15 years after wolf reintroduction. Biol Conserv 145:205–213. https://doi.org/10.1016/j.biocon.2011.11.005
Ritchie ME, Tilman D, Knops JMH (1998) Herbivore effects on plant and nitrogen dynamics in oak savanna. Ecology 79:165–177. https://doi.org/10.2307/176872
Rossow LJ, Bryant JP, Kielland K (1997) Effects of above-ground browsing by mammals on mycorrhizal infection in an early successional taiga ecosystem. Oecologia 110:94–98. https://doi.org/10.1007/s004420050137
Rousk J (2016) Biomass or growth? How to measure soil food webs to understand structure and function. Soil Biol Biochem 102:1–3. https://doi.org/10.1016/j.soilbio.2016.07.001
Ruijven J, De Deyn GB, Raaijmakers CE et al (2004) Interactions between spatially separated herbivores indirectly alter plant diversity. Ecol Lett 8:30–37. https://doi.org/10.1111/j.1461-0248.2004.00688.x
Sankaran M, Augustine DJ (2004) Large herbivores suppress decomposer abundance in a semiarid grazing ecosystem. Ecology 85:1052–1061. https://doi.org/10.1890/03-0354
Sardans J, Rivas-Ubach A, Penuelas J (2012) The elemental stoichiometry of aquatic and terrestrial ecosystems and its relationships with organismic lifestyle and ecosystem structure and function: a review and perspectives. Biogeochemistry 111:1–39. https://doi.org/10.1007/s10533-011-9640-9
Schon NL, Mackay AD, Minor MA (2012) Vulnerability of soil invertebrate communities to the influences of livestock in three grasslands. Appl Soil Ecol 53:98–107. https://doi.org/10.1016/j.apsoil.2011.11.003
Schrama M, Veen GF, Bakker ES et al (2013) An integrated perspective to explain nitrogen mineralization in grazed ecosystems. Perspect Plant Ecol Evol Syst 15:32–44. https://doi.org/10.1016/j.ppees.2012.12.001

Singer FJ, Schoenecker KA (2003) Do ungulates accelerate or decelerate nitrogen cycling? For Ecol Manag 181:189–204. https://doi.org/10.1016/s0378-1127(03)00133-6
Sitters J, Olde Venterink H (2015) The need for a novel integrative theory on feedbacks between herbivores, plants and soil nutrient cycling. Plant Soil 396:421–426. https://doi.org/10.1007/s11104-015-2679-y
Sitters J, Olde Venterink H (2018) A stoichiometric perspective of the effect of herbivore dung on ecosystem functioning. Ecol Evol 8:1043–1046. https://doi.org/10.1002/ece3.3666
Sitters J, Maechler MJ, Edwards PJ et al (2014) Interactions between C:N:P stoichiometry and soil macrofauna control dung decomposition of savanna herbivores. Funct Ecol 28:776–786. https://doi.org/10.1111/1365-2435.12213
Sitters J, Atkinson CL, Guelzow N et al (2015) Spatial stoichiometry: cross-ecosystem material flows and their impact on recipient ecosystems and organisms. Oikos 124:920–930. https://doi.org/10.1111/oik.02392
Sitters J, Bakker ES, Veldhuis MP et al (2017a) The stoichiometry of nutrient release by terrestrial herbivores and its ecosystem consequences. Front Earth Sci 5. https://doi.org/10.3389/feart.2017.00032
Sitters J, te Beest M, Cherif M et al (2017b) Interactive effects between reindeer and habitat fertility drive soil nutrient availabilities in arctic tundra. Ecosystems 20:1266–1277
Sorensen LI, Mikola J, Kytoviita M-M, Olofsson J (2009) Trampling and spatial heterogeneity explain decomposer abundances in a sub-arctic grassland subjected to simulated reindeer grazing. Ecosystems 12:830–842. https://doi.org/10.1007/s10021-009-9260-6
Stark S, Grellmann D (2002) Soil microbial responses to herbivory in an arctic tundra heath at two levels of nutrient availability. Ecology 83:2736–2744. https://doi.org/10.2307/3072011
Stark S, Strommer R, Tuomi J (2002) Reindeer grazing and soil microbial processes in two suboceanic and two subcontinental tundra heaths. Oikos 97:69–78. https://doi.org/10.1034/j.1600-0706.2002.970107.x
Stark S, Tuomi J, Strommer R, Helle T (2003) Non-parallel changes in soil microbial carbon and nitrogen dynamics due to reindeer grazing in northern boreal forests. Ecography (Cop) 26:51–59. https://doi.org/10.1034/j.1600-0587.2003.03336.x
Stark S, Julkunen-Tiitto R, Kumpula J (2007) Ecological role of reindeer summer browsing in the mountain birch (*Betula pubescens ssp czerepanovii*) forests: effects on plant defense, litter decomposition, and soil nutrient cycling. Oecologia 151:486–498. https://doi.org/10.1007/s00442-006-0593-y
Stark S, Männistö MK, Eskelinen A (2015) When do grazers accelerate or decelerate soil carbon and nitrogen in tundra? A test of theory on grazing effects in fertile and infertile habitats. Oikos 124:593–602. https://doi.org/10.1111/oik.01355
Sterner RW (1990) The ratio of nitrogen to phosphorus resupplied by herbivores - zooplankton and the algal competitive arena. Am Nat 136:209–229. https://doi.org/10.1086/285092
Sterner RW, Elser JJ (2002) Ecological stoichiometry. Princeton University Press, Princeton, NJ
Streitwolf-Engel R, Boller T, Wiemken A, Sanders IR (1998) Mycorrhizal fungal diversity determines plant biodiversity, ecosystem variability and productivity. Nature 396:69–72
Subalusky AL, Dutton CL, Rosi-Marshall EJ, Post DM (2015) The hippopotamus conveyor belt: vectors of carbon and nutrients from terrestrial grasslands to aquatic systems in sub-Saharan Africa. Freshw Biol 60:512–525. https://doi.org/10.1111/fwb.12474
Sun G, Zhu-Barker X, Chen D et al (2017) Responses of root exudation and nutrient cycling to grazing intensities and recovery practices in an alpine meadow: an implication for pasture management. Plant Soil 416:515–525. https://doi.org/10.1007/s11104-017-3236-7
Suominen O, Danell K, Bergstrom R (1999) Moose, trees, and ground-living invertebrates: indirect interactions in Swedish pine forests. Oikos 84:215–226. https://doi.org/10.2307/3546716
Szelecz I, Sorge F, Seppey CVW et al (2016) Effects of decomposing cadavers on soil nematode communities over a one-year period. Soil Biol Biochem 103:405–416. https://doi.org/10.1016/j.soilbio.2016.09.011

Trap J, Bonkowski M, Plassard C et al (2016) Ecological importance of soil bacterivores for ecosystem functions. Plant Soil 398:1–24. https://doi.org/10.1007/s11104-015-2671-6

Urabe J, Nakanishi M, Kawabata K (1995) Contribution of metazoan plankton to the cycling of nitrogen and phosphorus in Lake Biwa. Limnol Oceanogr 40:232–241

Valdés-Correcher E, Sitters J, Wassen M, Brion N, Olde Venterink H (2019) Herbivore dung quality affects plant community diversity. Sci Rep 9:5675. https://doi.org/10.1038/s41598-019-42249-z

van der Waal C, Kool A, Meijer SS et al (2011) Large herbivores may alter vegetation structure of semi-arid savannas through soil nutrient mediation. Oecologia 165:1095–1107. https://doi.org/10.1007/s00442-010-1899-3

van der Wal R (2006) Do herbivores cause habitat degradation or vegetation state transition? Evidence from the tundra. Oikos 114:177–186. https://doi.org/10.1111/j.2006.0030-1299.14264.x

van der Wal R, Bardgett RD, Harrison KA, Stien A (2004) Vertebrate herbivores and ecosystem control: cascading effects of faeces on tundra ecosystems. Ecography (Cop) 27:242–252. https://doi.org/10.1111/j.0906-7590.2004.03688.x

van Groenigen JW, Lubbers IM, Vos H et al (2014) Earthworms increase plant production: a meta-analysis. Sci Rep:1–7. https://doi.org/10.1038/srep06365

Van Haveren BP (1983) Soil bulk density as influenced by grazing intensity and soil type on a shortgrass prairie site. J Range Manag 36:586. https://doi.org/10.2307/3898346

Vandegehuchte ML, Van Der Putten WH, Duyts H et al (2016) Aboveground mammal and invertebrate exclusions cause consistent changes in soil food webs of two subalpine grassland types, but mechanisms are system-specific. Oikos 126:212–223. https://doi.org/10.1111/oik.03341

Veldhuis MP, Howison RA, Fokkema RW et al (2014) A novel mechanism for grazing lawn formation: large herbivore-induced modification of the plant-soil water balance. J Ecol 102:1506–1517. https://doi.org/10.1111/1365-2745.12322

Veldhuis MP, Gommers MI, Olff H, Berg MP (2018) Spatial redistribution of nutrients by large herbivores and dung beetles in a savanna ecosystem. J Ecol 106:422–433. https://doi.org/10.1111/1365-2745.12874

Verschoor BC, De Goede RGM, De Vries FW, Brussaard L (2001) Changes in the composition of the plant-feeding nematode community in grasslands after cessation of fertiliser application. Appl Soil Ecol 17:1–17. https://doi.org/10.1016/S0929-1393(00)00135-9

Wardle DA, Barker GM, Yeates GW et al (2001) Introduced browsing mammals in New Zealand natural forests: aboveground and belowground consequences. Ecol Monogr 71:587–614

Wardle DA, Bonner KI, Barker GM (2002) Linkages between plant litter decomposition, litter quality, and vegetation responses to herbivores. Funct Ecol 16:585–595

Wardle DA, Bardgett RD, Klironomos JN et al (2004) Ecological linkages between aboveground and belowground biota. Science 304:1629–1633. https://doi.org/10.1126/science.1094875

Wu X, Griffin JN, Sun S (2013) Cascading effects of predator-detritivore interactions depend on environmental context in a Tibetan alpine meadow. J Anim Ecol. https://doi.org/10.1111/1365-2656.12165

Xiao Z, Wang X, Koricheva J et al (2017) Earthworms affect plant growth and resistance against herbivores: a meta-analysis. Funct Ecol:1–11. https://doi.org/10.1111/1365-2435.12969

Yang Y, Wu L, Lin Q et al (2013) Responses of the functional structure of soil microbial community to livestock grazing in the Tibetan alpine grassland. Glob Chang Biol 19:637–648. https://doi.org/10.1111/gcb.12065

Yang Z, Nolte S, Wu J (2017) Tidal flooding diminishes the effects of livestock grazing on soil micro-food webs in a coastal saltmarsh. Agric Ecosyst Environ 236:177–186. https://doi.org/10.1016/j.agee.2016.12.006

Yates CJ, Norton DA, Hobbs RJ (2000) Grazing effects on plant cover, soil and microclimate in fragmented woodlands in South-Western Australia: implications for restoration. Austral Ecol 25:36–47. https://doi.org/10.1046/j.1442-9993.2000.01030.x

Zhao F, Ren C, Shelton S et al (2017) Grazing intensity influence soil microbial communities and their implications for soil respiration. Agric Ecosyst Environ 249:50–56. https://doi.org/10.1016/j.agee.2017.08.007

Zhou GY, Zhou XH, He YH et al (2017) Grazing intensity significantly affects belowground carbon and nitrogen cycling in grassland ecosystems: a meta-analysis. Glob Chang Biol 23:1167–1179. https://doi.org/10.1111/gcb.13431

Chapter 10
Effects of Grazing and Browsing on Tropical Savanna Vegetation

Frank van Langevelde, Claudius A. D. M. van de Vijver, Herbert H. T. Prins, and Thomas A. Groen

10.1 Introduction

Savannas cover an eighth of the world's land surface (Scholes and Archer 1997) and harbour around one fifth of the world's human population (Hoffmann et al. 2002). Savannas are characterized by a continuous grass layer and a discontinuous layer of trees and shrubs (Scholes and Archer 1997), that vary in cover (Sankaran et al. 2005), the determinants of which are poorly understood (Bond 2008; Lehmann et al. 2009).

The savanna tree to grass ratio is determined by several factors, including rainfall, soil type, herbivory, and fire occurrence and intensity (Van Langevelde et al. 2003; Bond et al. 2005). Savanna tree-grass dynamics are of particular interest in ecology, as sudden shifts can occur between a low biomass system state (i.e., tree-grass

F. van Langevelde (✉)
Resource Ecology Group, Wageningen University, Wageningen, The Netherlands

School of Life Sciences, Westville Campus, University of KwaZulu-Natal, Durban, South Africa
e-mail: Frank.vanlangevelde@wur.nl

C. A. D. M. van de Vijver
Graduate School Production Ecology and Resource Conservation, Wageningen University, Wageningen, The Netherlands
e-mail: Claudius.vandevijver@wur.nl

H. H. T. Prins
Animal Sciences Group, Wageningen University, Wageningen, The Netherlands
e-mail: Herbert.prins@wur.nl

T. A. Groen
Faculty of Geo-Information Science and Earth Observation, University of Twente, Enschede, The Netherlands
e-mail: t.a.groen@utwente.nl

I. J. Gordon, H. H. T. Prins (eds.), *The Ecology of Browsing and Grazing II*, Ecological Studies 239, https://doi.org/10.1007/978-3-030-25865-8_10

co-dominance) and a high biomass system state (i.e., tree dominance) (Briggs et al. 2002; Brown and Carter 1998; Roques et al. 2001; Van Auken 2000). As the two competing life forms trees and grasses co-occur under many environmental conditions in savanna ecosystems, savannas, therefore, provide an ideal ecosystem to investigate several ecologically important processes such as competition and facilitation (Van Langevelde et al. 2011).

In the past decade, the above-mentioned sudden shifts where, in this case, savannas containing both trees and grasses suddenly shift into ecosystems dominated by trees (so called bush encroachment), has become a serious point of concern (Brown and Carter 1998; Van Auken 2000; Groen et al. 2017). It has been suggested that the sudden shift from tree-grass co-existence towards tree dominance is the result of the positive feedback mechanism between fuel load (grass biomass) and fire intensity, which could lead the system towards either a savanna state with trees and grasses or a tree-dominated state (Van Langevelde et al. 2003; Bond and Keeley 2005). The occurrences of these "alternative stable states" have been analysed in detail for several ecosystems including shallow freshwater lakes (Carpenter et al. 1999), coral reefs (Nystrom et al. 2000), seegrasses (Van Langevelde and Prins 2007), mangroves (Huisman et al. 2009) and arid grasslands (Van Langevelde et al. 2016). An important feature of ecosystems having alternative stable states is the prediction that they are vulnerable to abrupt discontinuous changes, so called "catastrophic shifts".

Burning is a common phenomenon in savanna ecosystems (Bond et al. 2005; Carmona-Moreno et al. 2005; Groen et al. 2011; Van Langevelde et al. 2014; Lehmann et al. 2014), and can have profound effects on the tree-grass balance in savannas, depending on a variety of factors such as climatic conditions, tree height and fuel load (Govender et al. 2006). The damaging effect of fire on trees increases with fire intensity (De Ronde et al. 2004; Groen et al. 2008), which is determined by a number of factors, including grass biomass—more grass biomass resulting in more intense fires.

As mentioned before, herbivory is another determinant of savanna structure and dynamics. The effects of herbivory can be direct, for example through tree removal (Dublin et al. 1990) or suppression of seedling establishment (Loth et al. 2005; Prins and Van der Jeugd 1993). But the effects of herbivory can also be indirect, via removal of grass biomass which can reduce fire intensity and thus the effect of fire on trees. This can initiate a positive feedback where reduced fire intensity leads to less damage to trees and thus higher tree cover which consequently reduces grass cover through competition for resources.

Previous modelling studies predicted the occurrence of alternative stable states on the basis of the interaction of between herbivores and disturbances like fire (Dublin et al. 1990; Van Langevelde et al. 2003; D'Odorico et al. 2006; Groen et al. 2017). An important assumption in these models is that grazing and browsing pressure is constant. In human-managed systems this assumption might hold, but in natural systems it certainly does not. The population dynamics, and distribution for wild herbivore populations are largely determined by available resources (Owen-Smith and Mills 2006; Wato et al. 2016). When considering grazing and browsing in savannas, however, the indirect facilitative effect of grazers on browsers and vice versa, and their interactive effects on the positive feedback mechanism are so far unexplored. In this Chapter we review studies that model the dynamics of tropical savannas and the influence of grazing and browsing on these dynamics. We summarize this understanding in a simple model to capture the dynamics of trees, grasses, herbivores and fire in tropical savannas.

10.2 Review of Savanna Dynamics

Savanna vegetation occurs over a huge range of climate conditions: mean annual precipitation (MAP) alone ranges from 150 mm to more than 2800 mm (Sankaran et al. 2005; Lehmann et al. 2009). Soil properties (fertility, texture, depth) and irradiance also vary widely across savannas (Solbrig 1996). Additionally, large areas of savanna are subject to defoliation by herbivores and fire, two quite different forms of defoliation on plants with consequent effects on their (re)growth, which also vary substantially over the biome (Olff et al. 2002; Bond and Keeley 2005; Hempson et al. 2015). Ecologists have attempted to deal with the complexity of savannas by concentrating on the fundamental question how do trees and grasses co-exist without one dominating the other (Scholes and Archer 1997; Van Langevelde et al. 2011)? Here we will review two general explanations for tree-grass co-existence in savannas: niche separation and disturbances.

The mechanism of niche separation proposes that plants segregate along various environmental niche axes, including gradients of light, soil moisture and rooting depth (Sydes and Grime 1984). The classical explanation for long-term co-existence of trees and grasses in savannas, the two-layer hypothesis of Walter (1971), is based on niche separation by rooting depth of trees and grasses. Assuming that water is the most limiting factor for plant growth in savannas, this hypothesis states that because of differences in their rooting depth, trees and grasses differ in their access to water—trees and grasses both have access to the upper soil layers, where grasses are thought to be competitively superior, and trees have sole access to deeper soil layers. The two-layer hypothesis has a long history, and several subsequent models have employed it as a basis to explain shifts in trees and grasses in savannas (Walker et al. 1981; Van Langevelde et al. 2003).

Over the last decade, this two-layer hypothesis is debated (Scholes and Archer 1997; Higgins et al. 2000; Jeltsch et al. 2000) due to an accumulation of conflicting empirical evidence. Some field experiments did not find clear niche separation of roots (Seghieri 1995; Mordelet et al. 1997; Priyadarshini et al. 2014, 2015, 2016), whereas others appear to have done so (Weltzin and McPherson 1997; Schenk and Jackson 2002). Especially, in the drier areas with deep, sandy, free-draining soils where water penetration from rainfall does reach lower layers of the soil, allocation of root growth by trees to these layers has been observed (Schenk and Jackson 2002).

However, savannas can also occur on shallow soils with impermeable lower layers where grasses and trees are forced to share the same rooting space (Wiegand et al. 2005). Further, it appears that trees concentrate their rooting at shallow depths in certain environments even though the underlying substrate is not necessarily impermeable to deeper root penetration (Mordelet and Le Roux 2006). Presumably, advantages of competition by trees with grasses for soil resources outweigh advantages of deeper rooting under these conditions. Recently, the study of Holdo et al. (2018) confirmed that there are clear differences in rooting depth between grasses and trees across the MAP gradient, with grasses generally exhibiting shallower rooting profiles than trees. They also found that trees tended to become more shallow-rooted as a function of MAP, to the point that trees and grasses largely overlapped in terms of rooting depth at the wettest sites. Summarizing these results allow us to maintain the two-layer hypothesis as explanation for coexistence of trees and grasses.

Other explanations for the coexistence of trees and grasses in savannas recognise the fundamental role of disturbances (Jeltsch et al. 2000; Higgins et al. 2000; Van Wijk and Rodriguez-Iturbe 2002). These explanations are premised on the assumption that climate variation (drought) and defoliation (fire and herbivory) are disturbances that differ in their impact on different life-history stages, causing demographic bottlenecks on tree recruitment. Repeated negative impacts of these disturbances on the dominant life-form create opportunities for the establishment and persistence of the life-form that becomes out-competed (Warner and Chesson 1985). These explanations differ from the explanation based on niche separation in that they explicitly focus on population processes of trees.

The relative importance of these disturbances differs over the MAP gradient. Arid and semi-arid savannas have relatively high inter-annual variation in MAP, and consequently these systems frequently experience drought events which can kill adult plants (Fensham and Holman 1999), thereby opening resource space that may be re-occupied by grasses or tree seedlings (Jeltsch et al. 2000; Van Wijk and Rodriguez-Iturbe 2002). Trees can also be killed by other factors such as storms (which are rare around the equator) or very large herbivores such as elephant (*Loxodonta africana*) (e.g., Kohi et al. 2011; Pretorius et al. 2011). In humid savannas, empirical evidence has accumulated which indicates that fire and herbivory are important elements that structure the vegetation (Tomlinson et al. 2012), including the balance between trees and grasses, which suggested that these disturbances might play a fundamental role in ensuring co-existence (Sankaran et al. 2005; cf. **Smit and Coetsee** Chap. 13).

The effects of disturbances on savanna vegetation (fire, domestic livestock) can be separated over the rainfall gradient (Bucini and Hanan 2007). Water supply limits seedling recruitment (Jeltsch et al. 2000; Higgins et al. 2000; Van Wijk and Rodriguez-Iturbe 2002; Barbosa et al. 2014a, b) and precipitates adult mortality (Fensham and Holman 1999) in drier areas, while defoliation disturbances (herbivory and fire) limit adult recruitment in humid areas (Van Langevelde et al. 2003, 2014; Mourik et al. 2007; Bond 2008). Many empirical studies illustrate the large impact that grazers and browsers can have on savanna vegetation, both directly and indirectly (e.g., Prins and Van der Jeugd 1993; Loth et al. 2005; Sianga et al. 2017; Van der Waal et al. 2011a, b, 2016). The relative importance of fire versus herbivory is probably a function of plant shoot quality, because low-quality vegetation can sustain much lower large mammalian herbivore biomasses than does high-quality vegetation (Van Langevelde et al. 2008), and thus herbivory will increase in importance with increased forage quality while fire will increase in importance in areas with low forage quality (Bond and Keeley 2005; Hempson et al. 2015; Smit and Prins 2015). These in turn are likely to have major implications for savanna dynamics because of the different defoliation patterns associated with burning (non-selective defoliation, destroying leaves, stems, apical buds and in particular moribund, dead material) and herbivory (selective defoliation, removing live leaves mainly). Forage quality is positively correlated with soil fertility and negatively correlated with MAP (Olff et al. 2002) indicating that both MAP and soil fertility are major drivers of this disturbance regime. Hence, high rainfall savannas (>1000 mm) found on nutrient-poor soils might be mediated by fire. At intermediate MAP, forage quality could be sufficient to maintain large herbivores, meaning that these environments are often mediated by herbivory (Olff et al. 2002).

10.3 Modelling of Savanna Dynamics

To summarize the empirical studies on the dynamics of tropical savannas, we modelled savanna vegetation by including both niche separation as well as disturbances, i.e., fire and herbivory (Van Langevelde et al. 2003; Groen et al. 2017). The model simulates changes in grass and woody biomass as a function of rainfall, soil properties, fire and densities of both grazers and browsers. Compared to previous studies, we added herbivore populations that change depending on the available food (grass and browse). Grass biomass comprises both grass and herbaceous plants, while woody biomass comprises wood, twigs and leaves. We assume that trees and grasses compete for water, that trees have access to both top- (w_t) and sub- (w_s) soil layers and that grasses only have access to the topsoil layer (Schenk and Jackson 2002). We also assume that grasses are better competitors for water in the topsoil layer (Jackson et al. 1996). Water is often exclusively available during the wet season when the plants grow, whereas during the dry season growth reduces to almost zero (c.f. Priyadarshini et al. 2015). Including additional mortality due to drought is not included in the model and could quantitatively change the results. The model simulates with time steps of 1 year. All parameters with their description and units are listed in Table 10.1.

The rate of moisture recharge into the topsoil layer is the amount of infiltrated water (w_{in}) minus the loss of water to the subsoil layer (w_s):

$$w_t = w_{in} - w_s \tag{10.1}$$

Table 10.1 Overview of the variables and parameters used in the model

Symbols	Interpretation	Units	Values
H	Grass biomass	$g\ m^{-2}$	0–400
W	Woody biomass	$g\ m^{-2}$	0–1000
r_H	Water use efficiency of grasses	$g\ mm^{-1}$	1.0
r_W	Water use efficiency of trees	$g\ mm^{-1}$	0.5
θ_H	Rate of water uptake per unit grass biomass	$mm\ yr^{-1}\ g^{-1}$	0.9
θ_W	Rate of water uptake per unit woody biomass	$mm\ yr^{-1}\ g^{-1}$	0.5
d_H	Specific loss of grass biomass due to senescence	yr^{-1}	0.9
d_W	Specific loss of woody biomass due to senescence	yr^{-1}	0.4
d_G	Specific mortality rate of grazers	$m^2\ g^{-1}\ yr^{-1}$	0.02
d_B	Specific mortality rate of browsers	$m^2\ g^{-1}\ yr^{-1}$	0.02
c_H	Consumption coefficient of grass biomass by grazers	$m^2\ g^{-1}\ yr^{-1}$	0.02
c_W	Consumption coefficient of woody biomass by browsers	$m^2\ g^{-1}\ yr^{-1}$	0.02
e_H	Coefficient for consumption and conversion efficiency of grass biomass by grazers	$m^2\ g^{-1}\ yr^{-1}$	0–0.001
e_W	Coefficient for consumption and conversion efficiency of woody biomass by browsers	$m^2\ g^{-1}\ yr^{-1}$	0–0.001
k_H	Specific loss of grass biomass due to fire	yr^{-1}	0.1
k_W	Specific loss of woody biomass due to fire expressed per unit of energy	W^{-1}	0.01
n	Frequency of fire per year	yr^{-1}	0 or 1
a	Coefficient for the increase in fire intensity with grass biomass	$W\ m^{-1}\ g^{-1}$	0.5
α	Proportion of excess water that percolates to subsoil layer	–	0.4
β	Soil moisture content in the topsoil layer above which water starts to percolate to the subsoil layer	mm	200–500
w_{in}	Annual amount of infiltrated water	$mm\ m^{-2}\ yr^{-1}$	0–1000
w_t	Available amount of water in the topsoil layer	$mm\ m^{-2}\ yr^{-1}$	See Eq. (10.1)
w_s	Available amount of water in the subsoil layer	$mm\ m^{-2}\ yr^{-1}$	See Eq. (10.2)
G	Grazer density	$g\ m^{-2}$	0–30
B	Browser density	$g\ m^{-2}$	0–15

Parameter values are taken from Van Langevelde et al. (2003), and references therein

The amount of water that infiltrates to the subsoil layer is proportional to the amount of water that infiltrates in the topsoil:

$$\begin{aligned} w_s &= \alpha(w_{in} - \beta) \qquad & w_{in} > \beta \\ w_s &= 0 \qquad & w_{in} \le \beta \end{aligned} \tag{10.2}$$

where β is the amount of water in the topsoil layer above which water starts to percolate to the subsoil layer and α is the proportion of water that starts to percolate.

The proportion of water uptake from the topsoil layer per unit grass biomass is defined as (Walker et al. 1981)

$$U_H = \frac{\theta_H}{H\theta_H + W\theta_W + w_s} \tag{10.3}$$

where θ_H and θ_W are the water uptake rates of grasses and trees respectively, H and W are the amount of grass and woody biomass respectively, and w_s is the amount of water lost through percolation to the subsoil. Water is used by plants for production of new biomass, and the utilization of water by grasses for production of new grass biomass is expressed by the fraction of the total amount of water in the topsoil taken up by grasses ($U_H H$) times the amount of water in the topsoil (w_t) times the water use efficiency of the grasses (r_H). Total change in the amount of grass biomass in the model is expressed as

$$\frac{dH}{dt} = r_H w_t H \frac{\theta_H}{H\theta_H + W\theta_W + w_s} - d_H H - c_H GH - k_H nH \tag{10.4}$$

including mortality due to senescence ($d_H H$), grazing ($c_H GH$) and loss due to fire ($k_H nH$).

For woody biomass, the equation is similar except for the uptake of water, because trees also have access to the subsoil layer, and for the effect of fire on trees where grass biomass determines the severity of burning. Water uptake from the topsoil layer by trees is formulated in a similar fashion to that for grasses:

$$U_W = \frac{\theta_W}{H\theta_H + W\theta_W + w_s} \tag{10.5}$$

Because trees also have access to water in the subsoil layer, the total rate of change in woody biomass is:

$$\frac{dW}{dt} = r_W \left[w_t W \frac{\theta_W}{H\theta_H + W\theta_W + w_s} + w_s \right] - d_W W - c_W BW - k_W anHW \tag{10.6}$$

where the last term ($k_W anHW$) expresses the loss of tree biomass due to fire which is a function of fire frequency (n), the specific loss of biomass due to fire per unit heat released (k_W), the amount of grass biomass present (H) and the heat yield per g of grass biomass (a).

The effect of fire on trees is mostly limited to small trees, whereas large trees hardly suffer any damage from fires (De Ronde et al. 2004). Also the effect of browsing on trees might not be the same as trees grow out of reach for most browsers (De Knegt et al. 2008; Cameron and Du Toit 2007), and small trees are likely to suffer more from browsing than large trees, but it has been shown that mature trees also suffer from browsing (Goheen et al. 2007). We, therefore, modelled trees occurring in the most susceptible range to fire and herbivory, being up to 5 m (De Ronde et al. 2004).

Several studies showed a direct relation between vegetation and the density of herbivores (Kumar et al. 2002; Owen-Smith and Mills 2006; Hilbers et al. 2015; Wato et al. 2016, 2018), We modelled the consumption of plant biomass by the populations of grazing (G) and browsing (B) herbivores by assuming a linear numerical response. Increments in herbivore biomass occur as a result of the intake rate of plant biomass and the conversion efficiency of consumed plant biomass into herbivore biomass. We combined these properties into one parameter for grazers (e_H) and one for browsers (e_W), assuming that both are constants (Prins 1993). We assume a density dependent growth function resulting in the following equations describing herbivore dynamics:

$$\frac{dG}{dt} = e_H GH - d_G G^2 \tag{10.7}$$

$$\frac{dB}{dt} = e_W BW - d_B B^2 \tag{10.8}$$

10.4 Analysis of Savanna Dynamics

We analysed the stability of the equilibria of the model with extensive simulations. We simulated the model with parameter values as presented in Van Langevelde et al. (2003) and references within. For each combination of parameters we simulated with different initial densities of grass biomass to make sure that if alternative stable states exist the simulations started in the different domains of attraction. We varied water availability (w_{in}), the efficiency with which the grazers consume and convert grass biomass (e_H), the specific mortality rate of grazers (d_G) and fire frequency (n) over realistic ranges of values to analyse the behaviour of the model. We simulated the model with (1) constant grazer and browser populations (Eqs. 10.4 and 10.6); (2) a dynamic grazer population and constant browser population (Eqs. 10.4, 10.6 and 10.7); (3) a constant grazer population and dynamic browser population (Eqs. 10.4, 10.6 and 10.8); and (4) dynamic grazer and browser populations (Eqs. 10.4, 10.6, 10.7 and 10.8). When alternative stable states were found, the location of the separatrix was determined using simulations by initialising the model over a range of values of grass biomass. The separatrix indicates, at which level of grass biomass, the ecosystem switches from one stable state to another stable state, given fixed environmental conditions. By recording between which levels of initial H the system switched from one stable equilibrium to the other stable equilibrium the separatrix for several environmental conditions could be determined.

In Fig. 10.1, the effect of water availability (w_{in}) on grass and woody biomass for all analyses is presented, i.e., no herbivore population dynamics, only grazer population dynamics, only browser population dynamics and both grazer and browser population dynamics. When herbivore populations are not modelled dynamically, increasing the water availability (w_{in}) drives the system from a grassland system via

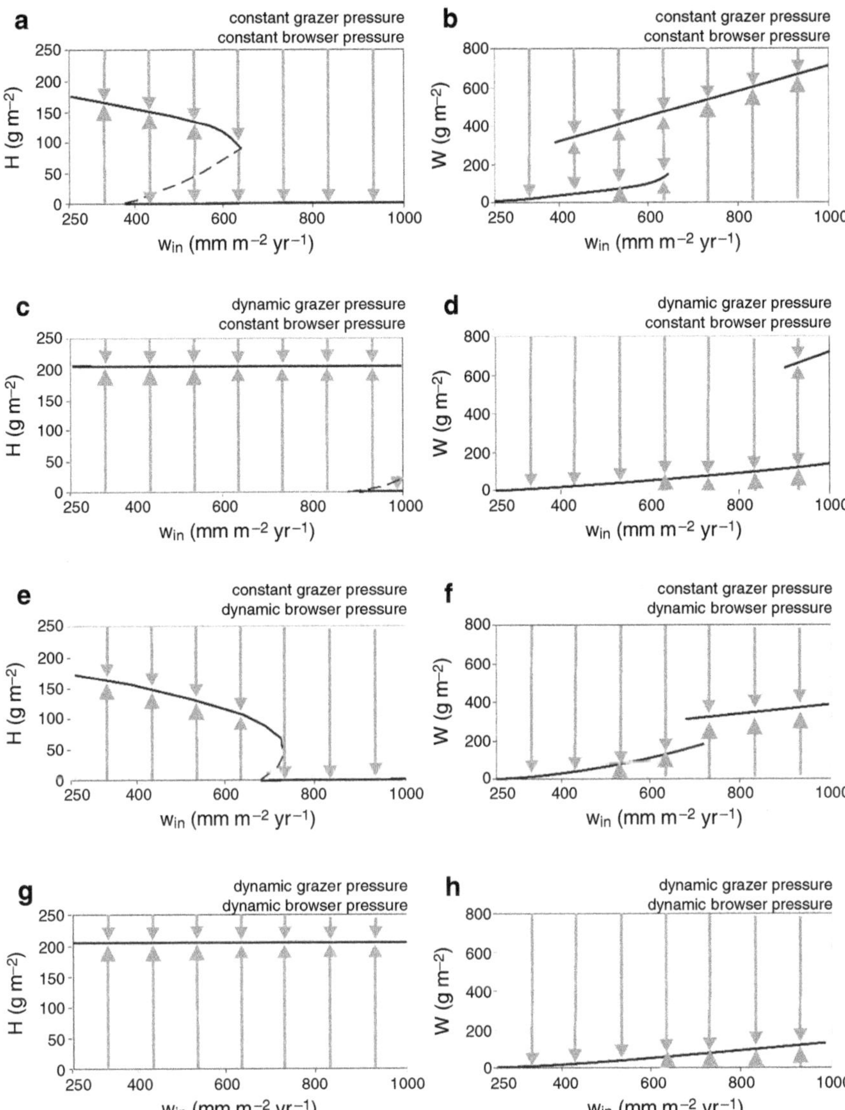

Fig. 10.1 Stable equilibria (solid lines) for grasses (H) and trees (W) over a precipitation range (250–1000 mm) with different situations of herbivore dynamics. (**a**) and (**b**) constant grazer and browser pressure; (**c**) and (**d**) constant browsers pressure and dynamic grazer pressure; (**e**) and (**f**) constant grazer pressure and dynamic browser pressure and (**g**) and (**h**) dynamic grazer and browser pressure. The separatrix (dashed line) is obtained by running simulations with different initial values for H. Which equilibrium for W is selected depends on the initial value in H, and therefore this separatrix cannot be shown. Grey arrows show to which equilibrium the system will develop, given initial grass biomass. Parameters: $\alpha = 0.4$; $a = 0.5$; $c_H = c_W = 0.02$; $d_G = 0.02$; $d_B = 0.02$; $d_H = 0.9$; $d_W = 0.4$; $e_H = e_W = 0.001$; $k_H = 0.1$; $k_W = 0.01$; $n = 1$; $r_H = 1$; $r_W = 0.5$; $\theta_W = 0.5$; $\theta_H = 0.9$; $\beta = 250$; $G = 15$ (in **a**, **b**, **e** and **f**); B=5 (in **a**–**d**)

tree-grass co-occurrence in savannas towards a woodland system (Fig. 10.1a, b). In regions of w_{in} with alternative stable states, both savannas and woodlands can occur. Alternative stable states occur in most analyses (Fig. 10.1a–f) except when both the grazer and browser population is dynamic (Fig. 10.1g, h). We simulated the latter even up to $w_{in} = 2500$ mm (data not shown), but this did not provide us with alternative stable states. In the model with constant grazer and browser numbers, alternative stable states occur at intermediate levels of w_{in}, while they occur at higher w_{in} levels when either the grazer or browser populations are simulated dynamically, and also the range of w_{in} where alternative stable states occur decreases. Finally, the domain of attraction of the savanna equilibrium is much larger when herbivore populations are dynamic than with constant herbivore numbers as the parameter ranges where savanna occurs increases at the expense of woodland and alternative stable states.

The effect of varying the efficiency with which the grazer population consumes and converts grass biomass (e_H) and the specific mortality rate of the grazer population (d_G) on grass and woody biomass are shown in Fig. 10.2. When grazers are modelled dynamically but browser numbers are kept constant, alternative stable states exist only at low values of e_H (Fig. 10.2a, b). At higher values of e_H, the system switches to woodland without grass biomass. When browsers are modelled dynamically as well, savanna is the only stable equilibrium (Fig. 10.2c, d). Changing d_G leads to alternative stable states at quite low values, while woodland occurs at low values of d_G and savanna at high values (Fig. 10.2e, f). Again when we include browser dynamics as well, the alternative stable states disappear and only savanna occurs. Here, the abundance of grass biomass increases with increasing d_G (Fig. 10.2g, h). The effect of e_H and d_G on the herbivore dynamics in the model is that they determine the speed of the response of the herbivore population to changes in forage availability. With high values of e_H, the grazer population grows fast with an increment in grass biomass. With high d_G, the grazer populations' mortality rate is high, meaning that the consumption of grass has to be high in order to maintain the population density, and with a decrease of grass biomass, a rapid decline in grazer density is to be expected.

To analyse the combined effect of water availability (w_{in}), soil water percolation threshold (β) and fire frequency (n), we simulated the model for a range of environmental conditions (Fig. 10.3). Figure 10.3a, b and c show that alternative stables states only occur in the presence of fire at the expense of woodland, and that the range increases with an increasing burning frequency (Van Langevelde et al. 2003). When including dynamic grazer or browser populations (Fig. 10.3d–i), the parameter range where alternative stable states occur decreases. When including both dynamic grazer and browser populations (Fig. 10.3j–l), alternative stable states and also woodland do not occur anymore within the ranges investigated. This concurs with the results in Fig. 10.1. Figure 10.3 also shows the interaction between w_{in} and β. The positive effect of water availability on tree dominance becomes less when β increases, because water is retained better in the topsoil layer. The interaction between β and w_{in} results in a diagonal edge between conditions where woodland occurs and where savanna occurs, or (when fire is included) where woodland occurs

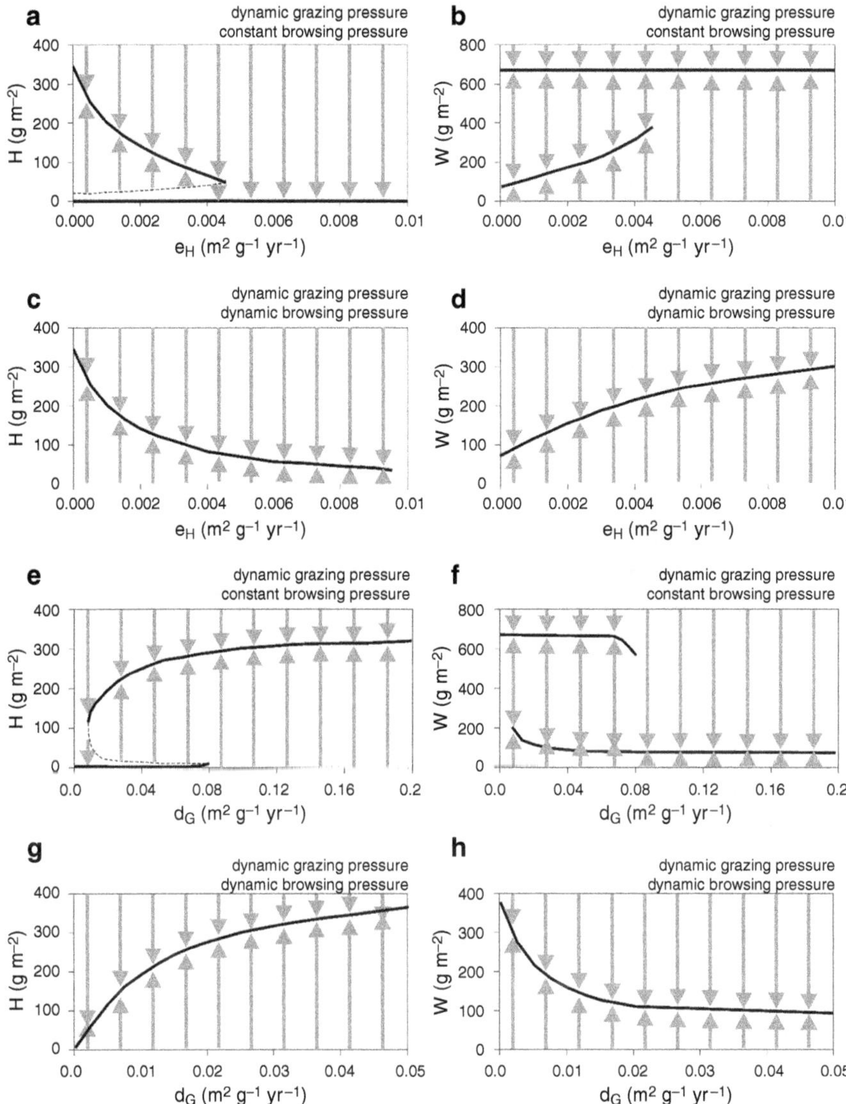

Fig. 10.2 Stable equilibria (solid lines) for grasses and trees with dynamic G and static B (**a**, **b**, **e** and **f**) and dynamic G and B (**c**, **d**, **g** and **h**) at varying values for the conversion coefficient of grass biomass to grazer biomass (e_H in **a–d**) and the specific grazer mortality rate (d_G in **e–h**). The separatrix (dashed line) is obtained by running simulations with different initial values for H. Parameters: $\alpha = 0.4$; $a = 0.5$; $c_H = c_W = 0.02$; $d_G = 0.02$ (in **a–d**); $d_B = 0.02$; $d_H = 0.9$; $d_W = 0.4$; $e_W = 0.001$; $e_H = 0.001$ (in **c–d**); $k_H = 0.1$; $k_W = 0.01$; $n = 1$; $r_H = 1$; $r_W = 0.5$; $\theta_W = 0.5$; $\theta_H = 0.9$; $\beta = 250$, $w_{in} = 940$; $B = 5$ (in **a**, **b**, **e** and **f**)

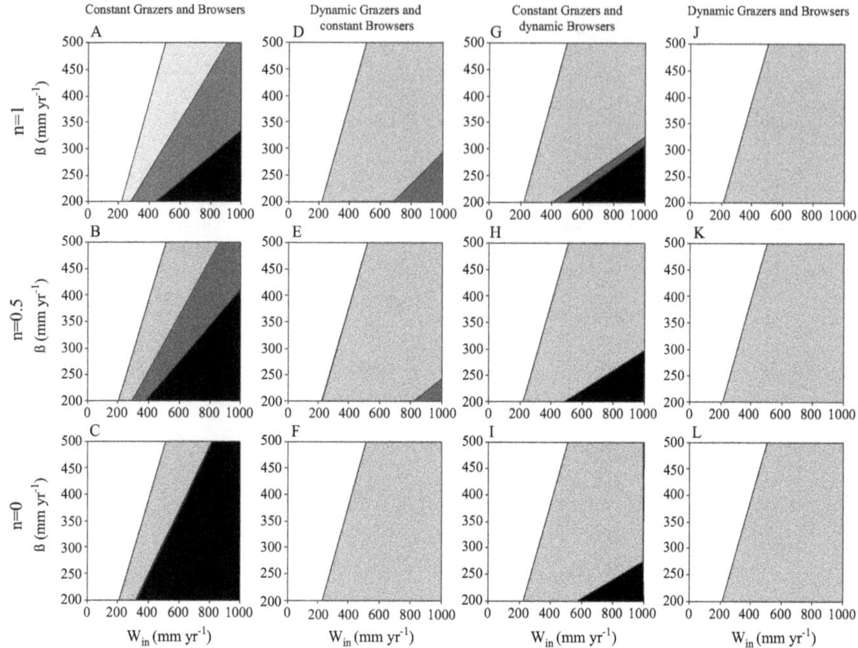

Fig. 10.3 Parameter plane showing the output for different settings of annual water infiltration (w_{in}), percolation threshold above which water starts to percolate to the subsoil (β) and fire frequency (n). Other parameters: $\alpha = 0.4$; $a = 0.5$; $c_H = c_W = 0.02$; $d_G = 0.02$; $d_B = 0.02$; $d_H = 0.9$; $d_W = 0.4$; $e_H = e_W = 0.001$; $k_H = 0.1$; $k_W = 0.01$; $r_H = 1$; $r_W = 0.5$; $\theta_W = 0.5$; $\theta_H = 0.9$; $B = 5$ (in **a–f**); $G = 22$ (in **a–c** and **g–i**). ☐ = grassland; ▒ = savanna; ▩ = alternative stable states; ■ = woodland

and where alternative stable states occur. The slope of this edge indicates the strength of this interaction; a steeper slope means that the effect of β on w_{in} becomes less strong. The slope becomes less steep when fire frequency increases.

Figure 10.4 shows the effect of changing grazer and browser response to biomass availability (e_H and e_W), when both grazers and browsers are modelled dynamically. At very low values of e_H and e_W alternative stable states can occur, but when either one of the two parameters increases, i.e., the population responds faster to increases in biomass, alternative stable states disappear, and either woodland or grassland is the resultant landscape. With increasing e_H, the grassland state is suppressed, and the resultant landscape is woodland. With increasing e_W, the woodland state is suppressed and the resultant landscape is savanna. When the mortality of grazers increases (c.f. Fig. 10.4c with 10.4d), the parameter space of e_H and e_W where woodland can occur decreases, because grazers are less capable of suppressing the woodland state. Also, when the mortality of browsers increases, the parameter space where savannas can occur decreases because the browsers are less capable of suppressing the savanna state. The borders between the different states display convex monotonously increasing curves, suggesting that the sensitivity of the

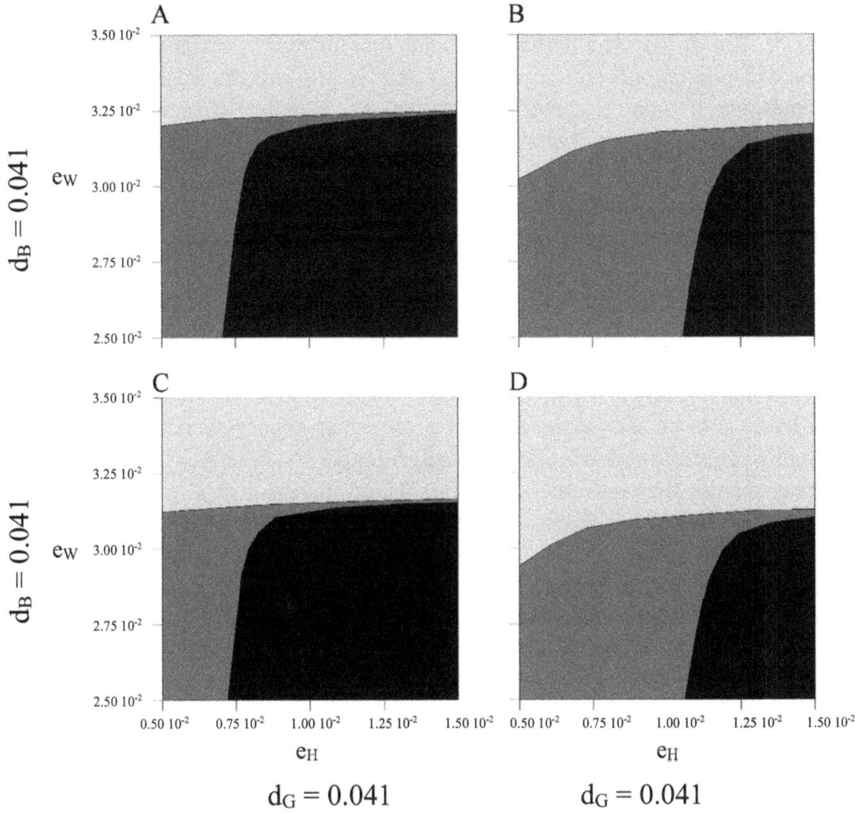

Fig. 10.4 Parameter plane showing the output for different settings of e_H and e_W, and different levels of mortality (d_G and d_B) when both grazer and browser populations respond dynamically. Other parameters $\alpha = 0.4$; $a = 0.5$; $c_H = c_W = 0.02$; $d_H = 0.9$; $d_W = 0.4$; $k_H = 0.1$; $k_W = 0.01$; $n = 1$; $\theta_W = 0.5$; $\theta_H = 0.9$; $r_H = 1$; $r_W = 0.5$; $w_{in} = 940$; $\beta = 250$; d_G and d_B as indicated in the figure. Shades indicate same states as in Fig. 10.3

system to changes in e_H is high at low values, but decreases once this value increases. Also, the system is much more sensitive to changes in e_W than to changes in e_G (compare scales) and to changes in d_G and in d_B (compare different values).

The results of the analysis show that including dynamic herbivore populations reduces the occurrence of alternative stable states and increases the occurrence of savanna when compared with constant herbivore numbers (Fig. 10.2). The decrease in parameter ranges where alternative stable states occur result in more resilient savannas as under these conditions sudden shifts from savanna to woodland can occur (i.e., bush encroachment). First, these results suggest that when herbivores respond dynamically to changes in food abundance, the occurrence of savannas is much more likely then under constant herbivore numbers. Second, they suggest that the risk of, so called, bush encroachment is reduced when herbivores can respond dynamically because the conditions under which alternative stable states occur decrease. Third, dynamic herbivore populations increase the domain of attraction of the savanna equilibrium considerably, indicating that the savanna equilibrium becomes much more resilient to perturbations. Finally, when both grazer and browser populations are dynamic, savanna as stable equilibrium prevails as the dominant type, and woodland no longer occurs as possible state.

As is explained by Van Langevelde et al. (2003), the positive feedback between fuel load (grass biomass) and fire intensity results under certain conditions in alternative stable states. High amounts of grass biomass result in intense fires, which lead to severe damage to trees. The negative effect of grazers on this positive feedback mechanism diminishes when the grazer population responds dynamically to the available amount of grass biomass. A decrease in grass biomass due to consumption triggers a negative feedback as this reduced grass biomass also leads to a decrease in grazer density through density dependent effects on mortality and fecundity. This allows the grass to recover from herbivory and to provide high fuel load for intense fires until the grazer population increases as a result or relatively high food availability. This ensures a more resilient presence of grass biomass in the system and the occurrence of savannas where otherwise woodland would occur.

An important factor that determines the effect of this negative feedback is the speed at which herbivores respond changes in grass biomass. We investigated the effect of the response of herbivores on the stability of savannas by varying the efficiency with which the grazers consume and convert grass biomass (e_H) and the specific mortality rate of the grazers (d_G). First, we found that the stability of savannas decreases with an increase in the response of grazer biomass to grass biomass (i.e., high e_H). Second, the stability of savannas increases with an increase in the specific mortality rate, i.e., a rapid response of density-dependent mortality to increasing grazer biomass (i.e., high d_G). The speed at which grazer populations respond to forage presence is not necessarily only a function of mortality or consumption and conversion efficiency. At a larger scale, grazers migrate to locations with more food when forage becomes scarce reducing the effect of the grazers on the vegetation (Drent and Prins 1987; McNaughton 1979; Van de Koppel et al. 2002; **Venter et al.** Chap. 5). We expected that such migratory behaviour reduces the risk that savannas collapse into a bush-encroached state. Moreover, grazer populations, both wildlife (Owen-Smith and Ogutu 2003) and livestock (Toulmin 1994), follow climatic variability (Illius and O'Connor 1999). Our results suggest that when pastoralists respond quickly to a decrease in forage availability and slowly to vegetation regeneration, the stability of the savannas increases and the risk on vegetation collapse decreases. However, it should be kept in mind that the response of herbivore populations depends on more than forage availability alone, like water accessibility, vegetation structure and predation risk (**Kiffner and Lee** Chap. 6). This means that savanna vegetation near water points can experience heavier grazing pressure than predicted by our model (as is for example described by Sianga et al. 2017). Alternatively, areas with high predation risk, or high tsetse densities, can experience less grazing than predicted by our model, resulting in more resilient savannas.

10.5 Future Avenues for Research

This Chapter demonstrates the importance of grazing and browsing in underpinning the dynamics of tropical savannas. The vegetation in savannas, characterized by the co-existence of trees and grasses, is determined by a combination of mean annual precipitation, soil characteristics and disturbances such as fire and herbivory. Current understanding about the dynamics of savanna vegetation is that both grazers and browsers can trigger shifts in states, from a mixture of grasses and trees to a tree-dominated state. These shifts occur when herbivores overexploit their foraging resources, especially heavy grazing pressure results in low amounts of fuel for ground fires which leads to fires that are not very intense and do not have a large damaging effects on woody biomass. Overexploitation is often the result of keeping large herbivores in a confined area (Boone et al. 2008; **Fynn et al.** Chap. 14). Limitations on movement may lead to overexploitation of the vegetation by both grazers and browsers and hence to shifts from tree-grass coexistence to tree-dominated vegetation. These shifts are less likely to occur when populations of grazers and browsers are free-ranging without movement limitations so that herbivores can migrate to mitigate for food shortage during the dry season. The model presented in this Chapter brings together numerous studies on the influence of grazing and browsing in tropical savanna vegetation and generates predictions to be tested by future studies.

One of the next steps is to model the dynamics of trees and grasses along a gradient away from water points, because herbivores are limited to the water source during the dry season (Wato et al. 2016), and hence have a large impact on the vegetation next to waterpoints, the so-called piosphere effect (Sianga et al. 2017). The vegetation along this gradient may largely differ depending on the composition of the herbivore community that can vary in body size and diet (especially grazers and less so for browsers as they are less water limited) and on the fire frequency. This gradient will differ in length and effects on the vegetation when considered in areas that differ in soil type and mean annual precipitation. More sandy soils and higher MAP will promote the cover of woody biomass, probably increasing the effects of herbivores (especially browsers) as their density also increase accordingly. Another topic that can be studied using the existing model is whether herbivore density (especially grazers) peaks at lower MAP than fire occurrence does on the same range of MAP. Data on herbivore density (Hempson et al. 2015) and fire occurrence (Lehmann et al. 2014) support this hypothesis.

Another expansion of the model would be to make it more spatially explicit, with differential run-on and run-off across the landscape: indeed our modelling and experimental work in West Africa points at the importance of this effect (Rietkerk et al. 2002a, b; Stigter and Van Langevelde 2004). Modelling done by Rietkerk et al. (2004) has, however, some intrinsic problems since plant growth was taken as a diffusion equation, which is mathematically convenient but may lead to wrong predictions. Also, in our model we have not taken seasonality into account: the work by Priyadarshini et al. (2015) shows that trees and grasses change their access to ground water over the year. The work of, for example, Van der Waal et al. (2009, 2011a, b, 2016) shows unexpected effects of nutrients on their scale on seedling establishment of woody species and their competitive ability with grasses. In response to herbivory, plants evolved defence mechanisms such as spines (Tomlinson et al. 2016) and chemical defences (Kohi et al. 2009), which reduces the effect of herbivory on plant growth. Plant defences are known to depend on nutrients in the soil, for example secondary metabolites are developed by plants in soils with low nutrient status, which would create an extra complexity in the current model. In a series of experiments, we explored the differences in responses of seedlings from savanna trees to several stresses, such as drought (Barbosa et al. 2014a, b; Tomlinson et al. 2012, 2013b, 2014), nutrient availability (Van der Waal et al. 2009; Tomlinson et al. 2013a; Barbosa et al. 2014a, b), herbivory (Tomlinson et al. 2016), competition with grasses (Tomlinson et al. 2019). This work clearly shows that plant responses differ largely between species (e.g., between evergreen and deciduous tree species), which has not included in modelling studies so far. These different responses may imply that the model results may only be valid for a selected group of plant species. Recognition of these different responses in models would improve our understanding of the effects of changes in climate and land use on vegetation dynamics. Finally, we now know that cascading effects of water availability or elephant browsing (Hilbers et al. 2015) may have unexplored consequences. We believe that much still has to be gained by modelling to explore vegetation dynamics, not only of savannas but of many other systems. Indeed, savannas are defined as stable associations of grasses and trees but much of the Palearctic is covered in taïga, which basically is the same formation yet hardly studied.

Nevertheless, our results provide insight in the interactive effects of both grazing and browsing on the dynamics in savanna ecosystems. An important conclusion of our modelling study is that the speed with which herbivores respond to changes in biomass availability has consequences for the occurrence of alternative stable states.

References

Barbosa ERM, Tomlinson KW, Carvalheiro LG, Kirkman K, de Bie S, Prins HHT, van Langevelde F (2014a) Short-term effect of nutrient availability and rainfall distribution on biomass production and leaf nutrient content of savanna tree species. PLoS One 9:e92619

Barbosa ERM, van Langevelde F, Tomlinson KW, Carvalheiro LMGR, Kirkman K, de Bie S, Prins HHT (2014b) Tree species from different functional groups respond differently to environmental changes during establishment. Oecologia 174:1345–1357

Bond WJ (2008) What limits trees in C-4 grasslands and savannas? Annu Rev Ecol Evol Syst 39:641–659

Bond WJ, Keeley JE (2005) Fire as a global 'herbivore': the ecology and evolution of flammable ecosystems. Trends Ecol Evol 20:387–394

Bond WJ, Woodward FI, Midgley GF (2005) The global distribution of ecosystems in a world without fire. New Phytol 165:525–538

Boone RB, Burnsilver SB, Worden JS, Galvin KA, Hobbs NT (2008) Large-scale movements of large herbivores livestock following changes in seasonal forage supply. In: Prins HHT, Van Langevelde F (eds) Resource ecology. Springer, Dordrecht, pp 187–206

Briggs JM, Knapp AK, Brock BL (2002) Expansion of woody plants in tallgrass prairie: a fifteen-year study of fire and fire-grazing interactions. Am Midl Nat 147:287–294

Brown JR, Carter J (1998) Spatial and temporal patterns of exotic shrub invasion in an Australian tropical grassland. Landsc Ecol 13:93–102

Bucini G, Hanan N (2007) A continental-scale analysis of tree cover in African savannas. Glob Ecol Biogeogr 16:593–605

Cameron EZ, Du Toit JT (2007) Winning by a neck: tall giraffes avoid competing with shorter browsers. Am Nat 169:130–135

Carmona-Moreno C, Belward A, Malingreau JP, Hartley A, Garcia-Alegre M, Antonovskiy M, Buchshtaber V, Pivovarov V (2005) Characterising interannual variation in global fire calendar using data from Earth observing satellites. Glob Chang Biol 11:1537–1555

Carpenter SR, Ludwig D, Brock WA (1999) Management of eutrophication for lakes subject to potentially irreversible change. Ecol Appl 9:751–771

D'Odorico P, Laio F, Ridolfi L (2006) A probabilistic analysis of fire-induced tree-grass coexistence in savannas. Am Nat 167:E79–E87

De Knegt H, Groen TA, van de Vijver CADM, Prins HHT, van Langevelde F (2008) Herbivores as architects of savannas: inducing and modifying spatial vegetation patterning. Oikos 117:543–554

De Ronde C, Trollope WSW, Parr CL, Brockett BH, Geldenhuys CJ (2004) Fire effects on flora and fauna. In: Goldhammer JG, De Ronde C (eds) Wildland fire management: handbook for Sub-Sahara Africa. Global Fire Monitoring Centre, Freiburg, pp 60–87

Drent RH, Prins HHT (1987) The herbivore as prisoner of its food supply. In: Van Andel J, Bakker J, Snaydon RW (eds) Disturbance in grasslands; species and population responses. Dr. W. Junk Publishing, Dortrecht, pp 133–149

Dublin HT, Sinclair ARE, McGlade J (1990) Elephants and fire as causes of multiple stable states in the serengeti-mara woodlands. J Anim Ecol 59:1147–1164

Fensham RJ, Holman RE (1999) Temporal and spatial patterns in drought-related tree dieback in Australian savanna. J Appl Ecol 36:1035–1050

Goheen JR, Young TP, Keesing F, Palmer TM (2007) Consequences of herbivory by native ungulates for the reproduction of a savanna tree. J Ecol 95:129–138

Govender N, Trollope LA, Wilgen BW (2006) The effect of fire season, fire frequency, rainfall and management on fire intensity in savanna vegetation in South Africa. J Appl Ecol 43:748–758

Groen TA, van Langevelde F, van de Vijver CADM, Govender N, Prins HHT (2008) Soil clay content and fire frequency affect clustering in savanna trees in South African savannas. J Trop Ecol 24:269–279

Groen TA, van Langevelde F, van de Vijver CADM, de Raad AL, de Leeuw J, Prins HHT (2011) A continental analysis of correlations between tree patterns in African savannas and human and environmental variables. J Arid Environ 75:724–733

Groen TA, van de Vijver CADM, van Langevelde F (2017) Do spatially homogenising and heterogenising processes affect transitions between alternative stable states? Ecol Model 365:119–128

Hempson GP, Archibald S, Bond WJ (2015) A continent-wide assessment of the form and intensity of large mammal herbivory in Africa. Science 350(6264):1056–1061

Higgins SI, Bond WJ, Trollope WSW (2000) Fire, resprouting and variability: a recipe for grass-tree coexistence in savanna. J Ecol 88:213–229

Hilbers JP, van Langevelde F, Prins HHT, Grant CC, Peel M, Coughenour MB, de Knegt HJ, Slotow R, Smit IPJ, Kiker GA, de Boer WF (2015) Modeling elephant-mediated cascading effects of water point closure. Ecol Appl 25:402–415

Hoffmann WA, Schroeder W, Jackson RB (2002) Positive feedbacks of fire, climate, and vegetation and the conversion of tropical savanna. Geophys Res Lett 29. https://doi.org/10.1029/2002GL015424

Holdo RM, Nippert JB, Mack MC (2018) Rooting depth varies differentially in trees and grasses as a function of mean annual rainfall in an African savanna. Oecologia 186:269–280

Huisman TJ, van Langevelde F, de Boer WF (2009) Local positive feedback and the persistence and recovery of fringe *Avicennia marina* (Forssk.) vierh. Mangroves. Wetl Ecol Manag 17:601–611

Illius AW, O'Connor TG (1999) On the relevance of nonequilibrium concepts to arid and semiarid grazing systems. Ecol Appl 9:798–813

Jackson RB, Canadel J, Ehleringer JR, Mooney HA, Sala OE, Schulze ED (1996) A global analysis of root distributions for terrestrial biomes. Oecologia 108:389–411

Jeltsch F, Weber GE, Grimm V (2000) Ecological buffering mechanisms in savannas: a unifying theory of long-term tree-grass coexistence. Plant Ecol 161:161–171

Kohi EM, de Boer WF, Slot M, van Wieren SE, Ferwerda JG, Grant CC, Heitkönig IMA, Knox N, van Langevelde F, Peel M, Slotow R, van der Waal C, Prins HHT (2009) Effects of simulated browsing on growth and leaf chemical properties in *Colophospermum mopane* saplings. Afr J Ecol 48:190–196

Kohi EH, de Boer WF, Peel MJS, Slotow R, van der Waal C, Heitkönig IMA, Skidmore A, Prins HHT (2011) African elephants *Loxodonta africana* amplify browse heterogeneity in African savannas. Biotropica 43:711–721

Kumar L, Rietkerk M, van Langevelde F, van de Koppel J, van Andel J, Hearne J, de Ridder N, Stroosnijder L, Skidmore AK, Prins HHT (2002) Relationship between vegetation growth rates at the onset of the wet season and soil type in the Sahel of Burkina Faso: implications for resource utilisation at large scales. Ecol Model 149:143–152

Lehmann CER, Ratman J, Hetley LB (2009) Which of these continents is not like the other? Comparisons of tropical savanna systems: key questions and challenges. New Phytol 181:508–511

Lehmann CE, Anderson TM, Sankaran M, Higgins SI, Archibald S, Hoffmann WA, Hanan NP, Williams RJ, Fensham RJ, Felfili J, Hutley LB, Ratnam J, San Jose J, Montes R, Franklin D, Russell-Smith J, Ryan CM, Durigan G, Hiernaux P, Haidar R, Bowman DMJS, Bond WJ (2014) Savanna vegetation-fire-climate relationships differ among continents. Science 343 (6170):548–552

Loth PE, De Boer WF, Heitkönig IMA, Prins HHT (2005) Germination strategy of the East African savanna tree *Acacia tortilis*. J Trop Ecol 21:509–518

McNaughton SJ (1979) Grassland-herbivore dynamics. In: Sinclair ARE, Norton-Griffiths M (eds) Serengeti. Dynamics of an ecosystem. The University of Chicago Press, Chicago, pp 46–81

Mordelet P, Le Roux X (2006) Lamto, structure, functioning, and dynamics of a savanna ecosystem. Ecological studies, vol 179. Springer, New York, pp 139–161

Mordelet P, Menaut JC, Mariotti A (1997) Tree and grass rooting patterns in an African humid savanna. J Veg Sci 8:65–70

Mourik AA, van Langevelde F, van Tellingen E, Heitkönig IMA, Gaigher I (2007) Stability of wooded patches in a South African nutrient-poor grassland: do nutrients, fire or herbivores limit their expansion? J Trop Ecol 23:529–537

Nystrom M, Folke C, Moberg F (2000) Coral reef disturbance and resilience in a human-dominated environment. Trends Ecol Evol 15:413–417

Olff H, Ritchie ME, Prins HHT (2002) Global environmental controls of diversity in large herbivores. Nature 415:901–904

Owen-Smith N, Mills MGL (2006) Manifold interactive influences on the population dynamics of a multispecies ungulate assemblage. Ecol Monogr 76:73–92

Owen-Smith N, Ogutu J (2003) Rainfall influences on ungulate populations dynamics. In: Du Toit JT, Rogers KH, Biggs HC (eds) The kruger experience: ecology and management of savanna heterogeneity. Island Press, Washington, pp 310–330

Pretorius Y, de Boer WF, van der Waal C, de Knegt HJ, Grant RC, Knox NM, Kohi EM, Mwakiwa E, Page BR, Peel MJS, Skidmore AK, Slotow R, van Wieren SE, Prins HHT (2011) Soil nutrient status determines how elephant utilize trees and shape environments. J Anim Ecol 80:875–883

Prins HHT (1993) Productivity of tropical herbivore assemblages: nature's basis for exploitation by man. Report number: 67 Published by: Wageningen Pers, pp 18

Prins HHT, Van der Jeugd HP (1993) Herbivore population crashes and woodland structure in East Africa. J Ecol 81:305–314

Priyadarshini KVR, Prins HHT, de Bie S, Heitkönig IMA, Woodborne S, Gort G, Kirkman K, Fry B, de Kroon H (2014) Overlap in nitrogen sources and redistribution of nitrogen between trees and grasses in a semi-arid savanna. Oecologia 174:1107–1116

Priyadarshini KVR, Prins HHT, de Bie S, Heitkonig IMA, Woodborne S, Gort G, Kirkman K, Ludwig F, Dawson TE, de Kroon H (2015) Seasonality of hydraulic redistribution by trees to grasses and changes in their water-source use that change tree–grass interactions. Ecohydrology 9:218–228

Priyadarshini KVR, de Bie S, Heitkonig IMA, Woodborne S, Gort G, Kirkman KP, Prins HHT (2016) Competition with trees does not influence root characteristics of perennial grasses in semi-arid and arid savannas in South Africa. J Arid Environ 124:270–277

Rietkerk M, Boerlijst M, van Langevelde F, HilleRisLambers R, van de Koppel J, Kumar L, Prins HHT, de Roos A (2002a) Self-organization of vegetation in arid ecosystems. Am Nat 160:524–530

Rietkerk M, Ouedraogo T, Kumar L, Sanou S, van Langevelde F, Kiema A, van de Koppel J, van Andel J, Hearne J, Skidmore AK, de Ridder N, Stroosnijder L, Prins HHT (2002b) Fine-scale spatial distribution of plants and resources on a sandy soil in the Sahel. Plant Soil 239:69–77

Rietkerk M, Dekker SC, De Ruiter PC, van de Koppel J (2004) Self-organized patchiness and catastrophic shifts in ecosystems. Science 305:1926–1929

Roques KG, O'Connor TG, Watkinson AR (2001) Dynamics of shrub encroachment in an African savanna: relative influence of fire, herbivory, rainfall and density dependence. J Appl Ecol 38:268–280

Sankaran M, Hanan NP, Scholes RJ, Ratnam J, Augustine DJ, Cade BS, Gignoux J, Higgins SI, Le Roux X, Ludwig F, Ardo J, Banyikwa F, Bronn A, Bucini G, Caylor KK, Coughenour MB, Diouf A, Ekaya W, Feral CJ, February EC, Frost PGH, Hiernaux P, Hrabar H, Metzger KL, Prins HHT, Ringrose S, Sea W, Tews J, Worden J, Zambatis N (2005) Determinants of woody cover in African savannas. Nature 438:846–849

Schenk HJ, Jackson RB (2002) Rooting depth, lateral root spreads and below-ground/above-ground allometries of plants in water-limited ecosystems. J Ecol 90:480–494

Scholes RJ, Archer SR (1997) Tree-grass interactions in savannas. Annu Rev Ecol Syst 28:517–544

Seghieri J (1995) The rooting patterns of woody and herbaceous plants in a savanna: are they complementary or in competition? Afr J Ecol 33:358–365

Sianga K, van Telgen M, Vrooman J, Fynn R, van Langevelde F (2017) Spatial refuges buffer landscapes against homogenization and degradation by large herbivore populations and facilitate vegetation heterogeneity. Koedoe 59:a1434

Smit IPJ, Prins HHT (2015) Predicting the effects of woody encroachment on mammal communities, grazing biomass and fire frequency in African savannas. PLoS One. https://doi.org/10.1371/journal.pone.0137857

Solbrig OT (1996) The diversity of the savanna ecosystem. In: Solbrig OT, Medina E, Silva JF (eds) Biodiversity and savanna ecosystem processes: a global perspective. Ecological studies, vol 121. Springer, Berlin, pp 1–27

Stigter JD, van Langevelde F (2004) Optimal harvesting in a two-species model under critical depensation. The case of optimal harvesting in semi-arid grazing systems. Ecol Model 179:153–161

Sydes CL, Grime JP (1984) A comparative study of root development using a simulated rock crevice. J Ecol 72:937–946

Tomlinson KW, Sterck FJ, Bongers F, da Silva DA, Barbosa ERM, Ward D, Bakker FT, van Kaauwen M, Prins HHT, de Bie S, van Langevelde F (2012) Biomass partitioning and root morphology of savanna trees across a water gradient. J Ecol 100:1113–1121

Tomlinson KW, van Langevelde F, Ward D, Bongers F, da Silva DA, Prins HHT, de Bie S, Sterck FJ (2013a) Deciduous and evergreen trees differ in juvenile biomass allometries because of differences in allocation to root storage. Ann Bot 112:575–587

Tomlinson KW, Poorter L, Sterck F, Borghetti F, Ward D, de Bie S, van Langevelde F (2013b) Leaf adaptations of evergreen and deciduous trees of semi-arid and humid savannas on three continents. J Ecol 101:430–440

Tomlinson KW, Poorter L, Bongers F, Borghetti F, Jacobs L, van Langevelde F (2014) Relative growth rate variation of evergreen and deciduous savanna tree species is driven by different traits. Ann Bot 114:315–324

Tomlinson KW, van Langevelde F, Ward D, Prins HHT, de Bie S, Vosman B, Sampaio EVSB, Sterck FJ (2016) Defence against vertebrate herbivores trades off into architectural and low nutrient strategies amongst savanna Fabaceae species. Oikos 125:126–136

Tomlinson KW, Sterck FJ, Barbosa ERM, de Bie S, Prins HHT, van Langevelde F (2019) Seedling growth of savanna tree species from three continents under grass competition and nutrient limitation in a greenhouse experiment. J Ecol 107:1051–1066

Toulmin C (1994) Tracking drought: options for destocking and restocking. In: Scoones I (ed) Living with uncertainty; new directions in pastoral development in Africa. Intermediate Technology Publications, London, pp 95–115

Van Auken OW (2000) Shrub invasions of north american semiarid grasslands. Annu Rev Ecol Syst 31:197–215

Van de Koppel J, Rietkerk M, Van Langevelde F, Kumar L, Klausmeier CA, Fryxell JM, Hearne JW, Van Andel J, De Ridder N, Skidmore AK, Stroosnijder L, Prins HHT (2002) Spatial heterogeneity and irreversible vegetation change in semiarid grazing systems. Am Nat 159:209–218

Van der Waal C, de Kroon H, de Boer WF, Heitkönig IMA, Skidmore AK, de Knegt HJ, van Langevelde F, van Wieren SE, Grant RC, Page BR, Slotow R, Kohi E, Mwakiwa E, Prins HHT (2009) Water and nutrients alter herbaceous competitive effects on tree seedlings in a semi-arid savanna. J Ecol 97:430–439

Van der Waal C, deKroon H, Heitkönig IMA, Skidmore AK, van Langevelde F, de Boer WF, Slotow R, Grant RC, Peel MJS, Kohi E, de Knegt HJ, Prins HHT (2011a) Scale of nutrient patchiness mediates resource partitioning between trees and grasses in a semi-arid savanna. J Ecol 99:1124–1133

Van der Waal C, Kool A, Meijer SS, Kohi E, Heitkönig IMA, de Boer WF, van Langevelde F, Grant RC, Peel MJS, Slotow R, de Knegt HJ, Prins HHT, de Kroon H (2011b) Large herbivores may alter vegetation structure of semi-arid savannas through soil nutrient mediation. Oecologia 165:1095–1107

Van der Waal C, de Kroon H, van Langevelde F, de Boer WF, Heitkönig IMA, Slotow R, Pretorius Y, Prins HHT (2016) Scale-dependent bi-trophic interactions in a semi-arid savanna: how herbivores eliminate benefits of nutrient patchiness to plants. Oecologia. https://doi.org/10.1007/s00442-016-3627-0

Van Langevelde F, Prins HHT (2007) Resilience and restoration of soft-bottom near-shore ecosystems. Hydrobiologia 591:1–4

Van Langevelde F, Van de Vijver CADM, Kumar L, Van de Koppel J, De Ridder N, Van Andel J, Skidmore AK, Hearne JW, Stroosnijder L, Bond WJ, Prins HHT, Rietkerk M (2003) Effects of fire and herbivory on the stability of savanna ecosystems. Ecology 84:337–350

Van Langevelde F, Drescher M, Heitkönig IMA, Prins HHT (2008) Modelling instantaneous intake rate of herbivores as function of forage quality and mass: effects on facilitative and competitive interactions between differently sized herbivores. Ecol Model 213:273–284

Van Langevelde F, Tomlinson K, Barbosa ERM, De Bie S, Prins HHT, Higgins SI (2011) Understanding tree-grass co-existence and impacts of disturbance and resource variability in savannas. In: Hill MJ, Hanan NP (eds) Ecosystem function in savannas. Measurement and modeling at landscape to global scales. CRC Press, Boca Raton, FL, pp 257–271

Van Langevelde F, de Groot C, Groen TA, Heitkönig IMA, Gaigher I (2014) Effect of patches of woody vegetation on the role of fire in tropical grasslands and savannas. Int J Wildland Fire 23:410–416

Van Langevelde F, Tessema ZK, de Boer WF, Prins HHT (2016) Soil seed bank dynamics under the influence of grazing as alternative explanation for herbaceous vegetation transitions in semi-arid rangelands. Ecol Model 337:253–261

Van Wijk MT, Rodriguez-Iturbe I (2002) Tree-grass competition in space and time: insights from a simple cellular automata model based on ecohydrological dynamics. Water Resour Res 38:18.11–18.15

Walker BH, Ludwig D, Holling CS, Peterman RM (1981) Stability of semi-arid savanna grazing systems. J Ecol 69:473–498

Walter H (1971) Ecology of tropical and subtropical vegetation. Oliver and Boyd, Edinburgh

Warner RR, Chesson PL (1985) Coexistence mediated by recruitment fluctuations: a field guide to the storage effect. Am Nat 125:769–787

Wato YA, Heitkönig IMA, van Wieren SE, Wahungu G, Prins HHT, van Langevelde F (2016) Prolonged drought results in starvation of African elephant (Loxodonta africana). Biol Conserv 203:89–96

Wato YA, Prins HHT, Heitkonig IMA, Wahungu G, Ngene S, Njumbi S, van Langevelde F (2018) Movement patterns of African elephants (Loxodonta africana) in a semi-arid savanna suggest that they have information on the location of dispersed water sources. Front Ecol Evol 6:167

Weltzin JF, McPherson GR (1997) Spatial and temporal soil moisture resource partitioning by trees and grasses in a temperate savanna, Arizona, USA. Oecologia 112:156–164

Wiegand K, Ward D, Saltz D (2005) Multi-scale patterns and bush encroachment in an arid savanna with a shallow soil layer. J Veg Sci 16:311–320

Chapter 11
Impacts of Browsing and Grazing Ungulates on Plant Characteristics and Dynamics

Autumn E. Sabo

11.1 Introduction

Ungulates occupy a wide variety of habitats, from forests and grasslands to deserts and tundra, and are on all continents except for Antarctica. Wild ruminants number approximately 214×10^6 (Pérez-Barbería 2017). Being primarily herbivores, they ingest vast amounts of forage: approximately 2.5% (dry matter) of their body mass daily (Prins and van Langevelde 2008). Given that ungulates are nearly ubiquitous across rural and natural landscapes, and that they consume huge quantities of plant material, they can dramatically affect vegetation communities.

Plant biologists view ungulates as one of the many disturbance factors that influence plants. Interacting with fire and drought, ungulates drive the dynamics of ecosystems across the globe, including savanna ecosystems as described in **van Langevelde et al.** (Chap. 10). In this Chapter, I will explore how ungulates affect lower levels of biological organization, from the morphology and chemistry of individual plants, through plant populations and vegetation communities. These effects, in turn, contribute to nutrient cycling and ecosystem patterns (as detailed elsewhere in this book).

11.2 Plant Architecture

11.2.1 Introduction

Edibility and ease of access result in different species of ungulates selecting specific plant types (e.g., Holechek 1984; Shipley 1999), and plant tissues (reviewed by Hanley et al. 2007). Each plant part serves a particular physiological function that is

A. E. Sabo (✉)
Penn State Beaver, Monaca, PA, USA
e-mail: axs182@psu.edu

I. J. Gordon, H. H. T. Prins (eds.), *The Ecology of Browsing and Grazing II*,
Ecological Studies 239, https://doi.org/10.1007/978-3-030-25865-8_11

Fig. 11.1 Abundant ungulates may consume nearly all of the foliage within their reach, which creates a visible browse line, as shown in this photo from northern Wisconsin, USA by Alison Paulson

crucial for plant survival and reproductive success. For example, leaves are the primary photosynthetic powerhouse of most species. Thus, when ungulates consume foliage, plants' abilities to transform solar energy into chemical energy are reduced. Damage to flowers and fruits directly diminishes short-term plant reproductive output (Russell et al. 2001). Browsing of buds, or apical or lateral meristems, alters plant repair responses and growth forms (reviewed by Russell et al. 2001), which may feedback to also modify photosynthetic and reproductive capacities.

Physical characteristics of plants partially determine their susceptibility and tolerance to ungulate damage. Some ungulates are deterred by spines and thorns (Cooper and Owen-Smith 1986). Thus, they may become a prominent feature of surviving plants in heavily browsed habitats, increasing in length with increased herbivory levels (reviewed by Hanley et al. 2007). A telltale sign of heavy browsing by ungulates is a lack of nearly all foliage and small twigs within their reach, a phenomenon known as a browse line (Fig. 11.1). Differential responses by woody and herbaceous plants to damage correspond, partially, to variation in lifespan, size, and biomass allocation, as well as physiological integration between plant parts (Haukioja and Koricheva 2000).

11.2.2 Woody Plants

In forest systems, most research regarding how ungulates impact plants has been conducted on trees, with little attention given to shrubs and herbaceous species, likely due to their disparities in commercial value. Unsuccessful regeneration by natural tree seedlings, and failed plantings because of browsing pressure, have generated substantial interest (reviewed by Gill 1992; Russell et al. 2001; Rooney and Waller 2003; Côté et al. 2004). In addition to stalled growth and increased mortality, browsing damage to terminal buds may alter tree growth forms and, consequently, stem quality (reviewed by Putman and Moore 1998; Côté et al. 2004). Stem quality is also affected in some regions by ungulates stripping bark, with levels varying between tree species and plant age and morphology (reviewed by Gill 1992). Tree rubbing (also known as fraying or thrashing) is a non-consumptive type of stem damage caused when deer scratch their velvet antlers on tree trunks (reviewed by Gill 1992).

Compared to single-trunked trees, shorter stature shrubs, with multiple stems, may be more susceptible to sustained browsing. Unlike many overstory trees that have the possibility of outgrowing ungulates' reach at maturity, shrubs may hold most of their edible-sized branches at a height continually accessible by ungulates. Similarly, the year-round availability of shrub stems may also increase their relative exposure to ungulate browsing compared to ephemeral herbaceous plants (Russell et al. 2001). In support, the abundances of shrubs have been shown to be inversely correlated with ungulate population density in many studies (e.g., McShea and Rappole 2000; Augustine and McNaughton 2004; Mudrak et al. 2009; Frerker et al. 2014).

11.2.3 Herbaceous Plants

Herbaceous vascular plants are often categorized into grasses and grass-like plants (e.g., sedges and rushes), or graminoids, and broad-leaved herbs, or forbs. Graminoids are considered to be well adapted to herbivore foraging due to their protected basal meristems, high shoot density, belowground nutrient reserves, and, in many species, short stature (Coughenour 1985). These common characteristics of graminoids allow for rapid, and substantial, regrowth following grazing damage (Haukioja and Koricheva 2000; but see Kirby 2001; Moser and Schütz 2006). On the other hand, one browsing event per growing season is more likely to be catastrophic for forbs, because new growth is typically generated at the end of shoots (Shipley 1999). Often, the showy flowers and large fruits of forbs are herbivore attractants, and fencing studies indicate that sexual plant parts may be preferred over the vegetative tissues (e.g., Hülber et al. 2005). While such preferences could relate to visibility, differences in chemical composition are also likely contributors.

11.3 Plant Chemistry

The chemistry of plants influences their digestibility and, subsequently, how highly they rank in preference by various herbivores (e.g., Hanley 1997; Bee et al. 2011). Differences in digestive system morphologies, body size and feeding behaviours (Holechek 1984) have been used to characterise ungulates based on their forage selection. While various systems have been promoted that include mixed feeders (Lamprey 1963), a browser-grazer continuum (McNaughton and Georgiadis 1986) and frugivores (Bodmer 1990), for simplicity, I discuss only the broadest categories of grazers and browsers (Lamprey 1963) in this Chapter. Grazers preferentially feed on graminoids that have low levels of potentially poisonous volatile oils but are difficult to digest. Browsers consume primarily forbs, shrubs and trees. These plants are often rich in defensive oils, proteins and phosphorous, but lower in cellulose than graminoids, which can make them easier to digest.

In addition to the graminoid vs. forb/tree/shrub categories discussed above, grouping plant species according to their photosynthetic pathways (C_3, C_4 and crassulacean acid metabolism [CAM]) also provides important information for understanding herbivory. For example, herbivores often prefer C_3 plants (reviewed by Heckathorn et al. 1999). Generally, plant species that utilize the typical C_3 pathway are higher in nitrogen and phosphorus, and lower in structural compounds such as lignin and silica, as compared to C_4 and CAM plants (reviewed by Heckathorn et al. 1999). Structural compounds appear to reduce digestibility in herbivores (reviewed by Augustine and McNaughton 1998); in particular, the ratios of lignin, or carbon, to nitrogen concentrations in leaves are negatively correlated with digestibility (e.g., Wardle et al. 2002; Cornelissen et al. 2004) and/or consumption (e.g., Kurokawa et al. 2010). Structural compounds contribute to physical properties of plants, such as leaf toughness, that appear to be important for determining ungulate preferences (e.g., Cingolani et al. 2005). Leaf toughness and specific leaf area are examples of functional traits being explored in cross-taxa analyses of ungulate impacts (e.g., Díaz et al. 2001).

Plant secondary compounds, including phenolics, are thought to provide resistance, or protection, from herbivory (reviewed by Augustine and McNaughton 1998; but see Carmona et al. 2011). However, avoidance of certain plant species by ungulates, based on high levels of secondary compounds, might only occur when well-defended species are of moderate to high abundance (Bryant et al. 1991). Potential reasons may be that secondary compounds only affect digestion at large doses, or that species must be dense for herbivores to learn which vegetation to avoid. The production of secondary chemicals may come at a metabolic cost to plants (Herms and Mattson 1992), particularly where nutrient availability is limited (reviewed by Augustine and McNaughton 1998).

Plants that are preferred by herbivores often have high levels of nitrogen (reviewed in Gill 1992), a key component of enzymes. As concentrations of nitrogen in plant tissues increase, photosynthetic rates and, thus, growth rates also increase

(Field and Mooney 1986; Reich et al. 1995). Hence, some chemical aspects of plants relate not only to herbivore preference but also to plants' abilities to recover following damage (reviewed by Augustine and McNaughton 1998).

11.4 Defoliation and Its Effects

11.4.1 Introduction

As discussed above, variation in ungulate damage is related to the physical structure and chemistry of plant species. Effects on plant survival, regrowth and reproduction are related to the timing and severity of herbivory, as well as the subsequent abilities of damaged vs. undamaged plants to attain necessary resources. The removal of leaf tissue diminishes plants' immediate photosynthetic capacities, but the long-term consequences vary according to plant traits and species, as well as the availability of resources to support recovery (Aarssen 1995).

11.4.2 Timing of Herbivory

The timing of ungulate damage can determine outcomes for individual plants (reviewed by Russell et al. 2001). Browsing of woody plants during the growing season, in temperate climates, appears to be more detrimental than damage that occurs on those same species while they are dormant (e.g., Canham et al. 1994). For forbs, because their susceptible tissues never grow completely out of the reach of ungulates, the timing of browsing has the potential to dramatically reduce, or completely eliminate, seed production for the year (reviewed by Russell et al. 2001). Impacts on reproduction are likely to be especially pronounced in short-lived plants and perennials that are only fleetingly aboveground (Augustine and McNaughton 1998; Augustine and DeCalesta 2003). For example, the fruit number of two lakeshore annuals declined drastically following simulated herbivory that was applied during late stages of development, although clipping plants early during their growth cycles had little effect (Gedge and Maun 1992). The severity and return interval of damage events may also influence tolerance, or regrowth potential (Bergelson and Crawley 1992).

11.4.3 Individual Plant Responses to Herbivory

Some species have lower tolerance to herbivory than others. Low tolerance relates to characteristics that inhibit fast regrowth: investing large quantities of nutrients and carbon into evergreen foliage (reviewed by Bloom et al. 1985); having low photosynthetic capacity (reviewed by Côté et al. 2004); lacking structures that allow for horizontal, vegetative spread (reviewed by Augustine and McNaughton 1998); having minimal reserves of belowground energy (reviewed by Augustine and McNaughton 1998); or failing to reallocate resources from belowground to photosynthetic tissues (Richards 1984). The ability of certain species to compensate for tissue loss, through increased growth, has been heavily researched but the hope of identifying herbivore optimization levels (where herbivory intensity maximizes net primary productivity, Dyer and Shugart 1992; McNaughton 1983) has rarely been realized (Belsky 1986). Belsky (1986) clarified this muddy concept of compensatory growth by introducing more exact terms including overcompensation, where damaged plants' biomasses exceed that of undamaged plants, and partial compensation, indicated by some regrowth of removed tissues (Belsky 1986).

Conclusive research on compensatory growth is difficult because all plant parts (above- and belowground tissues) and life stages (vegetative and reproductive, Russell et al. 2001) need to be thoroughly evaluated to avoid mistaken conclusions from the reallocation of plant resources rather than from actual compensatory growth (reviewed in Côté et al. 2004). For instance, root biomass is often reduced by ungulate herbivory (e.g., Ellison 1960; Crawley 1983; Richards 1984), so measuring only aboveground growth leads to misleading findings. Most robust evidence for overcompensation (reviewed by Côté et al. 2004) is from research on grasses, which may be better adapted than other taxa to counteract defoliation effects because of their basal meristems. However, some forbs also have well-protected meristems and show compensatory regrowth after grazing (e.g., *Plantago maritima* after goose grazing (Prins et al. 1980)). Following damage to shoot apical meristems, species with terminal leaders often develop into more bushy forms (reviewed by Côté et al. 2004) that can produce greater numbers of flowering stems (Mopper et al. 1991). Reduced fecundity in response to browsing is, however, more commonly observed (reviewed by Russell et al. 2001).

11.4.4 Plant Population Responses to Herbivory

Ungulate herbivory can reduce growth (Tanentzap et al. 2012; Koh et al. 2010), health (Switzenberg et al. 1955), cover (Rooney and Waller 2001; Habeck and Schultz 2015), and stature of preferred species (e.g., McNaughton 1984), even shifting populations to mostly prostrate genotypes (reviewed by Augustine and McNaughton 1998). Such changes may alter the environment and related resources, contributing to additional subsequent changes in plant populations (e.g., Heckel et al. 2010; Sabo et al. 2017). Variation in site-level availability of plant resources may likewise account for differences in herbivory tolerance between plant populations (Saunders and Puettmann 1999).

11.5 The Competitive Balance Between Plants

11.5.1 Introduction

Ungulate preferences for different species are related to plant structure and chemistry, as well as the temporal and spatial availability of preferred and less preferred species. Damaged plants may be less able to acquire resources including light, nutrients and water. Reductions in resources result in lower rates of vegetative growth and reproduction. When a species is unable to obtain the necessary amount of a resource in a given habitat, other species that either have lower requirements for that resource or that are better at acquiring it increase in relative abundance (in accordance with the resource ratio, or R*, competition model proposed by Tilman 1982). Eventually, selective foraging by ungulates can drive the competitive balance within communities, resulting in declines of preferred species that have insufficient resources and, potentially, long-term decreases in species richness and diversity.

11.5.2 Seasonal Availability of Browse

Long-distance migrations of ungulates may be a consequence of plant species' configurations (e.g., Prins and van Langevelde 2008), and growth phases varying in space and time (reviewed by Augustine and McNaughton 1998 and Hebblewhite et al. 2008). As landscapes become fragmented by land use change, historical patterns of transient feeding by native ungulates are interrupted (reviewed by Augustine and McNaughton 1998; Hebblewhite et al. 2008). Ungulate impacts may become more severe in fragmented areas (Reimoser 2003) because of reduced overall quantities of forage as land becomes developed, as well as higher quality or quantities of forage along patch edges (e.g., Takatsuki 2009).

Habitats with less distinct delineations of vegetation may also be structured by seasonal differences in ungulate preferences. For example, preferences for deciduous woody plants shift to evergreens during harsh seasons when deciduous foliage senesces. This can result in dominance by deciduous species where browsers are abundant during dry or cold leaf-off seasons (e.g., Owen-Smith and Cooper 1987; Ammer 1996).

11.5.3 Relative Abundance of Preferred Species

Even where the availability of preferred species does not differ between seasons, differences in the local abundance of plants are important for determining ungulate effects. If the abundance of a specific, preferred species is high, the overall detrimental impact from ungulate damage to the population of that species may be minor. This concept is called herbivore consumption saturation (Noy-Meir 1975). To extend this concept, as the abundance of ungulates increases, they eventually reach a population threshold after which the relative quantities of preferred plant species begin to decline noticeably (Tilghman 1989).

11.5.4 Changes in Competitive Relationships

The duration and intensity of ungulate pressure can influence community impacts. For example, when grazing was intense, but of short duration, sheep exhibited little selectivity for preferred forage species (reviewed by Augustine and McNaughton 1998). Similarly, herd membership, with its corresponding intraspecific competition, may also decrease selectivity (reviewed by Augustine and McNaughton 1998). Where selectivity is low, but biomass removed by grazing is high, preferred and/or less competitive species may actually increase following herbivory. Likewise, moderate densities of ungulates may reduce strong competitors within the plant community, allowing less robust species to increase (e.g., Schütz et al. 2003; Itô and Hino 2005; Takatsuki 2009).

Many studies focus on the effects of high, rather than moderate, densities of ungulates. Elevated ungulate populations reduce preferred plant species to the advantage of less preferred, more inaccessible, or more browse-tolerant species (e.g., Anderson et al. 2002; White 2012; Bradshaw and Waller 2016). Repeated grazing of grassland patches can lead to short, thick "grazing lawns" (McNaughton 1984). Similarly, in savannas, repeated browsing of trees may result in squat, resprouting trees or "browsing lawns" (Jachmann and Bell 1985). "Alternative successional trajectories" may develop whereby ungulates stall vegetation communities from transitioning to dominance by plant species that are characteristic of long time periods without other types of natural disturbances (Hidding et al. 2013).

Non-consumptive effects of ungulates on plant tissues can also drive community dynamics, e.g., trampling damage (reviewed in Persson et al. 2000), dispersal of seeds (Collins and Uno 1985; Myers and Harms 2009; Albert et al. 2015), and defecation (reviewed by Augustine and McNaughton 1998). In addition, herbivory may indirectly mediate important environmental conditions in the groundlayer. For example, abiotic factors such as light availability (Fig. 11.2, Takatsuki and Itô 2009; Sabo et al. 2017; Boulanger et al. 2018), soil fertility (Pastor et al. 1988; Augustine and Frank 2001; Chen et al. 2013; Sabo et al. 2017), soil moisture content (Chen et al. 2013), soil compaction (Heckel et al. 2010; Chen et al. 2013; Shelton et al. 2014; Kardol et al. 2014; Sabo et al. 2017), and bareground (Knight et al. 2009; DiTommaso et al. 2014), as well as biotic variables including pathogens (reviewed by Russell et al. 2001) and mycorrhizae (Rossow et al. 1997), may drive understory plant establishment, competitive ability and long-term success.

Fig. 11.2 Area accessible to white-tailed deer on the left side of the fence vs. inside the exclosure (on right). Photo from Kemp Natural Resources Area, WI, USA by Katie Frerker

11.5.5 Ungulate Effects on Species Richness and Diversity

The immediate impacts of ungulates on plant communities are typically changes in the cover or density of individual species. However, in time, ungulate pressure can lead to differential extirpation and colonization rates (Olff and Ritchie 1998), which result in altered species richness. Changes in the number of species may occur more quickly where plant generation times are shorter, e.g., grasslands vs. forest (Frank 2005).

Ungulates can increase overall species richness by reducing the dominance of strong competitors, dispersing seeds (e.g., Vellend et al. 2003), or creating microsites that are conducive to the establishment of incoming species (e.g., Frank 2005). In forests, there are many examples of decreased ungulate densities increasing tree richness (or vice versa, high densities of ungulates reducing tree richness), e.g., Tilghman (1989), Gill and Beardall (2001) and Habeck and Schultz (2015). Changes in the richness of forest herb communities in response to ungulates are, however, less well documented (Habeck and Schultz 2015; Sabo et al. 2019; but see Rooney and Dress 1997; Stockton et al. 2005; Wiegmann and Waller 2006). Additional long-term studies are needed to explore whether ungulates do, indeed, have limited effects on herb species richness, or if forest herb communities instead have long lag times before responding markedly to altered grazing pressure.

Scientists have found some evidence that ungulates facilitate invasion by non-native plant species (e.g., Vavra et al. 2007; Bartuszevige and Endress 2008; Knight et al. 2009; Duguay and Farfaras 2011; Castellano and Gorchov 2013; Frerker et al. 2014; Dávalos et al. 2015). Possible mechanisms for ungulates increasing exotic species are the same mechanisms described above for increasing overall species richness. Many non-native invasives are well-suited for colonizing bareground, the creation of which is a documented outcome of ungulate disturbance (Fahnestock and Knapp 1994; Knight et al. 2009; Chips et al. 2014). Lower preference by native ungulates for exotic species, compared to native plants, may contribute to invasions in some systems (e.g., Knight et al. 2009). There is, however, evidence of native herbivores suppressing invasive plants (Rossell et al. 2007, reviewed in Parker et al. 2006), and of exotic herbivores promoting invasions (reviewed in Parker et al. 2006). Therefore, trying to generalize about how species with exotic evolutionary histories contribute to ecosystem degradation (e.g., are they "drivers, back-seat drivers or passengers" Bauer 2012) is fraught with complexity, stemming from the range of community assembly factors, including ungulate effects, that vary between sites.

In order to concisely describe overall changes to communities, ecologists often combine abundance and richness into diversity indices. For example, in addition to ungulates reducing tree species richness, as discussed above, ungulates also contributed to decreases in tree diversity in many studies (reviewed by Gill and Beardall 2001; Kuiters and Slim 2002; Habeck and Schultz 2015). Effects can be diversity index and scale dependent (e.g., Frank 2005), however, so examining detailed species richness, abundance and demography data remains valuable.

11.5.6 Community Structure Effects

In addition to influencing specific species, ungulates also affect plants at broader levels of categorization. In areas with chronically large populations of ungulates, plant species with single shoots, which rely primarily on sexual reproduction, may decline while plants that spread vegetatively, from ground-level meristems, dominate. Once clonal plants are well established in large patches, they can inhibit the germination, or growth, of less vigorous species, becoming a self-perpetuating, relatively static "recalcitrant layer" (Royo and Carson 2006). Another group of species influenced by ungulate abundance is the legume family, which has an unusual symbiosis with nitrogen-fixing bacteria. Due to their high nitrogen concentrations, legumes are likely to decline in habitats with abundant grazers because ungulates prefer to forage on these species (e.g., Ritchie and Tilman 1995). Without legume litter to increase soil nutrient levels, late successional species, with higher nitrogen requirements, may be slow to colonize low fertility habitats. More directly, ungulates often suppress tree establishment and growth through herbivory. As detailed in the previous Chapter of this book (**van Langevelde** et al. Chap. 10), an extreme example is the maintenance of savannas *in lieu* of woodland/forest systems (e.g., Russell et al. 2001).

11.6 Resilience and Recovery of Browsed/Grazed Systems

11.6.1 Site-Level Conditions

Worldwide, many ecosystems have evolved under the influence of grazing and browsing ungulates and are, thus, resilient to certain levels of herbivory (Gunderson 2000). Historically, many plant communities depended on native grazers, or browsers, to support ecological processes, whilst the vegetation provided ungulates with essential food and cover. In these systems, predators fed on the ungulates, which prevented them from sustaining high population densities. "Non-equilibrium dynamics" result when additional stressors, such as drought or fire suppression, disrupt these natural feedbacks (Ellis and Swift 1988). The system may become strained to a point where even low population densities of ungulates impact the ecosystem, making passive recovery following ungulate reductions unlikely (Holling 1973).

Other than stressors like drought or fire, as well as site-level conditions can affect vegetation recovery. Low-productivity areas, where plant growth is limited by resource availability, may be less resilient to herbivory than high-productivity habitats (Danell et al. 1991), either because plant communities consist of slow-growing species, or because plants have limited access to resources for recovery. Even in ecosystems that receive sufficient precipitation and that are rich in nutrients, such as temperate deciduous forests, understory layer plants can experience resource limitations. For example, low light at the groundlayer (e.g., Sabo et al. 2017) may reduce opportunities for germination and growth.

11.6.2 Legacy Effects of Chronic Heavy Levels of Browsing

In forest systems, understory trees often experience more pronounced recovery than herbaceous species following ungulate removal (reviewed in Habeck and Schultz 2015). One potential reason may be the availability of seed rain from mature forest trees (Gill 1992). In addition, wide-ranging vertebrates are a common dispersal mode for the seeds of many trees species (reviewed in Habeck and Schultz 2015). In comparison, seed rain of preferred herbs may be negligible in systems that had chronically high densities of browsers, because herbs are unable to escape ungulate foraging by growing out of reach. Short-distance dispersal modes, including those provided by ants, are typical for many forest herbs, and fragmentation can slow even wind-dispersed seeds (McLachlan and Bazely 2001; Flinn and Vellend 2005; Soons et al. 2005). Reduced propagule input is one example of a "legacy effect" from sustained high densities of browsers (reviewed in Cuddington 2011). Additional symptoms could include loss of seed banks (DiTommaso et al. 2014), decreased pollinator populations (Vázquez and Simberloff 2003), and recalcitrant understory vegetation.

Recalcitrant layers develop after severe disturbances, such as chronically high densities of ungulates. The dense layers of rhizomatous ferns, ericaceous shrubs, graminoids or non-native invasive species block regeneration, or succession to shrub and forest communities. Possible mechanisms include above- or belowground competition, physical and chemical interference, and improved habitat for seed and seedling predators (reviewed by Royo and Carson 2006; Young and Peffer 2010). Reductions in invasive species abundance following decreased ungulate pressure (e.g., Kalisz et al. 2014; Dávalos et al. 2015) suggest that this is a community-level change that may be less "recalcitrant" than other types (Nuttle et al. 2014).

11.6.3 Ecological Models

Recalcitrant layers may be one cause of non-linear trajectories of system recovery following reduced ungulate pressure. Non-linear, and non-equilibrium, systems have been characterized by various ecological models. For example, ecological hysteresis acknowledges non-linear cause and effect relationships (May 1977; Laycock 1991; Rietkerk et al. 1996; Scheffer et al. 2001), and that active management may be required to generate habitats that resemble past conditions. State-and-transition models (Laycock 1991) accommodate alternate successional trajectories that may occur because of differential ungulate grazing/browsing pressure (or other disturbance factors). Catastrophe theory includes thresholds to predict the impacts of ungulate population densities on vegetation (Lockwood and Lockwood 1993). Despite continuous advances (reviewed by Briske et al. 2010), the hope that these types of models might serve as early warning systems to prevent severe ecosystem damage has been overly optimistic given the variability between systems and even sites of the same system type (Hastings and Wysham 2010).

11.7 Conclusions

Following decades of laboratory and field research into the impacts of ungulates on plants, useful generalizations across taxa and ecosystems remain largely elusive. The complexity of ungulate-vegetation dynamics demands continued, site-specific attention. Advances in restoration ecology, also focused narrowly on systems of interest, will become increasingly critical unless stakeholders decide to content themselves with alternative stable states (Suding et al. 2004).

References

Aarssen LW (1995) Hypotheses for the evolution of apical dominance in plants: implications for the interpretation of overcompensation. Oikos 74:149–156

Albert A, Auffret AG, Cosyns E et al (2015) Seed dispersal by ungulates as an ecological filter: a trait-based meta-analysis. Oikos 124:1109–1120. https://doi.org/10.1111/oik.02512

Ammer C (1996) Impact of ungulates on structure and dynamics of natural regeneration of mixed mountain forests in the Bavarian Alps. For Ecol Manag 88:43–53. https://doi.org/10.1016/S0378-1127(96)03808-X

Anderson CE, Chapman KA, White MA, Cornett MW (2002) Effects of browsing control on establishment and recruitment of eastern white pine (*Pinus strobus* L.) at Cathedral Grove, Lake Superior Highlands, Minnesota, USA. Nat Areas J 22:202–210

Augustine DJ, Decalesta D (2003) Defining deer overabundance and threats to forest communities: from individual plants to landscape structure. Ecoscience 10:472–486

Augustine DJ, Frank DA (2001) Effects of migratory grazers on spatial heterogeneity of soil nitrogen properties in a grassland ecosystem. Ecology 82:3149–3162. https://doi.org/10.1890/0012-9658(2001)082[3149,EOMGOS]2.0.CO;2

Augustine DJ, McNaughton SJ (1998) Ungulate effects on the functional species composition of plant communities: herbivore selectivity and plant tolerance. J Wildl Manag 62:1165–1183

Augustine DJ, McNaughton SJ (2004) Regulation of shrub dynamics by native browsing ungulates on East African rangeland. J Appl Ecol 41:45–58

Bartuszevige AM, Endress BA (2008) Do ungulates facilitate native and exotic plant spread?: Seed dispersal by cattle, elk and deer in northeastern Oregon. J Arid Environ 72:904–913

Bauer JT (2012) Invasive species: “back-seat drivers” of ecosystem change? Biol Invasions 14:1295–1304

Bee JN, Tanentzap AJ, Lee WG et al (2011) Influence of foliar traits on forage selection by introduced red deer in New Zealand. Basic Appl Ecol 12:56–63. https://doi.org/10.1016/j.baae.2010.09.010

Belsky AJ (1986) Does herbivory benefit plants? A review of the evidence. Am Nat 127:870–892

Bergelson J, Crawley MJ (1992) The effects of grazers on the performance of individuals and populations of scarlet gilia, *Ipomopsis aggregate*. Oecologia 90:435–444. https://doi.org/10.1007/BF00317703

Bloom AJ, Chapin FS III, Mooney HA (1985) Resource limitation in plants-an economic analogy. Annu Rev Ecol Syst 16:363–392. https://doi.org/10.1146/annurev.es.16.110185.002051

Bodmer RE (1990) Ungulate frugivores and the browser-grazer continuum. Oikos 57:319–325

Boulanger V, Dupouey J-L, Archaux F et al (2018) Ungulates increase forest plant species richness to the benefit of non-forest specialists. Glob Chang Biol 24:e485–e495. https://doi.org/10.1111/gcb.13899

Bradshaw L, Waller DM (2016) Impacts of white-tailed deer on regional patterns of forest tree recruitment. For Ecol Manag 375:1–11

Briske DD, Washington-Allen RA, Johnson CR et al (2010) Catastrophic thresholds: a synthesis of concepts, perspectives, and applications. Ecol Soc 15:37

Bryant JP, Provenza FD, Pastor J et al (1991) Interactions between woody plants and browsing mammals mediated by secondary metabolites. Annu Rev Ecol Syst 22:431–446

Canham CD, McAninch JB, Wood DM (1994) Effects of the frequency, timing, and intensity of simulated browsing on growth and mortality of tree seedlings. Can J For Res 24:817–825

Carmona D, Lajeunesse MJ, Johnson MT (2011) Plant traits that predict resistance to herbivores. Funct Ecol 25:358–367

Castellano SM, Gorchov DL (2013) White-tailed deer (*Odocoileus virginianus*) disperse seeds of the invasive shrub, Amur honeysuckle (*Lonicera maackii*). Nat Areas J 33:78–80

Chen D, Zheng S, Shan Y et al (2013) Vertebrate herbivore-induced changes in plants and soils: linkages to ecosystem functioning in a semi-arid steppe. Funct Ecol 27:273–281. https://doi.org/10.1111/1365-2435.12027

Chips MJ, Magliocca MR, Hasson B, Carson WP (2014) Quantifying deer and turkey leaf litter disturbances in the eastern deciduous forest: have nontrophic effects of consumers been overlooked? Can J For Res 44:1128–1132
Cingolani AM, Posse G, Collantes MB (2005) Plant functional traits, herbivore selectivity and response to sheep grazing in Patagonian steppe grasslands. J Appl Ecol 42:50–59
Collins SL, Uno GE (1985) Seed predation, seed dispersal, and disturbance in grasslands: a comment. Am Nat 125:866–872
Cooper SM, Owen-Smith N (1986) Effects of plant spinescence on large mammalian herbivores. Oecologia 68:446–455
Cornelissen JHC, Quested HM, Gwynn-Jones D et al (2004) Leaf digestibility and litter decomposability are related in a wide range of subarctic plant species and types. Funct Ecol 18:779–786
Côté SD, Rooney TP, Tremblay J-P et al (2004) Ecological impacts of deer overabundance. Annu Rev Ecol Evol Syst 35:113–147
Coughenour MB (1985) Graminoid responses to grazing by large herbivores: adaptations, exaptations, and interacting processes. Ann Mo Bot Gard 72:852–863
Crawley MJ (1983) Herbivory. The dynamics of animal—plant interactions. Blackwell Scientific Publications, Hoboken
Cuddington K (2011) Legacy effects: the persistent impact of ecological interactions. Biol Theory 6:203–210. https://doi.org/10.1007/s13752-012-0027-5
Danell K, Niemela P, Varvikko T, Vuorisalo T (1991) Moose browsing on Scots pine along a gradient of plant productivity. Ecology 72:1624–1633
Dávalos A, Nuzzo V, Blossey B (2015) Single and interactive effects of deer and earthworms on non-native plants. For Ecol Manag 351:28–35. https://doi.org/10.1016/j.foreco.2015.04.026
Díaz S, Noy-Meir I, Cabido M (2001) Can grazing response of herbaceous plants be predicted from simple vegetative traits? J Appl Ecol 38:497–508. https://doi.org/10.1046/j.1365-2664.2001.00635.x
DiTommaso A, Morris SH, Parker JD et al (2014) Deer browsing delays succession by altering aboveground vegetation and belowground seed banks. PLoS One 9:e91155
Duguay JP, Farfaras C (2011) Overabundant suburban deer, invertebrates, and the spread of an invasive exotic plant. Wildl Soc Bull 35:243–251. https://doi.org/10.1002/wsb.27
Dyer MI, Shugart HH (1992) Multi-level interactions arising from herbivory: a simulation analysis of deciduous forests utilizing Foret. Ecol Appl 2:376–386. https://doi.org/10.2307/1941872
Ellis JE, Swift DM (1988) Stability of African pastoral ecosystems: alternate paradigms and implications for development. Rangel Ecol Manag/J Range Manag Arch 41:450–459
Ellison L (1960) Influence of grazing on plant succession of rangelands. Bot Rev 26:1–78
Fahnestock JT, Knapp AK (1994) Plant responses to selective grazing by bison: interactions between light, herbivory and water stress. Vegetatio 115:123–131
Field C, Mooney HA (1986) The photosynthesis-nitrogen relationship in wild plants. In: On the economy of plant form and function: proceedings of the Sixth Maria Moors Cabot Symposium. Cambridge University Press, Cambridge, pp 57–77
Flinn KM, Vellend M (2005) Recovery of forest plant communities in post-agricultural landscapes. Front Ecol Environ 3:243–250. https://doi.org/10.1890/1540-9295(2005)003[0243,ROFPCI]2.0.CO;2
Frank DA (2005) The interactive effects of grazing ungulates and aboveground production on grassland diversity. Oecologia 143:629–634
Frerker K, Sabo A, Waller D (2014) Long-term regional shifts in plant community composition are largely explained by local deer impact experiments. PLoS One 9:e115843
Gedge KE, Maun MA (1992) Effects of simulated herbivory on growth and reproduction of two beach annuals, *Cakile edentula* and *Corispermum hyssopifolium*. Can J Bot 70:2467–2475
Gill RMA (1992) A review of damage by mammals in north temperate forests: 1. Deer. Forestry: Int J For Res 65:145–169

Gill RMA, Beardall V (2001) The impact of deer on woodlands: the effects of browsing and seed dispersal on vegetation structure and composition. Forestry: Int J For Res 74:209–218
Gunderson LH (2000) Ecological resilience—in theory and application. Annu Rev Ecol Syst 31:425–439
Habeck CW, Schultz AK (2015) Community-level impacts of white-tailed deer on understorey plants in North American forests: a meta-analysis. AoB Plants 7:plv119
Hanley TA (1997) A nutritional view of understanding and complexity in the problem of diet selection by deer (Cervidae). Oikos 79:209–218
Hanley ME, Lamont BB, Fairbanks MM, Rafferty CM (2007) Plant structural traits and their role in anti-herbivore defence. Perspect Plant Ecol Evol Syst 8:157–178
Hastings A, Wysham DB (2010) Regime shifts in ecological systems can occur with no warning. Ecol Lett 13:464–472
Haukioja E, Koricheva J (2000) Tolerance to herbivory in woody vs. herbaceous plants. Evol Ecol 14:551–562
Hebblewhite M, Merrill E, McDermid G (2008) A multi-scale test of the forage maturation hypothesis in a partially migratory ungulate population. Ecol Monogr 78:141–166. https://doi.org/10.1890/06-1708.1
Heckathorn SA, McNaughton SJ, Coleman JS et al (1999) C4 plants and herbivory. In: C4 Plant Biology. Academic Press, Cambridge, pp 285–312
Heckel CD, Bourg NA, McShea WJ, Kalisz S (2010) Nonconsumptive effects of a generalist ungulate herbivore drive decline of unpalatable forest herbs. Ecology 91:319–326. https://doi.org/10.1890/09-0628.1
Herms DA, Mattson WJ (1992) The dilemma of plants: to grow or defend. Q Rev Biol 67:283–335. https://doi.org/10.1086/417659
Hidding B, Tremblay J-P, Côté SD (2013) A large herbivore triggers alternative successional trajectories in the boreal forest. Ecology 94:2852–2860. https://doi.org/10.1890/12-2015.1
Holechek JL (1984) Comparative contribution of grasses, forbs, and shrubs to the nutrition of range ungulates. Rangelands 6:261–263
Holling CS (1973) Resilience and stability of ecological systems. Annu Rev Ecol Syst 4:1–23
Hülber K, Ertl S, Gottfried M et al (2005) Gourmets or gourmands?—diet selection by large ungulates in high-alpine plant communities and possible impacts on plant propagation. Basic and Applied Ecology 6:1–10
Itô H, Hino T (2005) How do deer affect tree seedlings on a dwarf bamboo-dominated forest floor? Ecol Res 20:121–128
Jachmann H, Bell RHV (1985) Utilization by elephants of the Brachystegia woodlands of the Kasungu National Park, Malawi. Afr J Ecol 23:245–258. https://doi.org/10.1111/j.1365-2028.1985.tb00955.x
Kalisz S, Spigler RB, Horvitz CC (2014) In a long-term experimental demography study, excluding ungulates reversed invader's explosive population growth rate and restored natives. PNAS 111:4501–4506. https://doi.org/10.1073/pnas.1310121111
Kardol P, Dickie IA, St. John MG et al (2014) Soil-mediated effects of invasive ungulates on native tree seedlings. J Ecol 102:622–631. https://doi.org/10.1111/1365-2745.12234
Kirby KJ (2001) The impact of deer on the ground flora of British broadleaved woodland. Forestry 74:219–229
Knight TM, Dunn JL, Smith LA et al (2009) Deer facilitate invasive plant success in a Pennsylvania forest understory. Nat Areas J 29:110–116
Koh S, Bazely DR, Tanentzap AJ et al (2010) *Trillium grandiflorum* height is an indicator of white-tailed deer density at local and regional scales. For Ecol Manag 259:1472–1479
Kuiters AT, Slim PA (2002) Regeneration of mixed deciduous forest in a Dutch forest-heathland, following a reduction of ungulate densities. Biol Conserv 105:65–74
Kurokawa H, Peltzer DA, Wardle DA (2010) Plant traits, leaf palatability and litter decomposability for co-occurring woody species differing in invasion status and nitrogen fixation ability. Funct Ecol 24:513–523. https://doi.org/10.1111/j.1365-2435.2009.01676.x

Lamprey HF (1963) Ecological separation of the large mammal species in the Tarangire Game Reserve, Tanganyika. Afr J Ecol 1:63–92
Laycock WA (1991) Stable states and thresholds of range condition on North American rangelands: a viewpoint. J Range Manag 44:427–433
Lockwood JA, Lockwood DR (1993) Catastrophe theory: a unified paradigm for rangeland ecosystem dynamics. J Range Manag 46:282–288
May RM (1977) Thresholds and breakpoints in ecosystems with a multiplicity of stable states. Nature 269:471
McLachlan SM, Bazely DR (2001) Recovery patterns of understory herbs and their use as indicators of deciduous forest regeneration. Conserv Biol 15:98–110. https://doi.org/10.1111/j.1523-1739.2001.98145.x
McNaughton SJ (1983) Compensatory plant growth as a response to herbivory. Oikos 40:329–336
McNaughton SJ (1984) Grazing lawns: animals in herds, plant form, and coevolution. Am Nat 124:863–886
McNaughton SJ, Georgiadis NJ (1986) Ecology of African grazing and browsing mammals. Annu Rev Ecol Syst 17:39–65
McShea WJ, Rappole JH (2000) Managing the abundance and diversity of breeding bird populations through manipulation of deer populations. Conserv Biol 14:1161–1170
Mopper S, Maschinski J, Cobb N, Whitham TG (1991) A new look at habitat structure: consequences of herbivore-modified plant architecture. In: Habitat structure. Springer, Dordrecht, pp 260–280
Moser B, Schütz M (2006) Tolerance of understory plants subject to herbivory by roe deer. Oikos 114:311–321
Mudrak EL, Johnson SE, Waller DM (2009) Forty-seven year changes in vegetation at the Apostle Islands: effects of deer on the forest understory. Nat Areas J 29:167–176
Myers JA, Harms KE (2009) Seed arrival, ecological filters, and plant species richness: a meta-analysis. Ecol Lett 12:1250–1260
Noy-Meir I (1975) Stability of grazing systems: an application of predator-prey graphs. J Ecol 63:459–481. https://doi.org/10.2307/2258730
Nuttle T, Ristau TE, Royo AA (2014) Long-term biological legacies of herbivore density in a landscape-scale experiment: forest understoreys reflect past deer density treatments for at least 20 years. J Ecol 102:221–228. https://doi.org/10.1111/1365-2745.12175
Olff H, Ritchie ME (1998) Effects of herbivores on grassland plant diversity. Trends Ecol Evol 13:261–265
Owen-Smith N, Cooper SM (1987) Palatability of woody plants to browsing ruminants in a South African savanna. Ecology 68:319–331. https://doi.org/10.2307/1939263
Parker JD, Burkepile DE, Hay ME (2006) Opposing effects of native and exotic herbivores on plant invasions. Science 311:1459–1461
Pastor J, Naiman RJ, Dewey B, McInnes P (1988) Moose, microbes, and the boreal forest. BioScience 38:770–777. https://doi.org/10.2307/1310786
Pérez-Barbería FJ (2017) Scaling methane emissions in ruminants and global estimates in wild populations. Sci Total Environ 579:1572–1580. https://doi.org/10.1016/j.scitotenv.2016.11.175
Persson I-L, Danell K, Bergström R (2000) Disturbance by large herbivores in boreal forests with special reference to moose. Ann Zool Fenn 37:251–263
Prins HHT, van Langevelde F (2008) Assembling a diet from different places. In: Resource ecology: spatial and temporal dynamics of foraging. Springer, Dordrecht, pp 129–158
Prins HHT, Ydenberg RC, Drent RH (1980) The interaction of Brent geese (*Branta bernicla*) and sea plantain (*Plantago maritima*) during spring staging: field observations and experiments. Acta Botanica Neerlandica 29:585–596
Putman RJ, Moore NP (1998) Impact of deer in lowland Britain on agriculture, forestry and conservation habitats. Mammal Rev 28:141–164
Reich PB, Walters MB, Kloeppel BD, Ellsworth DS (1995) Different photosynthesis-nitrogen relations in deciduous hardwood and evergreen coniferous tree species. Oecologia 104:24–30

Reimoser F (2003) Steering the impacts of ungulates on temperate forests. J Nat Conserv 10:243–252

Richards JH (1984) Root growth response to defoliation in two *Agropyron* bunchgrasses: field observations with an improved root periscope. Oecologia 64:21–25. https://doi.org/10.1007/BF00377538

Rietkerk M, Ketner P, Stroosnijder L, Prins HHT (1996) Sahelian rangeland development: a catastrophe? J Range Manag 49:512–519

Ritchie ME, Tilman D (1995) Responses of legumes to herbivores and nutrients during succession on a nitrogen-poor soil. Ecology 76:2648–2655

Rooney TP, Dress WJ (1997) Species loss over sixty-six years in the ground-layer vegetation of Heart's content, an old-growth forest in Pennsylvania, USA. Nat Areas J 17:297–305

Rooney TP, Waller DM (2001) How experimental defoliation and leaf height affect growth and reproduction in *Trillium grandiflorum*. J Torrey Bot Soc 128:393–399

Rooney TP, Waller DM (2003) Direct and indirect effects of white-tailed deer in forest ecosystems. For Ecol Manag 181:165–176

Rossell CR, Patch S, Salmons S (2007) Effects of deer browsing on native and nonnative vegetation in a mixed oak-beech forest on the Atlantic coastal plain. Northeast Nat 14:61–72

Rossow LJ, Bryant JP, Kielland K (1997) Effects of above-ground browsing by mammals on mycorrhizal infection in an early successional taiga ecosystem. Oecologia 110:94–98

Royo AA, Carson WP (2006) On the formation of dense understory layers in forests worldwide: consequences and implications for forest dynamics, biodiversity, and succession. Can J For Res 36:1345–1362

Russell FL, Zippin DB, Fowler NL (2001) Effects of white-tailed deer (*Odocoileus virginianus*) on plants, plant populations and communities: a review. Am Midl Nat 146:1–26

Sabo AE, Frerker KL, Waller DM, Kruger EL (2017) Deer-mediated changes in environment compound the direct impacts of herbivory on understorey plant communities. J Ecol 105:1386–1398

Sabo AE, Forrester JA, Burton JI et al (2019) Ungulate exclusion accentuates increases in woody species richness and abundance with canopy gap creation in a temperate hardwood forest. For Ecol Manag 433:386–395

Saunders MR, Puettmann KJ (1999) Effects of overstory and understory competition and simulated herbivory on growth and survival of white pine seedlings. Can J For Res 29:536–546

Scheffer M, Carpenter S, Foley JA et al (2001) Catastrophic shifts in ecosystems. Nature 413:591

Schütz M, Risch AC, Leuzinger E et al (2003) Impact of herbivory by red deer (*Cervus elaphus* L.) on patterns and processes in subalpine grasslands in the Swiss National Park. For Ecol Manag 181:177–188

Shelton AL, Henning JA, Schultz P, Clay K (2014) Effects of abundant white-tailed deer on vegetation, animals, mycorrhizal fungi, and soils. For Ecol Manag 320:39–49. https://doi.org/10.1016/j.foreco.2014.02.026

Shipley LA (1999) Grazers and browsers: how digestive morphology affects diet selection. In: Grazing behavior of livestock and wildlife. Idaho Forest, Wildlife and Range Experimental Station Bulletin 70

Soons MB, Messelink JH, Jongejans E, Heil GW (2005) Habitat fragmentation reduces grassland connectivity for both short-distance and long-distance wind-dispersed forbs. J Ecol 93:1214–1225

Stockton SA, Allombert S, Gaston AJ, Martin J-L (2005) A natural experiment on the effects of high deer densities on the native flora of coastal temperate rain forests. Biol Conserv 126:118–128. https://doi.org/10.1016/j.biocon.2005.06.006

Suding KN, Gross KL, Houseman GR (2004) Alternative states and positive feedbacks in restoration ecology. Trends Ecol Evol 19:46–53

Switzenberg DF, Nelson TC, Jenkins BC (1955) Effect of deer browsing on quality of hardwood timber in northern Michigan. For Sci 1:61–67

Takatsuki S (2009) Effects of sika deer on vegetation in Japan: a review. Biol Conserv 142:1922–1929

Takatsuki S, Itô TY (2009) Plants and plant communities on Kinkazan Island, northern Japan, in relation to sika deer herbivory. In: Sika Deer. Springer, Tokyo, pp 125–143

Tanentzap AJ, Kirby KJ, Goldberg E (2012) Slow responses of ecosystems to reductions in deer (Cervidae) populations and strategies for achieving recovery. For Ecol Manag 264:159–166. https://doi.org/10.1016/j.foreco.2011.10.005

Tilghman NG (1989) Impacts of white-tailed deer on forest regeneration in northwestern Pennsylvania. J Wildl Manag 53:524–532

Tilman D (1982) Resource competition and community structure. Princeton University Press, Princeton

Vavra M, Parks CG, Wisdom MJ (2007) Biodiversity, exotic plant species, and herbivory: the good, the bad, and the ungulate. For Ecol Manag 246:66–72

Vázquez DP, Simberloff D (2003) Changes in interaction biodiversity induced by an introduced ungulate. Ecol Lett 6:1077–1083

Vellend M, Myers JA, Gardescu S, Marks PL (2003) Dispersal of trillium seeds by deer: implications for long-distance migration of forest herbs. Ecology 84:1067–1072

Wardle DA, Bonner KI, Barker GM (2002) Linkages between plant litter decomposition, litter quality, and vegetation responses to herbivores. Funct Ecol 16:585–595. https://doi.org/10.1046/j.1365-2435.2002.00659.x

White MA (2012) Long-term effects of deer browsing: composition, structure and productivity in a northeastern Minnesota old-growth forest. For Ecol Manag 269:222–228

Wiegmann SM, Waller DM (2006) Fifty years of change in northern upland forest understories: identity and traits of "winner" and "loser" plant species. Biol Conserv 129:109–123

Young TP, Peffer E (2010) "Recalcitrant understory layers" revisited: arrested succession and the long life-spans of clonal mid-successional species. Can J For Res 40:1184–1188

Chapter 12
Impacts of Browsing and Grazing Ungulates on Faunal Biodiversity

Krisztián Katona and Corli Coetsee

12.1 Impact of Grazing and Browsing on Different Biodiversity Components

Noss (1990) proposed that biodiversity has three primary attributes (i.e., composition, structure and function), and that these attributes can be expanded into four levels of organisation—namely, genes-genome, population-species, community-ecosystem and regional-landscape. Seen within this framework, the group Ungulata, with its own high species diversity, has numerous direct and indirect effects on broader biodiversity. The high species diversity in the Ungulata is associated with high levels of diversity in morphology and physiology, which manifests in diverse diet types, e.g., ungulates may graze, browse or do both; within these groups they prefer certain plant parts or feed at specific heights within the vegetation strata (Gordon and Prins 2008). These preferences may vary seasonally. Dietary preferences, in turn, govern foraging behaviour, distribution and habitat use, social behaviour, and interactions with other species of herbivore taxa (Gordon 2003; Illius and Gordon 1987; McNaughton and Georgiadis 1986).

The most obvious effects of ungulates on biodiversity take place at the population-species level and result in changes in the abundance and richness of a range of plant and animal taxa. These interactions can be direct (e.g., ungulates consuming plants or animals) or indirect (e.g., ungulates competing with, or facil-

K. Katona (✉)
Szent István University Institute for Wildlife Conservation, Gödöllő, Hungary
e-mail: Katona.Krisztian@mkk.szie.hu

C. Coetsee
Scientific Services, South African National Parks, Skukuza, Mpumalanga, South Africa

School of Natural Resource Management, George Campus, Nelson Mandela University, George, South Africa
e-mail: Corli.Wigley-Coetsee@sanparks.org

I. J. Gordon, H. H. T. Prins (eds.), *The Ecology of Browsing and Grazing II*,
Ecological Studies 239, https://doi.org/10.1007/978-3-030-25865-8_12

itating, other animal species by changing vegetation dynamics thereby limiting, or increasing, available habitat or food for other species). Less intuitive are the impacts on biodiversity that come about at the genetic, ecosystem and landscape level. Ungulates (along with other vertebrate or invertebrate herbivores and pathogens) drive the evolution of anti-herbivore defences in plants and, in turn, have physiological and behavioural adaptations in response to plant defences (Hanley et al. 2007). Altered level of plant defences, brought about by ungulates, may affect other fauna such as invertebrates; for instance, whistling thorn (*Vachellia drepanolobium*) offers more rewards to ants when browsed than when unbrowsed (Huntzinger et al. 2004). Ungulates, and their actions, can also affect ecosystem processes at a much larger scale, often contributing to trophic cascades. As an example, in the Serengeti, wildebeest (*Connochaetes taurinus*) grazing decreases fuel loads which results in less intense fires and increased woody cover that may culminate in increased habitat for browsers (Holdo et al. 2009a, b). Browsing and grazing ungulates can also have effects on nutrient dynamics at landscape level; for instance, savanna ungulate browsing may accelerate soil nutrient cycling when pruning stimulates shoot regrowth which, in turn, is associated with increased consumption and decomposition (Fornara and Du Toit 2007, 2008).

In this chapter we focus on the impacts of browsing and grazing ungulates on faunal biodiversity at the population-species level. It is not intended to be an exhaustive review of the literature, but aims to increase understanding of the expansive scope of direct and indirect effects that ungulates may have on faunal diversity (Table 12.1). In addition to this, **Sitters and Andriuzzi** Chap. 9 provides information on indirect effects of ungulates on ecosystem processes and **Sabo** Chap. 11 addresses impacts of ungulates on plants and plant communities.

12.2 Direct Impact of Grazing and Browsing Ungulates on Animal Diversity

Grazing and browsing ungulates can have many direct effects on coexisting animal species; these include basic ecological processes, such as predation, scavenging or parasitism, but also cover incidental effects.

12.2.1 Impacting Biodiversity by Predating on Other Fauna

Although ungulates are herbivores and feed mainly on plants, they can, to some extent, predate on non-vertebrate animals. Predation by mammalian herbivores on phytophagous insects has mostly been neglected by ecologists (Gish et al. 2017). Intraguild predation by ungulates has been reported in Mediterranean savanna, where red deer (*Cervus elaphus*) and wild boar (*Sus scrofa*) incidentally ingest

Table 12.1 Overview of the issues covered; numbers in parentheses show the sections where more details or related examples are available

Behaviour	Effect type	Direct impact on animal species (2.1–2.2)	Direct impact on plant species (3.1)	Indirect impact on animal species (3.2–3.7)	Related management implications (4.)
Feeding	Grazing Browsing	Incidental feeding on plant-dwelling insects (2.1) Bird nest predation (2.1) (Providing food/habitat to symbiotic ungulate rumen microbiota) (2.2)	Landscape scale modification of the vegetation: Impeding forest regeneration; diminishing understory vegetation or grass cover Alteration of plant species composition through food selectivity	Establishing new or destroying available (micro) habitats (3.2–3.6) Competition: Reducing food availability (3.6, 3.7) Facilitation: Improving food quality or availability for other species (3.7)	Close-to-nature forestry and wildlife-friendly farming to conserve biodiversity and maintain less damage-susceptible habitat complexes ensuring niche segregation and resource partitioning of animal species Control of dense wild ungulate populations Conservation of the diverse composition of ungulate communities Maintaining adequate density and composition of grazing domestic ungulates Establishing landscape of fear by hunting and carnivore conservation resulting in spatially heterogenous ungulate impact Planning the timing and spatial distribution of livestock grazing
Moving	Trampling Seed dispersal Dispersing ectoparasites	Destroying bird nests (2.1) Ectoparasites occupy new areas and reach new hosts (2.2)	Forest gap establishment Decreasing grass cover Changing the vegetation distribution and plant species composition	Establishing new or destroying available micro-habitats or food resources (3.2, 3.3) Increased parasite load and risk of infections	
Defecating	Nutrient deposition Seed dispersal Dispersing endoparasites	Endoparasites occupy new areas and reach new hosts (2.2)	Changing the vegetation distribution and plant species composition	Modifying dung-beetle assemblages (3.2) Increased parasite load and risk of condition loss	
Wallowing	Changing the microtopography Seed dispersal	Killing or eating amphibian in adult or larval stages	Changing the patterns of the terrestrial and aquatic vegetation	Modifying amphibian and aquatic invertebrate communities (3.2, 3.3)	Establishing and maintenance of natural wallowing sites in habitat patches less-susceptible to ungulate impacts

(continued)

Table 12.1 (continued)

Behaviour	Effect type	Direct impact on animal species (2.1–2.2)	Direct impact on plant species (3.1)	Indirect impact on animal species (3.2–3.7)	Related management implications (4.)
Dying	Being predated Carcasses provide food for scavenger birds, mammals and arthropods Carcasses provide regeneration opportunity for plant recruitment	Increasing density and diversity of large carnivores and carrion-feeding animal taxa (2.2) Contaminated remains kill the consumers (2.2)	Changing the vegetation distribution and plant species composition	Direct or indirect competition for limited carcass resources among carnivorous mammals and scavenger birds Carcass provides feeding place for insect-eating birds (3.5)	Maintaining suitable density of prey and carcass-producing wild and domestic ungulates Leaving carcasses in situ only when appropriate Proper disposing of contaminated domestic carcasses

chestnut weevils (*Curculio elephas*) within acorns of holm oak (*Quercus ilex*) (Bonal and Muñoz 2007), and in warm-temperate old-growth forest where sika deer (*Cervus nippon centralis*) consume small moths of leaf miner (*Stigmella* spp.) (Yamazaki and Sugiura 2008). van Noordwijk et al. (2012) reported that 64% of caterpillar nests of the Glanville fritillary butterfly (*Melitaea cinxia*) were damaged after high intensity sheep grazing, mainly due to incidental ingestion of caterpillars.

Opportunistic consumption of animals, or animal products, by ungulates has been reported quite widely; such as the case of Abbott's duiker (*Cephalophus spadix*) catching a frog in Mwanihana Forest, Udzungwa Mountains, Tanzania (Rovero et al. 2005), red deer predation on ground nesting Manx shearwater (*Puffinus puffinus*) on the Isle of Rhum, Inner Hebrides (Furness 1988), white-tailed deer (*Odocoileus virginianus*) predation on Northern bobwhite (*Colinus virginianus*) eggs in south Georgia, USA (Ellis-Felege et al. 2008), cattle predation on grassland bird eggs and nestlings in Wisconsin, USA (Nack and Ribic 2005), swan (*Cygnus bewickii*) egg and chick predation by reindeer (*Rangifer tarandus*) in Yakutia, Russia (Degtyarev 2010), and white-tailed deer predation on nestlings of five grassland songbird species in North-Dakota, USA (Pietz and Granfors 2000). Around the Baltic sea, direct killing of nestlings, and crushing of eggs by trampling of cattle, lead to high extinction risk, over the long term, for sub-populations of southern dunlin (*Calidris alpina schinzii*) (Pakanen et al. 2016). The same research group found trampling rates of artificial nests to be very high, even at relatively low stocking rates of cattle, i.e., 0.83 head.ha^{-1} (Pakanen et al. 2011).

12.2.2 Influencing Biodiversity by Symbiosis or Being Parasitised or Predated

Ungulates, particularly ruminants, share their bodies (their gut) with countless microbiota through mutualistic symbiosis with important consequences for the health of the host individual (Bergmann et al. 2015; Hu et al. 2017). Although not strictly qualifying as animals, in a systematic sense, the interaction between ungulates and their microbiota has important consequences for biodiversity. These groups include Prokaryotes, i.e., Bacteria and Archae, of an estimated 7000 and 1500 species, respectively (Chaucheyras-Durand and Ossa 2014). Beside Prokaryotes, numerous species of Protozoa (eukaryotic single-cell, animal-like organisms showing mobility and heterotrophy) occupy the rumen, where they comprise up to half of the total microbial biomass (Tymensen et al. 2012). Although some work has shown that Archaea and Protozoa are remarkedly similar across groups of ruminants, many Archaea species have not been named, and we know little about bacterial species inhabiting ruminants and other ungulates (Henderson et al. 2015).

Ungulates also directly influence faunal diversity in a passive manner by providing habitat for endo- and ectoparasite invertebrates (Ablat et al. 2014; Davidson et al. 2014; Mukul-Yerves et al. 2014). Examples include rumen fluke (*Calicophoron daubneyi* and *Paramphistomum leydeni*) in fallow deer (*Dama dama*) in Ireland (O'Toole et al. 2014) and liver fluke (*Fasciola hepatica*) in red deer in the Highlands of Scotland (French et al. 2016). They can also change relative frequencies of endoparasites and thus affect other ungulate species (Odadi et al. 2011a).

Ungulates are important prey for many carnivore species. A global assessment of grey wolf (*Canis lupus*) diet variability, from 177 studies, among and within North America, Europe, and Asia, clearly pointed to the dominance of large (240–650 kg) and medium-sized (23–130 kg) wild ungulates in the diet (Newsome et al. 2016). From about 50,000 ungulate carcass records collected over 50 years in the Kruger National Park, only 6% mortality was not the result of predation (Owen-Smith and Mills 2008). López-Alfaro et al. (2015) found that the availability of ungulate meat (particularly elk, *Cervus elaphus*) governed the productivity of bear (*Ursus arctos*) populations in North America. In Cambodia, the Indochinese tiger (*Panthera tigris*) disappeared from one historically important site as a result of two decades of illegal hunting of both the tiger and its prey (red muntjac *Muntiacus muntjak*, sambar deer *Rusa unicolor*, wild cattle *Bos gaurus*, *B. javanicus*, O'Kelly et al. 2012).

Ungulates can also be predated on by raptors; for instance bald eagle (*Haliaeetus leucocephalus*) predate white-tailed deer (Duquette et al. 2011). Even after death, ungulates can improve the survival chances of predatory birds; ungulate carcasses are an important resource for golden eagles (*Aquila chrysaetos*) in Mediterranean Spain (Sánchez-Zapata et al. 2010). In Catalonian Pyrenees, Spain, Margalida et al. (2011), found that wild ungulates, primarily red deer, wild boar and chamois (*Rupicapra rupicapra*), are an important food resource for obligate scavenger vulture species. In Europe, vultures are also dependent on the carcasses of domestic animals, so the greater restrictions of sanitary legislations on the use of animal byproducts was suggested to lead to a decrease in vulture breeding success, and increases in mortality of younger age classes (Donázar et al. 2009). But, as a positive example, Margalida et al. (2017) concluded that the food shortage, caused by health policies, did not substantially affect either foraging behaviour or breeding output in bearded vultures (*Gypaetus barbatus*), in the Pyrenees. Meanwhile, across the Indian subcontinent, *Gyps* spp. vulture populations have collapsed, in the recent decades, because of the consumption of domestic and feral ungulate carcasses contaminated with residues of diclofenac, a drug widely used to treat pain, fever and inflammation in livestock (Taggart et al. 2007).

In the Greater Yellowstone Ecosystem, USA, the remains of ungulates, preyed upon by other carnivores, provide an important food source for scavenging canids, such as grey wolves and coyotes (*Canis latrans*) (Atwood and Gese 2008). In Slovenia, roe deer (*Capreolus capreolus*), preyed upon by lynx (*Lynx lynx*), are an important meat source for brown bears (*Ursus arctos*; acquired by kleptoparasitism) (Krofel et al. 2012). In south-eastern Australia, wild dogs (dingoes *Canis lupus dingo*, feral dogs and their hybrids *Canis lupus familiaris*) and foxes (*Vulpes vulpes*) also frequently visit carcasses of the exotic sambar deer, hunted by people throughout the year and left in the field (Forsyth et al. 2014). In Hungary, the high availability of carrion of wild ungulates, in areas of intensively managed rangeland, promoted the rapid expansion of golden jackal (*Canis aureus*) populations (Lanszki et al. 2015).

Ungulate carcasses function as essential food sources, but are also likely to create a particular microclimate. Studies in southern Norway indicated that the local abundance and diversity of *Coleoptera* spp. almost doubled around roe deer carcasses, highlighting the significant ecological role of ungulate carcasses in European boreal forest ecosystems (Melis et al. 2004).

12.3 Indirect Impact of Grazing and Browsing Ungulates on Animal Diversity

12.3.1 Modification of Resources and Habitat

The actions of grazing and browsing ungulates, in general, affect a wide range of other animals, including other mammals such as rodents, as well as birds, reptiles, amphibians and invertebrates. Here we document indirect effects brought about through habitat modification, which takes place, most often, through effects on vegetation. But ungulates can also modify landscapes directly; e.g., the wallowing of exotic water buffalo (*Bubalus bubalis*) in Australia has caused major landscape scale changes by affecting the functioning of tidal creeks and channels, with comcominant negative effects on native flora and fauna (Prins and Van Oeveren 2014).

Excessive utilization, by native ungulates, is often detrimental to native vegetation diversity; significantly impacting woody seedling recruitment, establishment and growth (e.g., Danell et al. 2003; Fornara and Du Toit 2008; Prins and Van der Jeugd 1993) or detrimental in association with fire (Staver et al. 2009; Staver and Bond 2014). Non-native ungulates may also affect vegetation negatively: Spear and Chown (2009) in their review of the effects of non-indigenous ungulates on biodiversity found that introduced ungulates often impact negatively on vegetation structure and composition and, in certain cases, lead to the extirpation of associated native species of fauna. However, moderate ungulate utilization is mostly beneficial to the diversity of native vegetation. For instance, Royo et al. (2010) documented that white-tailed deer, at moderate levels of abundance, promote herbaceous richness and abundance by preferentially browsing fast growing pioneer species that thrive following co-occurring disturbances (i.e., fire and gaps) in deciduous forest in West Virginia, USA. How the changes in vegetation, brought about by ungulates, both native and exotic, impact other fauna are not always clear and will be discussed in the following sections.

12.3.2 Interactions Between Ungulates and Invertebrates

The effects of ungulates on invertebrates depend on the group of invertebrates and their sensitivity to micro-climatic conditions, as well as the effects of ungulates on plant litter and the level of disturbance through trampling. Ungulates, and their actions, may impact negatively on invertebrates; such as the case of moose (*Alces alces*) browsing and ground-dwelling invertebrates in Swedish pine forest (Suominen et al. 1999), deer grazing and abundance of ants and richness of arthropods in Great Smoky Mountains National Park, USA (Lessard et al. 2012), and ungulate (i.e., red deer, wild boar) trampling and isopod and spider abundance in montane woodland in Germany (Mohr et al. 2005). Livestock trampling, even when very limited, decreased both land snail density and species richness in remnant forest on the North Island of New Zealand (Denmead et al. 2015). In African savannas, damage to seeds was more intense when large mammals were removed, as seed-feeding invertebrates increased (Shaw et al. 2002); in the Kruger National Park, the presence of meso-herbivores negatively affected the abundance of beetles and grasshoppers (Jonsson et al. 2010).

Ungulates, and their actions, are not always detrimental to invertebrates; Suominen et al. (2003) showed that Carabid beetles increased in plots grazed by reindeer in Lapland. Higher beetle diversity was found at intermediate levels of grazing and lower diversity at very high levels of grazing. In the Serengeti, the diversity of invertebrates was not strongly affected by grazing; spiders, ground beetles and ground-dwelling ants were more abundant in ungrazed sites, dung beetles numbers were unaffected by grazing and flies increased with grazing intensity (de Visser et al. 2015). The exclusion of white-tailed deer, which had historically occurred at high numbers (i.e., about 40 deer.km^{-2}) in Maryland, USA, had little effect on invertebrates (Duguay and Farfaras 2011). Ungulates (e.g., *Sus scrofa*), in the wetlands of Camargue in France, contributed to the dispersal of aquatic invertebrates after wallowing (Vanschoenwinkel et al. 2008). Likewise, litter arthropod species densities were highest on islands of Haida Gwaii (aka Queen Charlotte Islands), Canada, where deer have been present for >50 years (as opposed to islands where they had been present for a shorter period of time, or were absent); however, the understory invertebrate species density dramatically decreased with length of deer presence on the islands (Martin et al. 2010).

A large and diverse mammal fauna, and the associated dung, has been shown to be important for the maintenance of a large and diverse dung beetle fauna (e.g., Hanski and Cambefort 1991). In Southern Europe, the conservation of European dung beetle fauna could be enhanced by livestock grazing, as wild ungulate manure cannot support the regional dung beetle species pool (Jay-Robert et al. 2008). Akaba et al. (2014) showed that forest-dwelling dung beetle species gave way to grassland-dwelling species at higher sika deer (*Cervus nippon yesoensis*) densities in Japanese forests. Some scarce North American dung beetle species are specifically associated with deer dung; *Aphodius nemoralis* in northern coniferous forest and *A. zenkeri* in more southern woodlands (Stewart 2001).

12.3.3 Interactions Between Ungulates and Amphibians

Ungulates (e.g., peccaries *Tayassuidae*), through creating wallows, can increase habitat for amphibians (Ringler et al. 2015), and benefit amphibian population dynamics and species diversity (Beck et al. 2010). However, relationships between ungulates and amphibians can be complicated; Baruzzi and Krofel (2017) found that amphibian species richness in aquatic environments, created by ungulates, in Mediterranean-type systems in Italy, depended on the size of the water body and the level of disturbance. Although amphibian species richness decreased in small ponds regularly disturbed by wild boar and deer (*Dama dama* and *Capreolus capreolus*), richness increased in large ponds regularly used by ungulates. In the lowland wet forest of Costa Rica, more compacted litter, in plots with natural peccary (*Pecari tajacu* and *Tayassu pecari*) densities, had higher densities of amphibian and reptile individuals and more juveniles of anuran amphibian species (i.e., order Salientia) than plots where peccaries were excluded (Reider et al. 2013).

Based on long term data of British population of natterjack toad (*Bufo calamita*), grazing of the terrestrial habitat, by domestic livestock, was also found to positively influence the population growth rate (Buckley et al. 2014). Whereas some species (e.g., bufonids) may benefit from controlled grazing, Burton et al. (2009) suggested that fencing cattle from wetlands may be a prudent conservation strategy for other amphibian species (e.g., ranids).

12.3.4 Interactions Between Ungulates and Reptiles

Ungulates often have negative effects on reptiles (Beever and Brussard 2004). McCauley et al. (2006) showed that the removal of large native herbivores, as well as cattle, from East African savanna resulted in increases in snake numbers, concurrently with an increase in small mammal prey numbers. Pringle et al. (2007) found, using the same exclosures as above, that the removal of herbivores resulted in an increase in tree density and structural complexity with an associated increase in lizard density (and their main prey species of beetles). In a landscape, grazed by livestock, in south-eastern Australia, reptiles were found to be very closely associated with specific micro-habitats (Fischer et al. 2004), and landscape heterogeneity, with diverse habitats, was very important for reptile diversity. In systems where grazing, and management of grazing, have homogenized habitats, reptile diversity decreased. However, Brooks (1999) failed to identify a lasting effect of chronic high density of white-tailed deer on Northern redback salamander (*Plethodon cinereus*) abundance in a mixed oak forest in Massachusetts, USA.

Occasionally, reptile abundance is increased in the presence of ungulates; the exclusion of white-tailed deer in Ohio, USA, resulted in lower numbers of red-backed salamanders and garter snakes (*Thamnophis sirtalis*) (Greenwald et al. 2008). This study suggested that an increase in herpetofaunal abundance, in the presence of ungulates, was associated with an increase of their invertebrate prey, which may be linked to an increase in favourable plant species resulting from augmented nutrients when deer are present.

12.3.5 Interactions Between Ungulates and Birds

The presence of ungulates mostly leads to decreased bird density and composition through simplification of vegetation structure (Berger et al. 2001; Casey and Hein 1983; Holt et al. 2011; Martin et al. 2010; McShea and Rappole 2000). On the islands of Haida Gwaii, Canada, introduced Sitka black-tailed deer (*Odocoileus hemionus sitkensis*) has simplified vegetation (i.e., species composition and vegetation structure) in the understorey, with associated simplified invertebrate and songbird assemblages (Chollet et al. 2015; Martin et al. 2010). Following exclusion of native white-tailed deer in North American forests, the density and diversity of understorey woody plants increased, with a concomitant increase in numbers, but not diversity, of ground and intermediate canopy bird species (McShea and Rappole 2000).

In contrast to the results in forested areas, applying fire and grazing in North American grasslands created greater spatial heterogeneity in vegetation that provided greater variability in the grassland bird community (Fuhlendorf et al. 2006). Similarly, bird assemblages on sites intensively grazed by cattle (with associated higher landscape diversity) had higher species number and diversity, but lower densities than on the extensive sites (low grazing density) on Hungarian steppes and meadows (Báldi et al. 2005). Unfortunately, the higher species diversity masks the fact that several bird species, of conservation concern, decreased with increased grazing intensities. However, judicious grazing regimes may enhance rare bird densities (Faria and Morales 2017).

Interactions between ungulate carcasses and detritivorous invertebrates may also have consequences for birds; investigations in the Iberian Peninsula, Spain, revealed that 12 non-corvid passerine species (predominantly the white wagtail, *Motacilla alba*) opportunistically benefited from ungulate carrion, especially through predation on scavenging arthropods (Moreno-Opo and Margalida 2013). Dung, through providing resources to fungi, flies, beetles and other insects, also indirectly improves food opportunities to birds (Prins and Van Oeveren 2014 and references therein).

12.3.6 Interactions Between Ungulates and Rodents

The effects of ungulates on rodents are divergent; rodents generally benefit from the presence of ungulates (and *vice versa*) in North American grasslands (Davidson et al. 2010), but ungulates may impact negatively on numbers, and diversity, of rodents in African savannas (Keesing 1998; Pringle 2012; Pringle et al. 2010). In southern African savannas, Hagenah et al. (2009) found that murid rodent abundance, and species diversity, were higher in the absence of herbivores (including non-ungulates), and increased with herbaceous vegetation height that likely gave more cover from raptors. Pringle et al. (2010) suggest that, when larger herbivores are absent, there is reduced competition for resources that leads to higher numbers and diversity of rodents in African savannas, rather than less habitat disturbance or lower exposure to predators. In the boreal forests/subarctic tundra ecosystems in Finnish Lapland, there was a grazing-induced reduction by reindeer of small mammal abundance (den Herder et al. 2016). Experimental studies in European forests (in the Wytham Woods, UK) concluded that the negative responses of small mammal communities to intense deer browsing pressure can be rapidly reversed by an effective deer control measure (from a peak of 0.4–1.5 to 0.2 deer.ha^{-1}) (Bush et al. 2012).

12.3.7 Interactions Between Ungulates and Other Herbivorous Mammals

There is some debate, in the literature, as to whether smaller consumers competitively displace larger ones when resources are limited (Bagchi and Ritchie 2012) or whether larger size differences between consumers will lead to less intense competition and more co-existence (Ritchie and Olff 1999). Coexistence of sympatric ungulates can thus be regulated through competition and facilitation (Arsenault and Owen-Smith 2002; Gordon and Illius 1989; Latham 1999). Dietary studies of ungulates, inhabiting the same lowland habitat in Nepal (Wegge et al. 2006), suggested that the one-horned rhinoceros (*Rhinoceros unicornis*) facilitated feeding of smaller swamp deer (*Cervus duvauceli*) and hog deer (*Axis porcinus*), during the growing season, since feeding and trampling by rhinoceros stimulated high-quality grass. However, the smaller and more selective deer species outcompeted the rhinoceros, during the dry season, when food was more limited. Cattle grazing facilitated rabbits (*Oryctolagus cuniculus*), but not voles (*Microtus arvalis*), on floodplain grassland in the Netherlands (Bakker et al. 2009). Hippopotamus (*Hippopotamus amphibius*) grazing created structural vegetation heterogeneity and nutritious short lawn areas which facilitated subsistence of smaller herbivores (Kanga et al. 2013; Verweij et al. 2006). Van de Koppel and Prins (1998) further suggest that the interplay between competition and facilitation of large and smaller herbivores may drive transitions between grassland and woodland states in African savanna.

Murray and Illius (2000) found that resource competition between two meso-herbivores in African savanna (Serengeti, Tanzania) was achieved through the modification of the quantity and quality of their food sources. Wildebeest graze grass down to below the level at which topi (*Damaliscus lunatus*) can access grass; in turn, the narrower mouthed topi can reduce the leafy component to a level below that can be tolerated by wildebeest. Cameron and du Toit (2006) found that smaller foragers displace larger species, such as giraffe (*Giraffa camelopardalis*), from shared resources (e.g., giraffe browse high in canopy to avoid competition with smaller browsers). However, whether ungulates, that share space and resources, will compete depends on levels of forage production, browser population densities, habitat preferences and feeding preferences (Du Toit 1990). In a phenomenon called apparent competition, rarer ungulate species, that occur sympatrically with ungulate species that form the main prey of generalist predators, are subjected to disproportionally higher rates of predation, thereby potentially becoming even rarer (Johnson et al. 2013; Serrouya et al. 2011; Wittmer et al. 2010). Sound evidence for competition between ungulates, however, is extremely rare (Prins 2016). Behavioural observations by Bartos et al. (2002) revealed very few interspecific direct competitive interactions among four sympatric Cervids in a Czech forest.

According to the reviews by Prins (2016) and by Schieltz and Rubenstein (2016), negative effects of livestock grazing on wild ungulates are more prevalent than positive or neutral responses, but responses do vary by diet type and species. Foraging site use, food selection, foraging efficiency, health status (e.g., parasite loads, weight gain or fat stores) and demography of browsing and grazing wild ungulates were influenced by livestock presence in different situations. Grazers have been much less studied than browsers or mixed feeders, but several of the grazing species in Africa show a trend of positive responses with livestock, suggesting possible facilitation (Augustine et al. 2011; Odadi et al. 2011b), although numbers of wild ungulates may also be limited by livestock (Prins 2000; Young et al. 2005). Red deer preferentially grazed areas previously grazed by cattle on the Isle of Rhum, Scotland, and there were more calves per hind in those areas than in areas that were not grazed by cattle (Gordon 1988). On the other hand, in the Indian Trans-Himalaya, bharal (*Pseudois nayaur*), a grazing caprid, include more browse in their diet during winter, due to competition from livestock for grasses, resulting in suppressed fecundity and first-year survival for bharal (Suryawanshi et al. 2010). The endemic huemul deer (*Hippocamelus bisulcus*) in Argentina is endangered because of interspecific competition with domestic sheep for scarce winter resources (Vila et al. 2009). A recent review (Gordon 2018) suggests that human activities associated with livestock production (e.g., land use change, removal of predators, provision of water points) are the major factors affecting the outcome of the interactions between domestic and wild ungulate species.

There is also evidence of competition between native and introduced deer or caprid species across the globe, as suggested by their highly overlapping habitat and dietary requirements and expressed by examples of negative associations (e.g., spatial or habitat exclusion, opposing population trends). The widely introduced fallow deer is particularly effective as a competitor, meanwhile roe deer appear particularly vulnerable to competition from introduced deer species in Europe (Dolman and Wäber 2008; Ferretti 2011). Ferretti et al. (2015) found that the use of grassland by reintroduced red deer affected the quality of pasture for Apennine chamois (*Rupicapra pyrenaica*), with negative effects on the latter's foraging behavior and kid winter survival. Exotic deer do not always compete with native deer; Flueck (2010) in his review of interactions between native huemel deer and exotic red deer in Patagonia, found no evidence for exploitation competition.

12.4 Synthesis in Terms of Management Implications and Future Research

High ungulate diversity, with species distributed across body size classes and feeding guilds, can be directly linked to the heterogeneity of the ecosystems they occupy, regulated by a multitude of feedback loops between herbivores and plants, as observed in African savannas (Du Toit and Cumming 1999). Olff et al. (2002) propose that the highest diversity of large herbivore species should occur in areas of intermediate moisture and high nutrients because of the response of ungulates of differing body sizes to plant quality and quantity. In an evolutionary sense, high diversity of ungulates gives rise to high biodiversity; for instance of predator diversity (Sinclair et al. 2003) and floristic diversity (Du Toit and Cumming 1999).

However, ungulate effects on other faunal species are extremely variable and depend on the composition and abundance of species involved, for example, whether the ungulate species is a native species or replaces the functioning of a native species, the intactness of the ecosystem (i.e., in structure and function) and the scale. In certain parts of the world, wild ungulate species (e.g., white-tailed deer in much of North America), may become a threat to biodiversity if numbers increase unabated in the absence of predators (Horsley et al. 2003; Russell et al. 2001). However, grazing and browsing ungulates are essential for maintaining, or establishing, favourable conditions for various animal species and in the process, shape faunal assemblages (Gordon 2006; Pastor et al. 2006; Vera et al. 2006) (see Fig. 12.1).

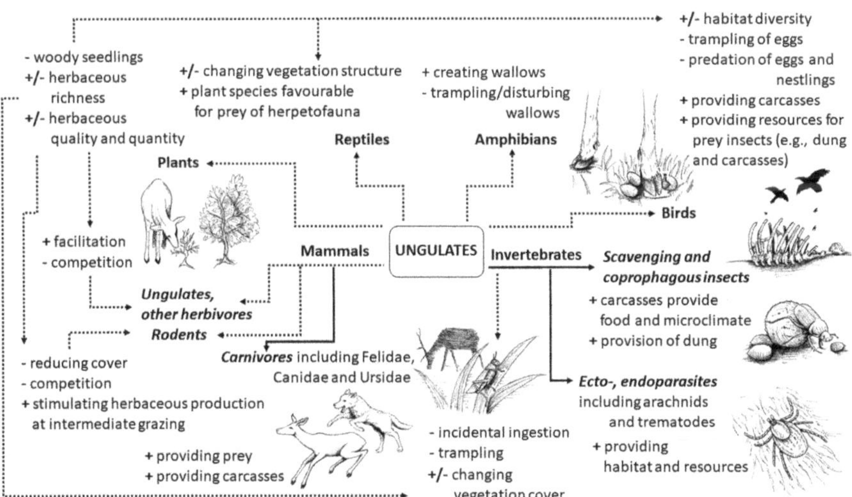

Fig. 12.1 Summary of the most important direct and indirect effects of ungulates on the diversity of different faunal taxa. The solid lines sign the mostly positive impact, while the dashed lines show variable effects on species groups, based on the most common results of the related studies (the drawings were made by Gergely Schally)

With such variable effects of ungulates on biodiversity, how can we manage for maximum biodiversity? Although some literature suggests that a moderate grazing intensity (i.e., intermediate grazing hypothesis) will maximise animal diversity (Milchunas et al. 1998; Suominen et al. 2003), this pattern is not always found (Kruess and Tscharntke 2002; Wardle et al. 2001). Fuhlendorf and Engle (2001) suggest that heterogeneity is the precursor to biological diversity at most levels of ecological organisation. For instance, applying variable fire and grazing regimes in North American grasslands creates greater spatial heterogeneity in vegetation composition and structure which leads to greater variability in the grassland bird community (Fuhlendorf et al. 2006). Thus, management that creates a shifting mosaic of vegetation composition and structure, using spatially and temporally discrete disturbances in grasslands, can be a useful tool in conservation. However, managing for heterogeneity is not only applicable to grasslands; managing within a heterogeneity paradigm has been widely applied in African protected areas (Chamaillé-Jammes et al. 2007; Owen-Smith et al. 2006; Parr and Brockett 1999; Rogers 2003; van Wilgen 2009). The significant influence of habitat management on ungulate impacts is also an emerging issue in Europe, especially in transition from even-aged to close-to-nature forest management (e.g., Reimoser and Gossow 1996; Reimoser 2003; Gerhardt et al. 2013).

The management involved in creating heterogeneity of landscape use of ungulates depends on the nature of the ungulate community. In the Northern Hemisphere (e.g., Europe, North America) free-ranging wild ungulate populations are common, and have the potential to heavily graze vegetation or heavily browse the forest understory, with associated decreases in vegetation and faunal diversity (Gordon 2006, and references therein). As these increases in ungulate populations, in some areas, are due to the loss of natural predators, predators should either be reintroduced, or their natural repatriation should be fostered; where this is impossible, or impractical, anthropogenic hunting should replace the missing predation process (e.g., hunting for fear, Cromsigt et al. 2013). A widely-known (although widely contested, e.g., Kauffman et al. 2010, 2013) example here would be the trophic cascade that resulted after grey wolves (*Canis lupus*) were reintroduced into Yellowstone National Park, USA. After the introduction of grey wolf, elk changed behaviour which appears to have contributed to the recovery of plant species such as aspen (*Populus tremuloides*), cottonwoods (*Populus* spp.) and willows (*Salix* spp.) (Ripple and Beschta 2003, 2007, 2012).

In the southern Hemisphere (e.g., Africa) ungulates are often found in fenced nature reserves and National Parks (Boone and Hobbs 2004). Fences often interrupt natural migration routes and limit access to water (Boone and Hobbs 2004); hence artificial water provision often follows the erection of fences (Smit et al. 2007). Water provision remains one of the most important intervention options available to managers to manipulate the densities and distribution of ungulates and other wildlife (Chamaillé-Jammes et al. 2007; Hilbers et al. 2015; Smit et al. 2007; Smit and Grant 2009). Apart from manipulating water provision, variable landscape use of ungulates can be encouraged through creating gradients of resources through the use of fire (Allred et al. 2011; Archibald et al. 2005; Archibald and Bond 2004).

Although livestock are often blamed for 'overgrazing', they can be sucessfully used to recreate processes important for healthy ecosystem functioning, when wildife is entirely absent, or specific functional types are lacking (Gordon and Duncan 1988; Gordon 2006). As different domestic species have very divergent effects on vegetation (e.g., horses keeping some patches of grass short and avoiding tall grass *vs.* cattle that graze less discriminately and will include some woody vegetation into their diet), practices that promote the favorable regulatory ungulate effects, through ensuring an appropriate range of browsing and grazing pressures, will lead to maximum biodiversity at all levels. Nevertheless, in most cases heavy grazing pressure by domestic or wild ungulates are not determined directly and quantitatively (Mysterud 2006). In order to continuously maintain favorable levels of different ungulate effects, implementing an adaptive management approach is recommended. This approach requires that the outcomes of interventions are monitored, with a set of population and environmental indices, which are used to continually revise the working hypothesis of the interventions (Apollonio et al. 2017).

Globally, rangelands are increasingly becoming invaded by woody vegetation, possibly due to CO_2 fertilization (Bond and Midgley 2000; Buitenwerf et al. 2012; Kgope et al. 2010; Stevens et al. 2017). The only savannas where this woody thickening has not taken place, over the last century, are in areas where elephants are present (Skowno et al. 2017; Stevens et al. 2016). Ungulates interact with megaherbivores to limit woody regeneration, but their importance is often underappreciated (O'Kane et al. 2014). Future research should aim to better understand how mixes of different ungulates, both wild and domestic, can be used to increase the health of ecosystems, in the face of the ecological challenges of our times (e.g., desertification, land desertion, woody encroachment and climate change). Venter et al. (2017), for instance, suggest that rewilding African rangelands, by using wildlife and livestock mixes to recreate mid-Holocene herbivory pressure and diversity, will shift ecosystems from being fire-dominated to herbivore-dominated, with associated increases in nutrient cycling, nutrient conservation and productivity. Studies on the joint effects of two, or more, herbivore species on the dynamics of temperate forests are also particularly relevant today, because of recent range expansions, and recolonization, by previously extirpated forest ungulates (Faison et al. 2016). As herbivory becomes a more important process in rewilded ecosystems, research on the interactions of ungulate herbivory and episodic disturbances (e.g., fire, timber harvesting, insect pests, diseases and drought) (Wisdom et al. 2006), and consequences for biodiversity become increasingly important.

References

Ablat A, Trudy R, Tohti A, Abduheni R, Canlin Z, Halik M (2014) Intestinal parasites of Tianshan red deer (*Cervus elaphus songaricus*) in Nanshan Mountains areas in Xinjiang. Acta Theriol Sin 34:87–92

Akaba S, Takafumi H, Yoshida T (2014) Impacts of high density of sika deer (*Cervus nippon yesoensis*) on dung beetle community. Jpn J Appl Entomol Zool 58:269–274. https://doi.org/10.1303/jjaez.2014.269

Allred BW, Fuhlendorf SD, Engle DM, Elmore RD (2011) Ungulate preference for burned patches reveals strength of fire–grazing interaction. Ecol Evol 1:132–144

Apollonio M, Belkin VV, Borkowski J, Borodin OI, Borowik T, Cagnacci F, Danilkin AA, Danilov PI, Faybich A, Ferretti F, Gaillard JM, Hayward M, Heshtaut P, Heurich M, Hurynovich A, Kashtalyan A, Kerley GIH, Kjellander P, Kowalczyk R, Kozorez A, Matveytchuk S, Milner JM, Mysterud A, Ozoliņš J, Panchenko DV, Peters W, Podgórski T, Pokorny B, Rolandsen CM, Ruusila V, Schmidt K, Sipko TP, Veeroja R, Velihurau P, Yanuta G (2017) Challenges and science-based implications for modern management and conservation of European ungulate populations. Mammal Res 62:209–217. https://doi.org/10.1007/s13364-017-0321-5

Archibald S, Bond WJ (2004) Grazer movements: spatial and temporal responses to burning in a tall-grass African savanna. Int J Wildland Fire 13:377–385

Archibald S, Bond WJ, Stock WD, Fairbanks DHK (2005) Shaping the landscape: fire–grazer interactions in an African savanna. Ecol Appl 15:96–109

Arsenault R, Owen-Smith N (2002) Facilitation versus competition in grazing herbivore assemblages. Oikos 97:313–318

Atwood TC, Gese EM (2008) Coyotes and recolonizing wolves: social rank mediates risk-conditional behaviour at ungulate carcasses. Anim Behav 75:753–762. https://doi.org/10.1016/j.anbehav.2007.08.024

Augustine DJ, Veblen KE, Goheen JR, Riginos C, Young TP (2011) Pathways for positive cattle-wildlife interactions in semiarid rangelands. Smithson Contrib Zool 632:55–71

Bagchi S, Ritchie ME (2012) Body size and species coexistence in consumer–resource interactions: a comparison of two alternative theoretical frameworks. Theor Ecol 5:141–151

Bakker ES, Olff H, Gleichman JM (2009) Contrasting effects of large herbivore grazing on smaller herbivores. Basic Appl Ecol 10:141–150

Báldi A, Batáry P, Erdős S (2005) Effects of grazing intensity on bird assemblages and populations of Hungarian grasslands. Agric Ecosyst Environ 108:251–263. https://doi.org/10.1016/j.agee.2005.02.006

Bartos L, Vankova D, Miller K, Siler J (2002) Interspecific competition between white-tailed, fallow, red, and roe deer. J Wildl Manag 66:522–527. https://doi.org/10.2307/3803185

Baruzzi C, Krofel M (2017) Friends or foes? Importance of wild ungulates as ecosystem engineers for amphibian communities. North-West J Zool 13:320–325

Beck H, Thebpanya P, Filiaggi M (2010) Do Neotropical peccary species (Tayassuidae) function as ecosystem engineers for anurans? J Trop Ecol 26:407–414

Beever E, Brussard P (2004) Community- and landscape-level responses of reptiles and small mammals to feral-horse grazing in the Great Basin. J Arid Environ 59:271–297. https://doi.org/10.1016/j.jaridenv.2003.12.008

Berger J, Stacey PB, Bellis L, Johnson MP (2001) A mammalian predator–prey imbalance: grizzly bear and wolf extinction affect avian neotropical migrants. Ecol Appl 11:947–960. https://doi.org/10.1890/1051-0761(2001)011[0947:AMPPIG]2.0.CO;2

Bergmann GT, Craine JM, Robeson MS II, Fierer N (2015) Seasonal shifts in diet and gut microbiota of the American bison (*Bison bison*). PLoS One 10:e0142409

Bonal R, Muñoz A (2007) Multi-trophic effects of ungulate intraguild predation on acorn weevils. Oecologia 152:533–540. https://doi.org/10.1007/s00442-007-0672-8

Bond WJ, Midgley GF (2000) A proposed CO2-controlled mechanism of woody plant invasion in grasslands and savannas. Glob Chang Biol 6:865–869

Boone RB, Hobbs NT (2004) Lines around fragments: effects of fencing on large herbivores. Afr J Range Forage Sci 21:147–158

Brooks R (1999) Residual effects of thinning and high white-tailed deer densities on northern redback salamanders in southern New England oak forests. J Wildl Manag 63:1172–1180. https://doi.org/10.2307/3802835

Buckley J, Beebee TJC, Schmidt BR (2014) Monitoring amphibian declines: population trends of an endangered species over 20 years in Britain. Anim Conserv 17:27–34. https://doi.org/10.1111/acv.12052

Buitenwerf R, Bond WJ, Stevens N, Trollope WSW (2012) Increased tree densities in South African savannas: >50 years of data suggests CO_2 as a driver. Glob Chang Biol 18:675–684

Burton EC, Gray MJ, Schmutzer AC, Miller DL (2009) Differential responses of postmetamorphic amphibians to cattle grazing in wetlands. J Wildl Manag 73:269–277. https://doi.org/10.2193/2007-562

Bush ER, Buesching CD, Slade EM, Macdonald DW (2012) Woodland recovery after suppression of deer: cascade effects for small mammals, wood mice (Apodemus sylvaticus) and bank voles (Myodes glareolus). PLoS One 7. https://doi.org/10.1371/journal.pone.0031404

Cameron EZ, du Toit JT (2006) Winning by a neck: tall giraffes avoid competing with shorter browsers. Am Nat 169:130–135

Casey D, Hein D (1983) Effects of heavy browsing on a bird community in deciduous forest. J Wildl Manag 47:829–836. https://doi.org/10.2307/3808620

Chamaillé-Jammes S, Valeix M, Fritz H (2007) Managing heterogeneity in elephant distribution: interactions between elephant population density and surface-water availability. J Appl Ecol 44:625–633

Chaucheyras-Durand F, Ossa F (2014) Review: The rumen microbiome: composition, abundance, diversity, and new investigative tools. Prof Anim Sci 30:1–12. https://doi.org/10.15232/S1080-7446(15)30076-0

Chollet S, Bergman C, Gaston AJ, Martin J-L (2015) Long-term consequences of invasive deer on songbird communities: going from bad to worse? Biol Invasions 17:777–790

Cromsigt JP, Kuijper DP, Adam M, Beschta RL, Churski M, Eycott A, Kerley GI, Mysterud A, Schmidt K, West K (2013) Hunting for fear: innovating management of human–wildlife conflicts. J Appl Ecol 50:544–549

Danell K, Bergström R, Edenius L, Ericsson G (2003) Ungulates as drivers of tree population dynamics at module and genet levels. For Ecol Manag 181:67–76

Davidson AD, Ponce E, Lightfoot DC, Fredrickson EL, Brown JH, Cruzado J, Brantley SL, Sierra-Corona R, List R, Toledo D, Ceballos G (2010) Rapid response of a grassland ecosystem to an experimental manipulation of a keystone rodent and domestic livestock. Ecology 91:3189–3200. https://doi.org/10.1890/09-1277.1

Davidson RK, Kutz SJ, Madslien K, Hoberg E, Handeland K (2014) Gastrointestinal parasites in an isolated Norwegian population of wild red deer (Cervus elaphus). Acta Vet Scand 56:59. https://doi.org/10.1186/s13028-014-0059-x

de Visser SN, Freymann BP, Foster RF, Nkwabi AK, Metzger KL, Harvey AW, Sinclair AR (2015) Invertebrates of the Serengeti: disturbance effects on arthropod diversity and abundance. Serengeti IV Sustain Biodivers Coupled Hum-Nat Syst 265

Degtyarev AG (2010) Monitoring of the Bewick's Swan in the tundra zone in Yakutia. Contemp Probl Ecol 3:90–99. https://doi.org/10.1134/S1995425510010157

den Herder M, Helle S, Niemela P, Henttonen H, Helle T (2016) Large herbivore grazing limits small-mammal densities in Finnish Lapland. Ann Zool Fenn 53:154–164

Denmead LH, Barker GM, Standish RJ, Didham RK (2015) Experimental evidence that even minor livestock trampling has severe effects on land snail communities in forest remnants. J Appl Ecol 52:161–170

Dolman PM, Wäber K (2008) Ecosystem and competition impacts of introduced deer. Wildl Res 35:202–214. https://doi.org/10.1071/WR07114

Donázar JA, Margalida A, Carrete M, Sánchez-Zapata JA (2009) Too sanitary for vultures. Science 326:664. https://doi.org/10.1126/science.326_664a

Du Toit JT (1990) Feeding-height stratification among African browsing ruminants. Afr J Ecol 28:55–61

Du Toit JT, Cumming DH (1999) Functional significance of ungulate diversity in African savannas and the ecological implications of the spread of pastoralism. Biodivers Conserv 8:1643–1661

Duguay JP, Farfaras C (2011) Overabundant suburban deer, invertebrates, and the spread of an invasive exotic plant. Wildl Soc Bull 35:243–251

Duquette JF, Belant JL, Beyer DE, Svoboda NJ, Albright CA (2011) Bald eagle predation of a white-tailed deer fawn. Northeast Nat 18:87–94. https://doi.org/10.1656/045.018.0108

Ellis-Felege SN, Burnam JS, Palmer WE, Sisson DC, Wellendorf SD, Thornton RP, Stribling HL, Carroll JP (2008) Cameras identify white-tailed deer depredating northern bobwhite nests. Southeast Nat 7:562–564

Faison EK, DeStefano S, Foster DR, Motzkin G, Rapp JM (2016) Ungulate browsers promote herbaceous layer diversity in logged temperate forests. Ecol Evol 6:4591–4602. https://doi.org/10.1002/ece3.2223

Faria N, Morales MB (2017) Population productivity and late breeding habitat selection by the threatened Little Bustard: the importance of grassland management. Bird Conserv Int:1–13. https://doi.org/10.1017/S0959270917000387

Ferretti F (2011) Interspecific aggression between fallow and roe deer. Ethol Ecol Evol 23:179–186. https://doi.org/10.1080/03949370.2011.554883

Ferretti F, Corazza M, Campana I, Pietrocini V, Brunetti C, Scornavacca D, Lovari S (2015) Competition between wild herbivores: reintroduced red deer and Apennine chamois. Behav Ecol 26:550–559. https://doi.org/10.1093/beheco/aru226

Fischer J, Lindenmayer DB, Cowling A (2004) The challenge of managing multiple species at multiple scales: reptiles in an Australian grazing landscape. J Appl Ecol 41:32–44

Flueck WT (2010) Exotic deer in southern Latin America: what do we know about impacts on native deer and on ecosystems? Biol Invasions 12:1909–1922. https://doi.org/10.1007/s10530-009-9618-x

Fornara DA, Du Toit JT (2007) Browsing lawns? Responses of Acacia nigrescens to ungulate browsing in an African savanna. Ecology 88:200–209

Fornara DA, Du Toit JT (2008) Community-level interactions between ungulate browsers and woody plants in an African savanna dominated by palatable-spinescent Acacia trees. J Arid Environ 72:534–545

Forsyth DM, Woodford L, Moloney PD, Hampton JO, Woolnough AP, Tucker M (2014) How does a carnivore guild utilise a substantial but unpredictable anthropogenic food source? Scavenging on hunter-shot ungulate carcasses by wild dogs/dingoes, red foxes and feral cats in South-Eastern Australia revealed by camera traps. PLoS One 9. https://doi.org/10.1371/journal.pone.0097937

French AS, Zadoks RN, Skuce PJ, Mitchell G, Gordon-Gibbs DK, Craine A, Shaw D, Gibb SW, Taggart MA (2016) Prevalence of liver fluke (Fasciola hepatica) in wild red deer (Cervus elaphus): Coproantigen ELISA is a practicable alternative to faecal egg counting for surveillance in remote populations. PLoS One 11:e0162420. https://doi.org/10.1371/journal.pone.0162420

Fuhlendorf SD, Engle DM (2001) Restoring heterogeneity on rangelands: ecosystem management based on evolutionary grazing patterns: we propose a paradigm that enhances heterogeneity instead of homogeneity to promote biological diversity and wildlife habitat on rangelands grazed by livestock. Bioscience 51:625–632. https://doi.org/10.1641/0006-3568(2001)051[0625:RHOREM]2.0.CO;2

Fuhlendorf SD, Harrell WC, Engle DM, Hamilton RG, Davis CA, Leslie DM (2006) Should heterogeneity be the basis for conservation? Grassland bird response to fire and grazing. Ecol Appl 16:1706–1716

Furness RW (1988) Predation on ground-nesting seabirds by island populations of red deer Cervus elaphus and sheep Ovis. J Zool 216:565–573

Gerhardt P, Arnold JM, Hackländer K, Hochbichler E (2013) Determinants of deer impact in European forests – a systematic literature analysis. For Ecol Manag 310:173–186. https://doi.org/10.1016/j.foreco.2013.08.030

Gish M, Ben-Ari M, Inbar M (2017) Direct consumptive interactions between mammalian herbivores and plant-dwelling invertebrates: prevalence, significance, and prospectus. Oecologia 183:347–352. https://doi.org/10.1007/s00442-016-3775-2

Gordon IJ (1988) Facilitation of red deer grazing by cattle and its impact on red deer performance. J Appl Ecol 25:1–9

Gordon IJ (2003) Browsing and grazing ruminants: are they different beasts? For Ecol Manag., Forest dynamics and ungulate herbivory: From leaf to landscape 181:13–21. https://doi.org/10.1016/S0378-1127(03)00124-5

Gordon IJ (2006) Restoring the functions of grazed ecosystems. In: Danell K, Duncan P, Bergstrom R, Pastor J (eds) Large herbivore ecology, ecosystem dynamics and conservation, Conservation biology series. Cambridge University Press, Cambridge, pp 449–467

Gordon IJ (2018) Review: livestock production increasingly influences wildlife across the globe. Animal 12:s372–s382. https://doi.org/10.1017/S1751731118001349

Gordon I, Duncan P (1988) Pastures new for conservation. New Sci 117:54–59

Gordon IJ, Illius AW (1989) Resource partitioning by ungulates on the Isle of Rhum. Oecologia 79:383–389. https://doi.org/10.1007/BF00384318

Gordon IJ, Prins HH (2008) The ecology of browsing and grazing, ecological studies. Springer, Berlin

Greenwald KR, Petit LJ, Waite TA (2008) Indirect effects of a keystone herbivore elevate local animal diversity. J Wildl Manag 72:1318–1321

Hagenah N, Prins HH, Olff H (2009) Effects of large herbivores on murid rodents in a South African savanna. J Trop Ecol 25:483–492

Hanley ME, Lamont BB, Fairbanks MM, Rafferty CM (2007) Plant structural traits and their role in anti-herbivore defence. Perspect Plant Ecol Evol Syst 8:157–178

Hanski I, Cambefort Y (eds) (1991) Dung beetle ecology. Princeton University Press, Princeton, NJ

Henderson G, Cox F, Ganesh S, Jonker A, Young W, Abecia L, Angarita E, Aravena P, Arenas GN, Ariza C (2015) Rumen microbial community composition varies with diet and host, but a core microbiome is found across a wide geographical range. Sci Rep 5:14567

Hilbers JP, van Langevelde F, Prins HHT, Grant CC, Peel MJS, Coughenour MB, de Knegt HJ, Slotow R, Smit IPJ, Kiker GA, de Boer WF (2015) Modeling elephant-mediated cascading effects of water point closure. Ecol Appl 25:402–415. https://doi.org/10.1890/14-0322.1

Holdo RM, Holt RD, Fryxell JM (2009a) Grazers, browsers, and fire influence the extent and spatial pattern of tree cover in the Serengeti. Ecol Appl 19:95–109

Holdo RM, Sinclair AR, Dobson AP, Metzger KL, Bolker BM, Ritchie ME, Holt RD (2009b) A disease-mediated trophic cascade in the Serengeti and its implications for ecosystem C. PLoS Biol 7:e1000210

Holt CA, Fuller RJ, Dolman PM (2011) Breeding and post-breeding responses of woodland birds to modification of habitat structure by deer. Biol Conserv 144:2151–2162. https://doi.org/10.1016/j.biocon.2011.05.004

Horsley SB, Stout SL, DeCalesta DS (2003) White-tailed deer impact on the vegetation dynamics of a northern hardwood forest. Ecol Appl 13:98–118

Hu X, Liu G, Shafer A, Wei Y, Zhou J, Lin S, Wu H, Zhou M, Hu D, Liu S (2017) Comparative analysis of the gut microbial communities in forest and alpine musk deer using high-throughput sequencing. Front Microbiol 8:572

Huntzinger M, Karban R, Young TP, Palmer TM (2004) Relaxation of induced indirect defenses of acacias following exclusion of mammalian herbivores. Ecology 85:609–614

Illius AW, Gordon IJ (1987) The allometry of food intake in grazing ruminants. J Anim Ecol 56:989–999

Jay-Robert P, Niogret J, Errouissi F, Labarussias M, Paoletti É, Luis MV, Lumaret J-P (2008) Relative efficiency of extensive grazing vs. wild ungulates management for dung beetle conservation in a heterogeneous landscape from Southern Europe (Scarabaeinae, Aphodiinae, Geotrupinae). Biol Conserv 141:2879–2887

Johnson HE, Hebblewhite M, Stephenson TR, German DW, Pierce BM, Bleich VC (2013) Evaluating apparent competition in limiting the recovery of an endangered ungulate. Oecologia 171:295–307. https://doi.org/10.1007/s00442-012-2397-6

Jonsson M, Bell D, Hjältén J, Rooke T, Scogings PF (2010) Do mammalian herbivores influence invertebrate communities via changes in the vegetation? Results from a preliminary survey in Kruger National Park, South Africa. Afr J Range Forage Sci 27:39–44

Kanga EM, Ogutu JO, Piepho H-P, Olff H (2013) Hippopotamus and livestock grazing: influences on riparian vegetation and facilitation of other herbivores in the Mara Region of Kenya. Landsc Ecol Eng 9:47–58

Kauffman MJ, Brodie JF, Jules ES (2010) Are wolves saving Yellowstone's aspen? A landscape-level test of a behaviorally mediated trophic cascade. Ecology 91:2742–2755. https://doi.org/10.1890/09-1949.1

Kauffman MJ, Brodie JF, Jules ES (2013) Are wolves saving Yellowstone's aspen? A landscape-level test of a behaviorally mediated trophic cascade: reply. Ecology 94:1425–1431

Keesing F (1998) Impacts of ungulates on the demography and diversity of small mammals in Central Kenya. Oecologia 116:381–389. https://doi.org/10.1007/s004420050601

Kgope BS, Bond WJ, Midgley GF (2010) Growth responses of African savanna trees implicate atmospheric [CO2] as a driver of past and current changes in savanna tree cover. Austral Ecol 35:451–463

Krofel M, Kos I, Jerina K (2012) The noble cats and the big bad scavengers: effects of dominant scavengers on solitary predators. Behav Ecol Sociobiol 66:1297–1304. https://doi.org/10.1007/s00265-012-1384-6

Kruess A, Tscharntke T (2002) Grazing intensity and the diversity of grasshoppers, butterflies, and trap-nesting bees and wasps. Conserv Biol 16:1570–1580. https://doi.org/10.1046/j.1523-1739.2002.01334.x

Lanszki J, Kurys A, Heltai M, Csányi S, Ács K (2015) Diet composition of the golden jackal in an area of intensive big game management. Ann Zool Fenn 52:243–255. https://doi.org/10.5735/086.052.0403

Latham J (1999) Interspecific interactions of ungulates in European forests: an overview. For Ecol Manag 120:13–21. https://doi.org/10.1016/S0378-1127(98)00539-8

Lessard J-P, Reynolds WN, Bunn WA, Genung MA, Cregger MA, Felker-Quinn E, Barrios-Garcia MN, Stevenson ML, Lawton RM, Brown CB, Patrick M, Rock JH, Jenkins MA, Bailey JK, Schweitzer JA (2012) Equivalence in the strength of deer herbivory on above and below ground communities. Basic Appl Ecol 13:59–66. https://doi.org/10.1016/j.baae.2011.11.001

López-Alfaro C, Coogan SCP, Robbins CT, Fortin JK, Nielsen SE (2015) Assessing nutritional parameters of brown bear diets among ecosystems gives insight into differences among populations. PLoS One 10:e0128088. https://doi.org/10.1371/journal.pone.0128088

Margalida A, Angels Colomer M, Sanuy D (2011) Can wild ungulate carcasses provide enough biomass to maintain avian scavenger populations? An empirical assessment using a bio-inspired computational model. PLoS One 6. https://doi.org/10.1371/journal.pone.0020248

Margalida A, Pérez-García JM, Moreno-Opo R (2017) European policies on livestock carcasses management did not modify the foraging behavior of a threatened vulture. Ecol Indic 80:66–73. https://doi.org/10.1016/j.ecolind.2017.04.048

Martin J-L, Stockton SA, Allombert S, Gaston AJ (2010) Top-down and bottom-up consequences of unchecked ungulate browsing on plant and animal diversity in temperate forests: lessons from a deer introduction. Biol Invasions 12:353–371

McCauley DJ, Keesing F, Young TP, Allan BF, Pringle RM (2006) Indirect effects of large herbivores on snakes in an African savanna. Ecology 87:2657–2663

McNaughton SJ, Georgiadis NJ (1986) Ecology of African grazing and browsing mammals. Annu Rev Ecol Syst 17:39–66

McShea WJ, Rappole JH (2000) Managing the abundance and diversity of breeding bird populations through manipulation of deer populations. Conserv Biol 14:1161–1170

Melis C, Teurlings I, Linnell JDC, Andersen R, Bordoni A (2004) Influence of a deer carcass on Coleopteran diversity in a Scandinavian boreal forest: a preliminary study. Eur J Wildl Res 50:146–149. https://doi.org/10.1007/s10344-004-0051-2
Milchunas DG, Lauenroth WK, Burke IC (1998) Livestock grazing: animal and plant biodiversity of shortgrass steppe and the relationship to ecosystem function. Oikos 83:65–74
Mohr D, Cohnstaedt LW, Topp W (2005) Wild boar and red deer affect soil nutrients and soil biota in steep oak stands of the Eifel. Soil Biol Biochem 37:693–700
Moreno-Opo R, Margalida A (2013) Carcasses provide resources not exclusively to scavengers: patterns of carrion exploitation by passerine birds. Ecosphere 4:1–15. https://doi.org/10.1890/ES13-00108.1
Mukul-Yerves JM, Zapata-Escobedo MR, Montes-Pérez RC, Rodríguez-Vivas RI, Torres-Acosta JF (2014) Gastrointestinal and ectoparasites in wildlife-ungulates undercaptive and free-living conditions in the Mexican tropic. Rev Mex Cienc Pecu 5:459–469. https://doi.org/10.22319/rmcp.v5i4.4017
Murray MG, Illius AW (2000) Vegetation modification and resource competition in grazing ungulates. Oikos 89:501–508
Mysterud A (2006) The concept of overgrazing and its role in management of large herbivores. Wildl Biol 12:129–141. https://doi.org/10.2981/0909-6396(2006)12[129:TCOOAI]2.0.CO;2
Nack JL, Ribic CA (2005) Apparent predation by cattle at grassland bird nests. Wilson Bull 117:56–62
Newsome TM, Boitani L, Chapron G, Ciucci P, Dickman CR, Dellinger JA, López-Bao JV, Peterson RO, Shores CR, Wirsing AJ, Ripple WJ (2016) Food habits of the world's grey wolves. Mammal Rev 46:255–269. https://doi.org/10.1111/mam.12067
Noss RF (1990) Indicators for monitoring biodiversity: a hierarchical approach. Conserv Biol 4:355–364
O'Kane CA, Duffy KJ, Page BR, Macdonald DW (2014) Model highlights likely long-term influences of mesobrowsers versus those of elephants on woodland dynamics. Afr J Ecol 52:192–208
O'Kelly HJ, Evans TD, Stokes EJ, Clements TJ, Dara A, Gately M, Menghor N, Pollard EHB, Soriyun M, Walston J (2012) Identifying conservation successes, failures and future opportunities; assessing recovery potential of wild ungulates and tigers in eastern Cambodia. PLoS One 7:e40482. https://doi.org/10.1371/journal.pone.0040482
O'Toole A, Browne JA, Hogan S, Bassière T, DeWaal T, Mulcahy G, Zintl A (2014) Identity of rumen fluke in deer. Parasitol Res 113:4097–4103. https://doi.org/10.1007/s00436-014-4078-3
Odadi WO, Jain M, Wieren SEV, Prins HHT, Rubenstein DI (2011a) Facilitation between bovids and equids on an African savanna. Evol Ecol Res 13:237–252
Odadi WO, Karachi MK, Abdulrazak SA, Young TP (2011b) African wild ungulates compete with or facilitate cattle depending on season. Science 333:1753–1755. https://doi.org/10.1126/science.1208468
Olff H, Ritchie ME, Prins HH (2002) Global environmental controls of diversity in large herbivores. Nature 415:901–904
Owen-Smith N, Mills MG (2008) Predator–prey size relationships in an African large-mammal food web. J Anim Ecol 77:173–183
Owen-Smith N, Kerley GIH, Page B, Slotow R, Van Aarde RJ (2006) A scientific perspective on the management of elephants in the Kruger National Park and elsewhere: elephant conservation. S Afr J Sci 102:389–394
Pakanen V-M, Luukkonen A, Koivula K (2011) Nest predation and trampling as management risks in grazed coastal meadows. Biodivers Conserv 20:2057–2073. https://doi.org/10.1007/s10531-011-0075-3
Pakanen V-M, Aikio S, Luukkonen A, Koivula K (2016) Grazed wet meadows are sink habitats for the southern dunlin (Calidris alpina schinzii) due to nest trampling by cattle. Ecol Evol 6:7176–7187. https://doi.org/10.1002/ece3.2369

Parr CL, Brockett BH (1999) Patch-mosaic burning: a new paradigm for savanna fire management in protected areas? Koedoe 42:117–130

Pastor J, Cohen Y, Hobbs NT (2006) The roles of large herbivores in ecosystem nutrient cycles. In: Danell K, Duncan P, Bergstrom R, Pastor J (eds) Large herbivore ecology, ecosystem dynamics and conservation, Conservation biology series. Cambridge University Press, Cambridge, pp 289–325

Pietz PJ, Granfors DA (2000) White-tailed deer (Odocoileus virginianus) predation on grassland songbird nestlings. Am Midl Nat 144:419–422

Pringle RM (2012) How to be manipulative intelligent tinkering is key to understanding ecology and rehabilitating ecosystems. Am Sci 100:30–37

Pringle RM, Young TP, Rubenstein DI, McCauley DJ (2007) Herbivore-initiated interaction cascades and their modulation by productivity in an African savanna. Proc Natl Acad Sci 104:193–197

Pringle RM, Palmer TM, Goheen JR, McCauley DJ, Keesing F (2010) Ecological importance of large herbivores in the Ewaso ecosystem. Smithson Contrib Zool 632:43–53

Prins HHT (2000) Competition between wildlife and livestock in Africa. In: Prins HHT, Grootenhuis JG, Dolan TT (eds) Wildlife conservation by sustainable use, Conservation biology series. Springer, Dordrecht, pp 51–80. https://doi.org/10.1007/978-94-011-4012-6_5

Prins HHT (2016) Interspecific resource competition in antelopes: search for evidence. In: Bro-Jorgensen J, Mallon DP (eds) Antelope conservation: from diagnosis to action, Conservation science and practice series. Wiley Blackwell, Oxford, pp 51–77

Prins HHT, Van der Jeugd HP (1993) Herbivore population crashes and woodland structure in East Africa. J Ecol 81:305–314

Prins HHT, Van Oeveren H (2014) Bovini as keystone species and landscape architects. In: Melletti M, Burton J (eds) Ecology, evolution and behaviour of wild cattle. Cambridge University Press, Cambridge, pp 21–29

Reider KE, Carson WP, Donnelly MA (2013) Effects of collared peccary (Pecari tajacu) exclusion on leaf litter amphibians and reptiles in a Neotropical wet forest, Costa Rica. Biol Conserv 163:90–98

Reimoser F (2003) Steering the impacts of ungulates on temperate forests. J Nat Conserv 10:243–252. https://doi.org/10.1078/1617-1381-00024

Reimoser F, Gossow H (1996) Impact of ungulates on forest vegetation and its dependence on the silvicultural system. For Ecol Manag., Ungulates in Temperate Forest Ecosystems 88:107–119. https://doi.org/10.1016/S0378-1127(96)03816-9

Ringler M, Hödl W, Ringler E (2015) Populations, pools, and peccaries: simulating the impact of ecosystem engineers on rainforest frogs. Behav Ecol 26:340–349

Ripple WJ, Beschta RL (2003) Wolf reintroduction, predation risk, and cottonwood recovery in Yellowstone National Park. For Ecol Manag 184:299–313

Ripple WJ, Beschta RL (2007) Restoring Yellowstone's aspen with wolves. Biol Conserv 138:514–519

Ripple WJ, Beschta RL (2012) Trophic cascades in Yellowstone: the first 15years after wolf reintroduction. Biol Conserv 145:205–213

Ritchie ME, Olff H (1999) Herbivore diversity and plant dynamics: compensatory and additive effects. Herbiv Plants Predat:175–204

Rogers KH (2003) Adopting a heterogeneity paradigm: implications for management of protected savannas. Kruger Exp Ecol Manag Savanna Heterog Isl Press Wash:41–58

Rovero F, Jones T, Sanderson J (2005) Notes on Abbott's duiker (Cephalophus spadix true 1890) and other forest antelopes of Mwanihana Forest, Udzungwa Mountains, Tanzania, as revealed by camera-trapping and direct observations. Trop Zool 18:13–23. https://doi.org/10.1080/03946975.2005.10531211

Royo AA, Collins R, Adams MB, Kirschbaum C, Carson WP (2010) Pervasive interactions between ungulate browsers and disturbance regimes promote temperate forest herbaceous diversity. Ecology 91:93–105

Russell FL, Zippin DB, Fowler NL (2001) Effects of white-tailed deer (Odocoileus virginianus) on plants, plant populations and communities: a review. Am Midl Nat 146:1–26
Sánchez-Zapata JA, Eguía S, Blázquez M, Moleón M, Botella F (2010) Unexpected role of ungulate carcasses in the diet of golden eagles Aquila chrysaetos in Mediterranean mountains. Bird Study 57:352–360. https://doi.org/10.1080/00063651003674946
Schieltz JM, Rubenstein DI (2016) Evidence based review: positive versus negative effects of livestock grazing on wildlife. What do we really know? Environ Res Lett 11:113003. https://doi.org/10.1088/1748-9326/11/11/113003
Serrouya R, McLellan BN, Boutin S, Seip DR, Nielsen SE (2011) Developing a population target for an overabundant ungulate for ecosystem restoration. J Appl Ecol 48:935–942
Shaw MT, Keesing F, Ostfeld RS (2002) Herbivory on Acacia seedlings in an East African savanna. Oikos 98:385–392
Sinclair ARE, Mduma S, Brashares JS (2003) Patterns of predation in a diverse predator–prey system. Nature 425:288–290
Skowno AL, Thompson MW, Hiestermann J, Ripley B, West AG, Bond WJ (2017) Woodland expansion in South African grassy biomes based on satellite observations (1990–2013): general patterns and potential drivers. Glob Chang Biol 23:2358–2369
Smit IPJ, Grant CC (2009) Managing surface-water in a large semi-arid savanna park: effects on grazer distribution patterns. J Nat Conserv 17:61–71
Smit IP, Grant CC, Devereux BJ (2007) Do artificial waterholes influence the way herbivores use the landscape? Herbivore distribution patterns around rivers and artificial surface water sources in a large African savanna park. Biol Conserv 136:85–99
Spear D, Chown SL (2009) Non-indigenous ungulates as a threat to biodiversity. J Zool 279:1–17
Staver AC, Bond WJ (2014) Is there a “browse trap”? Dynamics of herbivore impacts on trees and grasses in an African savanna. J Ecol 102:595–602
Staver AC, Bond WJ, Stock WD, Van Rensburg SJ, Waldram MS (2009) Browsing and fire interact to suppress tree density in an African savanna. Ecol Appl 19:1909–1919
Stevens N, Erasmus BFN, Archibald S, Bond WJ (2016) Woody encroachment over 70 years in South African savannahs: overgrazing, global change or extinction aftershock? Phil Trans R Soc B 371:20150437
Stevens N, Lehmann CE, Murphy BP, Durigan G (2017) Savanna woody encroachment is widespread across three continents. Glob Change Biol 23:235–244
Stewart AJA (2001) The impact of deer on lowland woodland invertebrates: a review of the evidence and priorities for future research. Forestry 74:259–270
Suominen O, Danell K, Bergström R (1999) Moose, trees, and ground-living invertebrates: indirect interactions in Swedish pine forests. Oikos 84:215–226
Suominen O, Niemelä J, Martikainen P, Niemelä P, Kojola I (2003) Impact of reindeer grazing on ground-dwelling Carabidae and Curculionidae assemblages in Lapland. Ecography 26:503–513
Suryawanshi KR, Bhatnagar YV, Mishra C (2010) Why should a grazer browse? Livestock impact on winter resource use by bharal Pseudois nayaur. Oecologia 162:453–462. https://doi.org/10.1007/s00442-009-1467-x
Taggart MA, Senacha KR, Green RE, Jhala YV, Raghavan B, Rahmani AR, Cuthbert R, Pain DJ, Meharg AA (2007) Diclofenac residues in carcasses of domestic ungulates available to vultures in India. Environ Int 33:759–765. https://doi.org/10.1016/j.envint.2007.02.010
Tymensen L, Barkley C, McAllister TA (2012) Relative diversity and community structure analysis of rumen protozoa according to T-RFLP and microscopic methods. J Microbiol Methods 88:1–6. https://doi.org/10.1016/j.mimet.2011.09.005
Van de Koppel J, Prins HH (1998) The importance of herbivore interactions for the dynamics of African savanna woodlands: an hypothesis. J Trop Ecol 14:565–576
van Noordwijk CGE, Flierman DE, Remke E, WallisDeVries MF, Berg MP (2012) Impact of grazing management on hibernating caterpillars of the butterfly Melitaea cinxia in calcareous grasslands. J Insect Conserv 16:909–920. https://doi.org/10.1007/s10841-012-9478-z

van Wilgen BW (2009) The evolution of fire management practices in savanna protected areas in South Africa. South Afr J Sci 105:343–349

Vanschoenwinkel B, Waterkeyn A, Vandecaetsbeek T, Pineau O, Grillas P, Brendonck L (2008) Dispersal of freshwater invertebrates by large terrestrial mammals: a case study with wild boar (Sus scrofa) in Mediterranean wetlands. Freshw Biol 53:2264–2273

Venter ZS, Hawkins H-J, Cramer MD (2017) Implications of historical interactions between herbivory and fire for rangeland management in African savannas. Ecosphere 8. https://doi.org/10.1002/ecs2.1946

Vera FWM, Bakker ES, Olff H (2006) Large herbivores: missing partners of western European light-demanding tree and shrub species? In: Danell K, Duncan P, Bergstrom R, Pastor J (eds) Large herbivore ecology, ecosystem dynamics and conservation, Conservation biology series. Cambridge University Press, Cambridge, pp 203–231

Verweij R, Verrelst J, Loth PE, Heitkönig IMA, Brunsting AMH (2006) Grazing lawns contribute to the subsistence of mesoherbivores on dystrophic savannas. Oikos 114:108–116

Vila AR, Borrelli L, Martínez L (2009) Dietary overlap between huemul and livestock in los Alerces National Park, Argentina. J Wildl Manag 73:368–373. https://doi.org/10.2193/2008-062

Wardle DA, Barker GM, Yeates GW, Bonner KI, Ghani A (2001) Introduced browsing mammals in New Zealand natural forests: aboveground and belowground consequences. Ecol Monogr 71:587–614

Wegge P, Shrestha AK, Moe SR (2006) Dry season diets of sympatric ungulates in lowland Nepal: competition and facilitation in alluvial tall grasslands. Ecol Res 21:698–706

Wisdom MJ, Vavra M, Boyd JM, Hemstrom MA, Ager AA, Johnson BK (2006) Understanding ungulate herbivory–episodic disturbance effects on vegetation dynamics: knowledge gaps and management needs. Wildl Soc Bull 34:283–292

Wittmer HU, Ahrens RN, McLellan BN (2010) Viability of mountain caribou in British Columbia, Canada: effects of habitat change and population density. Biol Conserv 143:86–93

Yamazaki K, Sugiura S (2008) Deer predation on leaf miners via leaf abscission. Naturwissenschaften 95:263–268. https://doi.org/10.1007/s00114-007-0318-z

Young TP, Palmer TM, Gadd ME (2005) Competition and compensation among cattle, zebras, and elephants in a semi-arid savanna in Laikipia, Kenya. Biol Conserv 122:351–359. https://doi.org/10.1016/j.biocon.2004.08.007

Chapter 13
Interactions Between Fire and Herbivory: Current Understanding and Management Implications

Izak P. J. Smit and Corli Coetsee

13.1 Introduction

Historically, studies on disturbances such as fire and herbivory have primarily focused on their effects separately. Although many studies continue in this way, it is increasingly appreciated that the interactive effects of these two processes are often different to the sum of the separate main effects (Fuhlendorf and Engle 2004). These interactive effects can be so strong that it has been argued that fire and herbivory should be viewed as a single disturbance in ecosystems that evolved with both processes active (Fuhlendorf and Engle 2004, i.e., pyric herbivory after 2001). Furthermore, while the effects of these disturbances were traditionally considered to be linear and unidirectional, current understanding explicitly highlights non-linearity, time-lags, possibility of multiple stable-states and several feedbacks (Archibald and Hempson 2016; Johnson et al. 2018; Rietkerk et al. 1996; Van Langevelde et al. 2003). This Chapter sets out to illustrate these interactions, and some feedbacks, typical of systems where both fire and herbivory are present.

Although the interaction between fire and herbivory (hereafter fire × herbivory) is becoming increasingly well established in the literature, it is a complex topic with very dynamic interplays with rainfall, soils, herbivore feeding guilds and population

I. P. J. Smit (✉)
Scientific Services, South African National Parks, Skukuza, Mpumalanga, South Africa

Centre for African Ecology, School of Animal, Plant and Environmental Sciences, University of the Witwatersrand, Wits, South Africa
e-mail: izak.smit@sanparks.org

C. Coetsee
Scientific Services, South African National Parks, Skukuza, Mpumalanga, South Africa

School of Natural Resource Management, George Campus, Nelson Mandela University, George, South Africa
e-mail: Corli.Wigley-Coetsee@sanparks.org

I. J. Gordon, H. H. T. Prins (eds.), *The Ecology of Browsing and Grazing II*, Ecological Studies 239, https://doi.org/10.1007/978-3-030-25865-8_13

sizes, and with multiple feedback loops. To complicate matters even further, these feedbacks are often context-specific and can be either positive or negative. Considering these complexities, a full review is beyond the scope of this Chapter, and perhaps even beyond current techniques to aid our comprehension. Instead, we provide a selection of relevant literature to illustrate concepts and observations common across many studies. As such, we start by providing a brief overview of how the paleo-ecological record suggests a long—and continuing—history of interaction between fire and herbivory (Sect. 13.2). This is followed by introducing conceptual models, in order to synthesize how grazing and browsing typically interact with fire (Sect. 13.3). We illustrate these conceptual models using a selection of case studies from across the world. This is followed by a discussion on implications of fire × herbivory, highlighting the importance of considering these processes together, rather than in isolation (Sect. 13.4). Section 13.5 provides a brief conclusion.

13.2 A Long and Dynamic History of Fire and Herbivory Interaction

Reconstructed herbivore and fire spatio-temporal patterns suggest that fire and herbivory have a long history of interaction spanning thousands of years (e.g., Gill et al. 2009; Pausas and Keeley 2009). The paleo-ecological record has been particularly useful in reconstructing these interactions, and various studies have indicated how decreases in herbivory (using dung fungus spores *Sporormiella* as proxy; see also **Rowan and Faith** Chap. 3) were often associated with periods of increases in fire frequency (using charcoal as proxy) and with associated changes in vegetation (using pollen as proxy) (Gill et al. 2009; Robinson et al. 2005). In a recent synthesis, Johnson et al. (2018) considered 14 studies that have tracked pollen, *Sporomiella* spores and charcoal over appropriate times-scales spanning extinctions of megafauna (and arrival of humans). At sites that were initially wooded or forested (but not where initially grass dominated), large increases in charcoal were observed once mega-herbivores went extinct, which was frequently followed by changes in vegetation. Although it is hard to ascertain whether this inverse relationship, between fire and herbivory, is causal as other factors can also play a role (e.g., arrival of humans lighting more fires, climate change, etc.), these patterns nonetheless seem consistent with prevailing patterns, and provide some circumstantial evidence for fire and herbivory as competing consumers of vegetation biomass over historical time frames.

Regardless of the cause(s), extinction of large mammal species has been documented from many parts of the world including Eurasia, Australia and even, although to a lesser degree, in Africa (Barnosky et al. 2004; Burney and Flannery 2005; c.f. **Rowan and Faith** Chap. 3). In North America all animals of >1000 kg, and over half of those >32 kg, became extinct at the end of the Pleistocene and early Holocene (Koch and Barnosky 2006). The disappearance of mega-herbivores at the end of the Pleistocene gave rise to no-analog, fire-driven plant communities in North America (Gill et al. 2009). Similarly, Rule et al. (2012) proposed that human-driven extinction of mega-fauna in Australia triggered replacement of rainforest by sclerophyll vegetation, as fuel build-up led to increased fire frequency and intensity. Johnson (2009) proposed, in his review of the ecological consequences of the late Quaternary mega-faunal extinctions, that vegetation reverted to more dense and uniform vegetation (c.f. **Saarinen** Chap. 2), and fire frequency increased due to accumulating plant material. Bakker et al. (2016) combined paleo-records with data from modern day herbivore exclosures, and came to a similar conclusion—i.e., high herbivore density and diversity result in less fires, and more open landscapes dominated by trees that are light-demanding and less preferred by herbivores. For much of the Holocene, it seems, therefore, that fire gained a competitive advantage over herbivory, as anthropogenic ignition frequency increased (Bowman et al. 2011), and native wildlife and mega-herbivores decreased (Johnson et al. 2018).

Although fire appear as the dominant vegetation consumption process during most of the Holocene ("flammable new worlds" of Johnson 2009), intensive live-stock farming, in more recent history (~400 years), resulted in fire decreasing again in some locations, as domestic livestock grazing reduced fire spread (Rius et al. 2009; Valese et al. 2014). Land abandonment, in certain parts of Europe in recent decades, may have shifted the balance towards fire ascendancy once again, this time due to changed agricultural practises, including destocking of domestic livestock (Archibald and Hempson 2016; Ascoli et al. 2009; Chauchard et al. 2007). It, therefore, remains clear that fire and herbivory remain in dynamic interplay.

In conclusion, there is circumstantial evidence that herbivory (sometimes interacting with climate and sometimes with anthropogenic management), contributed to, and interacted with, historic fire regimes, dynamically shaping vegetation patterns over time (Gill et al. 2009; Robinson et al. 2005; Rule et al. 2012; Williams et al. 2001).

13.3 Fire × Herbivory and Feedbacks: A Conceptual Model

Since Bond (2005) first suggested that, apart from a resource driven green world (where climate controls vegetation), the world could also be brown (where herbivory controls vegetation) or black (where fire controls vegetation), our understanding of 'consumer' controlled systems has greatly increased. Herbivory seems more likely to dominate in drier, higher nutrient conditions, as opposed to fire, which often dominates in higher rainfall and lower nutrient conditions (Bond 2005; Archibald and Hempson 2016). At intermediate conditions, either process may dominate (Archibald and Hempson 2016; see also later). In a worldwide biogeographical synthesis, Johnson et al. (2018) concluded that the evidence for vertebrate herbivores reducing fire regimes, by reducing fuel build-up, was strongest in environments with intermediate rainfall, warm temperatures and with grass-dominated ground vegetation.

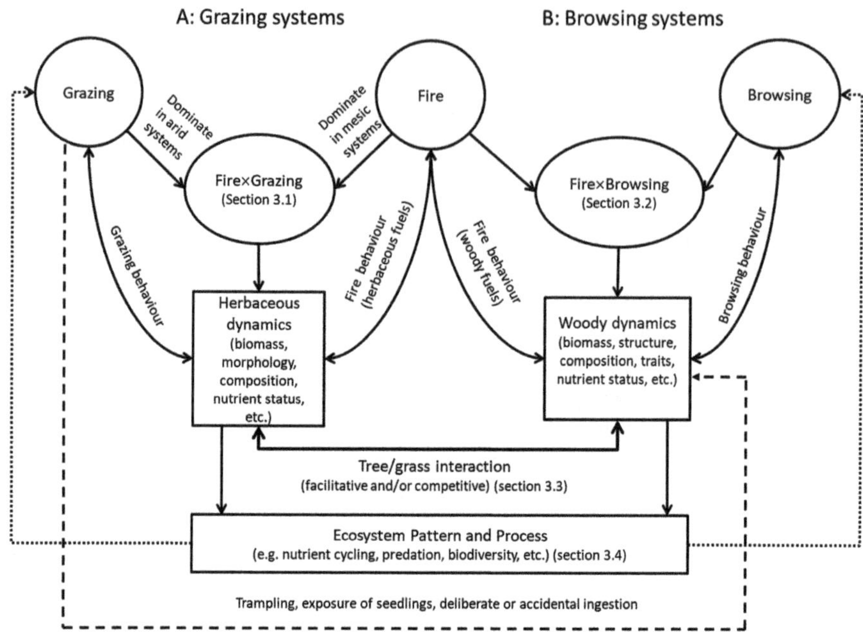

Fig. 13.1 Conceptual model of interactions and feedbacks between fire and herbivory in grazing (A), browsing (B) and mixed grazing/browsing systems (A + B) (inspired by Fuhlendorf et al. 2009)

In the paragraphs below, and summarised in Fig. 13.1, we propose a conceptual model, illustrating how (i) vertebrate grazing influences fire behaviour in grassy systems (typically grasslands; Sect. 13.3.1 and A in Fig. 13.1); (ii) vertebrate browsing influences fire behaviour in systems where fire spreads by means of woody vegetation (typically forests/shrubland; Sect. 13.3.2 and B in Fig. 13.1); and (iii) grazing and browsing can interact to influence fires in mixed tree/grass systems (typically savannas; Sect. 13.3.3; interaction between A and B in Fig. 13.1).

13.3.1 Fire × Grazing in Grass Dominated Systems

Managed grazing covers more than 25% of the global terrestrial surface, and as such, is the most dominant land use on earth (Asner et al. 2004). In these grass-dominated systems (Fig. 13.1, A), grazers and fire are competing for the herbaceous layer. Increased grazing will typically make the landscape less favourable for fires through both short- and long-term effects, as well as through changes to the abiotic landscape. Short-term effects refer to direct fuel removal, and changes in vertical structure brought about by herbivores (e.g., Kramer et al. 2003). The long-term compositional effects are due to the fact that tall, shade-tolerant flammable grasses, which typically vigorously resprout from stored underground reserves after a fire, tend to be sensitive to grazing (Archibald et al. 2005; Archibald and Hempson 2016; Hempson et al. 2015). Where fires, as opposed to grazers, are the dominant consumer, high-biomass, fire-adapted grass species are favoured, and this positive feedback results in a fire-dominated, often less preferred vegetation system. Interestingly, Leonard et al. (2010) suggest that light grazing of tall, tussock grasslands may increase fire potential due to animals selectively removing the live shoots leaving a high proportion of dry and flammable fuel. However, if these tall grasses are repeatedly defoliated through grazing, they tend to become out-competed by grazer-tolerant (but shade-intolerant) shorter, mostly stoloniferous, grasses (Archibald et al. 2005). These grasses are often more nutritious, reinforcing the grazing feedback (Hempson et al. 2015), and are less flammable due to reduced aerial cover, culminating in an herbivore-driven system. As such, persistent grazing not only reduces the fuel biomass in the short-term, but can also, in the longer-term, switch a system to a less flammable and possibly more nutritious state. As a third, less studied mechanism of herbivores influencing fires, herbivores can change the abiotic environment e.g., by creating a myriad of trails or bare patches, or influencing the litter layer (e.g., burying litter decreases fuel load) (Johnson et al. 2018).

African-wide analysis suggests that grazing will generally be the dominant consumer of grass for mean annual precipitation (MAP) <600 mm, whilst fire will typically dominate in areas above this threshold (Archibald and Hempson 2016). However, the areas with intermediate rainfall, around the threshold (500–800 mm MAP), can switch between herbivores or fire as dominant consumers (Hirota et al. 2011; Staver et al. 2011; Van Langevelde et al. 2003). In these systems, management actions, or natural phenomena such as disease, can possibly switch a system from being dominated by one process to the other. For example, Smit and Archibald (2019) showed fire frequency decreased once culling of mega-grazers terminated in the semi-arid savannas of the Kruger National Park, South Africa. The effect was especially pronounced in areas close to rivers, where grazing intensity is higher, and in lower rainfall areas where productivity is lower. Similarly, Waldram et al. (2008) showed that the removal of white rhinoceros (*Ceratotherium simum*), in a southern African mesic savanna, increased fuel loads and continuity, which resulted in larger, less patchy fires. In the Serengeti, rinderpest reduced wildebeest (*Connochaetes taurinus*) numbers resulting in a system that burned frequently, but once the wildebeest numbers recovered, annual fire extent halved (Dublin et al. 1990; Holdo et al. 2009).

Not all herbivory effects are similar, and herbivore traits (e.g., size and feeding guild) may influence the resulting herbaceous dynamics. For example, work from semi-arid southern African and north American savannas, has shown complicated interactions between fire and herbivory, and how this translates to herbaceous species composition and richness (Burkepile et al. 2013, 2016, 2017; Burns et al. 2009; Eby et al. 2014; Koerner et al. 2014). The magnitude and direction of effects on herbaceous species depended in the diversity of herbivores present (single vs. multiple), the suite of herbivores, the burning regime and the productivity of the site. For instance, at the southern African site, annually burnt areas attracted a diverse suite of herbivores; exclusion of larger herbivores (e.g., elephant, *Loxodonta africana*, zebra, *Equus quagga*, blue wildebeest, *Connochaetes taurinus*) increased plant abundance and cover, but not plant diversity, while excluding smaller (e.g., impala, *Aepyceros melampus*, warthog, *Phacochoerus africanus*, steenbok, *Raphicerus campestris*) as well as large species led to declines in plant diversity (Burkepile et al. 2016). On triennial burns, herbivore exclusion had no effect on plant richness or diversity (or cover), which could be attributed to little competitive exclusion due to the intermediate fire disturbance. Areas where fires were excluded attracted the least diverse suite of herbivores, but herbivore exclusion led to herbaceous competitive exclusion with reductions in plant richness and diversity, via increases in plant dominance and light limitation. In contrast, fire may affect herbivore species composition; Klop and Prins (2008), in their West African focussed review of how ungulate diversity and composition are affected by fire, climate and soil, showed highest grazing (but not browsing) ungulate species richness at intermediate fire frequencies.

In conclusion, high levels of grazing can reduce grass biomass and change the grass layer structure and composition, resulting in patches or landscapes with reduced flammability, with feedbacks to fire and grazing behaviour. There are, however, also some studies that have noted the opposite effect, e.g., Williamson et al. (2014) suggested that cattle grazing in the Australian Alps may have slightly increased fire severity, because more flammable woody shrubs have increased at the expense of grass under sustained grazing regimes.

13.3.2 Fire × Browsing Interaction in Woody Systems (Forests and Shrublands)

Browsers consume a relatively small fraction of net primary productivity in forests and shrublands (Hobbs 1996), and as such their direct effect on fuel accumulation, and hence fire behaviour, are limited as compared to grazers. Nonetheless, browsers can influence fire behaviour in woody systems, like forests and shrublands, either directly or indirectly (B in Fig. 13.1).

Browsing can influence fire patterns directly through reducing fuel loads and altering the vertical structure of vegetation. This process is very well documented in the Mediterranean where land abandonment, and the cessation of livestock farming, have led to a recovery of the fire-prone maquis vegetation with associated increases in fire risk (Lasanta et al. 2006). For example, Alberes cattle consume a high proportion of the most combustible species in Mediterranean scrublands (i.e., Ericaceous species), which reduces the forest fire hazard (Bartolomé et al. 2011). Similarly, Payoya goats (*Capra aegagrus hircus*) reduce the phytovolume of easily combustible undergrowth in pine forests in Spain, thereby reducing fire risk (Mancilla-Leytón et al. 2013). Browsers can also influence the vertical structure of fuel in forests, by decreasing the "bridging" fuels that may influence whether or not surface fires spread to crown fires (Hobbs 1996).

In turn, fire regimes have consequences for herbivores. Some work has shown that both browsing and grazing ungulates prefer to consume vegetation immediately post-fire (Cherry et al. 2018; Leverkus et al. 2018; Sachro et al. 2005; Vinton et al. 1993). Frequent fire may decrease concentrations of tannins and polyphenols making forage more digestible (e.g., Ferwerda et al. 2006), whilst in other systems fire suppression may decrease the forage quality for ungulates (Allred et al. 2011; Carlson et al. 1993; Fischer and Bradley 1987). For instance, fire suppression has led to the decline of Key deer (*Odocoileus virginianus clavium*), a subspecies of white-tailed deer restricted to the lower Florida Keys, as the pine community that the deer depends on has been negatively affected by a lack of recurrent fire (Carlson et al. 1993). However, frequent fire may also decrease nutritional value of vegetation; for instance, Ferwerda et al. (2006) have shown that frequent fire decreases nitrogen levels in mopane (*Colophospermum mopane*) woodlands in southern Africa. Season of fire can also influence the plant growth form and nutritional value. Griffin and Friedel (1984) found that winter fires maintained or increased pasture plants preferred by cattle, whereas summer fires decreased the grass component whilst increasing the non-preferred forb component.

Browsing can also indirectly influence fire patterns through restoration dynamics and post-fire vegetation succession. For example, browsers can retard regeneration of forests after fires, with very high consumption rates of emerging woody resprouts after burns, which can result in reduced forest recovery, and hence reduced fire frequency and intensity (Hobbs 1996). Fire suppression on the one hand, with heavy browsing pressure by ungulates on the other, has drastically altered species dominance and associated ecosystem function in North American deciduous forests (Brose et al. 2001; McEwan et al. 2011). For instance, the interaction between fire and appropriate levels of herbivory is important for the regeneration of aspen (*Populus tremuloides*) (Bailey and Whitham 2002; Krasnow and Stephens 2015). Work from Coconino National Forest, northern Arizona, USA, has shown that without elk (*Cervus elaphus*), aspen regenerates vigorously via asexual reproduction when intensely burned. However, when elk is present, they browse aspen ramets more intensely, in high-severity than in intermediate-severity burns, with the sum effect of reduced regeneration in high-severity burns. Royo et al. (2010) showed in eastern North American deciduous forests that white-tailed deer (*Odocoileus virginianus*), at moderate levels of abundance—as opposed to very high abundance—promote herbaceous richness (i.e., forbs) and abundance, by preferentially browsing fast growing pioneer species that thrive following co-occurring disturbances (i.e., fire and gaps). Similarly, Nuttle et al. (2013) found that woody diversity is higher under low browsing pressure of fire sites and forest gaps, but once browsing increases, these disturbances that are supposed to lead to increased woody diversity, actually lead to a decline in diversity, or are less effective in creating diversity.

Post-fire browsing selectivity can also influence competitive outcomes between plant species, which can subsequently influence flammability. For example, fire increased local browsing intensity and amplified per-unit effect of herbivores in an Australian Eucalypt forest, resulting in heavy browsing that created an understorey dominated by a less preferred, fire-resistant fern that suppresses the establishment of other plants, creating a state that is hard to reverse (Foster et al. 2015). In other landscapes, the opposite has been observed, where herbivory increased flammability. For example, cattle and European hare (*Lepus europaeus*), two exotic and widespread herbivores in Patagonia, inhibit forest recovery after fire, and favour a transition to a more flammable shrubland (Raffaele et al. 2011).

In conclusion, direct effects of fire × browsing interactions include changed fuel accumulation and vertical fuel structure, whilst indirect effects are often by means of herbivory influencing competitive effects in vegetation successional processes, influencing forest regeneration trajectories post-fire with implications for flammability of the system.

13.3.3 Grazing and Browsing Interaction with Fire in Savanna Systems

Heterogeneous tree/grass systems span a range of conditions from open grassy areas, dominated by grazers and often associated with higher fire frequency, to dense areas, typically with limited fire due to lower grass biomass, dominated by browsers (Klop and Prins 2008; Smit and Prins 2015). The mixture of trees and grasses means that the grazing processes described in the grassland systems (A in Fig. 13.1) and the browsing processes in forests/shrublands (B in Fig. 13.1) will interact, and have feedbacks across the grassy/grazer and woody/browser interfaces. There are various pathways in which grazing and browsing can interact to influence herbaceous and woody dynamics, and hence fire behaviour in mixed tree/grass systems.

Figure 13.2 (adapted from Van Langevelde et al. 2003) shows one, arguably over-simplified, scenario of how grazing and browsing may influence each other through direct and indirect effects on the fire regime. Firstly, if the level of grazing is high enough to reduce herbaceous biomass resulting in decreased fire frequency and intensity, this may favour woody vegetation through increased woody recruitment, allowing woody individuals to establish and grow tall enough to survive subsequent fires (escaping the "fire-trap"). Furthermore, reduced grass biomass will also reduce grass competition with trees (van der Waal et al. 2011a, b), allowing woody vegetation a competitive advantage (February et al. 2013). These processes result in higher woody cover, which may subsequently favour browsers. However, grazers may also reduce woody vegetation through trampling, exposing seedlings to browsing, and accidental, or even selective, ingestion of seedlings. Conversely, increased levels of browsing may reduce woody vegetation recruitment (Augustine and McNaughton 2004; Prins and Van der Jeugd 1993; Staver et al. 2009), and the cover of mature trees (Buechner and Dawkins 1961). Additionally, browsing can also indirectly influence surface fires, either through increasing fires by reducing the competitive effects of trees on grass (February et al. 2013), or increasing fires by decreasing tree clustering (tree clustering decreases fires by decreasing fuel load below trees according to Groen et al. 2008). The feedback between fire and browsers

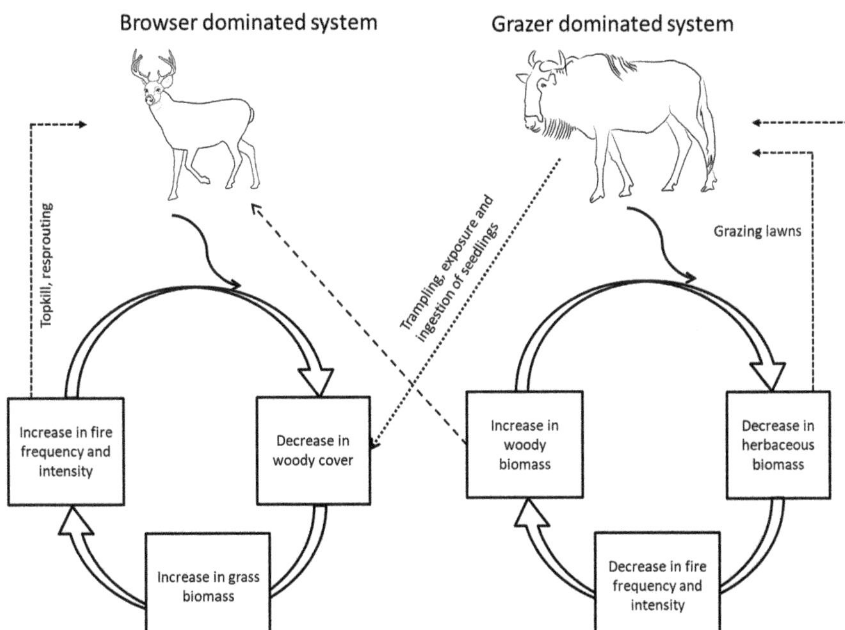

Fig. 13.2 Interaction and possible feedbacks between browsing and grazing herbivores, vegetation and fire (adapted from van Langevelde et al. 2003). Artwork by C. Coetsee

is complex, as increased fire frequency and intensity may reduce the total amount of forage available for browsers, but at the same time, browsers may be advantaged by fire maintaining browse within the browse height (Carlson et al. 1993). Mixed feeders, like elephants and impala, are also interesting to consider as that they can switch between grazing and browsing, and hence their feedback to fire can be either positive or negative. The combined effect of fire and elephants are a lethal combination for trees (Midgley et al. 2010; Shannon et al. 2011; Vanak et al. 2012), and some studies have argued that elephants are a net facilitator of fire, through removal of woody cover, favouring grass production and hence fire (Buechner and Dawkins 1961). However, a recent study has shown that increasing elephant density was associated with decreased fire frequency (Smit and Archibald 2019), likely due to elephants consuming and trampling large quantities of grass (Beekman and Prins 1989; Codron et al. 2006).

It is clear, from the above, that grazers, browsers and mixed feeders can influence fire, and hence savanna vegetation patterns, in complex ways through both positive and negative feedbacks. However, in general, it is agreed that large grazers affect fuel loads, and hence fire regimes, more than browsers (Johnson et al. 2018). Lehmann et al. (2011) suggest that if it was not for herbivory, the extent of denser woodland would have been much wider *in lieu* of more open African savannas. This is amply evidenced in long-term herbivory exclosures, in countries such as Kenya (Western and Maitumo 2004; Young et al. 1997) and South Africa (Asner et al. 2009; Levick et al. 2009; Wigley et al. 2014). Correspondingly, Doughty et al. (2016) proposed that South American savannas would be more similar to open African savannas had the many native mixed- and browsing mega-herbivores not gone extinct.

In conclusion, in mixed tree/grass and grazer/browser systems, increased grazing pressure can reduce fire frequency and intensity directly through fuel removal and compositional and structural changes to the herbaceous layer. Increased browsing may increase fire indirectly through reducing woody recruitment and decreasing the competitive effect of trees on grasses. However, there may also be other feedbacks working in opposing directions, and it is sometimes hard to ascertain the net effect. For example, grazing can also possibly reduce woody vegetation seedling establishment as a result of trampling, exposing seedlings to browsers (Pachzelt et al. 2015), and ingestion of seedlings. Also, grazing may lead to the formation of grazing lawns with preferred species, which may further favour grazers (Grant et al. 2019; Hempson et al. 2015; Venter et al. 2017), even though the total standing grass biomass may be reduced. Tall bunch grass areas may, however, remain important for some species, as well as being essential during drought periods when other grazing resources are depleted (Staver et al. 2019). Lastly, increased fires may provide some benefit to browsers, as resprouting from the base of the tree after aboveground parts are killed by fire (topkill), may increase the available browse within browse height.

13.3.4 Ecosystem Pattern and Process

Fire × herbivory will, through its influence on herbaceous and woody dynamics, influence ecosystem patterns and process, including biodiversity (bottom box in Fig. 13.1). Studies on various taxonomic groups have shown that fire and herbivory interaction can have species-specific positive or negative effects, with some species seemingly unaffected. For example, greater spatial heterogeneity, resulting from pyric herbivory, increases bird diversity and stability over time in tallgrass prairie (Hovick et al. 2015). Similarly, while different fire severity regimes had no effect on arthropod richness and abundance in north American aspen forest, intermediate-severity fire, combined with moderate elk herbivory, resulted in higher richness and abundance of arthropods, whilst severe fire and intense herbivory resulted in decreases (Bailey and Whitham 2002). However, the interaction between fire and herbivory does not always have consequences for biodiversity; for instance butterfly species richness seemingly does not respond to pyric herbivory in north American grasslands (Moranz et al. 2012).

We argue that the ecosystem patterns and processes, that are influenced through fire × herbivory, will also exert influence on grazing and browsing behaviour through various feedbacks. For example, it has been shown in African savannas that grazers may, at night, actively select more open and more frequently burned landscapes in order to lower their predation risk (Burkepile et al. 2013). As such, the spatio-temporal behaviour of herbivores is not only influenced by the availability of forage and water, but also the risk of foraging as influenced by their visibility as related to woody cover and grass structure, and the associated predation risk ("landscape of fear"). This can possibly have further trophic cascading effects, as herbivores can redistribute nutrients due to where they spend time as a result of avoiding areas of higher predation risk (Le Roux et al. 2018). In another study, Feng et al. (2012) showed how predation of moose (*Alces alces*) by grey wolf (*Canus lupus*) influences the flammability of a system. In the absence of wolves, moose numbers increase, and they selectively browse low-toxicity deciduous woody plants, thereby speeding succession towards high toxicity evergreens, such as spruce, culminating in increased overall landscape flammability. An example from Australia illustrates how the removal of an exotic predator, in the absence of any native predators, resulted in high abundance of native wallabies (*Wallabia bicolor*) in Booderee National Park, which may inhibit the successful recruitment of many plant species after fire-induced recruitment events (Dexter et al. 2013). As such, the complex interplay between fire, grazers, browsers and ecosystem processes, such as predation, are all interrelated, with various positive and negative feedbacks operating at a range of spatio-temporal scales, influencing vegetation dynamics and biodiversity.

13.3.5 Strength of Fire × Herbivore as Function of Size of Fire

The size of fires seems to be important for how both grazers and browsers subsequently influence fire behaviour (Archibald 2008; Archibald et al. 2005; Donaldson et al. 2018). In general, localised herbivore density, and hence their effect on fire behaviour, is higher in smaller resprouting grass/woody patches, as compared to large burnt areas where herbivory is less concentrated. For example, if relatively small areas in forests or shrublands burn, browsers may have a bigger concentration effect on the resulting post-fire fuel complexes (Hobbs 1996). Furthermore, with small fires, herbivores can possibly influence the successional trajectory by targeting preferred plant species—in contrast, for large fires, many preferred plant species will have the opportunity to escape herbivory, which may influence the ultimate flammability of the post-burn area. For example, large fires tend to disperse deer and cattle, but not elk, which results in positive effects on aspen regeneration and recruitment (Wan et al. 2014). Similarly, Donaldson et al. (2018) found that fires, less than 25 ha, attracted sufficient grazing herbivores in the Kruger National Park, South Africa, to shorten grass height, and to initiate positive feedbacks and maintain grazing lawns after only two seasons of burning, ultimately excluding repeat fires on those patches in the third year.

13.4 Management

From the conceptual models discussed above, as well as the wealth of studies, of which only some could be included here, it is clear that fire and herbivory need to be considered jointly by the managers of fire-prone landscapes (Smit and Archibald 2019). Management practices will depend on the objective(s) of the land manager, e.g., curbing bush encroachment, increasing land productivity, increasing biodiversity, increasing system resilience, reducing fire risk, etc. Various management tools such as (i) rewilding (reintroducing locally extinct fauna); (ii) mimicking historical fire and herbivory regimes through prescribed burning, livestock herbivory and mowing; and (iii) spatio-temporal manipulation of fire and herbivory, are increasingly employed in order to manipulate or restore fire and herbivory interactions, and to achieve specific objectives (Allred et al. 2014; Leverkus et al. 2018; Limb et al. 2011).

Rewilding, i.e., reintroducing locally extinct fauna or fauna that can fulfil the function of extinct fauna, is seen as a way to restore trophic processes and has been proposed as a potential way to reduce fires (Johnson et al. 2018). Venter et al. (2017) propose that the rewilding of African rangelands, with mid-Holocene herbivory pressure and diversity, may decrease the importance of fire and enhance nutrient cycling, conserve soil nutrient pools and benefit productivity. This will ultimately change African rangelands from a fire-dominated to an herbivore-dominated state (see also Holdo et al. 2009; Smit and Archibald 2019). Similarly, Johnson et al. (2018) argue that rewilding Australia, with medium-sized marsupials and rodents that dig for food and, in the process, mix litter with top soil, could reduce fire risk and at the same time also increase soil condition.

Where it is impractical to have natural fire and herbivory regimes, or to undertake rewilding, for example in small reserves or agricultural settings, management could aim to mimic these processes. This may include prescribed burning and herbivory by domestic ungulates, or even defoliating grass through mowing. Although domestic herbivores can, in some cases, be used as a substitute for native herbivores (e.g., cattle for bison; Doughty et al. 2010), in other cases livestock represent a different suite of functional traits from the original herbivore community (Archibald and Hempson 2016), which may lead to unintended cascading effects, e.g., decreases in soil carbon storage when livestock replaces native wildlife (Bagchi and Ritchie 2010).

Woody thickening is a major concern in many grassy systems worldwide (Saintilan and Rogers 2015 and references therein). Woody encroachment is important as it influences agricultural productivity (e.g., Anadón et al. 2014), as well as biodiversity (Smit and Prins 2015; Stanton et al. 2018). Various studies have shown that both fire and browsing are required to suppress woody vegetation, as an initial fire will reduce the vegetation height (top-kill), but follow-up browsing is important to slow browse plants' recruitment and growth and to keep plants within browse heights (Staver et al. 2009).

There is indirect evidence that agricultural practises, such as livestock grazing and mowing, have made possible the continued existence of many species threatened by the reduction in native herbivores by humans (Pykälä 2000). As partial surrogates, traditional livestock grazing and mowing have obscured the importance of natural disturbances to European biodiversity. Thus, the end of traditional animal husbandry, together with the suppression of natural disturbances, may cause even more adverse effects to biodiversity than is generally recognized (Pykälä 2000). Due to farm abandonment, and replacement with other land uses (e.g., forestry), destructive fires have increased—here rewilding has also been proposed as a solution (Dubinin et al. 2011; Navarro and Pereira 2015). In other areas, where native herbivores and livestock have disappeared, management fires are used to reduce fuel build up, and to reduce woody encroachment. However, this practise, of applying fire in isolation of herbivory, in order to reduce fuels, is arguably inappropriate, and may compromise long-term sustainability of the system (e.g., reduced soil fertility with increased fire; San Emeterio et al. 2016). Furthermore, fires may not be that effective for curbing shrub/bush encroachment, as fire, without guided herbivory, has proved largely ineffective in controlling shrub encroachment (Ascoli et al. 2013; Garbarino et al. 2013). As such, managers in places like southern Europe, are increasingly cognisant of the interactive effects of fire and herbivory, and are incorporating livestock grazing and browsing into fire prevention activities. It is argued that this reduces both the fuel loads, and the high cost of mechanical clearance of encroaching shrubs (Leytón and Vicente 2012; Mena et al. 2016; Varela et al. 2018).

Recent work in grasslands has promoted the idea that sequential positive and negative feedbacks between fire and grazing cause a shifting mosaic of disturbance patches in various post-fire and grazing-intensity successional stages across the landscape (e.g., Fuhlendorf and Engle 2004). These mosaics differ, *inter alia*, in nutritional and flammability qualities, and the resultant heterogenous landscape is argued to have high overall biodiversity and productivity (Engle et al. 2008; Fuhlendorf et al. 2006, 2009; Fuhlendorf and Engle 2001; Krook et al. 2007; **Fynn et al.** Chap. 14). Properly planned management fires can manipulate both herbivore and fire outcomes. For example, in the Kruger National Park, South Africa, an experiment was conducted to see whether a fire-driven, tallgrass system can, at least on a small spatial and temporal scale, be converted to a short grass, herbivore-driven system (Donaldson et al. 2018). This was successfully achieved by creating small fires, and ensuring the absence of large landscape fires in the surrounding region. These small fires attracted native grazers, which after 2 years of fire application, resulted in grazing lawns forming, excluding further fires. In another study (Allred et al. 2014), it was shown that spatial heterogeneity, created through patchy burning in the Great Plains grasslands, ensured stable livestock production through time as productivity was not related to precipitation, as opposed to livestock productivity decreasing with reduced rain when heterogeneity was absent (see also **Fynn et al.** Chap. 14). Research on East African savannas has shown that prescribed burning enhances cattle nutrition, but not when cattle shared grazing with native meso-herbivores (Odadi et al. 2017). However, a mosaic of burned and unburned areas, minimise fire-driven negative interactions between cattle and native meso-herbivores as cattle can use unburned refuge areas and burns draw wildlife away from important cattle foraging areas.

13.5 Conclusion

Herbivores can influence fire regimes directly through removal of fuels or indirectly through changing fuel structure, species composition (with different flammability characteristics), successional trajectories, or the creation of bare patches. Similarly, fires can influence herbivores by attracting them to recently burned areas. Fire and herbivory, or the lack thereof, are therefore, both drivers of, and responders to, each other, often through complex feedbacks. Generally, in systems where fires spread through grass fuels, an increase in herbivory will result in reduced fire frequency and intensity. The effect of herbivory is more variable in systems where fires spread through woody biomass, and this is mostly through indirect effects on species composition, which may ultimately alter the flammability. Considering the interaction and feedbacks between fire and herbivory, management should consider these processes simultaneously, in order to address specific concerns, e.g., reducing fire risk or curbing woody thickening. Actions like rewilding, and mimicking or manipulating fire and herbivore spatio-temporal patterns, may be important to achieve these outcomes.

Acknowledgements Brian van Wilgen for commenting on an earlier draft.

References

Allred BW, Fuhlendorf SD, Engle DM, Elmore RD (2011) Ungulate preference for burned patches reveals strength of fire–grazing interaction. Ecol Evol 1:132–144

Allred BW, Scasta JD, Hovick TJ, Fuhlendorf SD, Hamilton RG (2014) Spatial heterogeneity stabilizes livestock productivity in a changing climate. Agric Ecosyst Environ 193:37–41

Anadón JD, Sala OE, Turner BL, Bennett EM (2014) Effect of woody-plant encroachment on livestock production in North and South America. Proc Natl Acad Sci 111:12948–12953

Archibald S (2008) African grazing lawns—how fire, rainfall, and grazer numbers interact to affect grass community states. J Wildl Manag 72:492–501

Archibald S, Hempson GP (2016) Competing consumers: contrasting the patterns and impacts of fire and mammalian herbivory in Africa. Philos Trans R Soc B 371:20150309

Archibald S, Bond WJ, Stock WD, Fairbanks DHK (2005) Shaping the landscape: fire–grazer interactions in an African savanna. Ecol Appl 15:96–109

Ascoli D, Beghin R, Ceccato R, Gorlier A, Lombardi G, Lonati M, Marzano R, Bovio G, Cavallero A (2009) Developing an adaptive management approach to prescribed burning: a long-term heathland conservation experiment in north-west Italy. Int J Wildland Fire 18:727–735

Ascoli D, Lonati M, Marzano R, Bovio G, Cavallero A, Lombardi G (2013) Prescribed burning and browsing to control tree encroachment in southern European heathlands. For Ecol Manag 289:69–77

Asner GP, Elmore AJ, Olander LP, Martin RE, Harris AT (2004) Grazing systems, ecosystem responses, and global change. Annu Rev Environ Resour 29:261–299
Asner GP, Levick SR, Kennedy-Bowdoin T, Knapp DE, Emerson R, Jacobson J, Colgan MS, Martin RE (2009) Large-scale impacts of herbivores on the structural diversity of African savannas. Proc Natl Acad Sci 106:4947–4952
Augustine DJ, McNaughton SJ (2004) Regulation of shrub dynamics by native browsing ungulates on East African rangeland. J Appl Ecol 41:45–58
Bagchi S, Ritchie ME (2010) Introduced grazers can restrict potential soil carbon sequestration through impacts on plant community composition. Ecol Lett 13:959–968
Bailey JK, Whitham TG (2002) Interactions among fire, aspen, and elk affect insect diversity: reversal of a community response. Ecology 83:1701–1712
Bakker ES, Pagès JF, Arthur R, Alcoverro T (2016) Assessing the role of large herbivores in the structuring and functioning of freshwater and marine angiosperm ecosystems. Ecography 39:162–179. https://doi.org/10.1111/ecog.01651
Barnosky AD, Koch PL, Feranec RS, Wing SL, Shabel AB (2004) Assessing the causes of late pleistocene extinctions on the continents. Science 306:70–75
Bartolomé J, Plaixats J, Piedrafita J, Fina M, Adrobau E, Aixàs A, Bonet M, Grau J, Polo L (2011) Foraging behavior of Alberes cattle in a Mediterranean forest ecosystem. Rangel Ecol Manag 64:319–324
Beekman JH, Prins HHT (1989) Feeding strategies of sedentary large herbivores in East Africa, with emphasis on the African buffalo, Syncerus caffer. Afr J Ecol 27:129–147
Bond WJ (2005) Large parts of the world are brown or black: a different view on the 'Green World' hypothesis. J Veg Sci 16:261–266
Bowman DM, Balch J, Artaxo P, Bond WJ, Cochrane MA, D'antonio CM, DeFries R, Johnston FH, Keeley JE, Krawchuk MA (2011) The human dimension of fire regimes on Earth. J Biogeogr 38:2223–2236
Brose P, Schuler T, Van Lear D, Berst J (2001) Bringing fire back: the changing regimes of the Appalachian mixed-oak forests. J For 99:30–35
Buechner HK, Dawkins HC (1961) Vegetation change induced by elephants and fire in Murchison Falls National Park, Uganda. Ecology 42:752–766
Burkepile DE, Burns CE, Tambling CJ, Amendola E, Buis GM, Govender N, Nelson V, Thompson DI, Zinn AD, Smith MD (2013) Habitat selection by large herbivores in a southern African savanna: the relative roles of bottom-up and top-down forces. Ecosphere 4:1–19
Burkepile DE, Thompson DI, Fynn RW, Koerner SE, Eby S, Govender N, Hagenah N, Lemoine NP, Matchett KJ, Wilcox KR (2016) Fire frequency drives habitat selection by a diverse herbivore guild impacting top–down control of plant communities in an African savanna. Oikos 125:1636–1646
Burkepile DE, Fynn RW, Thompson DI, Lemoine NP, Koerner SE, Eby S, Hagenah N, Wilcox KR, Collins SL, Kirkman KP (2017) Herbivore size matters for productivity–richness relationships in African savannas. J Ecol 105:674–686
Burney DA, Flannery TM (2005) Fifty millennia of catastrophic extinctions after human contact. Trends Ecol Evol 20:395–401
Burns CE, Collins SL, Smith MD (2009) Plant community response to loss of large herbivores: comparing consequences in a South African and a North American grassland. Biodivers Conserv 18:2327–2342
Carlson PC, Tanner GW, Wood JM, Humphrey SR (1993) Fire in key deer habitat improves browse, prevents succession, and preserves endemic herbs. J Wildl Manag 57:914–928
Chauchard S, Carcaillet C, Guibal F (2007) Patterns of land-use abandonment control tree-recruitment and forest dynamics in Mediterranean mountains. Ecosystems 10:936–948
Cherry MJ, Chandler RB, Garrison EP, Crawford DA, Kelly BD, Shindle DB, Godsea KG, Miller KV, Conner LM (2018) Wildfire affects space use and movement of white-tailed deer in a tropical pyric landscape. For Ecol Manag 409:161–169

Codron J, Lee-Thorp JA, Sponheimer M, Codron D, Grant RC, de Ruiter DJ (2006) Elephant (Loxodonta africana) diets in Kruger National Park, South Africa: spatial and landscape differences. J Mammal 87:27–34

Dexter N, Hudson M, James S, MacGregor C, Lindenmayer DB (2013) Unintended consequences of invasive predator control in an Australian forest: overabundant wallabies and vegetation change. PLoS One 8:e69087

Donaldson JE, Archibald S, Govender N, Pollard D, Luhdo Z, Parr CL (2018) Ecological engineering through fire-herbivory feedbacks drives the formation of savanna grazing lawns. J Appl Ecol 55:225–235

Doughty CE, Wolf A, Field CB (2010) Biophysical feedbacks between the Pleistocene megafauna extinction and climate: the first human-induced global warming? Geophys Res Lett 37. https://doi.org/10.1029/2010GL043985

Doughty CE, Faurby S, Svenning J-C (2016) The impact of the megafauna extinctions on savanna woody cover in South America. Ecography 39:213–222

Dubinin M, Luschekina A, Radeloff VC (2011) Climate, livestock, and vegetation: what drives fire increase in the arid ecosystems of southern Russia? Ecosystems 14:547–562

Dublin HT, Sinclair ARE, McGlade J (1990) Elephants and fire as causes of multiple stable states in the Serengeti-Mara woodlands. J Anim Ecol 59(3):1147–1164

Eby S, Burkepile DE, Fynn RW, Burns CE, Govender N, Hagenah N, Koerner SE, Matchett KJ, Thompson DI, Wilcox KR (2014) Loss of a large grazer impacts savanna grassland plant communities similarly in North America and South Africa. Oecologia 175:293–303

Engle DM, Fuhlendorf SD, Roper A, Leslie DM Jr (2008) Invertebrate community response to a shifting mosaic of habitat. Rangel Ecol Manag 61:55–62

February EC, Higgins SI, Bond WJ, Swemmer L (2013) Influence of competition and rainfall manipulation on the growth responses of savanna trees and grasses. Ecology 94:1155–1164

Feng Z, Alfaro-Murillo JA, DeAngelis DL, Schmidt J, Barga M, Zheng Y, Tamrin MHBA, Olson M, Glaser T, Kielland K (2012) Plant toxins and trophic cascades alter fire regime and succession on a boreal forest landscape. Ecol Model 244:79–92

Ferwerda JG, Siderius W, van Wieren SE, Grant CC, Peel M, Skidmore AK, Prins HHT (2006) Parent material and fire as principle drivers of foliage quality in woody plants. For Ecol Manag 231(1-3):178–183

Fischer WC, Bradley AF (1987) Fire ecology of western Montana forest habitat types. Gen Tech Rep INT-223 Ogden UT US Dep Agric For Serv Intermt Res Stn 95:223

Foster CN, Barton PS, Sato CF, MacGregor CI, Lindenmayer DB (2015) Synergistic interactions between fire and browsing drive plant diversity in a forest understorey. J Veg Sci 26:1112–1123

Fuhlendorf SD, Engle DM (2001) Restoring heterogeneity on rangelands: ecosystem management based on evolutionary grazing patterns: we propose a paradigm that enhances heterogeneity instead of homogeneity to promote biological diversity and wildlife habitat on rangelands grazed by livestock. Bioscience 51:625–632

Fuhlendorf SD, Engle DM (2004) Application of the fire–grazing interaction to restore a shifting mosaic on tallgrass prairie. J Appl Ecol 41:604–614

Fuhlendorf SD, Harrell WC, Engle DM, Hamilton RG, Davis CA, Leslie DM (2006) Should heterogeneity be the basis for conservation? Grassland bird response to fire and grazing. Ecol Appl 16:1706–1716

Fuhlendorf SD, Engle DM, Kerby JAY, Hamilton R (2009) Pyric herbivory: rewilding landscapes through the recoupling of fire and grazing. Conserv Biol 23:588–598

Garbarino M, Lingua E, Weisberg PJ, Bottero A, Meloni F, Motta R (2013) Land-use history and topographic gradients as driving factors of subalpine Larix decidua forests. Landsc Ecol 28:805–817

Gill JL, Williams JW, Jackson ST, Lininger KB, Robinson GS (2009) Pleistocene megafaunal collapse, novel plant communities, and enhanced fire regimes in North America. Science 326:1100–1103

Grant RCC, Botha J, Peel MJS, Smit IPJ (2019) When less is more: heterogeneity in grass patch height support herbivores in counter intuitive ways. Afr J Range Forage Sci 36(1):1–8
Griffin GF, Friedel MH (1984) Effects of fire on central Australian rangelands. I Fire and fuel characteristics and changes in herbage and nutrients. Aust J Ecol 9:381–393
Groen TA, van Langevelde F, van de Vijver CA, Govender N, Prins HH (2008) Soil clay content and fire frequency affect clustering in trees in South African savannas. J Trop Ecol 24:269–279
Hempson GP, Archibald S, Bond WJ, Ellis RP, Grant CC, Kruger FJ, Kruger LM, Moxley C, Owen-Smith N, Peel MJS et al (2015) Ecology of grazing lawns in Africa. Biol Rev 90:979–994
Hirota M, Holmgren M, Van Nes EH, Scheffer M (2011) Global resilience of tropical forest and savanna to critical transitions. Science 334:232–235
Hobbs NT (1996) Modification of ecosystems by ungulates. J Wildl Manag 60:695–713. https://doi.org/10.2307/3802368
Holdo RM, Holt RD, Fryxell JM (2009) Grazers, browsers, and fire influence the extent and spatial pattern of tree cover in the Serengeti. Ecol Appl 19:95–109
Hovick TJ, Elmore RD, Fuhlendorf SD, Engle DM, Hamilton RG (2015) Spatial heterogeneity increases diversity and stability in grassland bird communities. Ecol Appl 25:662–672
Johnson CN (2009) Ecological consequences of late Quaternary extinctions of megafauna. Proc R Soc B 276:2509–2519
Johnson CN, Prior LD, Archibald S, Poulos HM, Barton AM, Williamson GJ, Bowman DM (2018) Can trophic rewilding reduce the impact of fire in a more flammable world? Philos Trans R Soc B 373:20170443
Klop E, Prins HHT (2008) Diversity and species composition of West African ungulate assemblages: effects of fire, climate and soil. Glob Ecol Biogeogr 17:778–787
Koch PL, Barnosky AD (2006) Late Quaternary extinctions: state of the Debate. Annu Rev Ecol Evol Syst 37:215–250
Koerner SE, Burkepile DE, Fynn RW, Burns CE, Eby S, Govender N, Hagenah N, Matchett KJ, Thompson DI, Wilcox KR (2014) Plant community response to loss of large herbivores differs between North American and South African savanna grasslands. Ecology 95:808–816
Kramer K, Groen TA, van Wieren SE (2003) The interacting effects of ungulates and fire on forest dynamics: an analysis using the model FORSPACE. For Ecol Manag 181:205–222
Krasnow KD, Stephens SL (2015) Evolving paradigms of aspen ecology and management: impacts of stand condition and fire severity on vegetation dynamics. Ecosphere 6:1–16
Krook K, Bond WJ, Hockey PA (2007) The effect of grassland shifts on the avifauna of a South African savanna. Ostrich-J Afr Ornithol 78:271–279
Lasanta T, González-Hidalgo JC, Vicente-Serrano SM, Sferi E (2006) Using landscape ecology to evaluate an alternative management scenario in abandoned Mediterranean mountain areas. Landsc Urban Plan 78:101–114
Le Roux E, Kerley GI, Cromsigt JP (2018) Megaherbivores modify trophic cascades triggered by fear of predation in an African savanna ecosystem. Curr Biol 28:2493–2499
Lehmann CE, Archibald SA, Hoffmann WA, Bond WJ (2011) Deciphering the distribution of the savanna biome. New Phytol 191:197–209
Leonard S, Kirkpatrick J, Marsden-Smedley J (2010) Variation in the effects of vertebrate grazing on fire potential between grassland structural types. J Appl Ecol 47:876–883
Leverkus SE, Fuhlendorf SD, Geertsema M, Allred BW, Gregory M, Bevington AR, Engle DM, Scasta JD (2018) Resource selection of free-ranging horses influenced by fire in northern Canada. Human–Wildlife Interact 12:10
Levick SR, Asner GP, Kennedy-Bowdoin T, Knapp DE (2009) The relative influence of fire and herbivory on savanna three-dimensional vegetation structure. Biol Conserv 142:1693–1700
Leytón JM, Vicente ÁM (2012) Biological fire prevention method: evaluating the effects of goat grazing on the fire-prone mediterranean scrub. For Syst 21:199–204
Limb RF, Fuhlendorf SD, Engle DM, Weir JR, Elmore RD, Bidwell TG (2011) Pyric–herbivory and cattle performance in grassland ecosystems. Rangel Ecol Manag 64:659–663
Mancilla-Leytón JM, Pino Mejías R, Martín Vicente A (2013) Do goats preserve the forest? Evaluating the effects of grazing goats on combustible Mediterranean scrub. Appl Veg Sci 16:63–73

McEwan RW, Dyer JM, Pederson N (2011) Multiple interacting ecosystem drivers: toward an encompassing hypothesis of oak forest dynamics across eastern North America. Ecography 34:244–256

Mena Y, Ruiz-Mirazo J, Ruiz FA, Castel JM (2016) Characterization and typification of small ruminant farms providing fuelbreak grazing services for wildfire prevention in Andalusia (Spain). Sci Total Environ 544:211–219

Midgley JJ, Lawes MJ, Chamaillé-Jammes S (2010) Savanna woody plant dynamics: the role of fire and herbivory, separately and synergistically. Aust J Bot 58:1–11

Moranz RA, Debinski DM, McGranahan DA, Engle DM, Miller JR (2012) Untangling the effects of fire, grazing, and land-use legacies on grassland butterfly communities. Biodivers Conserv 21:2719–2746

Navarro L, Pereira H (2015) Rewilding abandoned landscapes in Europe. In: Rewilding European landscapes. Springer, Cham

Nuttle T, Royo AA, Adams MB, Carson WP (2013) Historic disturbance regimes promote tree diversity only under low browsing regimes in eastern deciduous forest. Ecol Monogr 83:3–17

Odadi WO, Kimuyu DM, Sensenig RL, Veblen KE, Riginos C, Young TP (2017) Fire-induced negative nutritional outcomes for cattle when sharing habitat with native ungulates in an African savanna. J Appl Ecol 54:935–944

Pachzelt A, Forrest M, Rammig A, Higgins SI, Hickler T (2015) Potential impact of large ungulate grazers on African vegetation, carbon storage and fire regimes. Glob Ecol Biogeogr 24:991–1002

Pausas JG, Keeley JE (2009) A burning story: the role of fire in the history of life. Bioscience 59:593–601

Prins HHT, Van der Jeugd HP (1993) Herbivore population crashes and woodland structure in East Africa. J Ecol 81:305–314

Pykälä J (2000) Mitigating human effects on European biodiversity through traditional animal husbandry. Conserv Biol 14:705–712

Raffaele E, Veblen TT, Blackhall M, Tercero-Bucardo N (2011) Synergistic influences of introduced herbivores and fire on vegetation change in northern Patagonia. Argent J Veg Sci 22:59–71

Rietkerk M, Ketner P, Stroosnijder L, Prins HHT (1996) Sahelian rangeland development; a catastrophe? J Range Manag 49(6):512–519

Rius D, Vannière B, Galop D (2009) Fire frequency and landscape management in the northwestern Pyrenean piedmont, France, since the early Neolithic (8000 cal. BP). The Holocene 19:847–859

Robinson GS, Pigott Burney L, Burney DA (2005) Landscape paleoecology and megafaunal extinction in southeastern New York State. Ecol Monogr 75:295–315

Royo AA, Collins R, Adams MB, Kirschbaum C, Carson WP (2010) Pervasive interactions between ungulate browsers and disturbance regimes promote temperate forest herbaceous diversity. Ecology 91:93–105

Rule S, Brook BW, Haberle SG, Turney CS, Kershaw AP, Johnson CN (2012) The aftermath of megafaunal extinction: ecosystem transformation in Pleistocene Australia. Science 335:1483–1486

Sachro LL, Strong WL, Gates CC (2005) Prescribed burning effects on summer elk forage availability in the subalpine zone, Banff National Park. Can J Environ Manage 77:183–193

Saintilan N, Rogers K (2015) Woody plant encroachment of grasslands: a comparison of terrestrial and wetland settings. New Phytol 205:1062–1070

San Emeterio L, Múgica L, Ugarte MD, Goicoa T, Canals RM (2016) Sustainability of traditional pastoral fires in highlands under global change: effects on soil function and nutrient cycling. Agric Ecosyst Environ 235:155–163

Shannon G, Thaker M, Vanak AT, Page BR, Grant R, Slotow R (2011) Relative impacts of elephant and fire on large trees in a savanna ecosystem. Ecosystems 14:1372–1381

Smit IPJ, Archibald S (2019) Herbivore culling influences spatio-temporal patterns of fire in semiarid savanna. J Appl Ecol 56(3):711–721. https://doi.org/10.1111/1365-2644.13312

Smit IPJ, Prins HHT (2015) Predicting the effects of woody encroachment on mammal communities, grazing biomass and fire frequency in African savannas. PLoS One 10:e0137857
Stanton RAJ, Boone WW IV, Soto-Shoender J, Fletcher RJ Jr, Blaum N, McCleery RA (2018) Shrub encroachment and vertebrate diversity: a global meta-analysis. Glob Ecol Biogeogr 27:368–379
Staver AC, Bond WJ, Stock WD, Van Rensburg SJ, Waldram MS (2009) Browsing and fire interact to suppress tree density in an African savanna. Ecol Appl 19:1909–1919
Staver AC, Archibald S, Levin SA (2011) The global extent and determinants of savanna and forest as alternative biome states. Science 334:230–232
Staver AC, Botha J, Wigley-Coetsee C (2019) Grazer movements exacerbate grass declines during drought in an African savanna. J Ecol 107(3):1482–1491. https://doi.org/10.1111/1365-2745.13106
Valese E, Conedera M, Held AC, Ascoli D (2014) Fire, humans and landscape in the European Alpine region during the Holocene. Anthropocene 6:63–74
van der Waal C, de Kroon H, Heitkönig IM, Skidmore AK, van Langevelde F, de Boer WF, Slotow R, Grant RC, Peel MJS, Kohi EM et al (2011a) Scale of nutrient patchiness mediates resource partitioning between trees and grasses in a semi-arid savanna. J Ecol 99:1124–1133
van der Waal C, Kool A, Meijer SS, Kohi E, Heitkönig IM, de Boer WF, van Langevelde F, Grant RC, Peel MJS, Slotow R et al (2011b) Large herbivores may alter vegetation structure of semi-arid savannas through soil nutrient mediation. Oecologia 165:1095–1107
Van Langevelde F, Van De Vijver CA, Kumar L, Van De Koppel J, De Ridder N, Van Andel J, Skidmore AK, Hearne JW, Stroosnijder L, Bond WJ et al (2003) Effects of fire and herbivory on the stability of savanna ecosystems. Ecology 84:337–350
Vanak AT, Shannon G, Thaker M, Page B, Grant R, Slotow R (2012) Biocomplexity in large tree mortality: interactions between elephant, fire and landscape in an African savanna. Ecography 35:315–321
Varela E, Górriz-Mifsud E, Ruiz-Mirazo J, López-i-Gelats F (2018) Payment for targeted grazing: integrating local shepherds into wildfire prevention. Forests 9:464–483
Venter ZS, Hawkins H-J, Cramer MD (2017) Implications of historical interactions between herbivory and fire for rangeland management in African savannas. Ecosphere 8:e01946
Vinton MA, Hartnett DC, Finck EJ, Briggs JM (1993) Interactive effects of fire, bison (Bison bison) grazing and plant community composition in tallgrass prairie. Am Midl Nat 129(1):10–18
Waldram MS, Bond WJ, Stock WD (2008) Ecological engineering by a mega-grazer: white rhino impacts on a South African savanna. Ecosystems 11:101–112
Wan HY, Olson AC, Muncey KD, Clair SBS (2014) Legacy effects of fire size and severity on forest regeneration, recruitment, and wildlife activity in aspen forests. For Ecol Manag 329:59–68
Western D, Maitumo D (2004) Woodland loss and restoration in a savanna park: a 20-year experiment. Afr J Ecol 42:111–121
Wigley BJ, Fritz H, Coetsee C, Bond WJ (2014) Herbivores shape woody plant communities in the Kruger National Park: lessons from three long-term exclosures. Koedoe 56:1–12
Williams JW, Shuman BN, Webb T (2001) Dissimilarity analyses of late-Quaternary vegetation and climate in eastern North America. Ecology 82:3346–3362
Williamson GJ, Murphy BP, Bowman DM (2014) Cattle grazing does not reduce fire severity in eucalypt forests and woodlands of the Australian Alps. Austral Ecol 39:462–468
Young TP, Okello BD, Kinyua D, Palmer TM (1997) KLEE: a long-term multi-species herbivore exclusion experiment in Laikipia, Kenya. Afr J Range Forage Sci 14:94–102

Chapter 14
Managing Browsing and Grazing Ungulates

Richard W. S. Fynn, David J. Augustine, and Samuel D. Fuhlendorf

14.1 Introduction

Browsing and grazing ungulates comprise a diverse guild of herbivores, including domestic species such as goats (*Capra aegagrus*), sheep (*Ovis aries*), cattle (*Bos taurus*), donkeys (*Equus africanus*), horses (*Equus ferus*) and camels (*Camelus spp.*) and wild species, such as cervids (deer, moose), equids (zebra, wild horses), and bovids (wildebeest, impala, buffalo, bison, etc.). They occur across most of the earth's landscapes from intensively managed, planted pasture systems to extensive pastoral transhumance systems, and sedentary and migratory wildlife systems with little or no management (Fig. 14.1). The current distribution of individual ungulate species depends upon their historical range, prevailing environmental conditions, human population densities and associated socio-economic factors, as well as the objectives of management, such as biodiversity conservation, tourism, subsistence, commercial livestock products and game ranching (Gordon et al. 2004).

Domestic ungulate species play a major role in food production and the economies of most of the cultures of the world, yielding around 20% of the world's total food production (FAO 2013). Similarly, the livelihoods of millions of rural people are dependent upon livestock that range from subsistence systems to commercial ranching operations in developed countries. Across the globe wild ungulates such as

R. W. S. Fynn (✉)
Okavango Research Institute, University of Botswana, Maun, Botswana
e-mail: rfynn@ori.ub.bw

D. J. Augustine
USDA/ARS/NPA, Crops Research Laboratory, Fort Collins, CO, USA
e-mail: David.Augustine@ars.usda.gov

S. D. Fuhlendorf
Natural Resource Ecology and Management, Oklahoma State University, Stillwater, OK, USA
e-mail: sam.fuhlendorf@okstate.edu

I. J. Gordon, H. H. T. Prins (eds.), *The Ecology of Browsing and Grazing II*,
Ecological Studies 239, https://doi.org/10.1007/978-3-030-25865-8_14

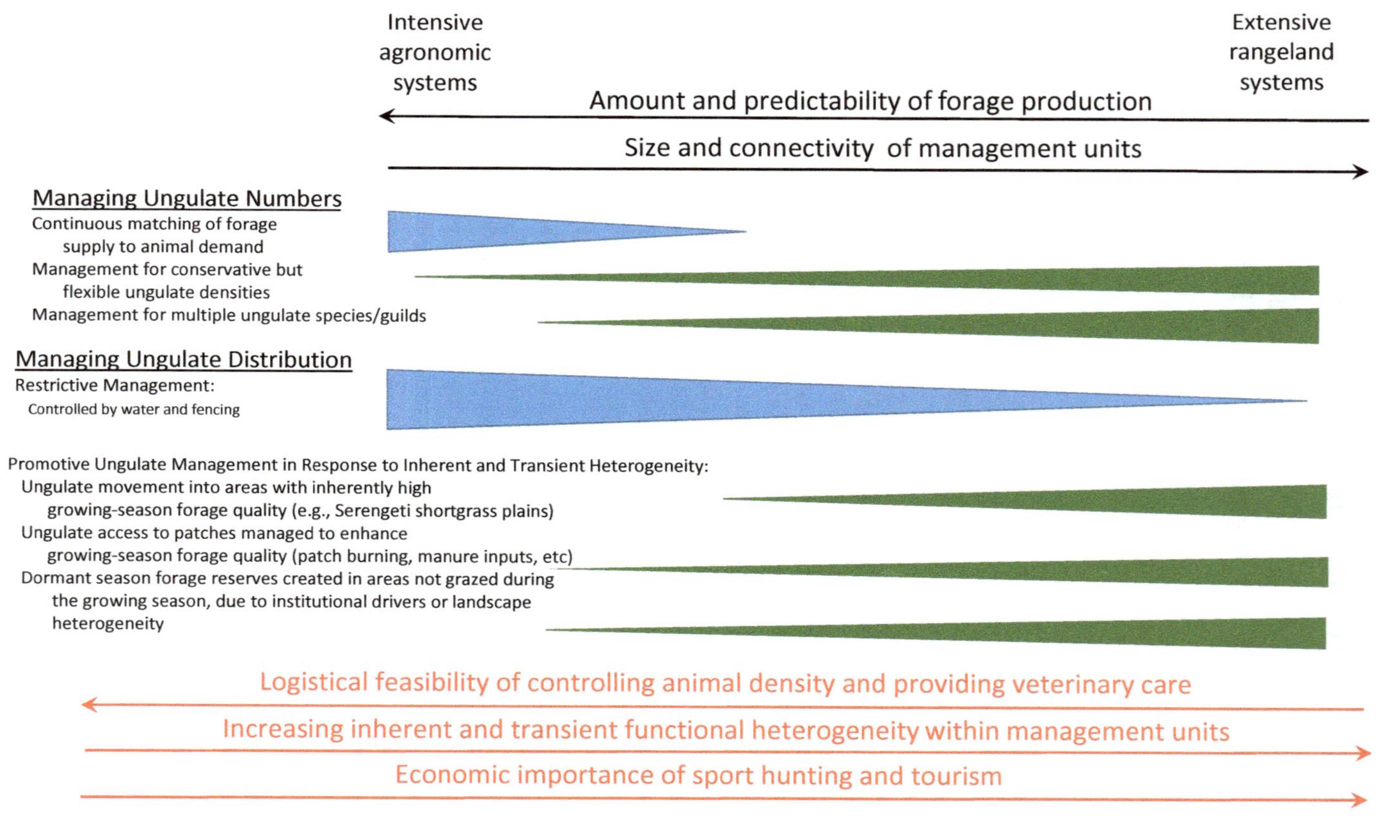

Fig. 14.1 Approaches to the management of ungulate abundance and mobility across a gradient from intensive agronomic systems to extensive rangeland systems, emphasizing the increasing importance of allowing ungulates to move in response to inherent and transient heterogeneity in extensive systems. Across this gradient, management emphasis shifts from restrictive to promotive management of ungulate foraging decisions in relation spatial heterogeneity

elephant (*Loxodonta africana* and *Elephas maximus*), wildebeest (*Connochaetes taurinus*), buffalo (*Syncerus caffer*), bison (*Bison bison*), elk (*Cervus elaphus*) and caribou (*Rangifer tarandus*) are important for tourism-based economies in many regions, including providing a prey base supporting iconic predators such as lion (*Panthera leo*), leopard (*Panthera pardus*), tiger (*Panthera tigris*) and grey wolf (*Canis lupus*). For example, wildlife tourism in Botswana contributed 11% to GDP in 2016 (WTTC 2017), in a country with a large beef industry as well as being the world's largest exporter of diamonds. Tourism in Tanzania is driven largely by its world-renowned wildlife, and is now the nation's leading economic sector, contributing 14% to GDP in 2014 (WTTC 2015). Yellowstone National Park, in the USA, attracted over 4.25 million visitors in 2016 (US National Park Service 2017). Apart from their economic importance, ungulates make up a large proportion of large mammal biodiversity and play an important role in shaping plant communities and ecosystem processes (Frank et al. 1998; Augustine et al. 2003; Augustine and McNaughton 2006; Fornara and Du Toit 2007; **Mishra et al.** Chap. 7). Their mobility requires conserving vast tracts of land, which is important for overall biodiversity conservation.

Having established the importance of browsing and grazing ungulates for livelihoods of people, and for biodiversity conservation, over a wide range of systems, from intensively-managed livestock systems to extensive migratory wildlife systems, and from arctic tundra to tropical savannas, an important question is whether there are some general concepts, and principles, that apply globally to their management? Here, we examine several concepts related to the ecology and productivity of ungulate populations and apply these concepts to their management. Central to the focus of this Chapter is the foundational role of resource heterogeneity in supporting ungulate populations, wild or domestic. The importance of resource heterogeneity for maintaining biodiversity, ecosystem processes and the productivity and stability of ungulate populations, is now well recognised (Illius and O'Connor 2000; Owen-Smith 2002, 2004; Augustine et al. 2003; Augustine and McNaughton 2006; Hobbs et al. 2008; Fuhlendorf et al. 2009, 2017; Fynn et al. 2015, 2016). In this Chapter we discuss the functional aspects of resource heterogeneity for ungulates, how this heterogeneity is distributed on ecological gradients, and temporally by disturbance, and the implications for management, at various scales.

14.2 Key Concepts Needed for the Management of Browsing and Grazing Ungulates

Ungulates require a balanced diet of energy, protein and minerals, which is rarely achievable at one given location (Prins and van Langevelde 2008). Moreover, overall intake requirements, and the components of a balanced diet, are not constant but vary seasonally with increasing demands for energy, protein and different minerals to support pregnancy and lactation (Murray 1995; Owen-Smith 2004; Prins and van Langevelde 2008). In addition, the quantity and quality of forage varies over both space and time, and is closely linked to the density and distribution of ungulates. For example, large declines in available plant biomass, and its protein and energy content, over the dry or winter seasons, often lead to deficits in nutrient intake relative to maintenance requirements (Ellis and Swift 1988; Prins 1996; Frank et al. 1998; Owen-Smith 2008; Parker et al. 2009). All systems can experience periodic droughts or other forms of extreme weather events that, depending on the duration and on ungulate densities, can result in acute deficits in nutrient intake relative to maintenance requirements, leading to mortality and population collapses; these are, however, most common in arid and semi-arid regions (Walker et al. 1987; Ellis and Swift 1988).

In this Section, we provide a brief review of how heterogeneity of functionally-different resources enables ungulates to adapt to seasonally varying physiological and biophysical constraints on meeting nutrient intake requirements. We also review how herbivory may interact with heterogeneity to promote biodiversity.

14.2.1 Inherent and Induced Sources of Functional Resource Heterogeneity for Ungulates

Functional resource heterogeneity refers to spatiotemporal variability in forage resources, whereby different patches in the landscape vary in their probability of meeting the specific nutritional needs of a grazing or browsing ungulate over time (between seasons or years). The functional nature of forage resources may be characterized by spatial or temporal differences in the biomass, structure/height, phenology, species composition and nutrient or chemical properties of forage. These differences may be driven by soil, geological, mean annual rainfall or hydrological variation fixed in space (inherent heterogeneity), or by patchy effects of rainfall, fire or grazing across space and time (induced heterogeneity). Inherent forms of functional heterogeneity are generally associated with ecological gradients arising from topographic, edaphic, and climatic variability, at various scales—including regional—or landscape-scale variation in soil moisture/salinity, across rainfall or flooding gradients (Bell 1970; McNaughton and Banyikwa 1995; Hopcraft et al. 2010; Fynn et al. 2015; Raynor et al. 2017), or altitudinal/temperature gradients in North America (Festa-Bianchet 1988; Frank et al. 1998), Fennoscandia (Albon and Langvatn 1992; Marell et al. 2006) and Asia (Omer et al. 2006). High-quality forages, supporting ungulate growth and reproduction, are mainly found, seasonally, in the less productive regions of ecological gradients, such as low-rainfall, saline regions and uplands, where grasses are shorter and more digestible, and support higher concentrations of minerals for pregnant and lactating animals (e.g., Murray 1995; Grant and Scholes 2006; Fynn et al. 2014). A key point is that important minerals, such as Ca, Na and P, are not at sufficient concentrations in forage across the broad landscape so animals must find specific sites where the concentration of these minerals is elevated (Murray 1995; Prins and van Langevelde 2008). This results in the migration of some species of ungulates to extensive saline regions at a regional scale (Murray 1995; Fynn et al. 2014), or movement on landscape catena's to saline patches (sodic sites) at a landscape scale (Grant and Scholes 2006).

More reliable dry/winter season forages are usually found in more productive regions of the gradient, such as lowlands, high rainfall areas and wetlands (Frank et al. 1998; Hopcraft et al. 2010; Fynn et al. 2015). Importantly, wetlands and high rainfall regions can provide green forage during the dry season when most landscapes support only dry, low-quality forage (Prins 1996; Hopcraft et al. 2010; Fynn et al. 2015). Thus, meeting maintenance requirements during resource-limited seasons, or extreme climatic events, requires access to forage resources in these more productive habitats (wetlands, lowlands and high-rainfall regions), that maintain adequate quantity and quality (Prins 1996; Owen-Smith 2002; Hopcraft et al. 2010; Fynn et al. 2015). When these areas are discrete, within the landscape, they are commonly known as reserve and key/buffer resources (Prins 1996; Illius and O'Connor 2000; Owen-Smith 2002; Hopcraft et al. 2010), and can be essential for reducing the rate of use of body stores and preventing mortality (Illius and O'Connor 2000; Owen-Smith 2002; Hopcraft et al. 2010). To summarize the salient points of the discourse so far, it appears that, in contrast to the commonly-held view of key resources being limited to dry season reserves of forage (Illius and O'Connor 2000), that there are two distinctly-different kinds of key resources, each acting on different demographic parameters: (1) high-quality wet season key resources, supporting animal growth and reproduction; and (2) adequate-quality dry season key resources, minimizing weight loss and mortality (see Selebatso et al. 2018).

Apart from functional heterogeneity being distributed in fixed locations across ecological gradients (inherent), it may also occur as transient (induced) heterogeneity through stochastic, patchy rainfall or fire events that stimulate growth of fresh, high-quality green forage, actively sought out by various ungulate species (Verlinden and Masogo 1997; Frank et al. 1998; Fryxell et al. 2005; Fuhlendorf et al. 2009; Sensenig et al. 2010; Mueller et al. 2011; Raynor et al. 2017). In addition, grazing and browsing by ungulates may remove older, low-quality grass or browse and stimulate fresh high-quality regrowth (Vesey-FitzGerald 1960; Fornara and du Toit 2007; Odadi et al. 2011), thereby increasing subsequent forage quality, and the weight gain of ungulates (Gordon 1988; Odadi et al. 2011). In some situations, ungulate foraging may even create grazing or browsing lawns supporting higher quality leafy forage (Augustine and McNaughton 2006; Verweij et al. 2006; Fornara and du Toit 2007; Anderson et al. 2010), where growth of new forage may be stimulated relative to ungrazed or unbrowsed areas (McNaughton 1985; Fynn et al. 2017; Fornara and du Toit 2007).

Functional heterogeneity of resources may also be facilitated by the diversity of plants in a landscape or the same patch. DNA metabarcoding has revealed a wide range of plant species, and a high diversity of taxonomically similar plants (e.g., grasses), in the diets of a wide range of ungulates (Kartzinel et al. 2015), indicating that ungulates rely on a varied diet to meet their daily nutritional requirements, among other needs (see Prins and van Langevelde 2008). A critical point, rarely recognised in the foraging ecology of ungulates, nor in optimal foraging theory, is that ungulates may select a diversity of plant species, not only for energy gain or for nutritional purposes, but also, for medicinal purposes. For example, different plant species may contain different macro- and micronutrients needed for specific physiological processes, or contain chemical compounds with anti-parasite properties (Lange et al. 2006; Shaik et al. 2006; Moreno et al. 2012; Mengistu et al. 2017). In addition, intake of forage may be stimulated by consumption of plant chemical compounds counteracting toxins in forage that retard intake; some plant species may contain compounds that nullify the effect of specific toxins (Provenza et al. 2003), or by foraging on different plant species, each containing different toxins that are complementary, a ungulate may obtain greater overall intake than by foraging on only one species containing only one of the toxins but which has a detrimental effect at higher doses (Provenza et al. 2003). Thus exposure of livestock to a greater diversity of plants has been shown to increase intake and stimulate milk production (Meuret and Dumont 1999; Provenza et al. 2003; Agreil et al. 2006).

Different plant species also allow diet breadth expansion over the annual cycle, as an adaptive foraging mechanism, to seasonal variation in the quantity and quality of individual plant species (Owen-Smith 2002). For example, grazers and browsers may shift their diets between different plant species, which meet different metabolic and nutritional needs, over the day's foraging activities, and over the annual cycle (Agreil et al. 2006; Owen-Smith 2002). The mechanistic basis for these diets shifts by browsers or grazers, over the annual cycle, is best understood by translating plant species-level selections into functionally different resource types, known as generic functional resources (Owen-Smith 2002). A combination of restricted intake, high-quality and staple functional resources are critical for animal growth and reproduction, while reserve, bridging and buffer functional resources help to maintain intake during the dry season, and may prevent population collapse during droughts (Owen-Smith 2002). For example, African buffalo were observed, during the dry season in Lake Manyara National Park, to increase protein intake by foraging on short, high-quality, *Cynodon dactylon*, lawns around the lake (restricted intake resource), but required taller sedge grasslands (reserve resource) to meet their overall intake requirements (Prins 1996).

Heterogeneity in forage resources, at landscape and regional scales, plays a crucial role in creating adaptive foraging options for ungulates over the seasonal cycle, and is a key consideration for how management actions influence the role of ungulates, and their productivity within the system (Fig. 14.1). A critical factor influencing ungulate access to inherent and induced forms of heterogeneity (adaptive foraging options), is the spatial scale available for movement, as heterogeneity may occur over hundreds of kilometres, either seasonally (Frank et al. 1998; Hopcraft et al. 2010; Fynn and Bonyongo 2011; Mueller et al. 2011), or stochastically and unpredictably within seasons (Verlinden and Masogo 1997; Fryxell et al. 2005; Mueller et al. 2011). The relationship between adaptive foraging options and spatial scale has major implications for management of ungulates, from the conservation of mobile wild ungulate species, to management of domestic ungulates on ranches, which will be discussed in later sections.

14.2.2 Anthropogenic Disruption of Functional Heterogeneity

A factor that plays an important role in creating heterogeneity in ecosystems, that support ungulates, is distance to permanent water, which if large enough (>15–20 km), creates herbivory gradients and so called piosphere effects (Sianga et al. 2017). For intensively managed rangelands, the provision of water sources, at high density across the landscape, as well as the installation of fencing, can enable maximal utilization of forage by domestic livestock species (Fig. 14.1; Fuhlendorf et al. 2017), or reduce spatial variation in vegetation utilisation in wildlife systems (Sianga et al. 2017); however, they may jeopardise ecological function, and other desired ecosystem services, such as landscape-scale hydrology, grassland structural heterogeneity and native plant and animal biodiversity (Fuhlendorf et al. 2017). Also in wildlife systems, for example, in regions with large elephant populations, maintaining large proportions of the landscape >15 km from permanent water (the maximum foraging distance of elephants from water—see Wato et al. 2018) facilitates woody vegetation structural heterogeneity and plant diversity, thereby promoting faunal diversity (Sianga et al. 2017). Similarly, placement of a high density of artificial water points across the landscape, in Kruger National Park, enabled the spread of water dependent ungulates to areas far from permanent water (perennial rivers) (Harrington et al. 1999). This may have led to increased lion densities in these back country areas, and an associated loss of refuges from predation for roan (*Hippotragus equinus*) and sable antelope (*Hippotragus niger*), possibly resulting in collapse of their populations.

Similar to the effect of artificial water eliminating predation refuges far from water in the landscape, heterogeneity in the age of forest stands in North America can play a critical role in the interactions and dynamics of moose (*Alces alces*) and woodland caribou (*Rangifer tarandus caribou*). Clear cutting of forest stands stimulates regrowth of trees and increases browse availability for moose, whereas old growth forests, with their abundant lichens, support woodland caribou (Bergerud and Elliot 1986; Rettie and Messier 2000; Wittmer et al. 2005). Widespread logging programs, that homogenize forest age and structure, can promote moose populations by increasing their food resource and, more insidiously, increasing the abundance of wolves (*Canis lupus*) in former caribou habitat, leading to increased predation pressure and declining caribou populations (Bergerud and Elliot 1986; Wittmer et al. 2005).

Prior to the European settlers disrupting natural processes in North American grasslands, interactions between fire and grazing, by the large mobile bison (*Bison bison*), created structural and compositional heterogeneity in the vegetation, as herds moved in response to burning, the distribution of water sources and the temporal pattern of precipitation (Fuhlendorf et al. 2009; Allred et al. 2011; Augustine and Derner 2014). Currently, Western-based approaches to management of rangelands prevents the maintenance of grassland structural heterogeneity, in extensive regions of North and South America, as well as portions of other continents (i.e., land is compartmentalized in small homogenous units, and fire and grazing are managed in a homogeneous manner); as a result undermining ungulate interactions with heterogeneity, with negative consequences for biodiversity (Derner et al. 2009; Fuhlendorf et al. 2009, 2017).

Another anthropogenic influence on functional heterogeneity is altering the density of ungulates, either by removing predators, increasing water availability or providing supplementary feeding. Research through the twentieth century clearly showed that the largest influence of mammalian ungulates on rangeland condition, and animal performance, depends on ungulate abundance, typically quantified, for livestock, in terms of stocking rate, or the number of animals per unit space and time (Heitschmidt and Taylor 1991; Holechek et al. 2004; Briske et al. 2008). The influence of stocking rate, or more generally grazing intensity, is identified as more important than the grazing management system (which alter the timing and distribution of animals) in determining rangeland condition (Briske et al. 2008). Perennial grasses require long recovery periods, after grazing, if they are to persist in the long-term, but this is not possible at heavy stocking rates, because of the rapid depletion of available forage, forcing premature utilization of rested paddocks (i.e., paddocks without grazing) (Holechek et al. 2004; Fynn et al. 2017). In heavily stocked wildlife areas, grassland is homogenized to a short grass state, eliminating taller grass forage reserves, leading to reduced rainfall-use efficiency of plants and/or population collapse of ungulates during drought (Walker et al. 1987; Irisarri et al. 2016). Similarly, increases in wild ungulate density, in response to predator extirpation, have substantial consequences for their effects on functional plant community composition (Augustine and McNaughton 1998; Ripple and Beschta 2004). Whilst this body of research has been focused mostly on domestic animals, and typically on small experimental units, it is clear that the impacts of grazing/browsing on grassland, savanna and woodland ecosystems is largely dependent on the number of ungulates per unit of space and time (Walker et al. 1987; Briske et al. 2008). The key message is that excessive ungulate densities can reduce grassland structural heterogeneity in landscapes, remove forage reserves, increase runoff and erosion during rainfall events, and degrade overall primary productivity (Walker et al. 1987; Augustine and McNaughton 1998; Ripple and Beschta 2004; Briske et al. 2008; Irisarri et al. 2016; Fynn et al. 2017).

14.3 Applying Key Concepts for Management of Browsing and Grazing Ungulates

Whether dealing with conservation strategies for wild ungulates, or the management of domestic ungulates by pastoralists or commercial ranchers, several key concepts stand out as important. Clearly incorporating inherent, and induced, heterogeneity at landscape and regional scales, into conservation and livestock management strategies, is needed if optimal conservation goals and livestock performance are to be realized. Conservation strategies globally, and in Africa in particular, have not been successful in maintaining the historic large populations, densities and diversity of various wild ungulates (Harris et al. 2009; Craigie et al. 2010; Fynn and Bonyongo 2011). One of the principal factors underlying declines in African wildlife is that functional heterogeneity of resources required by ungulates, over the annual cycle, is distributed on various ecological gradients, often at regional scales (see Sect. 14.2.1). However, protected areas mostly do not encompass the full extent of these gradients (Caro and Scholte 2007; Harris et al. 2009; Craigie et al. 2010; Fynn and Bonyongo 2011). With rapidly-growing human populations and settlements, agricultural expansion, and the associated construction of water sources, fences and roads, these ecological gradients are becoming fragmented and ungulate management is becoming increasingly restricted to smaller regions of these gradients (e.g., Sayre et al. 2013), thus offering less functional heterogeneity; i.e., heterogeneity declines with increasing management intensity, and smaller management units (left-hand side of the gradient in Fig. 14.1).

Those protected areas that do encompass sources of heterogeneity, at multiple spatial scales (e.g., the Serengeti-Mara ecosystem in Africa, and the Greater Yellowstone ecosystem in North America), still support productive and stable ungulate populations, and diverse ungulate communities (Fryxell et al. 2005; Frank et al. 1998). These examples, buttressed by ecological theory (e.g., Owen-Smith 2002, 2004; Fryxell et al. 2005; Hopcraft et al. 2010; Mueller et al. 2011), indicate that maintaining large areas available for the movement of nomadic and migratory populations should be a key management objective for conservation outcomes in rangelands. For less mobile ungulates, that utilize resource heterogeneity across smaller landscapes areas, such as moose (Mueller et al. 2011) or Greater kudu (*Tragelaphus strepsiceros*) (Owen-Smith 2002), conservation of extensive unfragmented landscapes is less critical (Mueller et al. 2011).

In the Serengeti-Mara system, the combination of the National Park, and various surrounding wildlife management areas, collectively achieve sufficient land area under protection to encompass the full rainfall gradient, the principal factor underlying functional resource heterogeneity (both inherent and induced) in the ecosystem (McNaughton 1985; McNaughton and Banyikwa 1995). In most regions of the globe, sufficiently large areas, for effective conservation of mobile ungulates, will only be possible if local communities of people are involved in the management of greater conservation regions using a multi-stakeholder approach. It is becoming increasingly recognized that protected areas alone are insufficient to conserve biodiversity and that implementation of conservation activities, across larger landscapes, including multiple land uses, within which protected areas are imbedded, is a key strategy for conservation (e.g., Miller et al. 2012; Sayer et al. 2013; Fynn et al. 2016; Fuhlendorf et al. 2018).

In this regard, there are several flaws in the current reliance on wilderness-based conservation strategy (or "fortress conservation"), such as (1) not accounting for the major historic influence of humans in shaping ecosystems; (2) being conceptually deficient in approaches to biodiversity conservation in landscapes now dominated by humans; and (3) being responsible for fragmentation of larger, more functional landscapes and ecosystems, by focussing conservation activities on small parts of these larger landscapes (Fynn et al. 2016; Fuhlendorf et al. 2018). Clearly, conservationists need innovative new approaches for the conservation of mobile ungulates, shifting emphasis from wilderness-based protected areas to approaches that promote connectivity across larger, more functional, landscapes in which protected areas are imbedded. Such an approach is possible with the understanding that conflict between wildlife, people and livestock is largely caused by an unfavourable spatio-temporal configuration of land use and livestock management, rather than primarily by human population growth; for example, changes from mobile pastoralism to sedentary ranching, can result in declining wildlife numbers, compared with pastoral areas, despite similar human population densities (Western et al. 2009; also see Keesing et al. 2018). Achieving productive populations of both wildlife and livestock, in larger pastoral landscapes, is possible if livestock management is based on ecological concepts aimed at promoting mobility and functional heterogeneity in the ecosystem (Fynn et al. 2016).

The theoretical basis for coexistence of livestock and wildlife, namely management approaches facilitating grassland structural heterogeneity, and a favourable spatio-temporal configuration of land use and livestock management across large landscapes (Western et al. 2009; Augustine et al. 2011; Fynn et al. 2016), is gaining empirical support (see Western et al. 2009; Shamhart et al. 2012; Tyrrell et al. 2017; Keesing et al. 2018). In a classic example of how ecological theory can be effectively applied for conservation of ungulate populations in human-dominated landscapes, grazing management committees, in a pastoral region of Kenya, set aside a core ungrazed area, providing a taller dry season reserve of grass (for both wildlife and livestock), surrounded by a heavily-grazed livestock area, that provides short higher-quality grass for short grass specialist wildlife species (Tyrrell et al. 2017). Impala, wildebeest and gazelles preferred foraging in the heavily-grazed livestock area during the wet season (facilitation of higher-quality short grass by livestock—see Gordon 1988; Fynn et al. 2016) but shifted their foraging to the ungrazed area during the dry season and drought periods (Tyrrell et al. 2017). This management approach creates structural heterogeneity similar to that observed on ecological gradients (Owen-Smith 2004; Hopcraft et al. 2010) and demonstrates the value of the structural heterogeneity concept for facilitating the co-existence of livestock and wildlife (also see Shamhart et al. 2012 for a North American example).

A key point from the Kenyan example is that local communities of people were able to maintain their traditional livestock-based livelihoods while conserving wild ungulate populations, which creates opportunities for additional income from tourism (community-based natural resource management—CBNRM). The effectiveness of biodiversity conservation in larger conservation landscapes will be determined by how well local communities are integrated into the management of, and benefit flows (ecosystem services and economic gains), from these larger landscapes; bottom up decision-making, and management processes, must replace traditional, top-down, command and control paradigms, where local people received few incentives to support conservation objectives, thereby fragmenting landscapes (Miller et al. 2012; Sayer et al. 2013; Fynn et al. 2016; Fuhlendorf et al. 2018).

The functional heterogeneity concept also provides a framework for planning water provision strategies, in protected areas, by ensuring that sufficient distance exists between water points to promote heterogeneity in vegetation structure and predation risk. Taking into account, for example, the maximum foraging range of around 15 km for bull elephants from water (and probably beyond the limits of most water-dependent ungulates) (O'Connor et al. 2007), artificial water points would have to be at least 45–50 km apart to create zones of refuge from herbivory of 15–20 km across. Such a strategy will promote reserves of forage for the dry season and droughts (Walker et al. 1987; Fynn et al. 2016), provide refuges from overuse for favoured plant species (O'Connor et al. 2007; Russell et al. 2011; Sianga et al. 2017), and refuges from predation for predation-sensitive ungulates (Harrington et al. 1999). Similarly, management of the scale and heterogeneity of clear felling of forests, will affect woodland caribou and moose habitat, and their interactions with wolves. In this system the coexistence of predators and their prey depends on a heterogeneous mosaic of old growth, low-predation-risk forests (low wolf density) favouring woodland caribou, and recently-felled forest with high browse availability favouring moose and their predators (e.g., Bergerud and Elliot 1986; Rettie and Messier 2000; Wittmer et al. 2005).

Clearly, management of natural ecosystems must consider the key processes contributing to vegetation heterogeneity, at various scales. Management can influence many variables underlying heterogeneity, i.e., protected area design relative to ecological gradients (Hopcraft et al. 2010; Fynn and Bonyongo 2011), the scale and heterogeneity of clear felling of forests (Rettie and Messier 2000), and the maintenance of large distance gradients from permanent water (Russell et al. 2011; Sianga et al. 2017). Additionally, induced heterogeneity can be created by using patchy fires regimes (Fuhlendorf et al. 2009), strategic grazing by livestock (Shamhart et al. 2012; Fynn et al. 2016; Tyrrell et al. 2017), and the creation of nutrient hotspots in landscapes through corralling of livestock (Muchiru et al. 2008; Augustine et al. 2011; Porensky and Veblen 2015). Creating greater heterogeneity in plant composition and structure across landscapes, not only increases the opportunity for adaptive foraging options for ungulates, but also increases niche diversity for biodiversity (Derner et al. 2009; Fuhlendorf et al. 2009, 2017; Fynn et al. 2016; Sianga et al. 2017).

Knowledge of the availability of various generic functional resource types (Owen-Smith 2002), in relation to physical traits (body size, specialized mouth/tongue anatomy) and feeding type (browser, mixed or grazer) of different ungulate species, provides a mechanistic framework for conservation and management of ungulates. For example, large-bodied tall grass grazers, such as buffalo, are unable to obtain sufficient intake on heavily grazed lawns (a restricted intake resource for buffalo—Prins 1996), whereas wildebeest and gazelles can (a high-quality/staple resource for these short grass specialists), explaining the spatial separation of buffalo vs. wildebeest and gazelles across grazing intensity gradients (Bhola et al. 2012). Thus the generic functional resource concept provides a useful framework for facilitating conservation management decisions; for example, whether an ungulate species could be successfully introduced to a particular region, or the effect that introducing livestock grazing to an area might have on various ungulate species.

Many pastoralists understand the distribution and role of functional heterogeneity across ecological gradients, and move their livestock seasonally and/or spatially, in a similar manner to the movement patterns of wild ungulates (e.g., right side of Fig. 14.1; Breman and de Wit 1983; Homewood 2008; Sayre et al. 2013; Fynn et al. 2015). Similar to the situation with wild ungulates, traditional movement patterns of livestock, along ecological gradients and across landscapes, have been greatly fragmented by human population growth, development, agriculture, privatization and fencing of land, formation of protected areas and colonial boundaries (Niamir-Fuller 1999; Homewood 2008; Western et al. 2009; Sayre et al. 2013). It is clear that pastoralists had insight into important ecological principles and have developed survival strategies for coping with variable and unpredictable rainfall (Breman and de Wit 1983; Homewood 2008; Sayre et al. 2013; Fynn et al. 2015).

Regional-scale pastoral movement strategies are likely to help livestock keepers cope with predicted increases in rainfall variability, and more frequent droughts associated with climate change (Ogutu and Owen-Smith 2003); large-scale movements allow access to drought (buffer) resources in high-rainfall regions, or in wetlands, and the ability to spatially track patchy rainfall events (Breman and de Wit 1983; Fryxell et al. 2005; Homewood 2008). With policy trends, in Africa, and elsewhere, leading to privatisation of land that is compartmentalised into smaller units (e.g., Western et al. 2009; Sayre et al. 2013), the ability of livestock keepers to adapt to increased variability under climate change is reduced, because they can no longer access the full range of functional resource types (generic functional resources), often distributed on regional-scale ecological gradients (see Sect. 14.2.1). With there being increased emphasis globally on development of 'climate-smart agriculture' (cf. Gordon et al. 2016), and resilient approaches to dealing with climate change, the evidence and concepts presented in this Chapter clearly indicate that policy actions need to be scaling up, rather than reducing the size of management units (see also Fryxell et al. 2005; Hobbs et al. 2008; Western et al. 2009).

The concepts of spatial scale of movement and adaptive foraging, in relation to functional heterogeneity, has implications for the management of livestock on commercial ranches as well. Rangeland science has been dominated by a paradigm advocating management that facilitates spatially uniform, and moderate, utilization of forage, which often results in uniformity of vegetation composition and structure at the landscape scale (Briske et al. 2008; Fuhlendorf et al. 2017; Sayre 2017). In order to facilitate more uniform utilization of rangelands, the rangeland management profession has promoted the use of an elaborate system of fencing and small paddocks (Sayre 2017), with the establishment and maintenance costs of these fences contributing to debt and capital depreciation, reducing profits (Knight et al. 2011). Another cost of the current paradigm is that, as described above, enclosing livestock in small paddocks restricts their ability to make foraging decisions in relation to functional heterogeneity on the property (including nutritional and medicinal resources). With these factors in mind, fundamentally different approaches to rangeland management are emerging, which aim to facilitate access to functional heterogeneity through use of fewer, much larger, paddocks and through seasonal movements between short grazed, high-quality grassland during the growing period and taller, ungrazed reserves of forage for the dry season (Fynn et al. 2017). Thus this approach more accurately simulates the natural seasonal movements of ungulates, and the ecological principles underlying their foraging ecology, than complex multi-paddock systems with their regular forced movements of livestock. Prevention of grassland maturation, and associated loss of quality, through sustained season-long grazing of priority paddocks increases animal production (the grazing facilitation concept explicitly addressed; see Gordon 1988; Odadi et al. 2011), while taller reserves of forage, in ungrazed paddocks (rested the entire growing season), reduce weight loss and mortality during the dry season and droughts (Walker et al. 1987). The purported advantages of managing for uniformity, therefore, are questionable given that rangeland heterogeneity enhances adaptive foraging options for livestock and, consequently, their nutrition and production (Owen-Smith 2002, 2004; Hobbs et al. 2008; Hopcraft et al. 2010).

Access to functional heterogeneity is related to the spatial area available for movement (Hobbs et al. 2008), with the consequence that negative density-dependent effects on animal performance (Sandland and Jones 1975) may be weaker at larger spatial scales (Wang et al. 2006; Hobbs et al. 2008). This key ecological principle is disregarded (with negative consequences for animal health and production) by ranching systems favouring an elaborate system of fencing and regular movement of livestock between small paddocks (see Fynn et al. 2017). Forage, and habitat selection, are highly complex and not always based on maximizing energy and protein intake. Ungulates also need to balance protein, energy and fibre intake (e.g., Prins 1996), counter toxins in forage (e.g., Provenza et al. 2003), deal with parasites or any other medicinal requirements (Lange et al. 2006; Shaik et al. 2006; Moreno et al. 2012; Mengistu et al. 2017) and other reasons (e.g., avoiding cold winds, heat, insects, etc.). Giving livestock greater freedom to forage in large paddocks facilitates adaptive foraging choices and habitat use decisions, with obvious benefits for livestock production and health (Provenza et al. 2003; Fynn et al. 2017), provided that stocking rates are set appropriately (Briske et al. 2008). Patchworks of fire-modified vegetation can provide a valuable alternative to the use of paddocks on ranches (Limb et al. 2011). In more arid regions, surface water management provides a useful alternative to fencing for controlling the distribution of livestock.

14.4 Conclusion

The concept of functional heterogeneity is applicable to the management of both wild and domestic browsing and grazing ungulates at the full range of scales, from landscapes to regions, and from intensive small stock systems in Europe (Meuret and Dumont 1999; Agreil et al. 2006) to extensive transhumance and wildlife systems in Africa (Breman and de Wit 1983; Homewood 2008; Fynn et al. 2015), Asia (Omer et al. 2006), North America (Festa-Bianchet 1988; Rettie and Messier 2000; Fuhlendorf et al. 2009) and Fennoscandia (Albon and Langvatn 1992; Marell et al. 2006). Facilitating access to, and maintaining, functional heterogeneity, therefore, should underlie the management and conservation of browsing and grazing ungulates, whatever the location, scale of management and animal type. It should be recognized, however, that the positive influence of functional heterogeneity is not independent of other key management factors. For example, negative density-dependent effects, due to excessive animal population size, can override the positive effects of, or even degrade, functional heterogeneity. Nevertheless, it is also true that the strength of the density-dependence effect increases with the decreasing area available for movement (decreasing management unit size; e.g., Wang et al. 2006; Hobbs et al. 2008). Consequently, appropriate management of browsing and grazing ungulates must explicitly consider their population size/stocking rate (increasingly relevant with decreasing management unit area), while facilitating access to functional heterogeneity, both spatially and temporally.

Acknowledgements The authors are indebted to Herbert Prins and Iain Gordon for their helpful comments, which improved this manuscript.

References

Agreil C, Meuret M, Fritz H (2006) Adjustment of feeding choices and intake by a ruminant foraging in varied and variable environments. In: Bels V (ed) Feeding in domestic vertebrates: from structure to behaviour. CABI Publishing, Cambridge, MA, pp 302–325

Albon SD, Langvatn R (1992) Plant phenology and benefits of migration in a temperate ungulate. Oikos 65:502–513

Allred BW, Fuhlendorf SD, Engle DM, Elmore RD (2011) Ungulate preference for burned patches reveals strength of fire–grazing interaction. Ecol Evol 1(2):132–144

Anderson TM, Hopcraft JGC, Eby S et al (2010) Landscape-scale analyses suggest both nutrient and antipredator advantages to Serengeti herbivore hotspots. Ecology 91:1519–1529

Augustine DJ, Derner JD (2014) Controls over the strength and timing of fire-grazer interactions in a semi-arid rangeland. J Appl Ecol 51:242–250

Augustine DJ, McNaughton SJ (1998) Ungulate effects on the functional species composition of plant communities: herbivore selectivity and plant tolerance. J Wildl Manag 62:1165–1183

Augustine DJ, McNaughton SJ (2006) Interactive effects of ungulate herbivores, soil fertility, and variable rainfall on ecosystem processes in a semi-arid savanna. Ecosystems 9:1242–1256

Augustine DJ, McNaughton SJ, Frank DA (2003) Feedbacks between soil nutrients and large herbivores in a managed savanna ecosystem. Ecol Appl 13:1325–1337

Augustine DJ, Veblen KE, Goheen JR et al (2011) Pathways for positive cattle-wildlife interactions in semi-arid rangelands. In: Georgiadis NJ (ed) Conserving wildlife in African landscapes: Kenya's Ewaso ecosystem. Smithsonian Institution Scholarly Press, Washington DC, pp 55–71

Bell R (1970) The use of the herb layer by grazing ungulates in the Serengeti. In: Watson A (ed) Animal populations in relation to their food resources. Blackwell, Oxford, pp 111–124

Bergerud AT, Elliot JP (1986) Dynamics of caribou and wolves in northern British Columbia. Can J Zool 64:1515–1519

Bhola N, Ogutu JO, Piepho H et al (2012) Comparitive changes in density and demography of large herbivores in the Masai-Mara and its surrounding human-dominated pastoral ranches in Kenya. Biodivers Conserv 21:1509–1530
Breman H, de Wit CT (1983) Rangeland productivity and exploitation in the Sahel. Science 221:1341–1347
Briske DD, Derner JD, Brown JR et al (2008) Rotational grazing on rangelands: reconciliation of perception and experimental evidence. Rangel Ecol Manag 61:3–17
Caro T, Scholte P (2007) When protection falters. Afr J Ecol 45:233–235
Craigie I, Baillie J, Balmford A et al (2010) Large mammal population declines in Africa's protected areas. Biol Conserv 143:2221–2228
Derner JD, Lauenroth WK, Stapp P et al (2009) Livestock as ecosystem engineers for grassland bird habitat in the Western Great Plains of North America. Rangel Ecol Manag 62:111–118
Ellis JE, Swift DM (1988) Stability of African pastoral ecosystems: alternate paradigms and implications for development. J Range Manag 41:450–459
FAO (2013) FAOSTAT database. Food and Agriculture Organization of the United Nations, Rome. http://www.fao.org/faostat/en/#data. Accessed 13 Mar 2015
Festa-Bianchet M (1988) Seasonal range selection in bighorn sheep: conflicts between forage quality, forage quantity and predator avoidance. Oecologia 75:580–586
Fornara DA, Du Toit JT (2007) Browsing lawns? Responses of *Acacia nigrescens* to ungulate browsing in an African savanna. Ecology 88:200–209
Frank DA, McNaughton SJ, Tracy BF (1998) The ecology of the earth's grazing ecosystems: comparing the Serengeti and Yellowstone. Bioscience 48:513–521
Fryxell JM, Wilmshurst JF, Sinclair ARE et al (2005) Landscape scale, heterogeneity, and the viability of Serengeti grazers. Ecol Lett 8:328–335
Fuhlendorf SD, Engle DM, Kerby J et al (2009) Pyric herbivory: rewilding landscapes through the recoupling of fire and grazing. Conserv Biol 23:588–598
Fuhlendorf SD, Fynn RWS, McGranahan D et al (2017) Heterogeneity as the basis for range management. In: Briske DD (ed) Rangeland systems: processes, management and challenges. Springer Nature, New York, pp 169–196
Fuhlendorf SD, Davis CA, Elmore RD et al (2018) Perspectives on grassland conservation efforts: should we rewild to the past or conserve for the future? Philos Trans R Soc B 373:20170438
Fynn RWS, Bonyongo MC (2011) Functional conservation areas and the future of Africa's wildlife. Afr J Ecol 49:175–188
Fynn RWS, Chase M, Roder A (2014) Functional habitat heterogeneity and large herbivore seasonal habitat selection in Northern Botswana. S Afr J Wildl Res 44:1–15
Fynn RWS, Murray-Hudson M, Dhliwayo M et al (2015) African wetlands and their seasonal use by wild and domestic herbivores. Wetl Ecol Manag 23:559–581
Fynn RWS, Augustine DJ, Peel MJS et al (2016) Strategic management of livestock to improve biodiversity conservation in African savannahs: a conceptual basis for wildlife-livestock co-existence. J Appl Ecol 53:388–397
Fynn RWS, Kirkman KP, Dames R (2017) Optimal grazing management strategies: evaluating key concepts. Afr J Range Forage Sci 34:87–98
Gordon IJ (1988) Facilitation of red deer grazing by cattle and its impact on red deer performance. J Appl Ecol 25:1–10
Gordon IJ, Hester AJ, Festa-Bianchet M (2004) The management of wild large herbivores to meet economic, conservation and environmental objectives. J Appl Ecol 41(6):1021–1031
Gordon IJ, Prins HHT, Squire G (eds) (2016) Food production and nature conservation: conflicts and solutions. Routledge, London
Grant CC, Scholes MC (2006) The importance of nutrient hot-spots in the conservation and management of large wild mammalian herbivores in semi-arid savannas. Biol Conserv 130 (3):426–437
Harrington R, Owen-Smith N, Viljoen PC et al (1999) Establishing the causes of the roan antelope decline in the Kruger National Park, South Africa. Biol Conserv 90:69–78

Harris G, Thirgood S, Hopcraft JGC et al (2009) Global decline in aggregated migrations of large terrestrial mammals. Endanger Species Res 7:55–76

Heitschmidt RK, Taylor CA (1991) Livestock production. In: Heitschmidt RK, Stuth JW (eds) Grazing management: an ecological perspective. Timber Press, Portland, OR, pp 161–177

Hobbs NT, Galvin KA, Stokes CJ et al (2008) Fragmentation of rangelands: implications for humans, animals, and landscapes. Glob Environ Chang 18:776–785

Holechek JL, Pieper RD, Herbel CH (2004) Range management, principles and practices, 5th edn. Prentice Hall, Upper Saddle River, NJ

Homewood K (2008) Ecology of African pastoral societies. James Currey, Oxford

Hopcraft JGC, Olff H, Sinclair ARE (2010) Herbivores, resources and risks: alternating regulation along primary environmental gradients in savannas. Trends Ecol Evol 25:119–128

Illius AW, O'Connor TG (2000) Resource heterogeneity and ungulate population dynamics. Oikos 89:283–294

Irisarri JGN, Derner JD, Porensky LM, Augustine DJ, Reeves JL, Mueller KE (2016) Grazing intensity differentially regulates ANPP response to precipitation in North American semiarid grasslands. Ecol Appl 26:1370–1380

Kartzinel TR, Chen PA, Coverdale TC et al (2015) DNA metabarcoding illuminates dietary niche partitioning by African large herbivores. Proc Natl Acad Sci USA 112:8019–8024

Keesing F, Ostfeld RS, Okanga S et al (2018) Consequences of integrating livestock and wildlife in an African savanna. Nat Sustain 1:566–573

Knight KB, Toombs TP, Derner JD (2011) Cross-fencing on private US rangelands: financial costs and producer risks. Rangelands 33:41–44

Lange K, Olcott DD, Miller JE (2006) Effect of *Sericea lespedeza* (Lespedeza cuneata) fed as hay, on natural and experimental Haemonchus contortus infections in lambs. Vet Parasitol 141:273–278

Limb RF, Fuhlendorf SD, Engle DM et al (2011) Pyric–herbivory and cattle performance in grassland ecosystems. Rangel Ecol Manag 64:659–663

Marell A, Hofgaard A, Danell K (2006) Nutrient dynamics of Reindeer forage species along snowmelt gradients at different ecological scales. J Basic Appl Ecol 7:13–20

McNaughton SJ (1985) Ecology of a grazing ecosystem: the Serengeti. Ecol Monogr 55:259–294

McNaughton SJ, Banyikwa FF (1995) Plant communities and herbivory. In: Sinclair ARE, Arcese P (eds) Serengeti II: dynamics, management, and conservation of an ecosystem. The University of Chicago Press, Chicago, pp 49–70

Mengistu G, Hoste H, Karonen M et al (2017) The in vitro anthelmintic properties of browse plant species against Haemonchus contortus is determined by the polyphenol content and composition. Vet Parasitol 237:110–116

Meuret M, Dumont B (1999) Advances in modelling animal-vegetation interactions and their use in guiding grazing management. In: Gagnaux D, Poffet JR (eds) 5th international livestock farming systems symposium – integrating animal science advances in the search of sustainability, Posieux, Switzerland, 19–20 August. EEAP Publication N° 97, pp 57–72

Miller JR, Morton LW, Engle DM et al (2012) Nature reserves as catalysts for landscape change. Front Ecol Environ 10:144–152

Moreno FC, Gordon IJ, Knox MR et al (2012) Anthelmintic efficacy of five tropical native plants against *Haemonchus contortus* and *Trichostrongylus colubriformis* in experimentally infected goats (*Capra hircus*). Vet Parasitol 187:237–243

Muchiru AN, Western DJ, Reid RS (2008) The role of abandoned pastoral settlements in the dynamics of African large herbivore communities. J Arid Environ 72:940–952

Mueller T, Olson KA, Dressler G et al (2011) How landscape dynamics link individual- to population-level movement patterns: a multispecies comparison of ungulate relocation data. Glob Ecol Biogeogr 20:683–694

Murray MG (1995) Specific nutrient requirements and migration of wildebeest. In: Sinclair ARE, Arcese P (eds) Serengeti II: dynamics, management, and conservation of an ecosystem. The University of Chicago Press, Chicago, pp 231–256

Niamir-Fuller M (1999) Managing mobility in African rangelands: legitimization of transhumance. IT Publications (FAO), London
O'Connor TG, Goodman PS, Clegg B (2007) A functional hypothesis of the threat of local extirpation of woody plant species by elephant in Africa. Biol Conserv 136:329–345
Odadi WO, Karachi MK, Abdulrazak SA et al (2011) African wild ungulates compete with or facilitate cattle depending on season. Science 333:1753–1755
Ogutu JO, Owen-Smith N (2003) ENSO, rainfall and temperature influences on extreme population declines among African savanna ungulates. Ecol Lett 6:412–419
Omer RM, Hester AJ, Gorden IJ et al (2006) Seasonal changes in pasture biomass, production and offtake under the transhumance system in northern Pakistan. J Arid Environ 60:641–660
Owen-Smith N (2002) Adaptive herbivore ecology. Cambridge University Press, Cambridge
Owen-Smith N (2004) Functional heterogeneity in resources within landscapes and herbivore population dynamics. Landsc Ecol 19:761–771
Owen-Smith N (2008) Effects of temporal variability in resources on foraging behaviour. In: Prins HHT, Van Langevelde F (eds) Resource ecology: spatial and temporal dynamics of foraging. Springer, Dordrecht, pp 159–181
Parker KL, Barboza PS, Gillingham MP (2009) Nutrition integrates environmental responses of ungulates. Funct Ecol 23:57–69
Porensky LM, Veblen KE (2015) Generation of ecosystem hotspots using short-term cattle corrals in an African savanna. Rangel Ecol Manag 68:131–141
Prins HHT (1996) Behaviour and ecology of the African buffalo: social inequality and decision making. Chapman & Hall, London
Prins HHT, van Langevelde F (2008) Assembling a diet from different places. In: Prins HHT, Van Langevelde F (eds) Resource ecology: spatial and temporal dynamics of foraging. Springer, Dordrecht, pp 129–155
Provenza FD, Villalba JJ, Dziba LE et al (2003) Linking herbivore experience, varied diets, and plant biochemical diversity. Small Rumin Res 49:257–274
Raynor EJ, Joern A, Skibbe A et al (2017) Temporal variability in large grazer space use in an experimental landscape. Ecosphere 8(1):e01674. https://doi.org/10.1002/ecs2.1674
Rettie WJ, Messier M (2000) Hierarchical habitat selection by woodland caribou: its relationship to limiting factors. Ecography 23:466–478
Ripple WJ, Beschta RL (2004) Wolves and the ecology of fear: can predation risk structure ecosystems. Bioscience 54:755–766
Russell BG, Letnic M, Fleming PJS (2011) Managing feral goat impacts by manipulating their access to water in the rangelands. Rangel J 33:143–152
Sandland RL, Jones RJ (1975) The relation between animal gain and stocking rate in grazing trials: an examination of published theoretical models. J Agric Sci-Cambridge 85:123–128
Sayer J, Sunderland T, Ghazoul J et al (2013) Ten principles for a landscape approach to reconciling agriculture, conservation, and other competing land uses. Proc Natl Acad Sci 110:8349–8356
Sayre NF (2017) The politics of scale: a history of rangeland science. The University of Chicago Press, Chicago
Sayre NF, McAllister RR, Bestelmeyer BT et al (2013) Earth stewardship of rangelands: coping with ecological, economic, and political marginality. Front Ecol Environ 11:348–354
Selebatso M, Maude G, Fynn RWS (2018) Adaptive foraging of sympatric ungulates in the Central Kalahari Game Reserve, Botswana. Afr J Wildl Res 48:023005
Sensenig RL, Demment MW, Laca EA (2010) Allometric scaling predicts preferences for burned patches in a guild of East African grazers. Ecology 91:2898–2907
Shaik SA, Terrill TH, Miller JE et al (2006) Sericea lespedeza hay as a natural deworming agent against gastrointestinal nematode infection in goats. Vet Parasitol 139:150–157
Shamhart J, King F, Proffitt KM (2012) Effects of a rest-rotation grazing system on wintering elk distributions at Wall Creek, Montana. Rangel Ecol Manag 65:129–136

Sianga K, van Telgen M, Vrooman J et al (2017) Spatial refuges buffer landscapes against homogenisation and degradation by large herbivore populations and facilitate vegetation heterogeneity. Koedoe 59(2):a1434. https://doi.org/10.4102/koedoe.v59i2.1434

Tyrrell P, Russell S, Western D (2017) Seasonal movements of wildlife and livestock in a heterogenous pastoral landscape: implications for coexistence and community based conservation. Glob Ecol Conserv 12:59–72

US National Park Service (2017) https://www.nps.gov/yell/learn/news/17002.htm. Accessed 15 Mar 2018

Verlinden A, Masogo R (1997) Satellite remote sensing of habitat suitability for ungulates and Ostrich in the Kalahari of Botswana. J Arid Environ 35:563–574

Verweij RJT, Verrelst J, Loth PE et al (2006) Grazing lawns contribute to the subsistence of mesoherbivores on dystrophic savannas. Oikos 114:108–116

Vesey-FitzGerald DF (1960) Grazing succession among East African game animals. J Mammal 41:161–172

Walker BH, Emslie RH, Owen-Smith RN et al (1987) To cull or not to cull: lessons from a southern African drought. J Appl Ecol 24:381–401

Wang G, Hobbs NT, Boone RB et al (2006) Spatial and temporal variability modify density dependence in populations of large herbivores. Ecology 87:95–102

Wato YA, Prins HHT, Heitkonig I et al (2018) Movement patterns of African elephants (Loxodonta africana) in a semi-arid savanna suggest that they have information on the location of dispersed water sources. Front Ecol Evol 6:167

Western D, Groom R, Worden J (2009) The impact of subdivision and sedentarization of pastoral lands on wildlife in an African savanna ecosystem. Biol Conserv 142:2538–2546

Wittmer HU, Sinclair ARE, McLellan BN (2005) The role of predation in the decline and extirpation of woodland caribou. Oecologia 144:257–267

WTTC (2015) Travel and tourism, economic impact, Tanzania. World Travel and Tourism Council, 2015 Annual Research: Key Facts

WTTC (2017) Travel and tourism, economic impact, Botswana. World Travel and Tourism Council, March 2017 Annual Research: Key Facts

Chapter 15
The Ecology of Browsing and Grazing in Other Vertebrate Taxa

Iain J. Gordon, Herbert H. T. Prins, Jordan Mallon, Laura D. Puk, Everton B. P. Miranda, Carolina Starling-Manne, René van der Wal, Ben Moore, William Foley, Lucy Lush, Renan Maestri, Ikki Matsuda, and Marcus Clauss

Since the publication of the "The Ecology of Browsing and Grazing" (Gordon and Prins 2008), a number of researchers have taken the approach outlined in the book to assess the impacts of differences in food and nutrient supply on the ecology of other vertebrate taxa. The approach may not work in all vertebrate taxa but understanding the similarities and differences between herbivorous vertebrate taxa provides ecologists with a broader canvas upon which to develop and test hypotheses about herbivore/plant interactions. In line with the slightly altered emphasis of the current book (The Ecology of Browsing and Grazing II), we also asked the authors of the

I. J. Gordon (✉)
James Cook University, Townsville, QLD, Australia
e-mail: Iain.gordon@jcu.edu.au

H. H. T. Prins
Animal Sciences Group, Wageningen University, Wageningen, The Netherlands
e-mail: Herbert.prins@wur.nl

J. Mallon
Beaty Centre for Species Discovery and Palaeobiology Section, Canadian Museum of Nature, Ottawa, ON, Canada
e-mail: jmallon@nature.ca

L. D. Puk
The University of Queensland, St. Lucia, QLD, Australia

E. B. P. Miranda
Universidade Estadual de Mato Grosso, Alta Floresta, Brazil

ONF Brasil Gestão Florestal, Cotriguaçu, Brazil

C. Starling-Manne
Universidade Federal do Rio de Janeiro, Rio de Janeiro, Brazil

R. van der Wal
Department of Ecology, Swedish University of Agricultural Sciences (SLU), Uppsala, Sweden

School of Biological Sciences, University of Aberdeen, Aberdeen, UK
e-mail: rene.van.der.wal@slu.se

I. J. Gordon, H. H. T. Prins (eds.), *The Ecology of Browsing and Grazing II*, Ecological Studies 239, https://doi.org/10.1007/978-3-030-25865-8_15

Sections in this Chapter to provide insights into the impacts that these different vertebrate taxa have on the ecosystems in which they exist. In this Chapter, the authors describe the findings from this research. As you will see, the depth of research on the ecology and impacts of the different herbivorous vertebrate taxa varies considerably and demonstrates the importance of further research endeavours, on herbivore/plant interactions, across the board. The taxa covered are:

- Dinosaurs (Jordan Mallon)
- Fish (Laura D. Puk)
- Reptiles (Everton B. P. Miranda and Carolina Starling-Manne)
- Birds (René van der Wal)
- Marsupials (Ben Moore and William Foley)
- Lagomorphs (Lucy Lush)
- Rodents (Renan Maestri)
- Primates (Ikki Matsuda and Marcus Clauss)

15.1 Dinosaurs

Jordan Mallon

The browser-grazer continuum is not one that readily applies to the non-avian dinosaurs (hereafter, simply 'dinosaurs'). Although grasses (Poaceae) had evolved by the Cretaceous (Prasad et al. 2005), they were not abundant and did not form

B. Moore
Hawkesbury Institute for the Environment, Western Sydney University, Richmond, NSW, Australia
e-mail: B.Moore@westernsydney.edu.au

W. Foley
Research School of Biology, The Australian National University, Canberra, ACT, Australia
e-mail: William.foley@anu.edu.au

L. Lush
School of Life and Environmental Sciences, University of Sydney, Sydney, NSW, Australia

R. Maestri
Departamento de Ecologia, Universidade Federal do Rio Grande do Sul, Porto Alegre, Brazil

I. Matsuda
Chubu University Academy of Emerging Sciences, Kasugai-shi, Aichi, Japan

Wildlife Research Center of Kyoto University, Sakyo-ku, Kyoto, Japan

Japan Monkey Centre, Inuyama, Aichi, Japan

Institute for Tropical Biology and Conservation, Universiti Malaysia Sabah, Sabah, Malaysia
e-mail: ikki-matsuda@isc.chubu.ac.jp

M. Clauss
Clinic for Zoo Animals, Exotic Pets and Wildlife Vetsuisse Faculty, University of Zürich, Zürich, Switzerland
e-mail: mclauss@vetclinics.uzh.ch

widespread grasslands until the Miocene (Potts and Behrensmeyer 1992). It, therefore, makes little sense to speak of dinosaurian 'grazers' *sensu stricto* (but see 'True Dinosaurian Grazers?' below). Rather, lycopods, ferns, sphenopsids, cycadophytes, ginkgos, conifers, and (non-poaecean) angiosperms made up the bulk of the plant material available for dinosaurian consumption (Gee 2011; Tiffney 2012). Ferns, in particular, likely filled the role of low growing, herbaceous colonizers for most of the Mesozoic (Wing et al. 1993; Collinson 1996). For this reason, it makes more sense to speak of herbivorous dinosaurs in terms of concentrate, intermediate, and bulk feeders (sensu Hofmann and Stewart 1972, Mallon and Anderson 2014a).

15.1.1 Size and Shape

Whether particular herbivorous dinosaurs were concentrate, intermediate, or bulk feeders would have been primarily influenced by their respective body sizes (Peters 1983). Dinosaur body mass spanned seven orders of magnitude (Benson et al. 2014), so these animals undoubtedly adopted a variety of feeding strategies. Small (10–100 kg) herbivores—including heterodontosaurids, small ornithopods, early thyreophorans, most pachycephalosaurs, and basal ceratopsians, among others—almost certainly concentrated on nutritious shoots, fruits, and seeds to fuel a relatively high metabolism (Weishampel 1984). These dinosaurs were obligatory bipeds, and possessed narrow, pointed beaks with which to selectively crop their food (Coe et al. 1987). Their teeth, and associated jaw mechanics, varied from the simple to the complex, indicative of corresponding variability in dietary fibre intake (Norman and Weishampel 1985; Nabavizadeh 2016) (Fig. 15.1a). Some derived clades developed rudimentary tooth batteries having a single, continuous occlusal surface (e.g., Norman et al. 2011). The jaw joint was depressed below the plane of occlusion, increasing the lever arm of the external mandibular adductor musculature, enabling a more powerful bite. This system was functionally equal, but mechanically opposite, to that of ungulates, where the jaw joint is positioned above the occlusal plane to increase the lever arm of the masseter musculature (Greaves 1995). One lineage, the Ornithopoda, is traditionally thought to have developed a 'pleurokinetic' skull having multiple, mobile intracranial joints, allowing the upper jaw to flex laterally during tooth occlusion, to accommodate the lower cheek teeth. Given the isognathous nature of the jaws (where the teeth occlude on both sides when the jaws are closed), this would have resulted in a transverse power stroke, functionally analogous to that of ungulates (Norman and Weishampel 1985). This hypothesis has recently received some push-back—Rybczynski et al. (2008) and Cuthbertson et al. (2012) showed that the secondary intracranial movements imposed by the pleurokinetic model could not be accommodated by the corresponding joints. Rather, minimal rotation of the lower jaw rami, about their long axes, would have produced a similar power stroke, and is mechanically more feasible (Nabavizadeh and Weishampel 2016).

At the opposite end of the size spectrum, megaherbivorous dinosaurs ($\geq 1 \times 10^3$ kg; sensu Owen-Smith 1988) included most sauropodomorphs, stegosaurs, ceratopsids, ankylosaurs, and iguanodontians. These forms were highly variable in

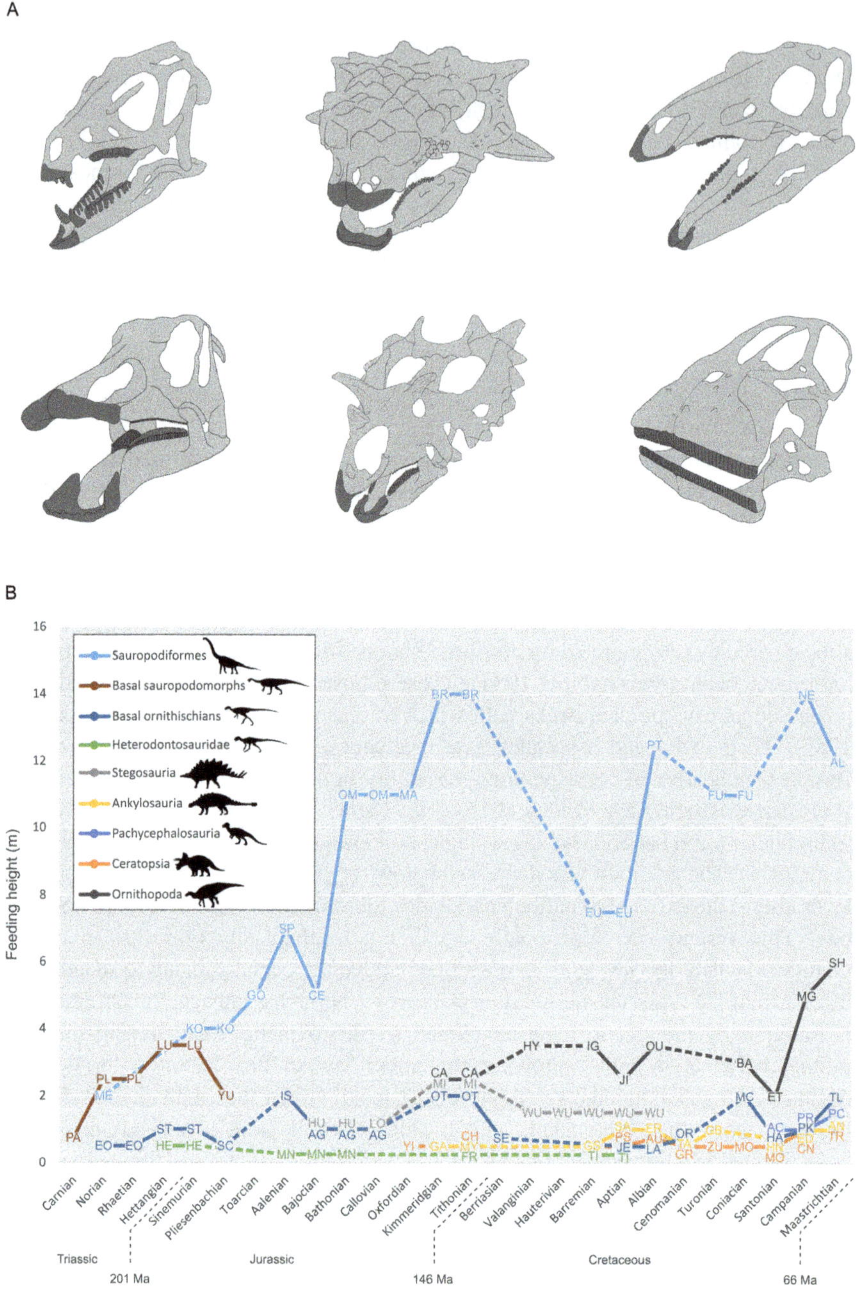

Fig. 15.1 Herbivorous dinosaur ecomorphology. A, Herbivorous dinosaur craniodental adaptations, with beaks and teeth in dark grey (clockwise from upper left): heterodontosaurid, ankylosaur, stegosaur, sauropod, ceratopsian, hadrosaur (skulls not to scale). B, Herbivorous dinosaur feeding heights through time. Sauropodiform feeding heights (assuming vertical necks) were calculated following Upchurch and Barrett (2000). The methodology of Mallon et al. (2013) was used for calculating bipedal feeding heights for basal sauropodomorphs, basal ornithischians,

morphology, and shared few ecomorphological features in common beyond large body size and minimally facultative quadrupedality. Stegosaurs and ankylosaurs retained primitively small, phylliform teeth, although the beaks of the latter were wider, and presumably predisposed to bulk feeding (Mallon and Anderson 2014a; Ősi et al. 2017). The exceptionally wide bellies of the ankylosaurs attest to lengthy gut-retention times and a preference for fibrous roughage (Bakker 1986).

Iguanodontians and ceratopsids both possessed complex dental batteries, but these functioned in quite different ways. Iguanodontian dental batteries, best exemplified by the hadrosaurids, consisted of hundreds of highly complex, tightly-spaced teeth, multiple rows of which contributed to the occlusal surface (Erickson et al. 2012; LeBlanc et al. 2016). This inclined surface was capable of both crushing and shearing functions. Tooth wear evidence further suggests some capacity for fore-aft grinding (Williams et al. 2009; Mallon and Anderson 2014b). This suite of functions would have allowed the hadrosaurids to rend all manner of plant types, substantiating their inferred role as herbivore generalists. By contrast, the tooth batteries of the ceratopsids were simpler, and the continuous occlusal surface of the teeth was limited to the vertical plane. Thus, ceratopsids were evidently more restricted in their diets, which is consistent with their narrow, selective beaks (Mallon and Anderson 2014a). In this sense, ceratopsids might be likened to the narrow-lipped black rhinoceros, *Diceros bicornis*, which selects for low-growing woody scrub (Owen-Smith 1988).

Largest of all were the sauropodomorphs, with some forms possibly approaching 90 tonnes (Benson et al. 2014). These were characterized by exceedingly long necks ending in proportionally tiny heads. The jaws were simple in construction, and the teeth were primitively small and leaf-shaped. Some derived forms (e.g., diplodocids)

Fig. 15.1 (continued) heterodontosaurids, pachycephalosaurs, basal ceratopsians, and ornithopods, and quadrupedal feeding heights for stegosaurs, ankylosaurs, and derived ceratopsians (ceratopsids). Dashed lines indicate missing data. Herbivorous theropod groups (e.g., Ornithomimosauria, Therizinosauria) are not shown due to space restrictions. Taxonomic abbreviations: AL, *Alamosaurus*; AC, *Acrotholus*; AG, *Agilisaurus*; AN, *Ankylosaurus*; AU, *Auroraceratops*; BA, *Bactrosaurus*; BR, *Brachiosaurus*; CA, *Camptosaurus*; CE, *Cetiosaurus*; CH, *Chaoyangsaurus*; CN, *Centrosaurus*; ED, *Edmontonia*; EO, *Eocursor*; ER, *Europelta*; ET, *Eotrachodon*; EU, *Euhelopus*; FR, *Fruitadens*; FU, *Futalongkosaurus*; GA, *Gargoyleosaurus*; GB, *Gobisaurus*; GO, *Gongxianosaurus*; GR, *Graciliceratops*; GS, *Gastonia*; HA, *Haya*; HE, *Heterodontosaurus*; HN, *Hungarosaurus*; HU, *Huayangosaurus*; HY, *Hypselospinus*; IG, *Iguanodon*; IS, *Isaberrysaura*; JE, *Jeholosaurus*; JI, *Jinzhousaurus*; KO, *Kotasaurus*; LE, *Leaellynasaura*; LO, *Loricatosaurus*; LU, *Lufengosaurus*; MA, *Mamenchisaurus*; MC, *Macrogryphosaurus*; ME, *Melanorosaurus*; MG, *Magnapaulia*; MI, *Miragaia*; MN, *Manidens*; MO, *Mosaiceratops*; MY, *Mymooropelta*; NE, *Neuquensaurus*; OM, *Omeiosaurus*; OR, *Orodromeus*; OT, *Othnielosaurus*; OU, *Ouranosaurus*; PA, *Panphagia*; PC, *Pachycephalosaurus*; PK, *Parksosaurus*; PL, *Plateosaurus*; PR, *Prenocephale*; PS, *Psittacosaurus*; PT, *Patagotitan*; SA, *Sauropelta*; SC, *Scutellosaurus*; SE, *Stenopelix*; SH, *Shantungosaurus*; SP, *Spinophorosaurus*; ST, *Stormbergia*; TA, *Talarurus*; TI, *Tianyulong*; TL, *Talenkauen*; TR, *Triceratops*; WU, *Wuerhosaurus*; YI, *Yinlong*; YU, *Yunnanosaurus*; ZU, *Zuniceratops*. Other abbreviations: Ma, Mega-annum. Silhouette credits: R. Amos, A. Farke, S. Hartman, N. Tamura, M. Taylor, E. Willoughby

had simple, peg-like teeth restricted to the front of the jaws; these would have been mechanically ideal for stripping branches of twigs and leaves (Young et al. 2012). Other contemporaneous sauropods (i.e., camarasaurids) had mouths full of robust, chisel-like teeth capable of delivering higher, sustained bite forces, probably for ingestion of harder foodstuffs (Button et al. 2014). *Nigersaurus taquietti* is distinctive in having an unusually broad, flat muzzle lined with a battery of pencil-like teeth (Sereno et al. 2007). This configuration was almost certainly an adaptation for bulk feeding, low to the ground.

Unfortunately, the ecology of most other sauropods is not so easily discernible. Most problematic has been the use of their long necks, and debate has waged over whether they were held horizontally or more nearly vertically. Proponents of the first view maintain that the cervical articulations prevented the neck from being elevated very much (Stevens and Parrish 2005a, b), that the blood pressure necessary to perfuse the brain, with the neck held in a vertical position, would have been dangerously high (Seymour 2009), or that the neck was mechanically more efficient while in a horizontal posture (Woodruff 2017). Proponents of the second view retort that the cervical vertebrae alone are not good indicators of neck posture in modern taxa (Taylor et al. 2009), that stresses distributed through an upright neck would not have been prohibitive (Christian 2010), and that partitioning of the forest strata would have made ecological sense in sauropod-dominated communities (Bakker 1978) (Fig. 15.1b), as it does among ruminants today (du Toit 1990). The debate is, as yet, unsettled, and the arguments grow increasingly nuanced.

15.1.2 Energetics

Dinosaur energetics are among the most difficult aspects of their biology to glean from the fossil record, but palaeontologists are anything but faint of heart. Although a range of educated assumptions must invariably be made, these serve to constrain interpretations within the realm of possibility. Dinosaur energetics are typically considered from the perspectives of supply and demand. On the supply side, Hummel et al. (2008) investigated the suitability of various Mesozoic (non-angiosperm) plants as sauropod fodder by subjecting samples of related living tissue to in vitro fermentation experiments, using gas production as a measure of metabolizable energy, while also quantifying other nutrients. Horsetails, *Equisetum* spp., proved to be the most nutritious, having high levels of both metabolizable energy and crude protein (the high silica content was likely of little concern to those dinosaurs that relied on gut processing). *Araucaria* spp., *Ginkgo* spp. and *Angiopteris* spp. were also likely dietary staples, whereas cycads, tree ferns, and podocarps proved to be poor sources of energy.

On the demand side, the work of Farlow (1976) and colleagues (Farlow et al. 2010) is instructive. It is possible to bracket the possible energy requirements of dinosaurs via consideration of how those requirements scale with body size in both reptilian ectotherms and mammalian endotherms (birds, although descended from dinosaurs, are likely poor models for dinosaur energetics because their physiology is fine-tuned to the aerial realm). Thus, a single adult *Brontosaurus excelsus*, weighing 3×10^4 kg, might have needed anywhere from 2×10^4 to 50×10^4 kJ of energy per day, depending on the animal's physiology (Farlow et al. 2010). By combining both supply and demand sides of the model, it is possible to further estimate the population density of *B. excelsus* on a Jurassic landscape, which varies from a few large adults.km^{-2} (endotherm model) to a few tens of large adults.km^{-2} (ectotherm model). Given that the most diverse sauropod communities of the Jurassic were situated in semi-arid climates at mid-latitudes (Noto and Grossman 2010; Mannion et al. 2011), primary productivity is unlikely to have been very high, and sauropods may have needed to seasonally migrate to find appropriate sustenance. Evidence for such migration, between lowland and upland environments, is preserved as oxygen isotope excursions in tooth enamel (Fricke et al. 2011).

15.1.3 True Dinosaurian Grazers?

In spite of the foregoing discussion, there do exist rare examples of grass-eating in the dinosaur fossil record. The first concerns a report of grass phytoliths found inside coprolites (fossil dung) attributed to Late Cretaceous titanosaurian sauropods from India (Prasad et al. 2005), based on their common association with diagnostic skeletal material. The variety of phytolith morphologies is suggestive of the presence of up to five different grass species, including relatives of modern rices, bamboos, and cool-season grasses. Other phytoliths, found within the same coprolites, are attributable to dicotyledons, conifers, and palms. Variation in phytolith preservation implies that some were purposefully ingested as forage, whereas others (particularly from the rarely preserved palms) were consumed incidentally.

A more recent example (Wu et al. 2017) concerns the discovery of grass phytoliths and silicified epidermis within the purported dental calculus (plaque) of the Early Cretaceous basal hadrosauroid *Equijubus normani* from China. These structures are assigned to Poaceae, based on the short-long cellular patterning and short-cell pairing of the epidermis, and on the equidimensional unlobed and slightly lobed morphologies of the phytoliths.

The presence of grass phytoliths in Cretaceous deposits is somewhat unexpected, as it was long held that these defensive structures evolved in response to later mammal grazing during the Miocene (Strömberg 2004). Could grass phytoliths have evolved in response to dinosaur herbivory instead? It seems unlikely; phytoliths are primarily thought to work by exacerbating tooth wear (Strömberg et al. 2016), which is hardly a concern for reptiles with unlimited tooth replacements, including dinosaurs. In the case of sauropods, these animals are unlikely to have done much oral processing, opting instead for hindgut fermentation (Hummel and Clauss 2011). The dental battery of *Equijubus normani*, and other iguanodontians, may have been well-suited for rending grass, but chance finds of gut and fecal contents in these animals indicates a quite varied diet, including conifer and angiosperm twigs and stems, bark, seeds, leaves, and even crustaceans (Currie et al. 1995;

Tweet et al. 2008; Chin et al. 2017). There is, as yet, no solid evidence that dinosaurs co-evolved with any group of angiosperms (Barrett and Willis 2001; Butler et al. 2009). Rather, grass phytoliths may have evolved in response to predation from the hypsodont gondwanatherian mammals or insects (Prasad et al. 2005).

15.2 Fishes

Laura D. Puk

Herbivorous fishes are found in both marine and freshwater ecosystems but are much more abundant in the marine environment. While in terrestrial ecosystems 'grazing' and 'browsing' refer to the feeding on monocotyl and dicotyl plants, in marine ecosystems, 'grazing' and 'browsing' refer to the removal of different kinds of algae (Green and Bellwood 2009). Marine grazing fishes remove small filamentous turf algae and marine browsing fishes remove larger macroalgae (Green and Bellwood 2009). In contrast to terrestrial ecosystems, 'real' grazing on angiosperms is rare and restricted to only one plant group: seagrasses (Beer 1989). In freshwater ecosystems the terminology is different yet again. The term 'grazer' is used for both plant—('macrophytes'; Fowler and Robson 1978) and algal—('periphyton', Power 1983) removing fishes, whereas the term 'browser' is not used at all.

Where a distinction between grazing and browsing fishes is made, they are faced with different challenges in terms of intake and digestion of their respective food source. Grazers feeding on turf algae ingest a low energy but relatively easily digestible and abundant resource. Browsers feeding on macroalgae on the other hand are faced with a resource that provides more energy, but exhibits chemical and structural defences, is difficult to digest, and patchily distributed. Both groups exhibit adaptations to their respective resources, including differing home range sizes (Nash et al. 2013; Welsh and Bellwood 2014), bite rates (Randall 1967; Burkepile and Hay 2010), and digestive mechanisms (Clements et al. 2009). Browsers are often larger in body size than grazers, potentially enabling them to take bigger, stronger bites (Bonaldo and Bellwood 2008; Lokrantz et al. 2008). They also have larger home ranges (Nash et al. 2013), which helps them locate their food and they take less frequent bites (Cardoso et al. 2009; Burkepile and Hay 2010), because their resource is more energy-dense. After ingestion, browsing fishes utilise fermentation to break down the macroalgal cell walls to access the nutrients (Clements et al. 2009); a mechanism not required by grazers.

While the importance of herbivorous fishes in marine ecosystems has been recognized for a long time, research in freshwater systems mainly focused on insects (Lodge et al. 1998) but is now recognizing the impact of herbivorous fishes on macrophytes (Lodge 1991; Lodge et al. 1998). Herbivorous fishes hold crucial functions in both freshwater and in marine systems, albeit of a different nature. In freshwater systems, they contribute significantly to seed dispersal in seasonally flooded forest ecosystems (Anderson et al. 2011; Horn et al. 2011), exhibiting some of the longest dispersal distances discovered in animals, including African hornbills and Asian elephants (Anderson et al. 2011). However, due to their removal of macrophytes, they can increase phytoplankton biomass and shift lakes towards a turbid state if certain conditions are met (Van Donk and Otte 1996). Grazing and browsing fishes exhibit their highest diversity in marine ecosystems, especially on coral reefs (Gaines and Lubchenco 1982), which has led to a research focus on these ecosystems. Here, herbivorous fishes are considered to provide a key ecosystem function because they control algae, which compete with the primary reef builder—corals.

On coral reefs, herbivorous fishes can consume up to 90% of the primary production (Carpenter 1986; Polunin and Klumpp 1992) and, therefore, play a major role in controlling the reef state. Grazers are more abundant than browsers (Choat et al. 2004), and coupled with their higher feeding rates (Burkepile and Hay 2010), they are responsible for consuming most of the primary production (Burkepile and Hay 2010). Coral reefs dominated by corals are considered 'healthy', but if grazing pressure is reduced, they often become dominated by algae (Hughes et al. 2007; Webster et al. 2015). Grazing pressure can be reduced due to overfishing, which is simulated experimentally by excluding fishes with cages (Bellwood et al. 2006; Webster et al. 2015), or due to coral bleaching events which kill vast amounts of corals, freeing up space for algal settlement (Hoegh-Guldberg 1999). Sudden increased availability of space reduces the grazing pressure per unit area on a reef (Williams et al. 2001; Mumby et al. 2007; Mumby 2009), because the existing fish community now has to distribute its feeding activity over the additional free substrate. Once a reef is dominated by algae a return to a coral-dominated state is difficult as macroalgae inhibit coral recruitment (Kuffner et al. 2006; Webster et al. 2015) and survival (Box and Mumby 2007; Webster et al. 2015). Algal-dominated reefs provide less ecosystem services to people compared to their healthy coral-dominated counterparts (Moberg and Folke 1999).

Shifts to algal-dominated coral reefs are projected to increase in the future due to climate change (Hoegh-Guldberg 1999). Higher temperatures cause a higher frequency of coral bleaching events and subsequent mass coral mortality (Hoegh-Guldberg 1999; Hughes et al. 2018). Additionally, algae experience a net positive physiological effect with rising temperatures, whereas corals experience a net negative physiological effect (Elfwing and Tedengren 2000). This could shift the competitive interaction further in favour of algae. However, grazing intensity is greater at higher temperatures (Smith 2008) and may counteract the increased primary production (O'Connor 2009). Higher production and herbivory can strengthen the producer-consumer interaction (O'Connor 2009) which may become a feature of future ecosystems. A strengthening of this interaction could increase the importance of herbivorous fish for the health of consumer-dominated ecosystems such as coral reefs even further.

While the state of coral reefs is heavily controlled by herbivorous consumers, other ecosystems are less dependent on consumers and instead producer-dominated (Connell et al. 2011). One example are kelp forests, common temperate marine ecosystems, where native herbivores (mostly invertebrates such as sea urchins) exhibit only minor control over the ecosystem state (Connell et al. 2011). However, rising temperatures cause a range expansion of tropical herbivorous fishes, which could have detrimental effects on kelp forests (Vergés et al. 2014). After a kelp die-off in Western Australia, tropical browsing fish (kyphosids) that had moved into higher latitudes consumed kelp and created a new barren ecosystem state dominated by turf algae (Bennett et al. 2015). Herbivorous fishes may, therefore, shift the production-dominated temperate marine systems towards consumption-dominated alternative states (Bennett et al. 2015). The strengthening of the producer-consumer interaction (O'Connor 2009), and the switch from production-dominated to consumer-dominated temperate systems (Bennett et al. 2015), suggest a stronger influence of grazers and browsers on future marine ecosystems. Only time will tell.

Similarly strong changes may occur in freshwater ecosystems. Increased grazing, for example through higher metabolic rates associated with increased temperatures, could destabilize the clear-water macrophyte-rich state of lakes and shift them to a phytoplankton-dominated turbid state (Van Donk and Otte 1996). Additionally, increased frequencies of droughts may decrease the number or intensity of floods in forest ecosystems that experience substantial seed dispersal by fishes (Horn et al. 2011). This could decrease the dispersal of plants, especially upstream (Horn et al. 2011). Overfishing can further intensify the problem as bigger fish are often targeted first but they are also the most effective long-distance dispersers (Anderson et al. 2011). Whether the role of fish as a disperser is important enough to make their disappearance noticeable in the forest community structure is currently unknown.

The potentially significant future ecosystem changes connected to herbivorous fishes highlight the importance of understanding their ecosystem function in detail. The study of the role of browsing and grazing fishes is still in its infancy and there is a lot to be learnt from studies on terrestrial ecosystems. However, fish herbivory is likely to differ from terrestrial herbivory because producers and consumers are structurally and functionally distinct from their terrestrial counterparts. Research should focus on understanding the drivers of food choice in herbivorous fishes, how climate change and habitat degradation influence interactions between herbivores and their ecosystems, and identify potential feedbacks.

15.3 Reptiles

Everton B. P. Miranda and Carolina Starling-Manne

Among reptiles, lizards and chelonians are the only modern groups that strictly, or mostly, feed on plant-matter (King 1996). Snakes are obligate carnivores, and crocodiles have evolved herbivory habits a few times in their evolutionary history (Kley et al. 2010; Fiorelli et al. 2016), but there is no modern herbivorous crocodile species. However, a large diversity of chelonian and lizard species is adapted to feed on plants, with several insular radiations that represent most of this diversity.

Given the physiological demands of digesting cellulose and high-fibre plant material, reptile species specialized for grazing and browsing are typically large-sized tropical or sub-tropical species (Pough 1973; Cooper and Vitt 2002; but see Espinoza et al. 2004). The latitude constraint is caused by temperature-dependent limitations to fermentation rates by bacteria in the gut, which require high temperatures to perform their activities (Cooper and Vitt 2002). Being ectotherms, reptiles do not maintain their body temperature stable, and would likely lose their microbial symbionts over cold winters, so full herbivore reptiles are usually restricted to tropical and subtropical climates. In terms of size, large ectotherms become two to three times smaller per each 10 °C of decrease in ambient temperature (Makarieva et al. 2005), due to metabolic constraints. There is also the need of having enough space to allow the fermentation of large amounts of fibre and cellulose (Pough 1973), so herbivory in ectotherms usually requires a large body size.

Consequently, many of the largest continental species of reptiles, and the megafauna in islands, are herbivorous lizards and chelonians (Hansen and Galetti 2009). Martin (1984) defined megafauna as animals above 44 kg, but throughout this section we will be using an operational concept of megafauna, which is animals whose evolution is constrained by the size of the land area each species inhabits (Burness et al. 2001; Hansen and Galetti 2009). Also, it is worth noting that reptiles are the vertebrates that achieve some of the highest densities and biomasses in terrestrial ecosystems, reaching up to around 600 $kg.ha^{-1}$ (Arce-Nazario and Carlo 2012; Lovich et al. 2018; Fig. 15.2). Since the grazing and browsing pressure of an herbivore is magnified by its density and biomass, in some ecosystems the ecological role of reptiles is also magnified.

15.3.1 Diversity

15.3.1.1 Insular

Some of the most conspicuous kinds of grazing and browsing reptiles are giant insular tortoises and iguanas (Gerlach 2014; Hastings et al. 2014). The colonization restrictions of islands normally results in a "reptile-only" megafauna (Hansen and Galetti 2009). In these ecosystems, tortoises have evolved, on multiple occasions, domed shells when grazing over grasslands and craning necks and saddleback shells when browsing over trees, shrubs, cacti, and palms. In the past, giant tortoises inhabited climates ranging from mediterranean to tropical climates, including archipelagos in the Mediterranean sea, and the Caribbean, Indic and Pacific oceans (Turtle Extinctions Working Group 2015). Today, there are extant island giant tortoise populations of only two species complexes, in the Galápagos Islands (*Chelonoidis niger* spp.) and in the Seychelles Islands (*Aldabracheys gigantea*; Hansen et al. 2010).

Another group of browsing and grazing reptiles are iguanas (Cooper and Vitt 2002). They represent a diverse group (Uetz et al. 2016), with three insular radiative adaptations in the Caribbean (nine species in the *Cyclura* genus), the islands of Fiji

Fig. 15.2 Densely packed green iguana (*Iguana iguana*) populations occur across a wide area of Neotropical forests. In this photo, several adult males display during the breeding season. http://www.thinkstockphotos.ca/image/stock-photo-iguanas-in-a-tree-in-mexico/464974882

and Tonga (six species in the *Brachylophus* and *Lapitiguana* genus, from which the two largest species are extinct), and the Galápagos archipelago (three terrestrial species from the *Conolophus* genus). As is the case with their continental counterparts, insular iguanas are mainly browsers, also being important for seed dispersal (Laurel et al. 2000; Meehan et al. 2002; Traveset et al. 2016). To complete the list, two other insular browsers are the monkey tailed lizard, *Corucia zebrata*, from the Solomon Islands, and the lacertid lizards, *Galotia* spp., from Mediterranean islands (Espinoza et al. 2004).

15.3.1.2 Continental

Continental browsing by lizards is restricted to iguanids in the Neotropics, such as the widespread green iguana, *Iguana iguana*, and by other species that are not so specialized on leaf or grass material, such as *Ctenosaurus*, *Sauromalus*, and *Dipsosaurus* (Vitt and Caldwell 2013). The only significant species outside the Neotropics are the mostly-grazing desert-specialists in the genus *Uromastyx* (Wilms et al. 2010), and the partial-browsers of the genus *Hydrosaurus*.

Large continental grazing chelonians (heavier than 44 kg) used to be found on all continents apart from Antarctica. As with most of the megafauna, they are now extinct in most of their historical range, except for the African continent (Martin 1984; Turtle Extinctions Working Group 2015). For herbivorous reptiles, the extinctions were caused mainly by direct and indirect effects of human action, including over-exploitation, and pressures exerted by invasive species (Turtle Extinctions Working Group 2015; Slavenko et al. 2016). Large continental grazing tortoises can only be found today in Africa (Turtle Extinctions Working Group 2015), likely due to their history of co-existence with hominids (Klein 1984). There are two widely known species of tortoises that fit this role, the leopard tortoise, *Stygmochelis pardalis*, and the spurred tortoise, *Centrochelys sulcata*. While the former may weigh up to 48 kg, and feeds almost exclusively on grasses (Branch et al. 1990), spurred tortoises are truly specialized grazers, that weight up to 96 kg (Lambert 1993), and are the last representatives of giant grazing tortoises on continental ecosystems. A few other smaller grazing tortoise species still inhabit continents, such as the Bolson tortoise, *Gopherus flavomarginatus*, from Mexico, the yellow-footed tortoise, *Chelonoidis denticulatus*, from South America and the Asian forest tortoise, *Manouria emys*, from Southeast Asia. Among the extinct giant species of continental tortoises are the *Chelonoidis lutzae* in South America, *Hesperotestudo* spp. in North America, Meiolanids (horned turtles) in Australia and the *Megalochelys* spp. in Asia (estimated weight: 1000 kg; Badam 1981)—being the latter the largest tortoises to have ever lived (Turtle Extinctions Working Group 2015).

15.3.1.3 Marine

Green sea turtles, *Chelonia mydas*, play a conspicuous role as a marine grazer, paralleling grazing ungulates in many aspects of their ecology (Christianen et al. 2014). They have some degree of endothermy which allows them to graze on water temperatures around 20 °C (Heath and Mcginnis 1980). Furthermore, two species of herbivorous reptiles are known as being associated with marine habitats. First, there is the marine iguana from the Galápagos Islands, which is a grazer, feeding on the rocky subtidal and intertidal zones of the archipelago (Shepherd and Hawkes 2005). Marine iguanas have tight schedules of basking and foraging—the second being usually restricted to only 1 h a day—which allows them to graze on the cold Galapagos waters. A second species, closely associated to marine ecosystems, is *Ctenosaura bakeri* (Köhler 2004), a browsing iguana, endemic to the mangroves of the Utila Island, in Honduras, in the Caribbean.

15.3.2 Effects on Vegetation Composition and Dynamics

Giant tortoises are highly influential on both insular and continental ecosystems, being best known as important seed dispersers (Falcon et al. 2018). They also play the, perhaps less appreciated, role of grazers and browsers on vegetation, affecting vegetation community composition and structure (Lovich et al. 2018).

Grazing herds of giant tortoises reportedly create and maintain a grassland ecosystem called tortoise turf on islands in the Indian Ocean (Gibson and Hamilton

1983; Cheke and Hume 2010). Composed of grasses, sedges and herbs, tortoise turf occurs in areas of high-tortoise concentration and, therefore, heavy grazing. Many of its characteristic plant species are dwarfs with highly specialized growth strategies (e.g., flowers and fruits produced at the base of the plants; Merton et al. 1976). This is a dominant ecosystem in the Seychelles and used to be widespread at the Mascarene Islands (Cheke and Hume 2010; Griffiths et al. 2010). Tortoise terraforming is, in its turn, important to other insular species of plants and animals (Griffiths et al. 2010, 2011, 2013). Many plants in the Mascarene archipelago have two kinds of foliage, from which their low-hanging leaves have adaptations to lower or no levels of herbivory by tortoises (Eskildsen et al. 2004). This phenomenon illustrates the high grazing pressure that can be exerted by large herds of tortoises. Tortoise-exclusion experiments are in operation on Aldabra, and are likely to show the importance of these keystone herbivores on the archipelago's vegetation dynamics in the near future. On the other hand, on some islands, where native giant tortoises have been extirpated, similar-species tortoise introductions are being conducted, as ecological replacements to restore important lost ecosystem functions. For example, that approach is rebuilding trophic interactions (e.g., herbivory and seed dispersal networks) on some islands of the Mascarene archipelago (Griffiths et al. 2010, 2013).

On the Galápagos Islands, giant tortoises are known to maintain plant communities with an upper strata, formed by arborescent cacti (Gibbs et al. 2010). If tortoises have been extirpated, these plant communities become encroached by woody vegetation (Hunter and Gibbs 2014). The Anthropocene extinctions, translocations and population fluctuations of tortoises in the Galapagos archipelago provided many ecological quasi-experiments to test the effects of these ecosystem engineers (Gibbs et al. 2008; Froyd et al. 2014). Tortoise reintroductions, after decades of absence, cause a marked rebuilding of cactus population, reducing clustering and increasing juvenile cactus recruitment (Gibbs et al. 2010). Intense grazing, soil disturbance, browsing, seed dispersal, pool forming and direct vegetation damage are some of the means by which giant tortoises restore degraded ecosystems in the Galápagos (Hamann 1993; Blake et al. 2012, 2013; Froyd et al. 2014).

The effects of grazing and browsing on vegetation composition and dynamics, by continental tortoise species, are still largely uninvestigated (Falcon et al. 2018). The last representative of giant tortoises in continental ecosystems, the spurred tortoise, is threatened by extinction in the wild, mainly due to collection for the pet trade and the bush meat industry (Garrigues and Cadi 2011; Petrozzi et al. 2018). The ecological effects of the removal of a potential ecosystem engineer are poorly understood. The same problem of lack of vegetation impact studies occurs with other large-sized tortoise species, such as the yellow-footed tortoise (up to 54 kg), the leopard tortoise (up to 48 kg) and the Asian forest tortoise (up to 37 kg). Hopefully, the reintroductions planned for some of those species will mirror the ecological quasi-experiments in Galápagos and in the Mascarenes. Thereby, we would gain invaluable insight into the ecological roles of the last representatives of large tortoises on continental vegetation dynamics, other than as seed dispersers.

Whereas there are many excellent studies on the foraging behavior of iguanids, studies of their role on vegetation dynamics have mainly focused on seed dispersal (Valido and Olesen 2007). As with tortoises, many iguanas play an important role increasing the germination rate and decreasing the germination time of several angiosperm species, some of the plants also threatened (Traveset 1990; Laurel et al. 2000; Moura et al. 2015; Traveset et al. 2016). The smaller size of iguanas, and their low metabolism (averaging just ~70 $kcal.day^{-1}.kg^{-1}$ for *Iguana iguana*; Lichtenbelt 1992), suggest they may influence vegetation dynamics through browsing and grazing to a lesser degree than do tortoises. However, by attaining extremely high densities, in both continental and insular ecosystems (up to 364 $ind.ha^{-1}$ in continental *Iguana iguana*; Rodda 1992), and by browsing or grazing over a small home range (Moura et al. 2015), they may be able to affect vegetation composition by: i.) reducing the leaf area of browsed species; ii.) increasing the amount of sunlight that reaches the undergrowth; and, iii.) affecting edaphic mosaics in the soil, by the repeated deposition of feces in latrines. It is hoped that these research gaps will be addressed in the future.

Marine, grazing reptiles, such as marine iguanas and green turtles, feed mostly on algae. Marine iguanas forage non-selectively on subtidal and intertidal zones, choosing abundant red and green algae species and avoiding brown algae (Shepherd and Hawkes 2005). While each adult consumes a fresh mass of just 35 g of algae per day, they can reach high densities, so populations of marine iguanas can crop a sizeable amount of fresh algal material (27–29 tons yearly for a single population in a 3 km^2 peninsula on the Galápagos Islands; Nagy and Shoemaker 1984). Together with decreased water temperature, during El Niño years, population numbers of marine iguanas are often limited by bottom-up effects on the algae cover of rocky substrata (Wikelski and Thom 2000). Green turtles, the only other herbivorous marine reptile, also feed almost exclusively on algae and sea grass. They prefer low-fibre and high nitrogen species, mainly *Thalassia* spp. (turtle grasses), although all algae grazed by green turtles are low-quality forage (Bjorndal 1980). Sea turtles graze on algae and sea grasses, keeping them fertilized with their dung (Moran and Bjorndal 2005; Hearne et al. 2018). Sea turtles further parallel the terrestrial ungulates by having complex space use relationships with their predators. The food resource of turtles could be at risk under conditions that their natural predators, namely sharks, are rare or absent due to overfishing (Heithaus et al. 2014).

15.3.3 Direct Impact of Domestic Herbivores

Habitat destruction resulting from introduced ungulates is a classical threat to grazing and browsing reptiles (Fig. 15.3). The transformation of forest ecosystems into grasslands, or the poor management of ungulates in natural grasslands, continue to negatively affect grazing and browsing reptiles all over the world. Uromastyx lizards—important ecosystem engineers (Williams et al. 1999)—are negatively impacted by heavy grazing pressure imposed by livestock (Cunningham 2000).

Fig. 15.3 A grassland grazed by introduced cattle on Santa Cruz Island, in Galápagos. In this photo, a cow and a giant tortoise, *Chelonoidis porteri*, share a field. https://www.alamy.com/stock-photo-wild-galapagos-giant-tortoise-geochelone-elephantopus-with-cow-on-164775731.html

Other grazing species, such as the tortoises of the genus Gopherus and the spurred tortoise, are also negatively affected by competition for forage resources with cattle, *Bos taurus*, (Grandmaison et al. 2010; Ureña-Aranda et al. 2015; Becerra-López et al. 2017; Petrozzi et al. 2018). Also, habitat loss in the form of forest fires, set by cattle-ranchers in tropical ecosystems, are frequent all over the world (Gibbons 1984; Wiewandt and García 2000; García and Gerber 2016; Tershy et al. 2016). Direct ecosystem conversion for beef and dairy cattle production is bound to continue (Fearnside 2018), whereas the harvest of native species (including reptiles), for food, remains largely unmanaged, unsustainable and irresponsible in most tropical countries (Fa et al. 2002; Fernandez et al. 2012; Terborgh and Peres 2017).

15.4 Concluding Remarks and Further Developments

Rather than biological oddities, large-sized herbivorous reptiles are fundamental components of some vertebrate communities of tropical and subtropical environments, all over the world (Miranda 2017). The low metabolic rate of reptiles allows them to exist at extremely high biomasses, as testified by early naturalists (Von Humboldt 1877; Leguat 2017), or in ecosystems that have not yet been destroyed by humans (Bourn et al. 1999; Mourão et al. 2000). Many of these important species have disappeared, with a higher rate of extinction for giant lizards and tortoises (Turtle Extinctions Working Group 2015; Slavenko et al. 2016). Ecological networks have been impoverished by the loss of these large herbivores, likely resulting in simpler, shorter and less resilient trophic networks (Malhia et al. 2016). For the remaining herbivorous reptilian species, their direct impacts on vegetation dynamics are still being discovered. Results coming from natural and planned ecological quasi-experiments on islands show the importance of large herbivorous reptiles for ecosystem composition, structure and function (Gibbs et al. 2010; Griffiths et al. 2013). Another interesting research avenue, yet to be explored, is investigating the role reptilian aquatic herbivores play for the ecosystems in which they exist. Some freshwater turtles, such as the Amazon giant river turtle, *Podocnemis expansa* (up to 65 kg), and the Central American river turtle, *Dermatemys mawii* (up to 22 kg), feed mostly on leaves. These species are seasonal or year-round browsers, and hold the potential to exert pressure on the submerged arboreal vegetation, and act as links between terrestrial and aquatic ecosystems.

Finally, the last decade has seen a multitude of translocation programs targeting threatened tortoises, turtles, and herbivorous lizards (Jones 2002; Attum et al. 2010; Nussear et al. 2012; Gibbs et al. 2014; Grant and Hudson 2015; Falcón and Hansen 2018). These conservation initiatives have been restoring not only threatened species populations, but also lost or diminished ecological interactions. Knowledge about the effects of those translocations on the vegetation dynamics of continental floras will provide information on rebuilding reptile-driven ecological interactions (*sensu* Genes et al. 2017) such as trampling, seed dispersal, grazing and browsing. While considered by many as "primitive" and dull, reptiles are significant players in the ecosystems they inhabit, and perform important functions as predators, pollinators, burrowers and frugivores. Acquiring knowledge on their role as grazers and browsers is a promising ecological research venture.

Acknowledgements
EBPM greatly appreciates the generous financial support of the following donors: Rufford Small Grants Foundation (18743-1 and 23022-2), Rainforest Biodiversity Group, Idea Wild, The Mamont Scholars Program of the Explorer's Club Exploration Fund, Cleveland Metroparks Zoo and the SouthWild.com Conservation Travel System. EBPM receives major logistical support from the Peugeot-ONF Brasil Carbon Sink Reforestation Project, based at Fazenda São Nicolau in the Municipality of Cotriguaçu, Mato Grosso, Brazil. This Project is a Peugeot initiative to fulfill some of the Kyoto Protocol directions and is run by the ONF Brasil enterprise. CSM's research is supported by a scholarship from The National Council for Scientific and Technological Development (CNPq), within the Ministry of Science of Brazil, and by the Young Talents for Conservation Award, by the Luísa Pinho Sartori Institute. We thank Bernardo Araujo and our reviewers Iain Gordon and Herbert Prins for reviewing the manuscript.

15.5 Birds

René van der Wal

15.5.1 *Grazing and Browsing Birds*

Foraging on plants is very widespread among birds. The most common form this takes is the selection of high energy and nutrient forage item such as seeds and fruits. Quite a number of species have specialised in this foraging strategy (see Whelan et al. 2015 for impacts of granivorous and frugivorous birds on ecosystems), but many more take these items as part of a much broader—often omnivorous—diet. Far fewer have adopted grazing or browsing as their main mode of foraging (see Lopes et al. 2016 for a classification of avian diet types), estimated to be the case for 101 species (80 grazers and 21 browsers—Table 15.1), and thus less than 1% of the Worlds' extant bird species (9993—Jetz et al. 2012). The distribution of herbivorous birds is strongly skewed towards a single phylogenetic order and family therein, namely the Anatidae or ducks, geese and swans (Table 15.1). All swans and geese

Table 15.1 Orders and families with bird species for which grazing or browsing is the main mode of foraging. The number of species per family that is primary grazing or browsing was estimated using https://en.wikipedia.org/wiki/aves and https://globalspecies.org/ntaxa/114863 as starting points, and consulting additional secondary sources (e.g., various Cornell Lab or Ornithology resources, Handbook of the Birds of the World https://www.hbw.com) or primary sources for further information or triangulation where information was insufficient. Phylogeny is after Jarvis et al. (2014)

Order	Family	Species (groups)	Primarily grazer	Primarily browser
Struthioniformes	Struthionidae	Ostriches	2	
Rheiformes	Rheidae	Rheas	2	
Cassuariiformes	Dromaiidae	Emu	1	
Anseriformes	Anseranatidae	Magpie goose	1	
	Anatidae	Ducks, geese and swans	51	
	Anhimidae	Screamers	3	
Galliformes	Phasianidae	Grouse		17
Opisthocomiformes	Opisthomidae	Hoatzin		1
Gruiformes	Gruidae	Cranes	6	
	Rallidae	Coots, moorhens	12	
Columbiformes	Columbidae	Pigeons and doves	1	
Psittaciformes	Strigopidae	Kakapo	1	
Passeriformes	Cotingidae	Plantcutters		3
Total number of species			80	21

are obligate herbivores, although occasionally consuming other food items too (e.g., insects, bones or even soil—Abrahams 2013). The extent to which plant-eating ducks consume plant material (other than seeds) varies considerably among species and the time of the year (Olsen 2015). About 20 ducks (e.g., whistling ducks, some diving ducks including pochards, and several dabbling ducks including wigeons, gadwall, *Anas strepera*, and Australian wood duck, *Chenonetta jubata*, can be considered primarily herbivorous, for at least part of the year (Table 15.1), but all will shift towards a more omnivorous diet during spring and summer to meet protein requirements for breeding and growth (Sedinger 1997).

Whilst many of the above Anatidae also consume seeds when present (and thus are in part, and to a various extent, also granivorous), grazing of grasses, sedges and forbs is the predominant mode of foraging. Grazing typically concerns the consumption of aboveground tissue of non-woody plants (notably leaves but also stems and seed heads, the latter particularly in the case of agricultural crops), but this regularly spills over in the extraction of belowground plant organs such as tubers and rhizomes (Fox and Abraham 2017). For some species groups, shorter or longer periods of the year are dedicated to such foraging for energy-rich storage organs—captured by the term 'grubbing' (Anderson et al. 2012)—which requires considerable strength, and hence is typically conducted by larger species (e.g., ostrich, cranes, larger-bodied geese, swans; Zhiheng and Clarke 2016). Several other aquatic families, or members thereof, such as screamers, coots and some moorhens, are primarily grazers—or consumers of belowground plant parts—too, as are several flightless members of the phylogenetically oldest clade (Palaeognathae) of extant birds (emu, *Dromaius novaehollandiae*, rheas), the common wood pigeon, *Columba palumbus*, and the large, flightless, and critically-endangered kakapo, *Strigops habroptilus*, whose diet (e.g., clubmosses, ferns and bark, and thus leaning towards being a browser—Best 1984) is as unusual as its appearance. Many more bird species consume fresh green shoots, flowers and buds (e.g., pheasants, turkeys, rails, turacos), or extract rhizomes, roots and bulbs (e.g., bustards, sandgrouse, tinamous), but only in limited amounts, and at certain times of the year, and are thus not genuine grazers.

Grouse is the family of birds with most genuine browsers (all 17 species—Table 15.1), often almost exclusively living from shrubs, or tree leaves and needles, outside the breeding season. Many also consume seeds, with the exception of the two sage-grouse, which lack a muscular gizzard and, therefore, consume only soft-tissue foods (mainly leaves of sagebrush). Within the breeding season, grouse diets are generally enriched with invertebrates (Sedinger 1997). Other bird species that forage primarily on buds, leaves and twigs are the three plantcutter—the only folivorous passerines (Bucher et al. 2003)—and the Amazonian-dwelling hoatzin, *Ophisthocomus hoazin*. The latter has even developed a crop and oesophagus with fermentation functionality akin to rumen-digestion (Grajal et al. 1989; Godoy-Vitorino et al. 2011).

With the exception of the Rattites (large flightless birds including ostriches, rheas and emu) and the (poorly flying) hoatzin—both of which can digest cellulose (Swart et al. 1993; Grajal et al. 1989)—the requirement to fly has prevented birds from developing relatively heavy but efficient digestive systems to deal with e.g., plant material (Caviedes-Vidal et al. 2007; but see Hunt et al. 2019 for limited evidence between flying ability and caecal length—greatest for grazers—within birds). This has two important consequences: firstly, a limited ability to digest cellulose and other highly recalcitrant plant parts (Prop and Vulink 1992; Durant 2003), meaning that large amounts of plant material need to be consumed; secondly, there is an even greater premium on plant quality than is the case for most other herbivores, as selection of the most nutritious and/or energy rich (depending on requirements and time of the year) food items—in itself a time-consuming process—is the only way to maximise daily nutrient and/or energy gains. Combined, many herbivorous birds have to forage for most of the day (Prop and Vulink 1992). For grazers, having to use a bill as feeding device brings the additional constraint of reduced bite size as compared to mammalian grazers. This may, in part, be compensated for by employing a greater bite frequency in order to achieve comparable instantaneous food intake rates (Steuer et al. 2015), or require further adaptation, in the form of yet longer daily foraging times (Van Gils et al. 2007). This puts in sharp focus the trade-off between plant quantity and quality, and with it a set of key behavioural adaptions (see below) to try and achieve the highest possible intake rates of high quality food items. Whilst none of these adaptations is unique to birds, and indeed trade-offs between quantity and quality rule foraging decisions of all herbivores (and beyond), birds' limited digestive ability of plant material tends to make conditions more extreme.

15.5.2 Ecosystem Impacts of Grazing and Browsing Birds

Many of the ecosystem impacts of birds observed today could be understood in the context of digestive constraints on herbivorous birds, whereby the ability to i) aggregate into groups to optimise foraging time and food finding (Stahl et al. 2001; Gyimes et al. 2010; Kułakowska et al. 2014); ii) undergo carefully scheduled movement and migration, to and from a limited number of sites, and thus requiring individuals, in a number of herbivorous bird species, to make decisions at larger spatial and temporal scales (Arzel et al. 2006; Van der Graaf et al. 2006; Shariatinajafabadi et al. 2014); and, iii) seek out, and subsequently utilise, new 'high quality' foraging opportunities (Van Eerden et al. 2005; Si et al. 2015; Fox and Madsen 2017; Buij et al. 2017), are key behavioural adaptations to those constraints.

For grazing or browsing birds, to exert large-scale impact on the vegetation, species need to be abundant and use a location for a reasonable length of time. As a consequence, herbivorous birds that are gregarious tend to have greatest impact on the ecosystems they inhabit (Wood et al. 2012; Kollars et al. 2017), though exceptions occur (e.g., ptarmigan controlling shrub architecture—Tape et al. 2010). Large aggregations concern mostly—but not exclusively (e.g., coots, cranes, wood pigeons)—"waterfowl", a loosely defined label that typically concerns ducks, geese and swans, and sometimes also rails. As the label suggests, waterfowl use both terrestrial and aquatic ecosystem components, and impacts on land are often inextricable connected to those occurring in nearby freshwater, or marine, ecosystems and vice versa (Hessen et al. 2017; Kollars et al. 2017).

15.5.3 Bird Density and Existence of Plant Refuge Determine Marine Ecosystem Impact

Impacts on marine ecosystems are restricted to the intertidal or shallow subtidal, and those zones are typically used seasonally, coinciding with spring or autumn migrations. For example, beds of seagrass, *Zostera spp*., are in high demand by brent geese, *Branta bernicla*, black swan, *Cygnus atratus,* mute swan, *Cygnus olor*, and American wigeon, *Anas americana*, but Zostera also occurs frequent in the diet of a further 20 waterfowl species (Kollars et al. 2017). Consumption of both above- and belowground components may severely reduce seagrass abundance, though accumulation of seeds in feeding pits, in autumn, can facilitate seedling recruitment in spring, thereby contributing to the persistence of seagrass beds (Zipperle et al. 2010). In some areas waterfowl may control the abundance and distribution of seagrasses (Kollars et al. 2017), in other areas (e.g., Gulf of Mexico where redhead duck, *Aythya americana*, forages extensive on seagrass beds that represent economically important fish nursing habitat—Kennedy et al. 2018), impacts appear to be limited. Critical to the extent of bird herbivory impact, in shallow marine, as well as comparable freshwater, systems (e.g., swans foraging on pondweeds, *Potamogeton* spp—Klaasen and Nolet 2007), is not only the density of birds, and duration of resource use, but also the existence of spatial refuges for plants (e.g., belowground plant parts that are out of reach—Santamaria and Rodríguez-Gironés 2008), and safe sites for seedling establishment to ensure plant regrowth and recruitment. Although recovery of previously endangered waterfowl, due to successful conservation efforts, combined with ongoing expansion of already highly successful species, may exert further impact on seagrass beds (Kollars et al. 2017), the attraction of most *Zostera*-consuming species to agricultural land for foraging (Fox and Madsen 2017) is likely to mitigate against greater bird grazing impacts on marine ecosystems.

15.5.4 Expanding Waterfowl Populations Contribute to Top-Down Control of Freshwater Ecosystems

The impacts of ducks, geese, swans and rails on freshwater ecosystems are manifold, but key are the accumulation of nutrients, bio-turbidity and the consumption of macrophytes (Wood et al. 2012; Bakker et al. 2018). A further significant, although less directly observable, impact, is the contribution by waterfowl to wetland biodiversity around the world, through the transport of a wide variety of aquatic plants and animals between wetlands and water bodies (Van Leeuwen et al. 2012) Extensive consumption of water plants, in lakes, by birds, does occur, but in most cases this is limited to very shallow, and often heavily vegetated, systems (e.g., swamps), although ponds may also be subject to intense waterfowl herbivory (and indeed kept apparently free of vegetation—Van Onsem and Triest (2018)). Within deeper lakes, there may be brief episodes of heavy grazing, for example when large numbers of waterfowl gather to moult during which they are flightless, and hence prone to predation (e.g., moult migration by greylag geese, *Anser anser*—Zijlstra et al. 1991). Whereas 3 decades ago, recognition of waterfowl shaping freshwater ecosystems, beyond single sites, was limited, the rapid increase in waterfowl abundance in recent years has changed this. In a review, Bakker et al. (2018) concluded that, currently, herbivorous birds remove 40–48% of plant biomass in aquatic ecosystems, well in excess of exploitation levels found in most terrestrial ecosystems. Impacts are manifold with, for example, the dramatic expansions of greylag geese being held responsible for the decline in reed, *Phragmites europaeus*, beds across Europe (Bakker et al. 2018), with knock-on effects on numerous species of insects and breeding birds. Moreover, the accumulation of nutrients, deposited in the water as uric acid and faeces, in no small part originating from food plant consumed in the terrestrial realm, is now viewed as default, and increasingly the main eutrophicating agent in lakes and ponds (Chaichana et al. 2010; Hessen et al. 2017). Whilst there is strong evidence for birds causing eutrophication, bioturbation and the reduction of aquatic plant abundance, and for bird density to be related to the extent of plant biomass suppression, this is also the case for other species groups (e.g., mammals, crustaceans, molluscs, fish, echinoderms—Wood et al. 2016 and elsewhere in this Chapter and Book). Hence, the current, and likely further, increasing impacts of herbivorous birds on freshwater ecosystems may best be viewed as contributing to a wider top-down control of primary productivity. It is likely that only part of the complexity of herbivorous bird impact on freshwater ecosystems has been revealed because of the large number of species interactions and mechanisms involved (e.g., influence of carp on habitat choice of water birds—Haas et al. 2007). Whilst the impact of birds on macrophytes, in water bodies, is thus beyond doubt, its impact on rivers is still deemed limited, and mostly local (Franklin et al. 2008).

15.5.5 Agriculture Drives Grazing Bird Abundance and Terrestrial Ecosystem Impacts

Numerous studies have revealed the intricate nature of impacts exerted by grazing and browsing birds on terrestrial ecosystems. Key to large-scale herbivore impacts are changes in wider land use, government policy and climate which, collectively, have allowed, notably waterfowl, to greatly increase in abundance (e.g., Jefferies et al. 2004; Van Eerden et al. 2005). Long-term changes in the quality of the vegetation on offer in grasslands and arable lands have, together with reductions in hunting practice and the formation of nature reserves, driven down over-winter mortality of waterfowl (Fox and Madsen 2017). Particularly dramatic population expansions have occurred in arctic-breeding geese, raising the status of e.g., the lesser snow goose, *Chen caerulescens*, to one of the world's most abundant grazing birds. The impact of this species on their breeding grounds has been evidenced particularly well in and around La Pérouse Bay, Hudson Bay area, subarctic Canada. Here, the relationship between snow geese and the subarctic tundra vegetation changed from one characterised by finely tuned herbivore-plant-soil feedbacks, which allowed geese to optimise food intake (Hik and Jefferies 1990), to a situation in which extensive loss in vegetation cover occurred, due to grubbing. The dramatic increase in the number of breeding and spring-staging lesser snow geese, together with a comparably low number of Ross's geese, *Chen rossii*, set in motion a positive feedback loop, wherein grubbing for belowground plant parts caused vegetation cover decline and soil exposure, which led to the development of hypersaline topsoil, in turn killing off extant vegetation (Srivastava and Jefferies 1996), and reducing plant recovery potential (Handa et al. 2002). This resulted in—for avian herbivory—exceptionally large-scale denudation of, notably, saltmarsh habitat, and to a lesser extent also freshwater habitat, in and around snow goose colonies, all along the 2000 km-long Hudson Bay coastline (Jano et al. 1998; Jefferies et al. 2006). Because recovery is slow (freshwater marsh) or absent (saltmarsh; Peterson et al. 2013), lack of vegetation causes geese to move away and establish colonies elsewhere, resulting in further goose impacts across the region.

Similarly long-lasting, and large-scale, disruptions of the vegetation can now be observed across the Canadian Arctic. For example, between 1988 and 2011, well north of the Hudson Bay in Nunavut, a five-fold increase in area of exposed peat habitat, from 269 to 1373 km^2 in a 36,370 km^2 large study area, and a 48% loss of wet sedge meadow, were observed (Conkin and Alisauskas 2017). Although goose impacts in large parts of the Canadian Arctic remain limited, goose numbers are very high, colonies numerous, and local grazing impacts strong and replicated over large areas. Hence, region-wide ecosystem impacts are likely, including population declines of other bird species e.g., shorebirds and passerines (Flemming et al. 2016). Other Nearctic goose populations are less large, but otherwise following suit (e.g., Canada goose *Branta canadensis*, greater snow goose), and impact staging and breeding grounds elsewhere. Hunting pressure has been increased significantly in an attempt to curtail goose populations (e.g., Lefebvre et al. 2017), and reduce the pressure on arctic ecosystems as well as on agricultural land further south used by geese outside the breeding season, and which fuel the population expansions. Hunting has stabilised the population of greater snow geese (to around one million birds—Lefebvre et al. 2017), but the, more than an order of magnitude, more abundant population of lesser snow geese continues to grow despite huge hunting efforts (Alisauskas et al. 2011).

Whilst Palearctic goose populations expanded almost as spectacularly, impacts on their Arctic breeding grounds are far less pronounced. Pink-footed goose impacts, which like snow geese are sufficiently powerful to dig for belowground plant parts (i.e., grubbing) early on in the breeding season (Anderson et al. 2016), do now occur throughout the productive parts of Svalbard, but with the initial disruption of the vegetation overlying organic soils recovering swiftly (Speed et al. 2010). Colony-breeding geese that do not have the physical ability to grub extensively, such as the barnacle goose, can locally also be influential—suppressing e.g., vascular plant and moss biomass and the carbon sequestration ability of tundra—but again recovery, when grazing pressure reduces, is rapid (Sjögersten et al. 2008, 2011), and hence is best not labelled "habitat degradation" (Van der Wal 2006). A fundamental difference with the coastal zone of NE Canada, subject to isostatic uplift (Jefferies et al. 2006), is the absence of saline subsoil; when exposed through grubbing, soils in the Hudson Bay turn hypersaline which kills the vegetation, and puts vegetation succession on long-term hold. Thus, waterfowl impacts on terrestrial ecosystems may be strong and long-lasting, but particularly so when amplified by specific abiotic conditions.

Extensive use of agricultural land—often superior in foraging quality to semi-natural habitats (e.g., Fox and Madsen 2017; Dokter et al. 2018)—by a growing number of grazing birds (Van Eerden et al. 2005), has attracted widespread disapproval among, notably, the farming community, leading to loci of conflict around specific species. Evidence for such grazing impacts is mixed (see Fox et al. 2017 for a review), but the conflicts remain real. Whilst overwintering species (of notably geese) were initially seen as most problematic by some, there is a shift towards a broader discontent about extremely rapidly growing populations of year-round bird species. Within Europe this concerns is expressed most prominently towards greylag geese (a species which has increased particularly rapidly—Fox and Madsen 2017), but also towards other expanding species originating, at least in part, from collections (e.g., Canada goose, Egyptian goose, *Alopochen aegyptiaca*, mute swan, barnacle goose, white-fronted goose, *Anser albifrons*). Increasingly, questions are being asked about how such summer herbivory may impact various ecosystem functions, including food stocks of migratory bird species arriving in autumn (e.g., Gyimesi et al. 2011). In many countries, resident waterfowl species have also penetrated into urban environments, and established sizeable populations. Their omnipresence, and diversity of impacts on society, ranging from fouling of lawns and pavements to aggressive encounters, air traffic accidents, and concerns about zoonotic diseases (Buij et al. 2017), has rendered several species subject to sometimes dramatic population control. The rapid expansion of, notably, geese, but also ducks, cranes, coots, wood pigeons and other species benefitting from almost unlimited feeding opportunities provided by farmland landscapes, and the resulting diversification of impacts, is leading to significant shifts in the dynamics of different interest groups' views on the future of herbivorous birds (Cusack et al. 2018). This is an area of growing academic, political and conservation interest, and is likely to spill over into other areas of society, across large areas of the northern hemisphere (Milton 2000; Fox and Madsen 2017).

15.6 Marsupials

Ben Moore and William Foley

Amongst marsupials, browsers and grazers are found almost exclusively among the Australasian order Diprotodontia (Hume 1999), with significant herbivory not reported, and certainly unstudied, in South American marsupial lineages. This assemblage evolved in the absence of eutherian browsers and grazers, to occupy a diversity of ecological niches, throughout the Australian continent (Tyndale-Biscoe 2005). However, the diversity, size range and impact of marsupial browsers and grazers is much reduced from that prior to the Pleistocene extinction of megafaunal browsers including the short-faced kangaroos, *Simosthenurus* spp., the chenopod specialist, *Procoptodon goliah*, the "marsupial tapir", *Palorchestes* spp., and the diprotodontines, *Euryzygoma* spp., *Diprotodon* spp. and *Zygomaturus* spp., (Johnson 2006; Prideaux et al. 2009). The only native Australian non-marsupial mammalian browsers and grazers are rodents, including the highly-specialised, vole-like grass-eating broad-toothed rat, *Mastacomys fuscus*, the swamp rat *Rattus lutreolus* and the arid-zone sticknest rats, *Leporillus* spp., (Breed and Ford 2007; Aplin and Ford 2014). Other omnivorous and primarily seed- and fruit-eating rodents (e.g., *Melmoys cervinipes*), also include grass and leaves in their diets, to varying extents (Breed and Ford 2007). In consequence, marsupial browsers and grazers account for most Australian native vertebrate herbivory. The direct impacts of marsupial herbivory are rarely quantified, except where economic impacts occur, or are perceived, and where species are judged to be overabundant, and detrimentally affecting vegetation condition or biodiversity values (e.g., Barnes and Hill 1992; Coulson 2007; Di Stefano et al. 2007).

15.6.1 Marsupial Grazing

Grazing, as the main dietary habit amongst marsupials, is restricted to some species in the family Macropodidae (kangaroos and wallabies), and to all members of the family Vombatidae (wombats). Arman and Prideaux (2015) classified the diets of 37 extant macropods, and identified only 3 browsers (diet comprising >70% dicot foliage; these were two arboreal tree kangaroos and the small terrestrial quokka) and 9 grazers (>70% grass; large kangaroos and wallabies), while the greatest number (19) were mixed feeders. Previous classifications, with a greater emphasis on dental morphology, had identified many of these mixed feeders as browsers (e.g., Sanson 1989).

Because of their perceived economic impact, the impacts of large and abundant grazing marsupials have attracted the most attention from ecologists. Several species of large kangaroos, in particular red kangaroos (*Osphranter rufus*, formerly *Macropus rufus*—here we adopt the taxonomy of Jackson (2015)) together with the euro/wallaroo, *Osphranter robustus*, and eastern grey, *Macropus giganteus*, and western grey, *M. fuliginosus*, kangaroos, often graze the same pastures as domestic livestock, and macropods including grey kangaroos can aggregate in crops, causing cereal production losses (Coulson 2007). Agile wallabies, *Notamacropus agilis*, and black-striped wallabies, *Notamacropus dorsalis*, are sometimes, locally, identified as pests of pasture and/or crops in northern Australia and Central Queensland, respectively (Baxter et al. 2001; Bedoya-Pérez et al. 2017).

There has been a long-standing debate about the extent to which these macropods and livestock, particularly sheep, compete for the same pasture resources. Demonstrating intraspecific competition is difficult, and the evidence, for competion between livestock and macropods, can only be considered insufficient to weak (e.g., dietary overlap with a lack of evidence that resources are limiting; Spear and Chown 2009, Prins 2016). In arid New South Wales, Dawson and Ellis (1996) found little dietary overlap occurred between sympatric euros and sheep, but a large manipulative experiment, in the same region, showed that the diets of red kangaroos and sheep overlapped considerably. This was true whether they grazed together or separately, but the degree of overlap varied with rainfall and available pasture biomass (Dawson and Ellis 1994; Edwards et al. 1995, 1996). Although sheep diets were affected, during dry times, by the presence of kangaroos (they consumed more chenopod shrubs), kangaroo diet was unaffected by the presence of sheep. It appears possible then that competition may occur when red kangaroos deprive sympatric sheep of pasture when pasture biomass is low, but moderate or strong support for either competition or facilitation between macropods and livestock, such as evidence of detrimental effects on individuals or populations of putatively competing species, is lacking (Edwards et al. 1995; Dawson 2012).

Australian rangelands are characterized by non-equilibrium dynamics, and at times both kangaroos and sheep can reduce pasture to an ungrazeable level (Short 1985), at which times food must become, at least transiently, limiting. Whilst primary producers can re- or destock, to rapidly alter grazing pressure as circumstances demand, the numerical response of kangaroo species, although high (Bayliss and Choquenot 2002), is sometimes insufficient to take advantage of the transient availability of good pasture. Red kangaroos, and perhaps, euros are exceptions in that they can travel long distances to take advantage of local rainfall events (Croft 1991; Clancy and Croft 1992).

Only weak evidence can be found for competition or facilitation between macropods and cattle. Consistent with competition, black-striped wallabies, *Notamacropus dorsalis*, show very strong dietary overlap with cattle in Central Queensland and consume economically very significant quantities of pasture (Baxter et al. 2001). In the Simpson Desert, removal of cattle resulted in a progressive increase in kangaroo numbers over a period of approximately 2 years compared to areas where cattle remained (Frank et al. 2016), but over the short duration of that study, kangaroo feed intake (estimated from dung deposition) remained a tiny fraction of that of cattle. More consistent with facilitation, time-series data, from across a large area of the South Australian pastoral zone, revealed that the presence of sheep and cattle had a positive effect on the population growth rate of red kangaroos, although this result was interpreted by Jonzen et al. (2005) as indicating that livestock are a surrogate for resource availability beyond rainfall. At a smaller scale, Payne and Jarman (1999) observed immediate disturbance impacts of cattle on grazing kangaroos in eastern Australia and found these to be minimal, with kangaroo groups distributed more closely to cattle groups than expected by chance.

A second aspect of the debate about the impact of kangaroo grazing on domestic livestock focusses on the amount of food consumed by kangaroos relative to sheep. The safe stocking rate of rangelands for sheep is usually expressed in terms of "dry sheep equivalents" (DSE) which is the amount of food required to support a "typical" 2-year-old ~ 45 kg non-lactating Merino sheep (Turner and Alcock 2000). This unit is then used to compare different livestock management options. Dawson and Munn (2007) note that kangaroos are regarded, by some, as eating twice as much food as sheep (i.e., 2 DSE) but values of around 0.7 are typically assumed by Departments of Agriculture, because the basal metabolic rate of marsupials is generally about 70% of that of eutherian mammals (Munn and Dawson 2003). Dawson and Munn (2007) bring together data from many sources, including from heart-rate telemetry (Munn et al. 2009), and observations of bite rates (McLeod 1996) to estimate feeding rates of free-living kangaroos. They conclude that, on rangelands, kangaroos can be considered to be equivalent to 0.4 DSE in the event that competition is occurring.

Although much research has been directed towards the implications for domestic stock of grazing by kangaroos, attention is turning to the effects of grazing on ecosystems and biodiversity, particularly by high-density kangaroo populations. Grazing marsupials can engineer environments by maintaining lawns—bare-nosed wombats, *Vombatus ursinus*, can exclude woody plant establishment, and maintain lawns, in Tasmanian wetlands (Roberts et al. 2011), and both bare-nosed and hairy-nosed wombats, *Lasiorhinus latifrons*, can maintain lawns, in well-drained areas, elsewhere, by closely and repeatedly cropping their preferred tough native perennial grasses (Hume 1999; Tyndale-Biscoe 2005; Kirkpatrick 2016). Observations have been reported of up to 80 wombats grazing by day in some Tasmanian locations (Temby 1998). Red kangaroos can also maintain grazing lawns in Central Australia (and depend upon them during drought), but initial establishment of these lawns is facilitated by removal of senescent biomass by cattle (Newsome 1997).

Experimental studies have shown that high densities of kangaroos, at multiple locations in south-eastern Australia, have significant negative effects on ground layer plants (Sluiter et al. 1997; McIntyre et al. 2010, 2015), tree seedlings and saplings (Noble 2001; Meers and Adams 2003; Allcock and Hik 2004; Stapleton et al. 2017), beetles (Barton et al. 2011), grassland-dependent reptiles (earless dragons, *Tympanocryptis pinguicolla*, Howland et al. 2014, and striped legless lizards, *Delma impar*, Howland et al. 2016a), and reintroduced eastern barred bandicoots, *Perameles gunnii*, (Winnard and Coulson 2008). Browsing damage by kangaroos can be particularly significant during droughts, when grass biomass is depleted (Couslon and Norbury 1988). For several other groups, including web-spinning spiders (Foster et al. 2015) and birds (Howland et al. 2016b), the results were more complex with some species of birds being affected and others not.

In semi-arid, low-productivity Australian rangelands and woodlands, where kangaroos exist at more moderate densities, the intensity of their grazing is substantially less than that of livestock or rabbits, *Oryctolagus cuniculus*, (Vandandorj et al. 2017). Structural equation and multiple linear regression models that have been designed to identify the relative impacts of different types of herbivory and abiotic factors, suggest that the impacts of kangaroos on plant species composition (Travers et al. 2018), plant species diversity (Eldridge et al. 2018), shrub and tree short-term recruitment and long-term regeneration (Tiver and Andrew 1997), soil health (Eldridge et al. 2017) and regulating and provisioning ecosystem services (Vandandorj et al. 2017), are correspondingly benign, again in contrast to past and present impacts of exotic herbivores, particularly sheep. Indeed, under low productivity, kangaroo grazing can increase native plant species richness (Eldridge et al. 2018). In contrast, Rees et al. (2017) experimentally excluded kangaroos from plots in the Strzelecki Desert, demonstrating that in the absence of dingoes (*Canis dingo*), which otherwise suppress their numbers, kangaroos can suppress pasture biomass and grass seed production. This in turn likely contributes to the decline of small graminivorous birds.

Solutions to the impacts of high kangaroo densities are varied and can involve restoring coarse woody debris into habitats to protect saplings and create microhabitats for ground flora (Stapleton et al. 2017), direct reduction of kangaroo numbers through shooting (from densities of ~2–4 ha^{-1} to as low as 0.4 ha^{-1}), and the possible re-introduction of apex predators such as the dingo (Letnic et al. 2009). Long-term studies indicate that grassland recovery, from heavy grazing pressure by kangaroos, can be a slow process, particularly if the grazing pressure from kangaroos is replaced by that of other herbivores, such as rabbits. Although direct reduction of kangaroo numbers is a simple long-term strategy, in some places there can be significant community and professional opposition to the process, and accompanying actions (Ben-Ami et al. 2014; McKinnon et al. 2018). Nonetheless, the four large kangaroo species are the focus of Kangaroo Management Plans, and permits for culling are assessed each year (Department of Parks and Wildlife, Western Australia 2013; Department of Environment, Water and Natural Resources, South Australia 2017; Department of Environment, Land, Water and Planning, Victoria 2017; Department of Environment and Heritage Protection, Queensland 2018). Southern hairy-nosed wombats and bare-nosed wombats are also occasionally culled under license, in response to perceived impacts on cereal crops and pasture (Marks 1998), although the impact of their digging and damage to fences is generally more problematic (Triggs 2009).

The potential also exists for competition between marsupial browsers and grazers and non-livestock herbivores species, most particularly rabbits. Strong, to very strong, experimental evidence has been presented to show that competition from rabbits negatively affects bare-nosed wombats (Cooke 1998) and red kangaroos (Cooke and Mutze 2018). Grazing by rabbits favours (often exotic) annual grasses and forbs over native perennial grasses, and these altered grasslands are less able to support the more protracted reproduction of marsupials (Tyndale-Biscoe 2005).

Both in Australia, and in its introduced range in New Zealand, the browsing common brushtail possum, *Trichosurus vulpecula*, sometimes feeds on pasture, particularly on clover, which can account for up to 30% of its diet (Harvie 1973). Possums have sometimes been considered to compete for pasture with livestock, with estimates of grazing equivalence as high as 0.072 stock units (SU; the New Zealand SU is for a 55 kg lactating ewe, so exceeds the Australian DSE) but a more convincing estimate is 0.01 SU (Cowan 2007). More strikingly, in Tasmania, the combined effects of possum and wallaby grazing have been suggested to reduce dry matter yields by up to 94% and 48% for improved and native pasture, respectively (Statham and Rayner 1995).

15.6.2 Marsupial Browsing

Marsupial browsing impacts can be attributed to browsing and mixed-feeding macropods, and a number of arboreal folivores. The arboreal folivores include several widespread and locally abundant species that specialize, to varying extents, on the foliage of eucalypts, and of these koalas, *Phascolarctos cinereus*, and common ringtail, *Pseudochirus peregrinus*, and common brushtail possums have all been linked to defoliation and mortality of trees, when they occur in high densities, although the impact of common brushtail possums has been most thoroughly investigated in its introduced range in New Zealand (Cowan and Waddington 1990; Pekelharing et al. 1998; Duncan et al. 2011).

In the absence of predation, or significant impacts of disease, koala populations can exhibit rapid population growth, resulting in population densities as high as 20 ha^{-1}, reported from Cape Otway, Victoria (Whisson et al. 2016). At high koala population densities, browsing is unsustainable, and results in widespread tree and koala mortality (Fig. 15.4) (either directly from starvation or from euthanasia of malnourished captured koalas by management agencies; Martin 1985a, b, Department of Environment, Land, Water and Planning, Victoria 2016) and loss of koala habitat, as well as posing a threat to certain vegetation classes and their associated biodiversity (Department of Sustainability and Environment, Victoria 2004). Localised events, like these, have been recorded repeatedly since the early twentieth century and occur most commonly in isolated patches of habitat on islands or surrounded by cleared land (Kershaw 1934; Menkhorst 2008). Koala overabundance is usually associated with *Eucalyptus viminalis* across a variety of soil types in coastal regions of Victoria and South Australia, although it can also develop in *E. ovata* and other species (Martin 1985a), and has been reported in association with *E. tereticornis* from northern coastal NSW (Frith 1978).

The population dynamics of koalas, in *E. viminalis* forest, have been well studied, and modelled (Todd et al. 2008; Whisson et al. 2016). Ramsey et al. (2016) estimated that the koala carrying capacity of the manna gum forest at Cape Otway is 5.3–8.3.ha^{-1}, but cautioned that this estimate has low precision, and may be too high to allow recovery of previously overbrowsed trees; an appropriate management

Fig. 15.4 Dense populations of arboreal browsing marsupials can cause severe overbrowsing of the forest canopy. In this forest at Cape Otway, Victoria, koalas have overbrowsed and killed all mature manna gum, *Eucalyptus viminalis*, trees, while the less preferred messmate stringybarks, *E. obliqua*, have survived. Photo Credit: Ben Moore

target for a "safe" population density is probably lower than this estimate. Most koalas show strong fidelity to their small home ranges (0.4–1.2 ha at Cape Otway; Whisson et al. 2016), even in the face of declining food resources, and despite adequate habitat connectivity making dispersal possible. A model of koala-manna gum dynamics, at Mt. Eccles in western Victoria, predicted the koala population increasing to peak abundance in 10–18 years before crashing—with the impact on tree mortality dependent upon the eventual maximum rate of decline of the koala population (Todd et al. 2008). Eucalypts have a remarkable capacity to replace foliage after defoliation or fire, but energy (starch) reserves can be depleted if repeated defoliation occurs over an extended period (Bamber and Humphreys 1965). A management strategy, suggested by Todd et al. (2008) on the basis of their modelling, requires the sterilization of a variable number of female koalas each year in order to maintain a fixed number of non-sterilised females. Chemical (and formerly, surgical) sterilization (Hynes et al. 2010) is a widely-implemented koala management strategy in South Australia and Victoria, as is translocation of koalas from overabundant populations.

Increasingly, koalas are now reaching high densities, in commercial plantations of *E. globulus* established in the 1990s in Victoria and South Australia (Natural Resources Kangaroo Island 2017, Department of Environment, Land, Water and Planning, Victoria 2018; Department of Environment, Water, and Natural Resources, Government of South Australia 2017). Although significant impacts on plantation productivity have not been claimed (nor measured), costs to the plantation industry are considerable, because harvesting operations must be modified, and significant reputational cost is incurred by forestry companies when koalas are killed or injured (e.g., ABC News 2013; HVP Plantations 2018). Marsupial browsing of eucalypt seedlings in Tasmanian hardwood plantations, is more obviously detrimental to productivity, because it reduces seedling growth and survival and affects tree form (Bulinski and McArthur 1999; Scott et al. 2002). Consequently, Tasmania has implemented intensive browser control measures and encouraged research into the problem (e.g., Miller et al. 2009; Close et al. 2010; Miller et al. 2011). While brushtail possums, pademelons and wallabies have all been implicated, Bulinski and McArthur (2003) suggested that the relative contribution of brushtail possums had previously been underestimated.

The second specialist folivore of eucalypts (in addition to the koala) is the greater glider, *Petauroides volans*, which does not appear to reach population densities that measurably affect tree canopy biomass. However, the two generalist browser species, that can feed to significant extents on eucalypt foliage, i.e., the common brushtail and ringtail possums, can both occur in very high densities, and can cause local canopy loss and eucalypt mortality (Loyn and Middleton 1980; Low 2002; Yugovic 2015).

Dramatic impacts of common brushtail possums, on native vegetation, have been reported when they have been introduced to islands, such as the Keppel Islands in Australia, and most famously, to New Zealand (Low 2002). Possums selectively browse on, and cause crown dieback of, numerous native forest trees and mistletoes in New Zealand (Sweetapple 2008; Sweetapple et al. 2016), and enormous efforts are expended to reduce and control possum populations. Holland et al. (2013) have produced a model to describe the impact of browsing by common brushtail possums on woody vegetation, and have parametrized, and run, this model for several NZ tree species (Holland et al. 2016). These models focus on tree mortality as an endpoint, and key input parameters are plant foliage cover, and an index of browse damage. These models are constructed such that the relationship of possum density to tree mortality is strongly non-linear, and high possum densities are required before browsing impacts become apparent. A strength of these models is the recognition, by the modellers, that browsing is selective at the level of browse species and individual plant (Windley et al. 2016). While many other modellers (e.g., Feng et al. 2009) average herbivore offtake evenly across plants in the landscape, the models of Holland and coworkers incorporate the recognition that herbivory is often concentrated on individual, preferred trees. Selective browsing, including that driven by strong differences in secondary chemistry among conspecific individual plants (e.g., Moore et al. 2010), explains why browsing impacts can become apparent even when overall herbivore densities appear to be too low to have an impact on foliar production at the ecosystem level. These models have not yet been applied to Australian eucalypts, although these trees might differ from the New Zealand examples used by Holland et al. (2016), in terms of leaf lifespan, carbohydrate storage and bud reserves to facilitate regrowth, and may show greater-between tree variability in herbivore preferences attributable to plant secondary metabolites.

As with grazing, browsing by marsupials can affect biodiversity, especially by limiting regeneration of tree species after disturbance (Allcock and Hik 2004). Overabundance of the mixed-feeding swamp wallaby, *Wallabia bicolor*, at Booderee National Park, in coastal NSW, suppresses regeneration of *Eucalyptus pilularis* seedlings, as well as other native and weedy trees, vines and shrubs, favouring the growth of bracken, *Pteridium esculentum*, (Dexter et al. 2013). Stutz et al. (2016) demonstrated the remarkable ability of these wallabies to locate and target small seedlings by olfaction, even when they are obscured by thick understorey. Exclosures were also used at Wilsons Promontory, Victoria, to demonstrate that secondary succession (from shrubland to forest) could only proceed in the absence of swamp wallabies, which were capable of pulling down and browsing saplings as tall as 2 m (Ashton and Chappill 1989). In different Victorian forests, swamp wallaby effects on forest regeneration, after harvesting, can range from minimal (Di Stefano et al. 2007) to substantial (Di Stefano 2005). Browsing can also, sometimes, threaten individual plant species such as the Tasmanian trees *Eucalyptus gunnii* (Calder and Kirkpatrick 2008) and the threatened *Eucalyptus morrisbyi*, in which one of two remaining populations is less able to resist brushtail possum herbivory, and suffers lower flowering as a consequence (Mann et al. 2012).

Both the burrowing, *Bettongia lesueurii*, and the rufous, *Aepyprymnus rufescens*, bettongs differ from the other, principally mycophagous, rat-kangaroos in consuming significant amounts of plant material. Although this is generally thought to comprise mostly roots and tubers (Claridge et al. 2007), foliage and branches can also be consumed, particularly in high-density populations (Bice and Moseby 2008; Linley et al. 2017). Woylies, *Bettongia penicillata ogilbyi*, too, include substantially higher amounts of plant food in their diet in a high-density, food-limited, fenced population than they do otherwise (Zosky et al. 2018). Anecdotally, bettong browsing has been suggested to have limited post-fire recruitment of eucalypt seedlings (Noble et al. 2007), and the decline of rufous bettongs was linked, again anecdotally, to increased recruitment of woody vegetation in western New South Wales (Rolls 1981). Noble et al. (2007) modelled the impact of browsing by burrowing bettongs, fire and rainfall on shrub population dynamics and suggested a potential for fire and browsing, either individually or in combination, to maintain low shrub densities.

As with grazing marsupials, the impacts of browsing marsupials may only become apparent in the absence of mechanisms that would otherwise suppress populations, or alter their foraging behavior (Pickett et al. 2005). This could include the cessation of hunting by humans, and a lack of predation by native owls (e.g., Kavanagh 1988), raptors and dingoes, due to human persecution, local extinction or exclusion from fenced reserves. The isolation of habitat patches, due to land clearing and the introduction of marsupial herbivores to islands, can also make dispersal impossible, thereby increasing population growth rates. For example, burrowing bettongs previously declined to extinction on mainland Australia, but now thrive in predator-exclusion reserves such as Arid Recovery in South Australia, where their densities are sufficient to reduce perennial plant species richness and reduce vegetation condition (Linley et al. 2017). These effects were not seen at lower bettong densities in the same habitat (Munro et al. 2009). Similarly, browsing by quokkas, *Settonix brachyurus*, acts, together with fire, to prevent tree seedling recruitment on Rottnest Island, Western Australia (Main 1992).

15.6.3 Conclusion

Australian mammalian herbivore communities have changed dramatically since the arrival of humans, which was followed by the loss of browsing and grazing megafauna. Not only the causes of these extinctions, but also their consequences continue to be debated, as researchers struggle to attribute evidence of changed vegetation to herbivory, fire and climate (Johnson 2006). Given the former diversity of large mammalian herbivores, one can imagine an impact on par with that of African herbivores today. A second wave of ecosystem change commenced with the invasion of Australia by Europeans 230 years ago. Since then, woody vegetation clearance, changes to fire regimes, provisioning of water, pasture improvement, and the introduction of exotic weedy and pasture species, have altered the environment in which marsupials forage. Australia's apex predator, the dingo, has been rendered functionally extinct in large parts of the country, while exotic mesopredators such as the fox have had devasting impacts on many smaller mammalian herbivores. Grazing herbivores now coexist alongside, and possibly compete, with exotic sheep and cattle, rabbits, camels, goats and pigs, which themselves alter the foraging environment encountered by native marsupial herbivores (e.g., Cooke 1998; Tyndale-Biscoe 2005). As they have disappeared from across the landscape, or as the ecosystems around them have been altered, the prospect of understanding the former ecological role of many marsupial herbivores has slipped from our grasp, but much also remains to be learnt about their contemporary impacts.

15.7 Lagomorphs

Lucy Lush

Lagomorphs are small- to medium-sized mammals that inhabit a wide variety of habitats worldwide, both as native or introduced species (Hutchings and Harris 1996; Trout 2003). The lagomorph Order includes hares, *Lepus* spp., rabbits, *Leporidae* spp., and pikas, *Ochotonidae* spp., many of which are classed as endangered in their native ranges, with introduced species often becoming pests (Cowan and Hartley 2008; Jennings 2008). Despite their small size, lagomorphs can dramatically shape vegetation structure and composition, both positively and negatively (Boag et al. 1990; Crawley 1990; Van der Wal et al. 2000a, b; Stahl et al. 2006). The way land is managed can equally affect the distribution and behaviour of lagomorphs (Petrovan et al. 2013; Lush et al. 2014). Understanding the determinants of lagomorph distribution and behaviour, and interactions between these species and the ecosystems they inhabit (Sinclair et al. 2000), could provide management solutions for both conservation and biosecurity.

15.7.1 Feeding Ecology of Lagomorphs

Lagomorphs are selective feeders and can be both grazers and browsers, depending on the habitat, season and food availability (Chapuis 1990; Schai-Braun et al. 2015). In general, lagomorphs graze on grasses and herbs but may browse on saplings and woody plants, particularly when forage resources are limited during winter (Homolka 1982; Rao et al. 2003). Some species of hares, such as the Mountain hare, *Lepus timidus*, are specialist browsers, preferentially feeding on deciduous tree species (Hjältén et al. 2004).

Lagomorphs are hind gut fermenters and, unusually, they perform coprophagy (re-ingesting the soft faeces initially produced after a feeding bout) that enables them to more effectively digest lower quality forage (in comparison to other mammalian herbivore species their size) (Kuijper et al. 2004). This allows their diet to be incredibly varied and provides the flexibility to adapt to changes in plant availability, management of agricultural land or competition from other herbivores.

Rabbits and pikas are central placed foragers, which results in grazing gradients, with reduced grazing intensity and increased dietary selectivity in grasslands further away from their burrows or talus (Huntly 1987; McIntire and Hik 2005; Bakker et al. 2005). This can lead to higher standing biomass, vegetation height and decreased plant nutrient concentration at greater distances from burrows (Bakker et al. 2005). Unlike rabbits, pikas also carry out haying (caching plants) during periods of high vegetation biomass, enabling them to survive through periods when vegetation is dormant (Huntly et al. 1986). In contrast, hares often select fields with taller vegetation that provides cover from predation, and less intensively managed agricultural land, with higher levels of fat in plant material, whereas rabbits select for more intensively livestock grazed pastures with nitrogen rich shorter grass (Bakker et al. 2005; Lush et al. 2014, 2017).

15.7.2 Impacts of Lagomorph Grazing and Browsing

The flexible feeding ecology of lagomorphs can affect, both positively and negatively, the localised impact of their grazing behaviour on the landscape. For example, rabbits and pikas usually graze close to their burrows, which creates high intensity grazing areas (Huntly et al. 1986; Cowan et al. 1989). Hares can also affect vegetation structure through selectively browsing on plant species (Rose and Platt 1992), reducing woody biomass (Pease et al. 1979) and removing seedlings (Wong and Hickling 1999). This can benefit other species, for example, hare browsing restricted the growth of shrubs such as, *Artemisia maritima* and *Atriplex portulacoides*, on salt-marshes that created preferential foraging habitat for Brent geese (Van der Wal et al. 2000b). High population densities of lagomorphs can, therefore, lead to fundamental alterations to grassland structure and composition (Boag et al. 1990; Crawley 1990), and can cause damage to native grasses and forbs, resulting in reduced species richness, reduction in the area of grassland swards and increased weeds (Mutze et al. 2016). Studies, using exclusion experiments, revealed that preferential grazing of certain plants by lagomorphs reduced plant diversity and vegetation growth (Gibbens et al. 1993; McIntire and Hik 2005). However, these effects can be confounded by grazing by other types of herbivores, weather conditions and long-term effects of grazing pressure.

European rabbits, *Oryctolagus cuniculus*, have become a significant pest species in many countries (Cowan 1987). In the UK, damage to crops have been estimated at £115 million annually (Smith et al. 2007), and, in Australia are attributed to the loss of A$206 million in agriculture (Gong et al. 2009). However, rabbits can provide commercial revenue for example, A$36 million in Australia from the sale of products as a result of shooting (Gong et al. 2009).

Rabbit grazing can also benefit the environment, providing an important mechanism for maintaining certain habitats such as, calcareous grasslands, heathland and sand dune grasslands (Lees and Bell 2008; Trout 2003). When rabbits were removed from these habitats, plant biodiversity reduced and indirectly affected numbers of important invertebrate species (Barham and Stewart 2005). Similarly, the burrowing and grazing activity of pikas, *Ochotona pallasi*, has improved soil nutrients in arid habitats, and subsequently increased grass productivity, creating higher species diversity and vegetation abundance (Wesche et al. 2007; Yu et al. 2017). Although, in contrast, other studies found they had little impact on grassland plant species richness, and pikas were negatively affected by high intensity livestock grazing (Komonen et al. 2003).

Being prey species, lagomorphs can, to some extent, be naturally controlled by predators, such as the red fox, *Vulpes vulpes*, and lynx, *Lynx lynx*. There has been a 10 year cyclic relationship between snowshoe hare populations, *Lepus americanus*, and lynx, in North America, through a combination of predation and forage quantity and quality (Krebs et al. 2001). Similarly, rabbits, in their native ranges, are declining due to changes in land use, habitat loss and introduced viruses, resulting in declines in lynx populations (Virgós et al. 2003; Lees and Bell 2008). The changes in population density of lagomorphs can, in turn, affect grazing pressure, plant growth and composition.

15.7.3 Effect of Management on Lagomorphs

Agricultural intensification, and changes in management practices, have altered the habitat and food availability for lagomorphs. The winter planting of agricultural crops has provided an important food source for hares and rabbits, when other natural forage resources are limited; inadvertently resulting in an increase in rabbit populations (Tapper and Barnes 1986). Removal of hedgerows, woodlands and the planting of larger, monocultural fields, reduced habitat diversity and structure, which has negatively impacted some lagomorphs, such as the brown hare, *Lepus europaeus*, that require a diverse variety of plants and habitat structure (Tapper and Barnes 1986). Increased livestock grazing reduces the available forage and cover for lagomorphs, and has been shown to have a major impact on lagomorphs' foraging distribution (jackrabbits, cottontails and brown hares), with higher numbers found on areas with moderate livestock grazing, as opposed to those areas that were heavily or lightly grazed (Milchunas et al. 1998; Karmiris and Nastis 2007). This could also create potential competition for resources between livestock and lagomorphs (Hulbert and Andersen 2001). Livestock grazing regimes, the application of fertiliser and planting different grass or crop species affects plant nutritional quality and, consequently, lagomorph diet (Bakker et al. 1983; Pavlů et al. 2006; Lush et al. 2017). In particular, the amount of nitrogen, crude fat and fibre available in plants have differing effects on lagomorph foraging distribution and body condition (Hackländer et al. 2002; Bakker et al. 2005; Lush et al. 2014).

15.7.4 Managing Lagomorph Grazing and Browsing

Biocontrol agents, such as, myxomatosis and rabbit haemorrhagic disease virus, were introduced in several countries to reduce the extremely high rabbit populations. Following the introduction of these agents, rabbit numbers in the UK reduced by 99.9% (Boag 1987), resulting in dramatic changes to the landscape, with increased grassland growth and regeneration of woodlands. The reduction in rabbit populations in the UK reduced plant diversity resulting in the extinction of some invertebrates, such as, the large blue butterfly, *Maculinea arion*, (Sumption and Flowerdew 1985). It also resulted in the decline of many predators that fed largely on rabbits such as, the stoat, *Mustela ermineu*, and buzzard, *Buteo buteo*, (Sumption and Flowerdew 1985); with knock on positive and negative effects on the environment and economy. Eradication, and control, of introduced rabbit populations, particularly on islands, has resulted in increased plant species richness and cover, although exotic plants often populate areas faster than native plants, highlighting the benefits of rabbit grazing in some situations (Schweizer et al. 2016). In Australia, the reduction of rabbits benefitted Australian agriculture by A$70 billion dollars (Cooke et al. 2013). Although, recently, increasing resistance to the virus, and different strains, have reduced the effectiveness of this control measure (Ross and Sanders 1984).

Less dramatic methods to control lagomorph populations and foraging behaviour can be mediated through changes in the structure and composition of available habitats (Boag 1987). Reduction of vegetation cover, removal of field boundaries and the use of set asides (removing areas of land from crop production creating strips of grassland at field edges), have reduced some crop losses from lagomorph grazing;

however, at high lagomorph densities these become ineffective (Trout 2003). Studies found that planting of older, larger tree saplings reduced lagomorph browsing in plantations (McArthur and Appleton 2004); similarly, reducing tree planting density, and planting seedlings in tall vegetation to reduce their visibility, lowered browsing pressure by hares (Rao et al. 2003). Less intensive farming practices, such as reducing livestock grazing and other inputs to pastures e.g., fertilisers that alter the nutritional value of vegetation, could benefit some lagomorph species (Lush et al. 2014). However, each lagomorph species has differing nutritional requirements and therefore, changes could be positive or negative. Further research, and regular monitoring, is required to assess the impact of different management practices on lagomorph species and measure the effectiveness of management interventions at a landscape scale.

15.8 Rodents

Renan Maestri

Rodents outnumber all other orders of mammals, in term of number of species (Wilson and Reeder 2005); approximately 42% of mammals are rodents, which corresponds to around 2400 species. Rodent radiations have occurred repeatedly across the globe, and they occupy all continents with the exception of Antarctica (Lacher et al. 2016), showing a parallelism in morphological and ecological features in each radiation (Wood 1947; Aplin and Ford 2014). The success of rodents in colonizing different environments is directly related to rodent's ability to exploit diverse food items, such as grasses, roots, leaves, fruits, seeds and insects. However, most rodents are herbivorous, or include some plant material in their diet, and have many adaptations for herbivory (Samuels 2009). This is apparent from the smallest rodents such as the pygmy jerboa, *Salpingotulus michaelis* (~4 g), that feed on desert-adapted leaves, to the largest living rodent, *Hydrochoerus hydrochaeris* (~60 kg), that feed on grasses, and even the largest extinct rodent, *Josephoartigasia monesi* (~1000 kg), is thought to have fed on soft vegetation and fruits (Rinderknecht and Blanco 2008).

A classification of rodents into browsing vs. grazing categories is uncommon compared to more traditional categories that divide rodents into chewing vs. gnawing for feeding habits, and into folivory (leaf eaters) vs. gramnivory (grass eaters), among others, for food preferences. A comparative assessment of adaptations for herbivory, considering these categories, reveals many morphological and physiological differences among living rodents, which are reflected in their impact on the ecosystems through feeding activities, reviewed below.

The herbivorous habit of rodents is related to their characteristic skull shape, with enlarged and ever-growing incisors that are adapted for biting (also gnawing) (Lacher et al. 2016). The check teeth, which are found behind the incisors and the diastema, are effective at chewing and grinding of plant material. Gnawing with the incisors and grinding with the check teeth are the main activities performed by rodents whilst feeding. Both functions take place in alternation: when the incisors are engaging, the jaw is positioned forward in a position that means that the check teeth do not meet each other; when the check teeth are positioned for grinding and chewing, the incisors are not positioned to allow gnawing (Vaughan et al. 2015). This difference in position relates to a trade-off between specialization for gnawing or chewing, and rodents can be roughly classified depending on their jaw musculature and skull adaptations for increased gnawing over chewing or vice versa (Cox et al. 2012).

Morphological adaptations for feeding, mainly related to the position and form of masseter muscles, can be used to segregate rodents in three non-monophyletic groups: sciuromorph, hystricomorph and myomorph (Simpson 1945). In the sciuromorph rodents the masseter lateralis and temporalis muscles are relatively large and the masseter lateralis extends onto the rostrum (Korth 1994). This leads to changes in feeding processes that, together, increase the gnawing abilities of these rodents. Greater gnawing abilities are associated with feeding on large seeds, nuts and roots. Examples of sciuromorph rodents are squirrels, beavers and pocket gophers. In the hystricomorph rodents it is the masseter medialis that is greatly developed and extended, passing through an enlarged infraorbital foramen before attaching to the rostrum (Wood 1965; Korth 1994). Rodents with this adaptation have an increased ability to perform varied movements with the jaw, resulting in an improved capacity for chewing and grinding at the same time (Cox et al. 2012). These adaptations allow increased processing efficiency when feeding on plant material, and most strictly grazing rodents have this morphotype. Examples of hystricomorph rodents are jerboas plus some Old-World rodents, the South American porcupines and many other caviomorph species; the largest rodent in the world, the South American capybara, *Hydrochoerus hydrochaeris*, is a grazing rodent with these features (Herrera 2013). The myomorph rodents have a combination of features from both the sciuromorph and the *hystricomorph* morphotypes, where the masseter lateralis extends onto the rostrum, as in sciuromorphs, and the masseter medialis passes inside the infraorbital foramen, as in hystricomorphs (Korth 1994). This combination of features produces a phenotype capable of performing both gnawing and chewing functions that leads to effective feeding on seeds, fruits, grains and plant material in general (Cox et al. 2012; Maestri et al. 2016).

Dietary shifts also trigger corresponding changes in the shape of the jawbone (e.g., Hautier et al. 2012; Maestri et al. 2016). A narrow angular process of the mandible is usually associated with the hystricomorph rodents, while a more robust angular process characterizes sciuromorphy. Even among hystricomorphs, those families feeding exclusively on grasses have a narrower and thinner angular process than those feeding on fruits and seeds (Hautier et al. 2011). Furthermore, a shorter, and curved, diastema, and a deeper ramus, characterize families that feed on fruits and seeds, as opposed to those that feed on grasses (Hautier et al. 2011). Other general differences in skull morphology occur according to whether a species feeds on grasses versus fruits and seeds (see Hautier et al. 2012), and still another set of morphological characteristics are found among rodents that feed on meat (e.g., Woollard et al. 1978; Rowe et al. 2016).

Herbivory in rodents can also be augmented by adaptations of the teeth (e.g., Willians and Kay 2001; Ma et al. 2016). Hypsodonty, one of the hallmarks of herbivory for mammals in general (**Saarinen** Chap. 2), is also a feature of herbivorous rodents. High crowned check teeth can increase the capacity for processing plant material consumed during grazing, and these type of teeth are found, for instance, among caviomorphs and several species of small rats and mice. A few studies of South American sigmodontines rodents have shown that the proportion of high-crowned teeth is greatest in species found in cold and dry and semiarid climates, while few high-crowned species are present in wet and hot climates (see Madden 2015). In addition, high teeth crowns, and larger check teeth, are positively correlated with the more seasonal and open environments (Madden 2015; Maestri et al. 2017).

Another related adaptation is the development of ever-growing check teeth in some caviomorph rodents such as the caviids and chinchillids. Nevertheless, relating hypsodonty to diet in rodents is limited by the scarcity of work on dietary composition for rodents in general (Madden 2015; Arregoitia 2016). Analyses of stomach contents using, for example, metabarcoding (Lopes et al. 2015), have revealed a great overlap in plant families consumed by rodent species. Intestinal physiology studies can contribute by comparing the concentration of types of bacteria between rodents and grazing ungulates, revealing, for example, a similarity in intestinal physiology between the capybara and bovids (Borges et al. 1996, see Herrera 2013 for a distinction between foregut and hindgut fermenters). Enamel microwear investigations are also a useful tool that helped to classify grazing and browsing mammals (Townsend and Croft 2008), and are applicable to distinguish food preferences (e.g., omnivore, frugivore, gramnivore) among rodents (Caporale and Ungar 2016). Multiple approaches are urgently needed to analyse rodent diet composition, so as to move away from the broad description of rodents as "opportunistic" and "omnivorous".

Given the diversity and wide geographical distribution of rodent species, and the propensity of its members to feed on many types of plant material, it is important to understand how the foraging behavior of rodents impacts ecosystems through grazing and browsing activities. A few studies have investigated the impacts of rodents on ecosystems through grazing (e.g., Howe et al. 2002; Bilodeau et al. 2014) and browsing (Ravolainen et al. 2014), yet most of the literature has focused on large mammals or rabbits (previous section **Katona and Coetsee** Chap. 12). Nevertheless, the impact of grazing rodents could be, at least, as great as that of rabbits at some regions, as in South America: Madden (2015) suggests that the invasion of South America by herbivorous caviomorph rodents (see also **Saarinen** Chap. 2) could have generated increased soil erosion resulting from their grazing activities. For example, there is evidence that the feeding behavior and burrowing activities of tuco-tucos (genus *Ctenomys*) reduces forage and leads to vegetation changes and habitat degradation (Jackson 1988). When these rodents cover large areas, they can alter the soil dynamics and the vegetation growth (Massoia 1970; Galiano et al. 2014). Similarly, the feeding behavior of vizcacha (genus *Lagostomus*), another caviomorph, can lead to reductions in vegetation cover and abundance and they even compete with livestock for food; they are considered a pest in some areas of Argentina (Jackson 1988).

The effects of grazing and browsing of small rats and mice might also impact ecosystems, but may go unnoticed due to the difficulty of measuring their impacts. Limited direct evidence exists on the role of small cricetid rodents in damaging pastures and being pests of cultivated plants (Jackson 1988). Nevertheless, damage of human cultivated crops, such as rice and cereal grains, by small rodents, has been estimated to cause a billions of dollars per year of damage worldwide (Lacher et al. 2016). Using experimental plots in England, Hulme (1994, 1996) compared the impact of small rodents and invertebrates on seedlings predation and the growth of grasslands. He showed that rodents and molluscs had similar negative effects on herbaceous seedlings, consuming about 30% within a given plot. Nevertheless, rodents had their greatest effect on the growth of grassland plants, being responsible for a 50% reduction in mean plant biomass, greater than the impact of molluscs or arthropods. Hagenah et al. (2009) found a proportionally higher effect of murid rodents on plant biomass in a South African savanna, after excluding large herbivores from experimental plots; the exclusion of large herbivores had the effect of increasing small rodent abundance (Hagenah et al. 2009; Luza et al. 2018). Howe and Brown (1999) compared the effects of herbivory between small browsing rodents and birds in a tallgrass praire, and found that the voles had negative effects on grass biomass and altered plant community structure, especially in low-density communities; while the effects of bird were more pronounced in high-density plantings. Browsing voles, such as *Microtus pennsylvanicus*, can even be considered a plague in prairie grasslands in North America, because they can cause drops in plant diversity (37% drop in Simpson's diversity index), greatly altering community composition (Howe et al. 2002).

Other studies have found contrasting results, showing that rodents have negligible top-down control on plant assemblages, which were mostly regulated by bottom-up processes, such as in native grasslands of North America (Báez et al. 2006). Similarly, grazing by lemmings had very weak effects on plant biomass at the Canadian arctic (Bilodeau et al. 2014). Therefore, while rodents have a pronounced effect on growth of grasses in some cases, the literature on the effects of browsing and grazing by rodents shows varying overall impacts on ecosystems. Furthermore work is needed, not just focusing specifically on rodents but also comparing the effects of rodents with those of large mammals, to reach a consensus about how great are the ecosystem impacts of grazing and browsing by rodents.

15.9 Primates

Ikki Matsuda and Marcus Clauss

Non-human primates (hereafter referred to as 'primates') cover various trophic niches, from nearly exclusive folivory to frugivory, gummivory, insectivory and omnivory; although the majority of species are folivorous, frugivorous, or both

(Campbell et al. 2011). One exceptional species, the gelada baboon, *Theropithecus gelada*, was thought to be graminivorous (a 'grazer'); however, a recent study showed that geladas also consume a substantial amount of non-grass (non-monocot) foods, implying that they are not strict grazers as previously thought (Fashing et al. 2014). Folivorous primates that exhibit hindgut and foregut fermentation such as colobines (foregut fermenters: Asia and Africa), lemurs (hindgut fermenters: Africa) and gorillas (hindgut fermenters: Africa) are widely distributed in the Old World, and hindgut fermenting howlers are found in the New World (Mittermeier et al. 2013). Recent studies on the feeding ecology of some primates, technically classified as folivorous, have reported high levels of fruit and/or seed consumption in response to local habitat conditions (Campbell et al. 2011). Thus, classifying primates as strictly folivorous is not a simple matter.

15.9.1 Anatomical Adaptations to Diet

In parallel to the dichotomy of hindgut and foregut fermenters seen in mammals, primates generally have two contrasting digestive strategies: high intake with fast throughput for low digestive efficiency or low intake with slow throughput for high digestive efficiency (Clauss et al. 2008). Folivorous primates that exhibit foregut fermentation are mostly limited to the high digestive efficiency strategy (Clauss et al. 2008). Another key digestive strategy in folivorous primates entails fine-tuning of salivary protein composition. Howler monkeys, the most folivorous New World primates, show continuous expression of tannin-binding salivary proline-rich proteins; this allows them to consume a diet with variable tannin content (Espinosa Gómez et al. 2015). Such salivary tannin defences have been demonstrated in a few primate species, including omnivorous baboons and macaques (Espinosa-Gómez et al. 2018). On the other hand, as do a great variety of other grazing mammals, graminivorous geladas completely lack proline-rich proteins, and a capacity to bind tannins, demonstrating their narrower dietary niche compared to that of other baboon species (Mau et al. 2009). Given that colobines generally have large salivary glands, they may use a similar strategy to howlers (Kay et al. 1976; Matsuda et al. 2017b); however, this hypothesis awaits further testing.

The microbial community in the gastrointestinal tract is believed to play an important role in facilitating the consumption of hard-to-digest foods, such as leaves of trees and grasses. As a result of recent developments in sequencing technology, gut microbiota analysis, based on large amplicon libraries of 16S ribosomal RNA (rRNA) genes and mostly faecal DNA, has increasingly been conducted on folivorous primates (Ley et al. 2008). In folivorous primates, the distal gut microbiome varies even within a species according to diet and/or living conditions (Amato et al. 2013; Clayton et al. 2017); gut microbial diversity in captive primates is generally reduced compared with that in their wild counterparts. A recent study suggested that gastrointestinal distress in folivorous primates, such as infestation with or disease caused by parasites, may be associated with an imbalance between the types of organism present in their natural microflora, especially that of the gut (Amato et al. 2016).

Nutritional studies have revealed that both hindgut and foregut fermenters generally prefer leaves rich in protein and lower in fibre (Ganzhorn et al. 2016). However, there are some folivorous primates that do not display a strong preference, indicating that a preference for protein depends on the overall protein availability in the environment; the preference for protein is only clearly demonstrated in environments with a low average protein content (Ganzhorn et al. 2016). Other factors that may affect dietary selection in folivorous primates are mechanical toughness and leaf digestibility. However, little information is available regarding primates' dietary choices, and studies evaluating a variety of nutritional and mechanical factors with diet digestibility are particularly lacking. One study providing new insights into in vitro digestibility, toughness and nutrients of leaves shows that the preferred leaves of foregut fermenting proboscis monkeys not only contain more protein and less fibre but are also less tough and more digestible than the alternatives (Matsuda et al. 2017b).

15.9.2 Behavioural Adaptations to Diet

The way a primate rests could be related to their digestive physiology. Among folivorous primates, this is most evident in colobines, that have a characteristic long resting period (over 70% of daylight hours) sitting with a vertical posture (Matsuda et al. 2017a). This might be because the position of the digestive chamber, and the need to frequently eructate digestive gases, force colobines to assume a posture that reduces pressure on the thorax and respiratory organs. In contrast to ruminants that are characterised by a sorting mechanism in their forestomach that operates based on the density of different-sized food particles (Lechner-Doll et al. 1991), with smaller particles generally having a higher density than larger ones (Clauss et al. 2009), experiments with captive colobines have shown no evidence for a forestomach sorting mechanism (Schwarm et al. 2009; Matsuda et al. 2015, 2019). Passage studies in general suggest that, as a group, primates do not pass solutes faster than particles, i.e., lack the physiological or mechanical ability to wash particulate digesta with fluids in the digestive tract that has evolved in all other mammalian herbivore clades (Müller et al. 2011). An interesting peculiarity is that a non-obligatory rumination-like behaviour has been demonstrated in one colobine species, the proboscis monkey (Matsuda et al. 2011; Matsuda et al. 2014).

As with most primates, folivorous primates live in groups varying in their sex-composition and -disperal, e.g., one-male-multi-female or multi-female-multi-male groups with both-sex-dispersal or female-philopatric/male-dispersal. As leaves are generally abundant and evenly distributed in their habitats, socioecological models indicate that food competition within folivorous primate groups can be assumed to be weak/absent and that populations and groups are not constrained by the availability of food (Wrangham 1980; Janson and Goldsmith 1995; Sterck et al.

1997). Increasing day range with increasing group size has been used as a behavioural indicator of food competition within primate groups, and this hypothesis on weak/absent of food competition is supported by the fact that a flat relationship exists between group size and day range across various folivorous primates (e.g., Yeager and Kirkpatrick 1998). However, many folivorous primates, despite this assumed lack of feeding competition within their groups, often live in relatively small groups, though larger groups are expected to have lower costs of predation due to better detection of predators. This inconsistency is referred to as the 'folivore paradox' (Steenbeek and van Schaik 2001). Indeed, the results contradict the assumption of leaves as ubiquitous or non-patchy resources in some colobines, e.g., red colobus, *Procolobus rufomitratus*, (Snaith and Chapman 2005) and gray langurs, *Semnopithecus entellus*, (Sayers and Norconk 2008). Infanticide has been suggested as a mechanism that could regulate group sizes in groups with a high proportion of males, although this theory alone cannot explain the folivore paradox in all cases (reviewed by Snaith and Chapman 2007).

15.9.3 Conservation of Primates

Nearly half of all primate species are threatened with extinction as a result of habitat destruction and poaching, and folivorous primates are no exception (Estrada et al. 2017). A fundamental challenge facing primate conservation today is a lack of knowledge regarding the status of endangered populations (Wich and Marshall 2016). Such information is essential in order to develop effective, long-term management plans for conservation. Therefore, understanding the determinants of abundance in folivorous primates is of utmost importance (Chapman and Peres 2001). It has been suggested that the biomass of folivorous primates, especially colobines, is related to the protein-to-fibre ratio of the leaves in their habitats (Chapman et al. 2002). In line with this protein-to-fibre model, the effect of global change on folivorous populations has been examined; a decline in colobine abundance may be explained by the fact that the fibre concentration in their consumed leaves has increased and protein content has decreased over the past 30 years (Rothman et al. 2015).

Contrary to the earlier reports that colobine monkeys are primarily seed-eaters (e.g., Sun et al. 2007), an impact of colobines on their ecosytems as seed dispersers has recently been reported in several taxa such as *Nasalis*, *Presbytis*, *Rhinopithecus* and *Trachypithecus* (Tsuji et al. 2017; Chen et al. 2018; McConkey 2018; Thiry et al. 2019). Even if they are not comparatively efficient endozoochorous and epizoochorous seed dispersers in forest ecosystems, their high abundance and biomass could make them quantitatively significant in seed dispersal (Matsuda et al. 2013).

Available data showing the impact of climate change on food quality and colobine abundance are very limited, and as colobines have been considered as seed dispersers only very recently, the available data testing its impact on their living-ecosystem are also scarce. Thus, this remains a highly relevant topic that can aid, not only in a basic understanding of colobine population and behavioural ecology, but also in the development of conservation strategies for colobine species. Further work on the importance of colobines in ecological processes (e.g., seed dispersal, nutrient cycling) and their reduced populations on the dynamics of ecosystems is clearly needed.

References

ABC News (2013) Woodchip giant Australian Bluegum plantations stripped of environmental certification over koala deaths

Abrahams PW (2013) Geophagy and the involuntary ingestion of soil. In: Selinus O (ed) Essentials of medical geology. Springer, Dordrecht

Alisauskas RT, Rockwell RF, Dufour KW, Cooch EG, Zimmerman G, Drake KL, Leafloor JO, Moser TJ et al (2011) Harvest, survival, and abundance of midcontinent lesser snow geese relative to population reduction efforts. Wildl Monogr 179:1–42

Allcock KG, Hik DS (2004) Survival, growth, and escape from herbivory are determined by habitat and herbivore species for three Australian woodland plants. Oecologia 138:231–241

Amato KR et al (2013) Habitat degradation impacts black howler monkey (*Alouatta pigra*) gastrointestinal microbiomes. ISME J 7:1344–1353

Amato KR et al (2016) Using the gut microbiota as a novel tool for examining colobine primate GI health. Glob Ecol Conserv 7:225–237

Anderson JT, Nuttle T, Rojas JSS et al (2011) Extremely long-distance seed dispersal by an overfished amazonian frugivore. Proc R Soc B Biol Sci 278:3329–3335. https://doi.org/10.1098/rspb.2011.0155

Anderson HB, Godfrey TG, Woodin SJ, Van der Wal R (2012) Finding food in a highly seasonal landscape: where and how pink footed geese (Anser brachyrhynchus) forage during the Arctic spring. J Avian Biol 43:415–422

Anderson HB, Speed JDM, Madsen J, Pedersen ÅØ, Tombre I, Van der Wal R (2016) Late snow melt moderates herbivore disturbance of the Arctic tundra. Ecoscience 23:29–39

Aplin K, Ford F (2014) Murine rodents: late but highly successful invaders. In: Prins HHT, Gordon IJ (eds) Invasion biology and ecological theory: insights from a continent in transformation. Cambridge University Press, Cambridge, pp 196–240

Arce-Nazario JA, Carlo TA (2012) Iguana iguana invasion in Puerto Rico: facing the evidence. Biol Invasions 14(9):1981–1984

Arman SD, Prideaux GJ (2015) Dietary classification of extant kangaroos and their relatives (Marsupialia: Macropodoidea). Austral Ecol 40:909–922

Arregoitia LDV (2016) Rethinking omnivory in rodents. Preprint https://doi.org/10.20944/preprints201609.0017.v1

Arzel C, Elmberg J, Guillemain M (2006) Ecology of spring-migrating Anatidae: a review. J Ornithol 147:167–184

Ashton DH, Chappill JA (1989) Secondary succession in post-fire scrub dominated by Acacia verticillata (L'Herit) Willd at Wilsons promontory, Victoria. Aust J Bot 37:1–18

Attum O et al (2010) Retention rate of hard-released translocated Egyptian tortoises Testudo kleinmanni. Endanger Species Res 12(1):11–15. https://doi.org/10.3354/esr00271

Badam GL (1981) Colossochelys atlas, a Giant tortoise from the upper Siwaliks of North India. Bull Deccan Coll Res Inst 40:149–153

Báez S, Collins SL, Lightfoot D, Koontz TL (2006) Bottom-up regulation of plant community structure in an aridland ecosystem. Ecology 87:2746–2754

Bakker RT (1978) Dinosaur feeding behaviour and the origin of flowering plants. Nature 274:661–663
Bakker RJ (1986) The dinosaur heresies. Kensington Publishing, New York
Bakker JP, de Leeuw J, Van Wieren SE (1983) Micro-patterns in grassland vegetation created and sustained by sheep-grazing. Vegetatio 55:153–161
Bakker ES, Reiffers RC, Olff H, Gleichman JM (2005) Experimental manipulation of predation risk and food quality: effect on grazing behaviour in a central-place foraging herbivore. Oecologia 146:157–167. https://doi.org/10.1007/s00442-005-0180-7
Bakker ES, Veen C, Ter Heerdt G, Huig N, Sameel J (2018) High grazing pressure of geese threatens conservation and restoration of reed belts. Front Plant Sci:12. https://doi.org/10.3389/fpls.2018.01649
Bamber RK, Humphreys FR (1965) Variations in sapwood starch levels in some Australian forest species. Aust For 29:15–23
Barham DF, Stewart AJA (2005) Differential indirect effects of excluding livestock and rabbits from chalk Heath on the associated leafhopper (Hemiptera: Auchenorrhyncha) Fauna. J Insect Conserv 9:351–361. https://doi.org/10.1007/s10841-005-0517-x
Barnes A, Hill GJE (1992) Estimating kangaroo damage to winter wheat crops in the Bungunya district of southern Queensland. Wildl Res 19:417–427
Barrett PM, Willis KJ (2001) Did dinosaurs invent flowers? Dinosaur–angiosperm coevolution revisited. Biol Rev 76:411–447
Barton PS, Manning AD, Gibb H, Wood JT, Lindenmayer DB, Cunningham SA (2011) Experimental reduction of native vertebrate grazing and addition of logs benefit beetle diversity at multiple scales. J Appl Ecol 48:943–951
Baxter GS, Moll EJ, Lisle AT (2001) Pasture grazing by black-striped wallabies (Macropus dorsalis) in Central Queensland. Wildl Res 28:269–276
Bayliss P, Choquenot D (2002) The numerical response: rate of increase and food limitation in herbivores and predators. Philos Trans R Soc Lond B Biol Sci 357:1233–1248
Becerra-López JL et al (2017) Effect of climate change on halophytic grasslands loss and its impact in the viability of Gopherus flavomarginatus. Nat Conserv 21:39–55
Bedoya-Pérez M, Lawes M, McMahon C (2017) Final report: protecting vulnerable land from high wallaby densities. Charles Darwin University and Meat and Livestock Australia, North Sydney
Beer S (1989) Photosynthesis and photorespiration in marine phytoplankton. Aquat Bot 34:105–130. https://doi.org/10.1016/0304-3770(89)90052-1
Bellwood DR, Hughes TP, Hoey AS (2006) Sleeping functional group drives coral-reef recovery. Curr Biol 16:2434–2439. https://doi.org/10.1016/j.cub.2006.10.030
Ben-Ami D, Boom K, Boronyak L, Townend C, Ramp D, Croft DB, Bekoff M (2014) The welfare ethics of the commercial killing of free-ranging kangaroos: an evaluation of the benefits and costs of the industry. Anim Welf 23:1–10
Bennett S, Wernberg T, Harvey ES et al (2015) Tropical herbivores provide resilience to a climate-mediated phase shift on temperate reefs. Ecol Lett 18:714–723. https://doi.org/10.1111/ele.12450
Benson RB, Campione NE, Carrano MT et al (2014) Rates of dinosaur body mass evolution indicate 170 million years of sustained ecological innovation on the avian stem lineage. PLoS Biol 12(5):e1001853. https://doi.org/10.1371/journal.pbio.1001853
Best HA (1984) The foods of kakapo on Stewart Island as determined from their feeding sign. N Z J Ecol 7:71–83
Bice JK, Moseby KE (2008) Diet of the re-introduced greater bilby (Macrotis lagotis) and burrowing bettong (Bettongia lesueur) in the arid recovery reserve, northern South Australia. Aust Mammal 30:1–12
Bilodeau F, Gauthier G, Fauteux D, Berteaux D (2014) Does lemming winter grazing impact vegetation in the Canadian Arctic? Polar Biol 37:845–857
Bjorndal KA (1980) Nutrition and grazing behavior of the green turtle Chelonia mydas. Mar Biol 56 (2):147–154

Blake S et al (2012) Seed dispersal by Galápagos tortoises. J Biogeogr 39(11):1961–1972. https://doi.org/10.1111/j.1365-2699.2011.02672.x
Blake S et al (2013) Vegetation dynamics drive segregation by body size in Galapagos tortoises migrating across altitudinal gradients. J Anim Ecol 82(2):310–321
Boag B (1987) Reduction in numbers of the wild rabbit (Oryctolagus cuniculus) due to changes in agricultural practices and land use. Crop Prot 6:347–351
Boag B, Macfarlane Smith WH, Griffiths DW (1990) Effects of grazing by wild rabbits (Oryctolagus cuniculus) on the growth and yield of oilseed and fodder rape (Brasslca napus sub. Sp. oleifera). Crop Prot 9:155–159
Bonaldo RM, Bellwood DR (2008) Size-dependent variation in the functional role of the parrotfish Scarus rivulatus on the great barrier reef, Australia. Mar Ecol Prog Ser 360:237–244. https://doi.org/10.3354/meps07413
Borges PA, Dominguez-Bello MG, Herrera EA (1996) Digestive physiology of wild capybara. J Comp Physiol B 166:55–60
Bourn D et al (1999) The rise and fall of the Aldabran giant tortoise population. Proc R Soc Lond Ser B Biol Sci 266(1424):1091–1100. https://doi.org/10.1098/rspb.1999.0748
Box SJ, Mumby PJ (2007) Effect of macroalgal competition on growth and survival of juvenile Caribbean corals. Mar Ecol Prog Ser 342:139–149. https://doi.org/10.3354/meps342139
Branch WR, Baard E, De Villiers A (1990) Some exceptionally large southern African chelonians. Afr J Herpetol 37(1):53–54
Breed B, Ford F (2007) Native mice and rats. CSIRO Publishing, Collingwood, Vic
Bucher EH, Tamburini D, Abril A, Torres P (2003) Folivory in the white-tipped plantcutter Phytotoma rutila: seasonal variations in diet composition and quality. J Avian Biol 34:211–216
Buij R, Melman TCP, Loonen MJJE, Fox AD (2017) Balancing ecosystem function, services and disservices resulting from expanding goose populations. Ambio 46:301–318
Bulinski J, McArthur C (1999) An experimental field study of the effects of mammalian herbivore damage on Eucalyptus nitens seedlings. For Ecol Manag 113:241–249
Bulinski J, McArthur C (2003) Identifying factors related to the severity of mammalian browsing damage in eucalypt plantations. For Ecol Manag 183:239–247
Burkepile DE, Hay ME (2010) Impact of herbivore identity on algal succession and coral growth on a Caribbean reef. PLoS One. https://doi.org/10.1371/journal.pone.0008963
Burness GP, Diamond J, Flannery T (2001) Dinosaurs, dragons, and dwarfs: the evolution of maximal body size. Proc Natl Acad Sci 98(5):14518–14523
Butler RJ, Barrett PM, Kenrick P et al (2009) Diversity patterns amongst herbivorous dinosaurs and plants during the cretaceous: implications for hypotheses of dinosaur/angiosperm co-evolution. J Evol Biol 22:446–459
Button DJ, Rayfield EJ, Barrett PM (2014) Cranial biomechanics underpins high sauropod diversity in resource-poor environments. Proc R Soc B 281:20142114. https://doi.org/10.1098/rspb.2014.2114
Calder JA, Kirkpatrick JB (2008) Climate change and other factors influencing the decline of the Tasmanian cider gum (Eucalyptus gunnii). Aust J Bot 56:684–692
Campbell CJ, Fuentes A, MacKinnon KC, Bearder SK, Stumpf RM (2011) Primates in perspective. Oxford University Press, New York
Caporale SS, Ungar PS (2016) Rodent incisor microwear as a proxy for ecological reconstruction. Palaeogeogr Palaeoclimatol Palaeoecol 446:225–233
Cardoso SC, Soares MC, Oxenford HA, Côté IM (2009) Interspecific differences in foraging behaviour and functional role of Caribbean parrotfish. Mar Biodivers Rec 2:e148. https://doi.org/10.1017/S1755267209990662
Carpenter RC (1986) Partitioning Herbivory and its effects on coral reef algal communities. Ecol Monogr 56:345–363. https://doi.org/10.2307/1942551
Caviedes-Vidal E, McWhorter TJ, Lavin SR, Chediack JG, Tracy CR, Karasov WH (2007) The digestive adaptation of flying vertebrates: high intestinal paracellular absorption compensates for smaller guts. PNAS 104:19132–19137

Chaichana R, Leah R, Moss B (2010) Birds as eutrophicating agents: a nutrient budget for a small lake in a protected area. Hydrobiologia 646:111–121
Chapman CA, Peres CA (2001) Primate conservation in the new millennium: the role of scientists. Evol Anthropol: Issues News Rev 10:16–33
Chapman CA, Chapman LJ, Bjorndal KA, Onderdonk DA (2002) Application of protein-to-fiber ratios to predict colobine abundance on different spatial scales. Int J Primatol 23:283–310
Chapuis JL (1990) Comparison of the diets of two sympatric lagomorphs, Lepus europaeus (Pallas) and Oryctolagus cuniculus (L.) in an agroecosystem of the Ile-de-France. Zeitschrift fur Saugetierkd 55:176–185
Cheke A, Hume JP (2010) Lost land of the dodo: the ecological history of Mauritius, Réunion and Rodrigues. Yale University Press, London
Chen Y, Chen H, Zhang Y, Yao H, Yang W, Zhao Y, Ruan X, Xiang Z (2018) First evidence of epizoochorous seed dispersal by golden snub-nosed monkeys (Rhinopithecus roxellana) in temperate forest. Plant Ecol 219:417–427
Chin K, Feldmann RM, Tashman JN (2017) Consumption of crustaceans by megaherbivorous dinosaurs: dietary flexibility and dinosaur life history strategies. Sci Rep 7(1):11163. https://doi.org/10.1038/s41598-017-11538-w
Choat JH, Robbins WD, Clements KD (2004) The trophic status of herbivorous fishes on coral reefs. Mar Biol 145:445–454. https://doi.org/10.1007/s00227-004-1341-7
Christian A (2010) Some sauropods raised their necks—evidence for high browsing in *Euhelopus zdanskyi*. Biol Lett 6:823–825
Christianen MJA et al (2014) Habitat collapse due to overgrazing threatens turtle conservation in marine protected areas. Proc R Soc Lond B Biol Sci 281(1777):20132890
Clancy TF, Croft DB (1992) Population dynamics of the common wallaroo (Macropus robustus erubescens) in arid New South Wales. Wildl Res 19:1–16
Claridge AW, Seebeck JH, Rozse R (2007) Bettongs, potoroos and the musky rat-kangaroo. CSIRO Publishing, Collingwood, Victoria
Clauss M, Streich WJ, Nunn CL, Ortmann S, Hohmann G, Schwarm A, Hummel J (2008) The influence of natural diet composition, food intake level, and body size on ingesta passage in primates. Comp Biochem Physiol A Mol Integr Physiol 150:274–281
Clauss M, Fritz J, Bayer D, Nygren K, Hammer S, Hatt JM, Südekum KH, Hummel J (2009) Physical characteristics of rumen contents in four large ruminants of different feeding type, the addax (*Addax nasomaculatus*), bison (*Bison bison*), red deer (*Cervus elaphus*) and moose (*Alces alces*). Comp Biochem Physiol A Mol Integr Physiol 152:398–406
Clayton JB et al (2017) Associations between nutrition, gut microbiome, and health in a novel nonhuman primate model. bioRxiv:177295
Clements KD, Raubenheimer D, Choat JH (2009) Nutritional ecology of marine herbivorous fishes: ten years on. Funct Ecol 23:79–92. https://doi.org/10.1111/j.1365-2435.2008.01524.x
Close DC, Paterson S, Corkrey R, McArthur C (2010) Influences of seedling size, container type and mammal browsing on the establishment of Eucalyptus globulus in plantation forestry. New For 39:105–115
Coe MJ, Dilcher DL, Farlow JO, Jarzen DM, Russell DA (1987) Dinosaurs and land plants. In: Friis EM, Chaloner WG, Crane PR (eds) Origins of angiosperms and their biological consequences. Cambridge University Press, Cambridge, pp 225–258
Collinson ME (1996) 'What use are fossil ferns?' ± 20 years on: with a review of the fossil history of extant pteridophyte families and genera. In: Camus JM, Gibby M, Johns RJ (eds) Pteridology in perspective. Kew, Royal Botanic Gardens, London, pp 349–394
Conkin J, Alisauskas RT (2017) Conversion of tundra to exposed peat habitat by snow geese (Chen caerulescens caerulescens) and Ross's geese (C. rossii) in the Central Canadian Arctic. Polar Biol 40:563–576
Connell SD, Russell BD, Irving AD (2011) Can strong consumer and producer effects be reconciled to better forecast "catastrophic" phase-shifts in marine ecosystems? J Exp Mar Bio Ecol 400:296–301. https://doi.org/10.1016/j.jembe.2011.02.031

Cooke BD (1998) Did introduced rabbits Oryctolagus cuniculus (L.) displace common wombats Vombatus ursinus (Shaw) from paert of their range in South Australia? In: Wells RT, Pridmore PA (eds) Wombats. Surrey Beatty, Chipping Norton, pp 262–270

Cooke BD, Mutze GJ (2018) How introduced rabbits Oryctolagus cuniculus limit the abundance of red kangaroos Macropus Rufus and other native grazers in Australia. Food Webs 15:e00079

Cooke B, Chudleigh P, Simpson S, Saunders G (2013) The economic benefits of the biological control of rabbits in Australia, 1950-2011. Aust Econ Hist Rev 53:91–107. https://doi.org/10.1111/aehr.12000

Cooper WE, Vitt LJ (2002) Distribution, extent, and evolution of plant consumption by lizards. J Zool 257(4):487–517

Coulson GM (2007) Exploding kangaroos: assessing problems and setting targets. In: Lunney D, Eby P, Hutchings P, Burgin S (eds) Pest or guest: the zoology of overabundance. Royal Zoological Society of New South Wales, Sydney

Couslon GM, Norbury GL (1988) Ecology and management of western grey kangaroos (Macropus fuliginosus) at Hattah-Kulkyne National Park. Arthur Rylah Institute for Environmental Research, Conservation, Forests & Lands, National Parks and Wildlife Division, Melbourne

Cowan DP (1987) Group living in the European rabbit (Oryctolagus cuniculus): mutual benefit or resource localization? J Anim Ecol 56:779–795

Cowan P (2007) How many possums make a cow? N Z J Ecol 31:261–262

Cowan DP, Hartley FG (2008) Order lagomorpha: rabbit. In: Harris S, Yalden D (eds) Mammals of the British Isles handbook, 4th edn. The Mammal Society, London, pp 201–210

Cowan PE, Waddington DC (1990) Suppression of fruit production of the endemic forest tree, Elaeocarpus dentatus, by introduced marsupial brushtail possums, Trichosurus vulpecula. N Z J Bot 28:217–224

Cowan DP, Hardy AR, Vaughan JP, Christie WG (1989) Rabbit ranging behaviour and its implications for the management of rabbit populations. In: Putman RJ (ed) Mammals as pests. The Mammal Society, London, pp 178–185

Cox PG, Rayfield EJ, Fagan MJ, Herrel A, Pataky TC, Jeffery N (2012) Functional evolution of the feeding system in rodents. PLoS One 7:e36299

Crawley MJ (1990) Rabbit grazing, plant competition and seedling recruitment in acid grassland. J Appl Ecol 27(3):803–820

Croft DB (1991) Home range of the red kangaroo Macropus rufus. J Arid Environ 20:83–98

Cunningham P (2000) Daily activity pattern and diet of a population of the Spinytailed lizard, Uromastyx aegyptius microlepis, during summer in the United Arab Emirates. Zool Middle East 21(1):37–46

Currie J, Koppelhus EB, Muhammad AF (1995) "Stomach" contents of a hadrosaur from the Dinosaur Park Formation (Campanian, Upper Cretaceous) of Alberta, Canada. In: Sun AY, Wang Y (eds) Sixth symposium of Mesozoic terrestrial ecosystems and biota. China Ocean, Beijing, pp 111–114

Cusack JJ, Duthie AD, Sarobidy O, Rocío R, Pozo A, Mason THE, Månsson J, Nilsson L, Tombre IM, Eythórsson E, Madsen J, Tulloch A, Hearn RD, Redpath S, Bunnefeld N (2018) Time series analysis reveals synchrony and asynchrony between conflict management effort and increasing large grazing bird populations in northern Europe. Conserv Lett 12:e12450

Cuthbertson RS, Tirabasso A, Rybczynski N et al (2012) Kinetic limitations of intracranial joints in *Brachylophosaurus canadensis* and *Edmontosaurus regalis* (Dinosauria: Hadrosauridae), and their implications for the chewing mechanics of hadrosaurids. Anat Rec 295:968–979

Dawson TJ (2012) Kangaroos. CSIRO Publishing, Collingwood, VIC

Dawson TJ, Ellis BA (1994) Diets of mammalian herbivores in Australian arid shrublands - seasonal effects on overlap between red kangaroos, sheep and rabbits on dietary niche breadths and selectivities. J Arid Environ 26:257–271

Dawson TJ, Ellis BA (1996) Diets of mammalian herbivores in Australian arid, hilly shrublands: seasonal effects on overlap between euros (hill kangaroos), sheep and feral goats, and on dietary niche breadths and electivities. J Arid Environ 34:491–506

Dawson TJ, Munn AJ (2007) How much do kangaroos of differing age and size eat relative to domestic stock?: implications for the arid rangelands. In: Dickman C, Lunney D, Burgin S (eds)

Animals of arid Australia: out on their own. Royal Zoological Society of New South Wales, Mosman, NSW, pp 96–101

Department of Environment and Heritage Protection, Queensland (2018) Overview of the Queensland macropod industry. Version 1.5

Department of Environment, Land, Water and Planning, Victoria (2016) Koalas at Cape Otway: Managing Koalas at Cape Otway (Factsheet)

Department of Environment, Land, Water and Planning, Victoria (2017) Kangaroo Population Management. Melbourne

Department of Environment, Land, Water and Planning, Victoria (2018) Koalas in blue gum plantations

Department of Environment, Water, and Natural Resources, Government of South Australia (2017) South Australian Commercial Kangaroo Management Plan 2018–2022. Adelaide

Department of Parks and Wildlife, Western Australia (2013) Management Plan for the Commercial Harvest of Kangaroos in Western Australia 2014–2018. Perth

Department of Sustainability and Environment, Victoria (2004) Victoria's Koala Management Strategy. Melbourne

Dexter N, Hudson M, James S, MacGregor C, Lindenmayer DB (2013) Unintended consequences of invasive predator control in an Australian Forest: overabundant wallabies and vegetation change. PLoS One 8:e69087

Di Stefano J (2005) Mammalian browsing damage in the Mt. Cole state Forest, southeastern Australia: analysis of browsing patterns, spatial relationships and browse selection. New For 29:43–61

Di Stefano J, Anson JA, York A, Greenfield A, Coulson G, Berman A, Bladen M (2007) Interactions between timber harvesting and swamp wallabies (Wallabia bicolor): space use, density and browsing impact. For Ecol Manag 253:128–137

Dokter AM, Fokkema W, Ebbinge BS, Olff H, Van der Jeugd H, Nolet BA (2018) Agricultural pastures challenge the attractiveness of natural saltmarsh for a migratory goose. J Appl Ecol 55:2707–2718

du Toit JT (1990) Feeding-height stratification among African browsing ruminants. Afr J Ecol 28:55–61

Duncan RP, Holland EP, Pech RP, Barron M, Nugent G, Parkes JP (2011) The relationship between possum density and browse damage on kamahi in New Zealand forests. Austral Ecol 36:858–869

Durant D (2003) The digestion of fibre in herbivorous Anatidae – a review. Wild 54:7–24

Edwards G, Dawson T, Croft D (1995) The dietary overlap between red kangaroos (Macropus rufus) and sheep (Ovis aries) in the arid rangelands of Australia. Austral Ecol 20:324–334

Edwards GP, Croft DB, Dawson TJ (1996) Competition between red kangaroos (Macropus rufus) and sheep (Ovis aries) in the arid rangelands of Australia. Aust J Ecol 21:165–172

Eldridge DJ, Delgado-Baquerizo M, Travers SK, Val J, Oliver I (2017) Do grazing intensity and herbivore type affect soil health? Insights from a semi-arid productivity gradient. J Appl Ecol 54:976–985

Eldridge DJ, Delgado-Baquerizo M, Travers SK, Val J, Oliver I, Dorrough JW, Soliveres S (2018) Livestock activity increases exotic plant richness, but wildlife increases native richness, with stronger effects under low productivity. J Appl Ecol 55:766–776

Elfwing T, Tedengren M (2000) A comparison of production effects between corals and macroalgae at increased seawater temperature. Proc Ninth Int Coral Reef Symp 2:1139–1142

Erickson GM, Krick BA, Hamilton M, Bourne GR, Norell MA, Lilleodden E, Sawyer WG (2012) Complex dental structure and wear biomechanics in hadrosaurid dinosaurs. Science 338:98–101

Eskildsen LI, Olesen JM, Jones CG (2004) Feeding response of the Aldabra giant tortoise (Geochelone gigantea) to island plants showing heterophylly. J Biogeogr 31(11):1785–1790

Espinosa Gómez F, Santiago García J, Gómez Rosales S, Wallis IR, Chapman CA, Morales Mávil J, Canales Espinosa D, Hernández SL (2015) Howler monkeys (Alouatta palliata mexicana) produce tannin-binding salivary proteins. Int J Primatol 36:1086–1100

Espinosa-Gómez FC, Serio-Silva JC, Santiago-García JD, Sandoval-Castro CA, Hernández-Salazar LT, Mejía-Varas F, Ojeda-Chávez J, Chapman CA (2018) Salivary tannin-binding proteins are a pervasive strategy used by the folivorous/frugivorous black howler monkey. Am J Primatol 80: e22737. https://doi.org/10.1002/ajp.22737

Espinoza RE, Wiens JJ, Tracy CR (2004) Recurrent evolution of herbivory in small, cold-climate lizards: breaking the ecophysiological rules of reptilian herbivory. Proc Natl Acad Sci 101 (48):16819–16824

Estrada A et al (2017) Impending extinction crisis of the world's primates: why primates matter. Sci Adv 3:e1600946

Fa JE, Peres CA, Meeuwig J (2002) Bushmeat exploitation in tropical forests: an intercontinental comparison. Conserv Biol 16(1):232–237

Falcón W, Hansen DM (2018) Island rewilding with giant tortoises in an era of climate change. Philos Trans R Soc Lond B Biol Sci 373(1761):20170442

Falcon, W., Hansen, D. M. and Moll, D. (2018) Frugivory and seed dispersal by chelonians: a review and synthesis, bioRxiv, p. 379933

Farlow JO (1976) A consideration of the trophic dynamics of a Late Cretaceous large-dinosaur community (Oldman formation). Ecology 57:841–857

Farlow JO, Coroian ID, Foster JR (2010) Giants on the landscape: modelling the abundance of megaherbivorous dinosaurs of the Morrison formation (late Jurassic, western USA). Hist Biol 22:403–429

Fashing PJ, Nguyen N, Venkataraman VV, Kerby JT (2014) Gelada feeding ecology in an intact ecosystem at Guassa, Ethiopia: variability over time and implications for theropith and hominin dietary evolution. Am J Phys Anthropol 155:1–16

Fearnside PM (2018) Challenges for sustainable development in Brazilian Amazonia. Sustain Dev 26(2):141–149

Feng ZL, Liu RS, DeAngelis DL, Bryant JP, Kielland K, Stuart Chapin F, Swihart R (2009) Plant toxicity, adaptive herbivory, and plant community dynamics. Ecosystems 12:534–547

Fernandez FA et al (2012) How sustainable is the use of natural resources in Brazil? Natureza & Conservação 10(1):77–82

Fiorelli LE et al (2016) A new late cretaceous crocodyliform from the western margin of Gondwana (La Rioja Province, Argentina). Cretac Res 60:194–209

Flemming SA, Calvert A, Nol E, Smith PA (2016) Do hyperabundant Arctic-nesting geese pose a problem for sympatric species? Environ Rev 24:1–10

Foster C, Barton P, Wood J, Lindenmayer D (2015) Interactive effects of fire and large herbivores on web-building spiders. Oecologia 179:237–248

Fowler MC, Robson TO (1978) The effects of food preference and stocking rates of grass carp (Ctenopharyngodon idella VAL.) on mixed plant communities. Aquat Bot 5:261–276

Fox AD, Abraham KF (2017) Why geese benefit from the transition from natural vegetation to agriculture. Ambio 46:S188–S197

Fox AD, Madsen J (2017) Threatened species to super-abundance: the unexpected international implications of successful goose conservation. Ambio 46:179–187

Fox AD, Elmberg J, Tombre IM, Hessel R (2017) Agriculture and herbivorous waterfowl: a review of the scientific basis for improved management. Biol Rev 92:854–877. https://doi.org/10.1111/brv.12258

Frank ASK, Wardle GM, Greenville AC, Dickman CR (2016) Cattle removal in arid Australia benefits kangaroos in high quality habitat but does not affect camels. Rangel J 38:73–84

Franklin P, Dunbar M, Whitehead P (2008) Flow controls on lowland river macrophytes: a review. Sci Total Environ 400:369–378. https://doi.org/10.1016/j.scitotenv.2008.06.018

Fricke HC, Hencecroth J, Hoerner ME (2011) Lowland–upland migration of sauropod dinosaurs during the late Jurassic epoch. Nature 480:513–515

Frith HJ (1978) Wildlife conservation. Angus and Robertson, Sydney

Froyd CA et al (2014) The ecological consequences of megafaunal loss: Giant tortoises and wetland biodiversity. Ecol Lett 17(2):144–154. https://doi.org/10.1111/ele.12203

Gaines SD, Lubchenco J (1982) A unified approach to marine plant-herbivore interactions. II. Biogeography. Annu Rev Ecol Syst 13:111–138
Galiano D, Kubiak BB, Overbeck GE, de Freitas TRO (2014) Effects of rodents on plant cover, soil hardness, and soil nutrient content: a case study on tuco-tucos (Ctenomys minutus). Acta Theriol 59:583–587
Ganzhorn JU et al (2016) The importance of protein in leaf selection of folivorous primates. Am J Primatol 79:1–13
García M, Gerber GP (2016) Conservation and management of Cyclura iguanas in Puerto Rico. Herpetol Conserv Biol 11:61–67
Garrigues L, Cadi A (2011) Re-introduction of African spurred tortoise in north Ferlo, Senegal. In: Soorae P (ed) Global re-introduction perspectives: more case studies from around the globe. IUCN/SSC Re-introduction Specialist Group & Environment Agency-ABU DHABI, Abu Dhabi, pp 94–97
Gee CT (2011) Dietary options for the sauropod dinosaurs from an integrated botanical and paleobotanical perspective. In: Klein N, Remes K, Gee CT, Sander PM (eds) Biology of the sauropod dinosaurs: understanding the life of giants. Indiana University Press, Bloomington, pp 34–56
Genes L et al (2017) Credit of ecological interactions: a new conceptual framework to support conservation in a defaunated world. Ecol Evol 7(6):1892–1897. https://doi.org/10.1002/ece3.2746
Gerlach J (2014) Western Indian Ocean tortoises: ecology, diversity, evolution, conservation, Palaeontology. Siri Scientific Press, Rocjdale
Gibbens RP, Havstad KM, Billheimer DD, Herbel CH (1993) Creosotebush vegetation after 50 years of lagomorph exclusion. Oecologia 94:210–217. https://doi.org/10.1007/BF00341319
Gibbons J (1984) Iguanas of the South Pacific. Oryx 18(2):82–91
Gibbs JP, Marquez C, Sterling EJ (2008) The role of endangered species reintroduction in ecosystem restoration: tortoise–Cactus interactions on Española Island, Galápagos. Restor Ecol 16(1):88–93. https://doi.org/10.1111/j.1526-100X.2007.00265.x
Gibbs JP, Sterling EJ, Zabala FJ (2010) Giant tortoises as ecological engineers: a long-term quasi-experiment in the Galápagos Islands. Biotropica 42(2):208–214. https://doi.org/10.1111/j.1744-7429.2009.00552.x
Gibbs JP et al (2014) Demographic outcomes and ecosystem implications of giant tortoise reintroduction to Española Island, Galapagos. PLoS One 9(10):e110742
Gibson CWD, Hamilton J (1983) Feeding ecology and seasonal movements of giant tortoises on Aldabra atoll. Oecologia 56(1):84–92
Godoy-Vitorino F, Goldfarb KC, Karaoz U, Leal S, Garcia-Amado MA, Hugenholtz P, Tringe SG, Brodie EL, Dominguez-Bello MG (2011) Comparative analyses of foregut and hindgut bacterial communities in hoatzins and cows. ISME J 6:531–541
Gong W, Sinden J, Braysher M, Jones R (2009) The economic impacts of vertebrate pests in Australia. Invasive Animals Cooperative Research Centre, Canberra
Gordon IJ, Prins HHT (2008) The ecology of browsing and grazing. Springer, Berlin
Grajal A, Strahl SD, Parra R, Dominguez MG, Neher A (1989) Foregut fermentation in the hoatzin, a neotropical leaf-eating bird. Science 245:1236–1238
Grandmaison DD, Ingraldi MF, Peck FR (2010) Desert tortoise microhabitat selection on the Florence military reservation, south-Central Arizona. J Herpetol 44(4):581–590
Grant TD, Hudson RD (2015) West Indian iguana Cyclura spp reintroduction and recovery programmes: zoo support and involvement. Int Zoo Yearb 49(1):49–55
Greaves WS (1995) Functional predictions from theoretical models of the skull and jaws of reptiles and mammals. In: Thomason JJ (ed) Functional morphology in vertebrate paleontology. Cambridge University Press, Cambridge, pp 99–115
Green AL, Bellwood DR (2009) Monitoring functional groups of herbivorous reef fishes as indicators of coral reef resilience – a practical guide for coral reef managers in the Asia Pacific region. Switzerland, IUCN working group on Climate Change and Coral Reefs, Gland, p 70

Griffiths CJ et al (2010) The use of extant non-indigenous tortoises as a restoration tool to replace extinct ecosystem engineers. Restor Ecol 18(1):1–7. https://doi.org/10.1111/j.1526-100X.2009.00612.x
Griffiths CJ et al (2011) Resurrecting extinct interactions with extant substitutes. Curr Biol 21:762–765. https://doi.org/10.1016/j.cub.2011.03.042
Griffiths CJ et al (2013) Assessing the potential to restore historic grazing ecosystems with tortoise ecological replacements. Conserv Biol 27(4):690–700. https://doi.org/10.1111/cobi.12087
Gyimes A, Van Rooij AP, Nolet BA (2010) Nonlinear effects of food aggregation on interference competition in mallards. Behav Ecol Sociobiol 64:1897–1904
Gyimesi A, De Vries PP, De Boer T, Nolet BA (2011) Reduced tuber banks of fennel pondweed due to summer grazing by waterfowl. Aquat Bot 36:24–28
Haas K, Köhler U, Diehl S, Köhler P, Dietrich S, Holler S, Jaensch A, Niedermaier M, Vilsmeier J (2007) Influence of fish on habitat choice of water birds: a whole system experiment. Ecology 88:2915–2925
Hackländer K, Tataruch F, Ruf T (2002) The effect of dietary fat content on lactation energetics in the European hare (Lepus europaeus). Physiol Biochem Zool 75:19–28. https://doi.org/10.1086/324770
Hagenah N, Prins HHT, Olff H (2009) Effects of large herbivores on murid rodents in a south African savanna. J Trop Ecol 25:483–492
Hamann O (1993) On vegetation recovery, goats and giant tortoises on Pinta Island, Galápagos, Ecuador. Biodivers Conserv 2(2):138–151. https://doi.org/10.1007/BF00056130
Handa IT, Harmsen R, Jefferies RL (2002) Patterns of vegetation change and the recovery potential of degraded areas in a coastal marsh system of the Hudson Bay lowlands. J Ecol 90:86–99
Hansen DM, Galetti M (2009) The forgotten megafauna. Science (New York, NY) 324 (5923):42–43. https://doi.org/10.1126/science.1172393
Hansen DM et al (2010) Ecological history and latent conservation potential: large and giant tortoises as a model for taxon substitutions. Ecography 33(2):272–284. https://doi.org/10.1111/j.1600-0587.2010.06305.x
Harvie AE (1973) Diet of the opossum (Trichosurus vulpecula Kerr) on farmland northeast of Waverley, New Zealand. Proc N Z Ecol Soc 20:48–52
Hastings AK et al (2014) Domination by reptiles in a terrestrial food web of the Bahamas prior to human occupation. J Herpetol 48(3):380–388. https://doi.org/10.1670/13-091R1
Hautier L, Lebrun R, Saksiri S, Michaux J, Vianey-Liaud M, Marivaux L (2011) Hystricognathy vs sciurognathy in the rodent jaw: a new morphometric assessment of hystricognathy applied to the living fossil Laonastes (Diatomyidae). PLoS One 6:e18698
Hautier L, Lebrun R, Cox PG (2012) Patterns of covariation in the masticatory apparatus of hystricognathous rodents: implications for evolution and diversification. J Morphol 273:1319–1337
Hearne EL et al (2018) Effects of green turtle grazing on seagrass and macroalgae diversity vary spatially among seagrass meadows. Aquat Bot 152:10–15
Heath ME, Mcginnis SM (1980) Body temperature and heat transfer in the Green Sea turtle, Chelonia mydas. Coastal Mar Sci 1980(4):767–773
Heithaus MR et al (2014) Seagrasses in the age of sea turtle conservation and shark overfishing. Front Mar Sci 1:28. https://doi.org/10.3389/fmars.2014.00028
Herrera EA (2013) Capybara digestive adaptations. In: Capybara. Springer, New York, NY, pp 97–106
Hessen DO, Tombre IM, Van Geest G, Alfsnes K (2017) Global change and ecosystem connectivity: how geese link fields of central Europe to eutrophication of Arctic freshwaters. Ambio 46:40–47
Hik DS, Jefferies RL (1990) Increases in the net above-ground primary production of a salt-marsh forage grass: a test of the predictions of the herbivore optimization model. J Ecol 78:180–195
Hjältén J, Danell K, Ericson L (2004) Hare and vole browsing preferences during winter. Acta Theriol (Warsz) 49:53–62

Hoegh-Guldberg O (1999) Climate change, coral bleaching and the future of the world's coral reefs. Mar Freshw Res 50:839–866. https://doi.org/10.1071/MF99078
Hofmann RR, Stewart DRM (1972) Grazer or browser: a classification based on the stomach-structure and feeding habits of east African ruminants. Mammalia 36:226–240
Holland EP, Pech RP, Ruscoe WA, Parkes JP, Nugent G, Duncan RP (2013) Thresholds in plant-herbivore interactions: predicting plant mortality due to herbivore browse damage. Oecologia 172:751–766
Holland EP, Gormley AM, Pech RP (2016) Species- and site-specific impacts of an invasive herbivore on tree survival in mixed forests. Ecol Evol 6:1954–1966
Homolka M (1982) The food of Lepus europaeus in a meadow and woodland complex. Folia Zool 31:243–253
Horn MH, Correa SB, Parolin P et al (2011) Seed dispersal by fishes in tropical and temperate fresh waters: the growing evidence. Acta Oecol 37:561–577. https://doi.org/10.1016/j.actao.2011.06.004
Howe HF, Brown JS (1999) Effects of birds and rodents on synthetic tallgrass communities. Ecology 80:1776–1781
Howe HF, Brown JS, Zorn-Arnold B (2002) A rodent plague on prairie diversity. Ecol Lett 5:30–36
Howland B, Stojanovic D, Gordon IJ, Manning AD, Fletcher D, Lindenmayer DB (2014) Eaten out of house and home: impacts of grazing on ground-dwelling reptiles in Australian grasslands and grassy woodlands. PLoS One 9:e105966
Howland BW, Stojanovic D, Gordon IJ, Fletcher D, Snape M, Stirnemann IA, Lindenmayer DB (2016a) Habitat preference of the striped legless lizard: implications of grazing by native herbivores and livestock for conservation of grassland biota. Austral Ecol 41:455–464
Howland BW, Stojanovic D, Gordon IJ, Radford J, Manning AD, Lindenmayer DB (2016b) Birds of a feather flock together: using trait-groups to understand the effect of macropod grazing on birds in grassy habitats. Biol Conserv 194:89–99
Hughes TP, Rodrigues MJ, Bellwood DR et al (2007) Phase shifts, herbivory, and the resilience of coral reefs to climate change. Curr Biol 17:360–365. https://doi.org/10.1016/j.cub.2006.12.049
Hughes TP, Anderson KD, Connolly SR, Heron SF, Kerry JT, Lough JM, Baird AH, Baum JK, Berumen ML, Bridge TC, Claar DC, Eakin CM, Gilmour JP, Graham NAJ, Harrison H, Hobbs J-PA, Hoey AS, Hoogenboom M, Lowe RJ, McCulloch MT, Pandolfi JM, Pratchett M, Schoepf V, Torda G, Wilson SK (2018) Spatial and temporal patterns of mass bleaching of corals in the Anthropocene. Science 359(6371):80–83
Hulbert IA, Andersen R (2001) Food competition between a large ruminant and a small hindgut fermentor: the case of the roe deer and mountain hare. Oecologia 128:499–508. https://doi.org/10.1007/s004420100683
Hulme PE (1994) Seedling herbivory in grassland: relative impact of vertebrate and invertebrate herbivores. J Ecol 82:873–880
Hulme PE (1996) Herbivores and the performance of grassland plants: a comparison of arthropod, mollusk and rodent herbivory. J Ecol 84:43–51
Hume ID (1999) Marsupial Nutrition. Cambridge University Press, Cambridge
Hummel J, Clauss M (2011) Sauropod feeding and digestive physiology. In: Remes K, Gee CT, Sander PM (eds) Biology of the sauropod dinosaurs: understanding the life of giants. Indiana University Press, Bloomington, pp 11–33
Hummel J, Gee CT, Südekum KH et al (2008) In vitro digestibility of fern and gymnosperm foliage: implications for sauropod feeding ecology and diet selection. Proc R Soc B 275:1015–1021
Hunt A, Al-Nakkash L, Lee AH, Smith HF (2019) Phylogeny and herbivory are related to avian cecal size. Sci Rep 9:4243
Hunter EA, Gibbs JP (2014) Densities of ecological replacement herbivores required to restore plant communities: a case study of giant tortoises on Pinta Island, Galápagos. Restor Ecol 22 (2):248–256
Huntly NJ (1987) Influence of refuging consumers (pikas: Ochotona princeps) on subalpine meadow vegetation. Ecology 68:274–283. https://doi.org/10.2307/1939258

Huntly NJ, Smith AT, Ivins BL (1986) Foraging behavior of the Pika (Ochotona princeps), with comparisons of grazing versus haying. J Mammal 67:139–148

Hutchings MR, Harris S (1996) The current status of the brown hare (Lepus europaeus) in Britain. 1–76. https://doi.org/10.1007/s13398-014-0173-7.2

HVP Plantations (2018) Koalas. https://www.hvp.com.au/hvp-environment-conservation/koalas/. Accessed 9 July 2018

Hynes EF, Handasyde KA, Shaw G, Renfree MB (2010) Levonorgestrel, not etonogestrel, provides contraception in free-ranging koalas. Reprod Fertil Dev 22:913–919

Jackson JE (1988) Terrestrial mammalian pests in Argentina—an overview. Proc Vertebr Pest Conf 13:196–198

Jackson SM (2015) Taxonomy of Australian mammals. CSIRO Publishing, Collingwood

Jano AP, Jefferies RL, Rockwell RF (1998) The detection of vegetational change by multitemporal analysis of LANDSAT data: the effects of goose foraging. J Ecol 86:93–99

Janson CH, Goldsmith ML (1995) Predicting group size in primates: foraging costs and predation risks. Behav Ecol 6:326–336

Jarvis ED et al (2014) Whole-genome analyses resolve early branches in the tree of life of modern birds. Science 346:1320–1331

Jefferies RL, Rockwell RF, Abraham KE (2004) Agricultural food subsidies, migratory connectivity and large-scale disturbance in arctic coastal systems: a case study. Integr Comp Biol 44:130–139

Jefferies RL, Jano AP, Abraham KF (2006) A biotic agent promotes large-scale catastrophic change in the coastal marshes of Hudson Bay. J Ecol 94:234–242

Jennings NV (2008) Order lagomorpha: Brown hare. In: Harris S, Yalden D (eds) Mammals of the British Isles handbook, 4th edn. The Mammal Society, London, pp 210–220

Jetz W, Thomas GH, Joy JB, Hartmann K, Mooers AO (2012) The global diversity of birds in space and time. Nature 491:444–448

Johnson CN (2006) Australia's mammal extinctions: a 50,000 year history. Cambridge University Press, Cambridge

Jones CG (2002) Reptiles and amphibians. In: Perrow M, David A (eds) Handbook of ecological restoration. Cambridge University Press, New York, pp 355–375

Jonzen N, Pople AR, Grigg GC, Possingham HP (2005) Of sheep and rain: large-scale population dynamics of the red kangaroo. J Anim Ecol 74:22–30

Karmiris IE, Nastis AS (2007) Intensity of livestock grazing in relation to habitat use by brown hares (Lepus europaeus). J Zool 271:193–197. https://doi.org/10.1111/j.1469-7998.2006.00199.x

Kavanagh RP (1988) The impact of predation by the powerful owl, Ninox strenua, on a population of the greater glider, Petauroides volans. Aust J Ecol 13:445–450

Kay RNB, Hoppe P, Maloiy GMO (1976) Fermentative digestion of food in the colobus monkey Colobus polykomos. Experientia 32:485–487

Kennedy MA, Heck KL, Michot TC (2018) Impacts of wintering redhead ducks (Aythya americana) on seagrasses in the northern Gulf of Mexico. J Exp Mar Biol Ecol 506:42–48

Kershaw JA (1934) The koala on Wilsons promontory. Victorian Nat 51:76–77

King GM (1996) Reptiles and herbivory. Springer Science & Business Media, New York

Kirkpatrick J (2016) Pattern and process in riparian vegetation of the treeless high country of Australia. In: Capon S, James C, Reid M (eds) Vegetation of Australian riverine landscapes: biology, ecology and management. CSIRO Publishing, Melbourne, pp 201–220

Klaasen M, Nolet BA (2007) The role of herbivorous water birds in aquatic systems through interactions with aquatic macrophytes, with special reference to the Bewick's swan – fennel pondweed system. Hydrobiologia 584:205–213

Klein RG (1984) Mammalian extinctions and stone age people in Africa. In: Martin PS, Klein RG (eds) Quaternary extinctions: a prehistoric revolution. University of Arizona Press, Tucson, pp 553–573

Kley NJ et al (2010) Craniofacial morphology of Simosuchus clarki (Crocodyliformes: Notosuchia) from the late cretaceous of Madagascar. J Vertebr Paleontol 30(1):13–98
Kollars NM, Henry AK, Whalen MA, Boyer KE, Cusson M, Eklöf JS, Hereu CM, Jorgensen P, Kiriakopolos SL, Reynolds PL, Tomas F, Turner MS, Ruesink JL (2017) Meta-analysis of reciprocal linkages between temperate seagrasses and waterfowl with implications for conservation. Front Plant Sci 8:2119
Köhler G (2004) Conservation status of spiny-tailed iguanas (genus Ctenosaura), with special emphasis on the Utila Iguana (C. bakeri). Iguana 11:207–211
Komonen M, Komonen A, Otgonsuren A (2003) Daurian pikas (Ochotona daurica) and grassland condition in eastern Mongolia. J Zool 259:281–288
Korth WW (1994) The tertiary record of rodents in North America. Springer Science, New York
Krebs CJ, Boonstra R, Boutin S, Sinclair ARE (2001) What drives the 10-year cycle of snowshoe hares? Bioscience 51:25. https://doi.org/10.1641/0006-3568(2001)051[0025,WDTYCO]2.0.CO;2
Kułakowska KA, Kułakowski TM, Inglis R, Smith GC, Haynes PJ, Prosser P, Thorbek P, Sibly RM (2014) Using an individual-based model to select among alternative foraging strategies of woodpigeons: data support a memory-based model with a flocking mechanism. Ecol Model 280:89–101
Kuffner IB, Walters LJ, Becerro MA et al (2006) Inhibition of coral recruitment by macroalgae and cyanobacteria. Mar Ecol Prog Ser 323:107–117. https://doi.org/10.3354/meps323107
Kuijper DJP, van Wieren SE, Bakker JP (2004) Digestive strategies in two sympatrically occurring lagomorphs. J Zool 264:171–178. https://doi.org/10.1017/S0952836904005722
Lacher TE, Murphy WJ, Rogan J, Smith AT, Upham NS (2016) Evolution, phylogeny, ecology, and conservation of the clade Glires: Lagomorpha and Rodentia. In: Wilson DE, Lacher JTE, Mittermeier RA (eds) Handbook of mammals of the world, volume 6: lagomorphs and rodents. Lynx Ediciones, Barcelona, pp 15–26
Lambert MRK (1993) On growth, sexual dimorphism, and the general ecology of the African spurred tortoise, Geochelone sulcata. Chelonian Conserv Biol 4:37–46
Laurel MH, Richard E, Amy L (2000) Germination rates of seeds consumed by two species of rock iguanas (Cyclura spp.) in the Dominican Republic. Caribb J Sci 36(1–2):149–151
LeBlanc AR, Reisz RR, Evans DC, Bailleul AM (2016) Ontogeny reveals function and evolution of the hadrosaurid dinosaur dental battery. BMC Evol Biol 16:152. https://doi.org/10.1186/s12862-016-0721-1
Lechner-Doll M, Kaske M, Engelhardt WV (1991) Factors affecting the mean retention time of particles in the forestomach of ruminants and camelids. In: Tsuda T, Sasaki Y, Kawashima R (eds) Physiological aspects of digestion and metabolism in ruminants. Academic, San Diego, pp 455–482
Lees AC, Bell DJ (2008) A conservation paradox for the 21st century: the European wild rabbit Oryctolagus cuniculus, an invasive alien and an endangered native species. Mamm Rev 38:304–320. https://doi.org/10.1111/j.1365-2907.2008.00116.x
Lefebvre J, Gauthier G, Giroux J-F, Reed A, Reed ET, Bélanger L (2017) The greater snow goose Anser caerulescens atlanticus: managing an overabundant population. Ambio 46:262–274
Leguat F (2017) The voyage of François Leguat of Bresse to Rodriguez, Mauritius, Java, and the Cape of Good Hope: volume I. Hakluyt Society, London
Letnic M, Koch F, Gordon C, Crowther MS, Dickman CR (2009) Keystone effects of an alien top-predator stem extinctions of native mammals. Proc Biol Sci 276:3249–3256
Ley RE et al (2008) Evolution of mammals and their gut microbes. Science 320:1647–1651
Lichtenbelt WDM (1992) Digestion in an ectothermic herbivore, the green iguana (Iguana iguana): effect of food composition and body temperature. Physiol Zool 65(3):649–673
Linley GD, Moseby KE, Paton DC (2017) Vegetation damage caused by high densities of burrowing bettongs (Bettongia lesueur) at arid recovery. Aust Mammal 39:33–41
Lodge DM (1991) Herbivory on freshwater macrophytes. Aquat Bot 41:195–224. https://doi.org/10.1016/0304-3770(91)90044-6

Lodge DM, Cronin G, van Donk E, Froelich AJ (1998) Impact of Herbivory on plant standing crop: comparisons among biomes, between vascular and nonvascular plants, and among freshwater herbivore taxa. In: Jeppesen E et al (eds) The structuring role of submerged Marcophytes in lakes. Springer, New York, pp 149–174

Lokrantz J, Nyström M, Thyresson M, Johansson C (2008) The non-linear relationship between body size and function in parrotfishes. Coral Reefs 27:967–974. https://doi.org/10.1007/s00338-008-0394-3

Lopes CM, Barba M, Boyer F, Mercier C, Filho PJSS, Heidtmann LM, Galiano D, Kubiak BB, Langone PQ, Garcias FM, Giely L, Coissac E, Freitas TRO, Taberlet P (2015) DNA metabarcoding diet analysis for species with parapatric vs sympatric distribution: a case study on subterranean rodents. Heredity 114:525–536

Lopes LE, Fernandes AM, Medeiros MCI, Marini MA (2016) A classification scheme for avian diet types. J Field Ornithol 87:309–322

Lovich JE et al (2018) Where have all the turtles gone, and why does it matter? Bioscience XX (X):1–11. https://doi.org/10.1093/biosci/biy095

Low T (2002) The new nature: winners and losers in wild Australia. Penguin, Melbourne

Loyn RH, Middleton WGD (1980) Eucalypt decline and wildlife in rural areas. In: Old KM, Kile GA, Ohmart CP (eds) Eucalypt dieback in forests and woodlands. Forest Research CSIRO, Melbourne

Lush L, Ward AI, Wheeler P (2014) Opposing effects of agricultural intensification on two ecologically similar species. Agric Ecosyst Environ 192:61–66. https://doi.org/10.1016/j.agee.2014.03.048

Lush L, Ward AI, Wheeler P (2017) Dietary niche partitioning between sympatric brown hares and rabbits. J Zool 303:36–45. https://doi.org/10.1111/jzo.12461

Luza AL, Trindade JPP, Maestri R, Duarte LDS, Hartz SM (2018) Rodent occupancy in grassland paddocks subjected to different grazing intensities in South Brazil. Perspect Ecol Conserv. https://doi.org/10.1016/j.pecon.2018.06.006

Ma H, Ge D, Shenbrot G, Pisano J, Yang Q, Zhang Z (2016) Hypsodonty of Dipodidae (Rodentia) in correlation with diet preferences and habitats. J Mamm Evol. https://doi.org/10.1007/s10914-016-9352-y

Madden RH (2015) Hypsodonty in mammals: evolution, geomorphology and the role of earth surface processes. Cambridge University Press, Cambridge

Maestri R, Patterson BD, Fornel R, Monteiro LR, Freitas TRO (2016) Diet, bite force and skull morphology in the generalista rodent morphotype. J Evol Biol 29:2191–2204

Maestri R, Monteiro LR, Fornel R, Freitas TRO, Patterson BD (2017) Geometric morphometrics meets metacommunity ecology: environment and lineage distribution affects spatial variation in shape. Ecography 41:90–100

Main AR (1992) Management to retain biodiversity in the face of uncertainty. In: Hobbs RJ (ed) Biodiversity in Mediterranean ecosystems in Australia. Surrey Beatty & Sons, Chipping Norton, NSW

Makarieva AM, Gorshkov VG, Li B-L (2005) Gigantism, temperature and metabolic rate in terrestrial poikilotherms. Proc R Soc B Biol Sci 272(1578):2325–2328. https://doi.org/10.1098/rspb.2005.3223

Malhia Y et al (2016) Megafauna and ecosystem function from the Pleistocene to the Anthropocene. Proc Natl Acad Sci 113(4):838–846

Mallon JC, Anderson JS (2014a) Implications of beak morphology for the evolutionary paleoecology of the megaherbivorous dinosaurs from the Dinosaur Park formation (upper Campanian) of Alberta, Canada. Palaeogeogr Palaeoclimatol Palaeoecol 394:29–41

Mallon JC, Anderson JS (2014b) The functional and palaeoecological implications of tooth morphology and wear for the megaherbivorous dinosaurs from the Dinosaur Park formation (upper Campanian) of Alberta, Canada. PLoS One 9(6):e98605. https://doi.org/10.1371/journal.pone.0098605

Mallon JC, Evans DC, Ryan MJ, Anderson JS (2013) Feeding height stratification among the herbivorous dinosaurs from the Dinosaur Park formation (upper Campanian) of Alberta, Canada. BMC Ecol 13(1):14. https://doi.org/10.1186/1472-6785-13-14
Mann AN, O'Reilly-Wapstra JM, Iason GR, Sanson G, Davies NW, Tilyard P, Williams D, Potts BM (2012) Mammalian herbivores reveal marked genetic divergence among populations of an endangered plant species. Oikos 121:268–276
Mannion PD, Upchurch P, Carrano MT et al (2011) Testing the effect of the rock record on diversity: a multidisciplinary approach to elucidating the generic richness of sauropodomorph dinosaurs through time. Biol Rev 86:157–181
Marks CA (1998) Review of the humaneness of destruction techniques used on the common wombat Vombatus ursinus in Victoria. In: Wells RT, Pridmore PA (eds) Wombats. Surrey Beatty and Sons, Sydney, pp 287–297
Martin PS (1984) Prehistoric overkill: the global model. In: Martin PS, Klein RG (eds) Quaternary extinctions: a prehistoric revolution. University of Arizona Press, Tucson, pp 354–403
Martin RW (1985a) Overbrowsing, and decline of a population of the koala, Phascolarctos cinereus, in Victoria. 1. Food preference and food tree defoliation. Aust Wildl Res 12:355–365
Martin RW (1985b) Overbrowsing, and decline of a population of the koala, Phascolarctos cinereus, in Victoria. 2. Population condition. Aust Wildl Res 12:367–375
Massoia E (1970) Mamiferos que contribuyen a deteriorar suelos y pasturas en la Republica Argentina. IDIA 276:14–17
Matsuda I, Murai T, Clauss M, Yamada T, Tuuga A, Bernard H, Higashi S (2011) Regurgitation and remastication in the foregut-fermenting proboscis monkey (Nasalis larvatus). Biol Lett 7:786–789
Matsuda I, Higashi S, Otani Y, Tuuga A, Bernard H, Corlett RT (2013) A short note on seed dispersal by colobines: the case of the proboscis monkey. Integr Zool 8:395–399
Matsuda I et al (2014) Faecal particle size in free-ranging primates supports 'rumination' strategy in the proboscis monkey (Nasalis larvatus). Oecologia 174:1127–1137
Matsuda I et al (2015) Excretion patterns of solute and different-sized particle passage markers in foregut-fermenting proboscis monkey (Nasalis larvatus) do not indicate an adaptation for rumination. Physiol Behav 149:45–52
Matsuda I, Chapman CA, Shi Physilia CY, Mun Sha JC, Clauss M (2017a) Primate resting postures: constraints by foregut fermentation? Physiol Biochem Zool 90:383–391
Matsuda I, Clauss M, Tuuga A, Sugau J, Hanya G, Yumoto T, Bernard H, Hummel J (2017b) Factors affecting leaf selection by foregut-fermenting proboscis monkeys: new insight from in vitro digestibility and toughness of leaves. Sci Rep 7:42774
Matsuda I, Espinosa-Gómez FC, Ortmann S, Sha JCM, Osman I, Nijboer J, Schwarm A, Ikeda T, Clauss M (2019) Retention marker excretion suggests incomplete digesta mixing across the order primates. Physiol Behav 208:112558. https://doi.org/10.1016/j.physbeh.2019.112558
Mau M, Südekum KH, Johann A, Sliwa A, Kaiser TM (2009) Saliva of the graminivorous Theropithecus gelada lacks proline-rich proteins and tannin-binding capacity. Am J Primatol 71:663–669
McArthur C, Appleton R (2004) Effect of seedling characteristics at planting on browsing of Eucalyptus globulus by rabbits. Aust For 67:25–29. https://doi.org/10.1080/00049158.2004.10676202
McConkey KR (2018) Seed dispersal by Primates in Asian habitats: from species, to communities, to conservation. Int J Primatol 39:466–492
McIntire EJB, Hik DS (2005) Influences of chronic and current season grazing by collared pikas on above-ground biomass and species richness in subarctic alpine meadows. Oecologia 145:288–297. https://doi.org/10.1007/s00442-005-0127-z
McIntyre S, Stol J, Harvey J, Nicholls A, Campbell M, Reid A, Manning AD, Lindenmayer D (2010) Biomass and floristic patterns in the ground layer vegetation of box-gum grassy eucalypt woodland in Goorooyarroo and mulligans flat nature reserves, Australian Capital Territory. Cunninghamia 11:9–357

McIntyre S, Cunningham R, Donnelly C, Manning A (2015) Restoration of eucalypt grassy woodland: effects of experimental interventions on ground-layer vegetation. Aust J Bot 62:570–579

McKinnon M, Ahmad M, Bongers M, Chevalier R, Telfer I, Van Dorssen C (2018) Media coverage of lethal control: a case study of kangaroo culling in the Australian Capital Territory. Hum Dimens Wildl 23:90–99

McLeod SR (1996) The foraging behaviour of the arid zone herbivores, the red kangaroo (*Macropus rufus*) and the sheep (*Ovis aries*) and its role in their competitive interactions, population dynamics and life-history strategies. University of New South Wales, Sydney

Meehan HJ, McConkey KR, Drake DR (2002) Potential disruptions to seed dispersal mutualisms in Tonga, Western Polynesia. J Biogeogr 29(5–6):695–712. https://doi.org/10.1046/j.1365-2699.2002.00718.x

Meers T, Adams R (2003) The impact of grazing by eastern grey kangaroos (*Macropus giganteus*) on vegetation recovery after fire at Reef Hill Regional Park, Victoria. Ecol Manag Restor 4:126–132

Menkhorst P (2008) Hunted, marooned, re-introduced, contracepted: a history of koala management in Victoria. In: Lunney D, Munn A, Meikle W (eds) Too Close for comfort: contentious issues in human-wildlife encounters. Royal Zoological Society of New South Wales, Mosman, pp 73–92

Merton LFH, Bourn DM, Hnatiuk RJ (1976) Giant tortoise and vegetation interactions on Aldabra atoll—part 1: inland. Biol Conserv 9(4):293–304

Milchunas DG, Lauenroth WK, Burke IC (1998) Livestock grazing: animal and plant biodiversity of Shortgrass steppe and the relationship to ecosystem function. Oikos 83:65. https://doi.org/10.2307/3546547

Miller AM, O'Reilly-Wapstra JM, Potts BM, McArthur C (2009) Non-lethal strategies to reduce browse damage in eucalypt plantations. For Ecol Manag 259:45–55

Miller AM, O'Reilly-Wapstra JM, Potts BM, McArthur C (2011) Repellent and stocking guards reduce mammal browsing in eucalypt plantations. New For 42:301–316

Milton K (2000) Ducks out of water: nature conservation as boundary maintenance. In: Knight J (ed) Natural enemies: people-wildlife conflicts in anthropological perspective. Berg, Oxford, pp 229–246

Miranda EBP (2017) The plight of reptiles as ecological actors in the tropics. Front Ecol Evol 5:159

Mittermeier RA, Rylands AB, Wilson DE (2013) Handbook of the mammals of the world, vol 3. Primates, Lynx Edicions, Barcelona

Moberg F, Folke C (1999) Ecological goods and services of coral reef ecosystems. Ecol Econ 29:215–233. https://doi.org/10.1016/S0921-8009(99)00009-9

Moore BD, Lawler IR, Wallis IR, Beale CM, Foley WJ (2010) Palatability mapping: a koala's eye view of spatial variation in habitat quality. Ecology 91:3165–3176

Moran KL, Bjorndal KA (2005) Simulated green turtle grazing affects structure and productivity of seagrass pastures. Mar Ecol Prog Ser 305:235–247

Moura ADA et al (2015) Can green iguanas compensate for vanishing seed dispersers in the Atlantic forest fragments of north-East Brazil? J Zool 295(3):189–196

Mourão G et al (2000) Aerial surveys of caiman, marsh deer and pampas deer in the Pantanal wetland of Brazil. Biol Conserv 92(2):175–183

Müller DWH, Caton J, Codron D, Schwarm A, Lentle R, Streich WJ, Hummel J, Clauss M (2011) Phylogenetic constraints on digesta separation: variation in fluid throughput in the digestive tract in mammalian herbivores. Comp Biochem Physiol A 160:207–220

Mumby PJ (2009) Phase shifts and the stability of macroalgal communities on Caribbean coral reefs. Coral Reefs 28:761–773. https://doi.org/10.1007/s00338-009-0506-8

Mumby PJ, Hastings A, Edwards HJ (2007) Thresholds and the resilience of Caribbean coral reefs. Nature 450:98–101. https://doi.org/10.1038/nature06252

Munn A, Dawson T (2003) Energy requirements of the red kangaroo (*Macropus rufus*): impacts of age, growth and body size in a large desert-dwelling herbivore. J Comp Physiol B 173:575–582

Munn A, Dawson T, McLeod S, Croft D, Thompson M, Dickman C (2009) Field metabolic rate and water turnover of red kangaroos and sheep in an arid rangeland: an empirically derived dry-sheep-equivalent for kangaroos. Aust J Zool 57:23–28

Munro NT, Moseby KE, Read JL (2009) The effects of browsing by feral and re-introduced native herbivores on seedling survivorship in the Australian rangelands. Rangel J 31:417–426

Mutze G, Cooke B, Jennings S (2016) Density-dependent grazing impacts of introduced European rabbits and sympatric kangaroos on Australian native pastures. Biol Invasions 18:2365–2376. https://doi.org/10.1007/s10530-016-1168-4

Nabavizadeh A (2016) Evolutionary trends in the jaw adductor mechanics of ornithischian dinosaurs. Anat Rec 299:271–294

Nabavizadeh A, Weishampel DB (2016) The predentary bone and its significance in the evolution of feeding mechanisms in ornithischian dinosaurs. Anat Rec 299:1358–1388

Nagy KA, Shoemaker VH (1984) Field energetics and food consumption of the Galápagos marine Iguana, Amblyrhynchus cristatus. Physiol Zool 57(3):281–290

Nash KL, Graham NAJ, Bellwood DR (2013) Fish foraging patterns, vulnerability to fishing, and implications for the management of ecosystem function across scales. Ecol Appl 23:1632–1644

Natural Resources Kangaroo Island, Department for Environment and Water, South Australia (2017) Unexpected rise in koala numbers on KI. http://www.naturalresources.sa.gov.au/kangarooisland/news/170829-unexpected-rise-in-koala-numbers Accessed 9 July 2018

Newsome AE (1997) Reproductive anomalies in the red kangaroo in Central Australia explained by aboriginal traditional knowledge and ecology. In: Saunders NR, Hinds LA (eds) Marsupial biology: recent research, new perspectives. University of New South Wales Press, Sydney, pp 229–236

Noble JC (2001) Regulating Callitris populations: a tale of two pineries. In: Dargavel J, Hart D, Libbis B (eds) Perfumed pineries: environmental histories of Australia's cypress pines. Centre for Resource and Environmental Studies, Australian National University, Canberra

Noble JC, Hik DS, Sinclair ARE (2007) Landscape ecology of the burrowing bettong: fire and marsupial biocontrol of shrubs in semi-arid Australia. Rangel J 29:107–119

Norman DB, Weishampel DB (1985) Ornithopod feeding mechanisms: their bearing on the evolution of herbivory. Am Nat 126:151–164

Norman DB, Crompton AW, Butler RJ et al (2011) The lower Jurassic ornithischian dinosaur *Heterodontosaurus tucki* Crompton & Charig, 1962: cranial anatomy, functional morphology, taxonomy, and relationships. Linnean Soc 163:182–276

Noto CR, Grossman A (2010) Broad-scale patterns of late Jurassic dinosaur paleoecology. PLoS One 5(9):e12553. https://doi.org/10.1371/journal.pone.0012553

Nussear KE et al (2012) Translocation as a conservation tool for Agassiz's desert tortoises: survivorship, reproduction, and movements. J Wildl Manag 76(7):1341–1353. https://doi.org/10.1002/jwmg.390

O'Connor MI (2009) Warming strengthens an herbivore-plant interaction. Ecology 90:388–398. https://doi.org/10.1890/08-0034.1

Olsen AM (2015) Exceptional avian herbivores: multiple transitions toward herbivory in the bird order Anseriformes and its correlation with body mass. Ecol Evol 5:5016–5032

Ősi A, Prondvai E, Mallon J et al (2017) Diversity and convergences in the evolution of feeding adaptations in ankylosaurs (Dinosauria: Ornithischia). Hist Biol 29:539–570

Owen-Smith RN (1988) Megaherbivores. Cambridge University Press, Cambridge

Pavlů V, Hejcman M, Pavlů L et al (2006) Effect of continuous grazing on forage quality, quantity and animal performance. Agric Ecosyst Environ 113:349–355. https://doi.org/10.1016/j.agee.2005.10.010

Payne AL, Jarman PJ (1999) Macropod studies at Wallaby Creek. X. Responses of eastern grey kangaroos to cattle. Wildl Res 26:215–225

Pease JL, Vowles RH, Keith LB (1979) Interaction of snowshoe hares and Woody vegetation. J Wildl Manag 43:43–60

Pekelharing CJ, Frampton CM, Suisted PA (1998) Seasonal variation in the impacts of brushtailed possums (Trichosurus vulpecula) on five palatable plant species in New Zealand beech (Nothofagus) forest. N Z J Ecol 22:141–148
Peters RH (1983) The ecological implications of body size. Cambridge University Press, Cambridge
Peterson SL, Rockwell RF, Witte CR, Koons DN (2013) The legacy of destructive snow goose foraging on supratidal marsh habitat in the Hudson Bay lowlands. Arct Antarct Alp Res 45:575–583
Petrovan SO, Ward AI, Wheeler PM (2013) Habitat selection guiding Agri-environment schemes for a farmland specialist, the brown hare. Anim Conserv 16:344–352. https://doi.org/10.1111/acv.12002
Petrozzi F et al (2018) Exploring the main threats to the threatened African spurred tortoise Centrochelys sulcata in the west African Sahel. Oryx 52(3):544–551
Pickett KN, Hik DS, Newsome AE, Pech RP (2005) The influence of predation risk on foraging behaviour of brushtail possums in Australian woodlands. Wildl Res 32:121–130
Polunin NV, Klumpp D (1992) Algal food supply and grazer demand in a very productive coral-reef zone. J Exp Mar Bio Ecol 164:1–15. https://doi.org/10.1016/0022-0981(92)90132-T
Potts R, Behrensmeyer AK (1992) Late Cenozoic terrestrial ecosystems. In: Behrensmeyer AK, Damuth JD, DiMichele WA et al (eds) Terrestrial ecosystems through time. Chicago University Press, Chicago, pp 419–541
Pough FH (1973) Lizard energetics and diet. Ecology 54(4):837–844
Power ME (1983) Grazing responses of tropical freshwater fishes to different scales of variation in their food. Environ Biol Fish 9:103–115. https://doi.org/10.1007/bf00690856
Prasad V, Strömberg CA, Alimohammadian H et al (2005) Dinosaur coprolites and the early evolution of grasses and grazers. Science 310:1177–1180
Prideaux GJ, Ayliffe LK, DeSantis LRG, Schubert BW, Murray PF, Gagan MK, Cerling TE (2009) Extinction implications of a chenopod browse diet for a giant Pleistocene kangaroo. Proc Natl Acad Sci U S A 106:11646–11650
Prins HHT (2016) Interspecific resource competition in antelopes: search for evidence. In: Bro-Jorgensen J, Mallon DP (eds) Antelope conservation: from diagnosis to action. Wiley, Hoboken, NJ, pp 51–77
Prop J, Vulink T (1992) Digestion by barnacle geese in the annual cycle: the interplay between retention time and food quality. Funct Ecol 6:180–189
Ramsey DSL, Tolsma AD, Brown GW (2016) Towards a habitat condition assessment method for guiding the management of overabundant koala populations. Victoria Arthur Rylah Institute for Environmental Research, Department of Environment, Land, Water and Planning, Heidelberg
Randall JE (1967) Food habits of reef fishes of the West Indies. Stud Trop Oceanogr 5:665–847
Rao SJ, Iason GR, Hulbert IAR et al (2003) The effect of sapling density, heather height and season on browsing by mountain hares on birch. J Appl Ecol 40:626–638. https://doi.org/10.1046/j.1365-2664.2003.00838.x
Ravolainen VT, Brathen KA, Yoccoz NG, Nguyen JK, Ims RA (2014) Complementary impacts of small rodents and semi-domesticated ungulates limit tall shrub expansion in the tundra. J Appl Ecol 51:234–241
Rees JD, Kingsford RT, Letnic M (2017) In the absence of an apex predator, irruptive herbivores suppress grass seed production: implications for small granivores. Biol Conserv 213:13–18
Rinderknecht A, Blanco RE (2008) The largest fossil rodent. Proc R Soc B 275:1637
Roberts C, Kirkpatrick JB, McQuillan PB (2011) Tasmanian lentic wetland lawns are maintained by grazing rather than inundation. Austral Ecol 36:303–309
Rodda GH (1992) The mating behavior of Iguana iguana. Smithson Contrib Zool 534:1–38
Rolls EC (1981) A million wild acres: 200 years of man and an Australian forest. GHR Press, Sydney

Rose AB, Platt KH (1992) Snow tussock (Chionochloa) population responses to removal of sheep and European hares, Canterbury, New Zealand. N Z J Bot 30:373–382. https://doi.org/10.1080/0028825X.1992.10412917
Ross J, Sanders MF (1984) The development of genetic resistance to myxomatosis in wild rabbits in Britain. J Hyg (Lond) 92:255–261. https://doi.org/10.1017/S0022172400064494
Rothman JM, Chapman CA, Struhsaker TT, Raubenheimer D, Twinomugisha D, Waterman PG (2015) Long-term declines in nutritional quality of tropical leaves. Ecology 96:873–878
Rowe KC, Achmadi AS, Esselstyn JA (2016) Repeated evolution of carnivory among indo-Australian rodents. Evolution 70:653–665
Rybczynski N, Tirabasso A, Bloskie P et al (2008) A three-dimensional animation model of *Edmontosaurus* (Hadrosauridae) for testing chewing hypotheses. Palaeontol Electr 11(2):9A. http://palaeo-electronica.org/2008_2/132/index.html
Samuels JX (2009) Cranial morphology and dietary habits of rodents. Zool J Linnean Soc 156:864–888
Sanson GD (1989) Morphological adaptations of teeth to diets and feeding in the Macropodoidea. In: Grigg G, Jarman PJ, Hume ID (eds) Kangaroos, Wallabies and Rat-Kangaroos. Surrey Beatty and Sons, Sydney, pp 151–168
Santamaria L, Rodríguez-Gironés MA (2008) Hiding from swans: optimal burial depth of sago pondweed tubers foraged by Bewick's swans. J Ecol 90:303–315
Sayers K, Norconk MA (2008) Himalayan Semnopithecus entellus at Langtang National Park, Nepal: diet, activity patterns, and resources. Int J Primatol 29:509–530
Schai-Braun SC, Reichlin TS, Ruf T et al (2015) The European hare (Lepus europaeus): a picky herbivore searching for plant parts rich in fat. PLoS One 10:1–16. https://doi.org/10.1371/journal.pone.0134278
Schwarm A, Ortmann S, Wolf C, Streich WJ, Clauss M (2009) Passage marker excretion in red kangaroo (Macropus rufus), collared peccary (Pecari tajacu) and colobine monkeys (Colobus angolensis, C. polykomos, Trachypithecus johnii). J Exp Zool A Ecol Genet Physiol 311:647–661
Schweizer D, Jones HP, Holmes ND (2016) Literature review and meta-analysis of vegetation responses to goat and European rabbit eradications on islands. Pac Sci 70:55–71. https://doi.org/10.2984/70.1.5
Scott SL, McArthur C, Potts BM, Joyce K (2002) Possum browsing - the downside to a eucalypt hybrid developed for frost tolerance in plantation forestry. For Ecol Manag 157:231–245
Sedinger JS (1997) Adaptations to and consequences of an herbivorous diet in grouse and waterfowl. Condor 99:314–326
Sereno PC, Wilson JA, Witmer LM et al (2007) Structural extremes in a Cretaceous dinosaur. PLoS One 2(11):e1230. https://doi.org/10.1371/journal.pone.0001230
Seymour RS (2009) Raising the sauropod neck: it costs more to get less. Biol Lett 5:317–319
Shariatinajafabadi M, Wang T, Skidmore AK, Toxopeus AG, Kölzsch A, Nolet BA, Exo K-M, Griffin L, Stahl J, Cabot D (2014) Migratory herbivorous waterfowl track satellite-derived green wave index. PLoSOne 9:e108331
Shepherd SA, Hawkes MW (2005) Algal food preferences and seasonal foraging strategy of the marine iguana, Amblyrhynchus cristatus, on Santa Cruz, Galapagos. Bull Mar Sci 77(1):51–72
Short J (1985) The functional-response of kangaroos, sheep and rabbits in an arid grazing system. J Appl Ecol 22:435–447
Si Y, Xin Q, Prins HHT, De Boer WF, Gong P (2015) Improving the quantification of waterfowl migration with remote sensing and bird tracking. Sci Bull 60:1984–1993
Simpson GG (1945) The principles of classification and a classification of mammals. Bull Am Mus Nat Hist 85:1–350
Sinclair ARE, Krebs CJ, Fryxell JM et al (2000) Testing hypotheses of trophic level interactions: a boreal forest ecosystem. Oikos 89:313–328. https://doi.org/10.1034/j.1600-0706.2000.890213.x
Sjögersten S, Van der Wal R, Woodin SJ (2008) Habitat sensitivity determines herbivory controls over CO2 fluxes in a warmer arctic. Ecology 89:2103–2116

Sjögersten S, Van der Wal R, Loonen MJJE, Woodin SJ (2011) Recovery of ecosystem carbon fluxes and storage from herbivory. Biogeochemistry 106:357–370

Slavenko A et al (2016) Late quaternary reptile extinctions: size matters, insularity dominates. Glob Ecol Biogeogr 25(11):1308–1320. https://doi.org/10.1111/geb.12491

Sluiter IRK, Allen GG, Morgan DG, Walker IS (1997) Vegetation responses to stratified kangaroo grazing pressure at Hattah-Kulkyne National Park, 1992–96. Flora and Fauna technical report no. 149. Department of Natural Resources and Environment, East Melbourne

Smith TB (2008) Temperature effects on herbivory for an indo-Pacific parrotfish in Panamá: implications for coral–algal competition. Coral Reefs 27:397–405. https://doi.org/10.1007/s00338-007-0343-6

Smith GC, Prickett AJ, Cowan DP (2007) Costs and benefits of rabbit control options at the local level. Int J Pest Manag 53:317–321

Snaith TV, Chapman CA (2005) Towards an ecological solution to the folivore paradox: patch depletion as an indicator of within-group scramble competition in red colobus monkeys (Piliocolobus tephrosceles). Behav Ecol Sociobiol 59:185–190

Snaith TV, Chapman CA (2007) Primate group size and interpreting socioecological models: do folivores really play by different rules? Evol Anthropol: Issues News Rev 16:94–106

Spear D, Chown SL (2009) Non-indigenous ungulates as a threat to biodiversity. J Zool 279:1–17

Speed JDM, Cooper EJ, Jónsdóttir IS, Van der Wal R, Woodin SJ (2010) Plant community properties predict vegetation resilience to herbivore disturbance in the Arctic. J Ecol 98:1002–1013

Srivastava DS, Jefferies RL (1996) A positive feedback: herbivory, plant growth, salinity, and the desertification of an Arctic salt-marsh. J Ecol 84:31–42

Stahl J, Tolsma PH, Loonen MJJE, Drent RH (2001) Subordinates explore but dominants profit: resource competition in high Arctic barnacle goose flocks. Anim Behav 61:257–264

Stahl J, Van Der Graaf AJ, Drent RH, Bakker JP (2006) Subtle interplay of competition and facilitation among small herbivores in coastal grasslands. Funct Ecol 20:908–915. https://doi.org/10.1111/j.1365-2435.2006.01169.x

Stapleton JP, Ikin K, Freudenberger D (2017) Coarse woody debris can reduce mammalian browsing damage of woody plant saplings in box-gum grassy woodlands. Ecol Manage Restor 18:223–230

Statham M, Rayner RJ (1995) Loss of pasture and crop to native animals in Tasmania. In: Proceedings of 10th annual vertebrate Pest conference. Hobart, Department of Primary Industries and Fisheries, pp 171–176

Steenbeek R, van Schaik CP (2001) Competition and group size in Thomas's langurs (Presbytis thomasi): the folivore paradox revisited. Behav Ecol Sociobiol 49:100–110

Sterck EHM, Watts DP, van Schaik CP (1997) The evolution of female social relationships in nonhuman primates. Behav Ecol Sociobiol 41:291–309

Steuer P, Hummel J, Grosse-Brinkhaus C, Südekum K-H (2015) Food intake rates of herbivorous mammals and birds and the influence of body mass. Eur J Wildl Res 61:91–102

Stevens KA, Parrish JM (2005a) Neck posture, dentition, and feeding strategies in Jurassic sauropod dinosaurs. In: Tidwell V, Carpenter K (eds) Thunder-lizards: the sauropodomorph dinosaurs. Indiana University Press, Bloomington, pp 212–232

Stevens KA, Parrish JM (2005b) Digital reconstructions of sauropod dinosaurs and implications for feeding. In: Curry Rogers KA, Wilson JA (eds) The sauropods: evolution and paleobiology. University of California Press, Berkeley, pp 178–200

Strömberg CA (2004) Using phytolith assemblages to reconstruct the origin and spread of grass-dominated habitats in the great plains of North America during the late Eocene to early Miocene. Palaeogeogr Palaeoclimatol Palaeoecol 207:239–275

Strömberg CA, Di Stilio VS, Song Z (2016) Functions of phytoliths in vascular plants: an evolutionary perspective. Funct Ecol 30:1286–1297

Stutz RS, Banks PB, Proschogo N, McArthur C (2016) Follow your nose: leaf odour as an important foraging cue for mammalian herbivores. Oecologia 182:643–651

Sumption KJ, Flowerdew JR (1985) The ecological effects of the decline in rabbits Oryctolagus cuniculus L. due to myxomatosis. Mamm Rev 15:151–186. https://doi.org/10.1111/j.1365-2907.1985.tb00396.x
Sun IF, Chen Y-Y, Hubbell SP, Wright SJ, Noor NSM (2007) Seed predation during general flowering events of varying magnitude in a Malaysian rain forest. J Ecol 95:818–827
Swart D, Mackie RI, Hayes JP (1993) Fermentative digestion in the ostrich (Struthio camelus var. domesticus), a large avian species that utilizes cellulose 1. S Afr J Anim Sci 23:127–135
Sweetapple PJ (2008) Spatial variation in impacts of brushtail possums on two Loranthaceous mistletoe species. N Z J Ecol 32:177–185
Sweetapple PJ, Nugent G, Whitford J, Latham MC, Pekelharing K (2016) Long-term response of temperate canopy trees to removal of browsing from an invasive arboreal herbivore in New Zealand. Austral Ecol 41:538–548
Tape KD, Lord R, Marshall H-P, Ruess RW (2010) Snow-mediated ptarmigan browsing and shrub expansion in arctic Alaska. Ecoscience 17:186–193
Tapper SC, Barnes RFW (1986) Influence of farming practice on the ecology of the brown hare (Lepus europaeus). J Appl Ecol 23:39–52
Taylor MP, Wedel MJ, Naish D (2009) Head and neck posture in sauropod dinosaurs inferred from extant animals. Acta Palaeontol Pol 54:213–220
Temby ID (1998) The law and wombats in Australia. In: Wells RT, Pridmore PA (eds) Wombats. Surrey Beatty and Sons, Sydney, pp 305–311
Terborgh J, Peres CA (2017) Do Community-managed forests work? A biodiversity perspective. Land 6(22):1–7
Tershy B et al (2016) The biogeography of threatened insular iguanas and opportunities for invasive vertebrate management. Herpetol Conserv Biol 11:222–236
Thiry V, Bhasin O, Stark DJ, Beudels-Jamar RC, Drubbel RV, Nathan SKSS, Goossens B, Vercauteren M (2019) Seed dispersal by proboscis monkeys: the case of Nauclea spp. Primates. https://doi.org/10.1007/s10329-019-00736-x
Tiffney BH (2012) Land plants as a source of food and environment in the age of dinosaurs. In: Brett-Surman MK, Holts TR Jr, Farlow JO (eds) The complete dinosaur, 2nd edn. Indiana University Press, Bloomington, pp 568–587
Tiver F, Andrew MH (1997) Relative effects of herbivory by sheep, rabbits, goats and kangaroos on recruitment and regeneration of shrubs and trees in eastern South Australia. J Appl Ecol 34:903–914
Todd CR, Forsyth DM, Choquenot D (2008) Modelling the effects of fertility control on koala-forest dynamics. J Appl Ecol 45:568–578
Townsend KEB, Croft DA (2008) Diets of notoungulates from the Santa Cruz formation, Argentina: new evidence from enamel microwear. J Vertebr Paleontol 28:217–230
Travers SK, Eldridge DJ, Dorrough J, Val J, Oliver I (2018) Introduced and native herbivores have different effects on plant composition in low productivity ecosystems. Appl Veg Sci 21:45–54
Traveset A (1990) Ctenosaura similis gray (Iguanidae) as a seed disperser in a central American deciduous forest. Am Midl Nat 123(2):402–404
Traveset A et al (2016) Galápagos land iguana (Conolophus subcristatus) as a seed dispersers. Integr Zool 11(3):207–213
Triggs B (2009) Wombats, 2nd edn. CSIRO Publishing, Collingwood
Trout RC (2003) Rabbits in the farmland ecosystem. In: Tattersall F, Manly WJ (eds) Conservation and conflict. Westbury Publishing, Otley, West Yorkshire, pp 198–210
Tsuji Y, Ningsih JIDP, Kitamura S, Widayati KA, Suryobroto B (2017) Neglected seed dispersers: endozoochory by Javan lutungs (Trachypithecus auratus) in Indonesia. Biotropica 49:539–545
Turner BW, Alcock D (2000) The dry sheep equivalent - redefining a 'standard'. Asian Australas J Anim Sci 13:215–215
Turtle Extinctions Working Group et al (2015) Turtles and tortoises of the world during the rise and global spread of humanity: first checklist and review of extinct Pleistocene and Holocene chelonians. 69(5). https://doi.org/10.3854/crm.5.000e.fossil.checklist.v1.2015

Tweet JS, Chin K, Braman DR et al (2008) Probable gut contents within a specimen of *Brachylophosaurus canadensis* (Dinosauria: Hadrosauridae) from the Upper Cretaceous Judith River Formation of Montana. PALAIOS 23:624–635

Tyndale-Biscoe H (2005) Life of marsupials. CSIRO Publishing, Collingwood

Uetz P, Freed P Hošek J (2016) The reptile database, Species Number. http://www.reptile-database.org/db-info/SpeciesStat.html. Accessed 19 Oct 2017

Upchurch P, Barrett PM (2000) The evolution of sauropod feeding mechanisms. In: Sues H-D (ed) Evolution of herbivorous in terrestrial vertebrates. Cambridge University Press, Cambridge, pp 79–122

Ureña-Aranda CA et al (2015) Using range-wide abundance modeling to identify key conservation areas for the micro-endemic bolson tortoise (Gopherus flavomarginatus). PLoS One 10(6): e0131452

Valido A, Olesen JM (2007) The importance of lizards as frugivores and seed dispersers. In: Dennis AJ et al (eds) Seed dispersal: theory and its application in a changing world. CAB International, Wallingford, pp 124–147

Van der Graaf AJ, Stahl J, Klimkowska A, Bakker JP, Drent RH (2006) Surfing on a green wave – how plant growth drives spring migration in the Barnacle Goose Branta leucopsis. Ardea 94:567–577

Van der Wal R (2006) Do herbivores cause habitat degradation or vegetation state transition? Evidence from the tundra. Oikos 114:177–186

Van der Wal R, Egas M, Van Der Veen A, Bakker J (2000a) Effects of resource competition and herbivory on plant performance along a natural productivity gradient. J Ecol 88:317–330

Van der Wal R, Van Wijnen H, Van Wieren S et al (2000b) On facilitation between herbivores: how Brent geese profit from brown hares. Ecology 81:969–980

Van Donk E, Otte A (1996) Effects of grazing by fish and waterfowl on the biomass and species composition of submerged macrophytes. Hydrobiologia 340:285–290. https://doi.org/10.1007/BF00012769

Van Eerden MR, Drent RH, Stahl J, Bakker JP (2005) Connecting seas: western Palaearctic continental flyway for water birds in the perspective of changing land use and climate. Glob Chang Biol 11:894–908

Van Gils JA, Gymesi A, Van Lith B (2007) Avian herbivory: an experiment, a field test, and an allometric comparison with mammals. Ecology 88:2926–2935

Van Leeuwen CHA, Van der Velde G, Van Groenendael JM, Klaassen M (2012) Gut travellers: internal dispersal of aquatic organisms by waterfowl. J Biogeogr 39:2031–2040

Van Onsem S, Triest L (2018) Turbidity, waterfowl herbivory, and propagule banks shape submerged aquatic vegetation in ponds. Front Plant Sci 9:1514

Vandandorj S, Eldridge DJ, Travers SK, Delgado-Baquerizo M (2017) Contrasting effects of aridity and grazing intensity on multiple ecosystem functions and services in Australian woodlands. Land Degrad Dev 28:2098–2108

Vaughan TA, Ryan JN, Czaplewski NJ (2015) Mammalogy, 6th edn. Jones & Bartlett, Burlington

Vergés A, Steinberg PD, Hay ME et al (2014) The tropicalization of temperate marine ecosystems: climate-mediated changes in herbivory and community phase shifts. Proc Biol Sci 281:1–10. https://doi.org/10.1098/rspb.2014.0846

Virgós E, Cabezas-díaz S, Malo A et al (2003) Factors shaping European rabbit abundance in continuous and fragmented populations of Central Spain. Acta Theriol (Warsz) 48:113–122

Vitt LJ, Caldwell JP (2013) Herpetology: an introductory biology of amphibians and reptiles. Academic, Cambridge

Von Humboldt A (1877) Personal narrative of travels to the equinoctial regions of the new continent during the years 1799–1804. AMS Press, New York

Webster FJ, Babcock RC, Van Keulen M, Loneragan NR (2015) Macroalgae inhibits larval settlement and increases recruit mortality at Ningaloo reef, Western Australia. PLoS One. https://doi.org/10.1371/journal.pone.0124162

Weishampel DB (1984) Interactions between Mesozoic plants and vertebrates: fructifications and seed predation. N Jb Geol Paläont Abh 167:224–250
Welsh JQ, Bellwood DR (2014) Herbivorous fishes, ecosystem function and mobile links on coral reefs. Coral Reefs 33:303–311. https://doi.org/10.1007/s00338-014-1124-7
Wesche K, Nadrowski K, Retzer V (2007) Habitat engineering under dry conditions: the impact of pikas (Ochotona pallasi) on vegetation and site conditions in southern Mongolian steppes. J Veg Sci 18:665–674. https://doi.org/10.1111/j.1654-1103.2007.tb02580.x
Whelan CJ, Şekercioğ ÇH, Wenny DG (2015) Why birds matter: from economic ornithology to ecosystem services. J Ornithol 156:227–238
Whisson DA, Dixon V, Taylor ML, Melzer A (2016) Failure to respond to food resource decline has catastrophic consequences for koalas in a high-density population in southern Australia. PLoS One 11:12
Wich SA, Marshall AJ (2016) An introduction to Primate Conservation. Oxford University Press, New York
Wiewandt TA, García M (2000) Mona Island iguana Cyclura cornuta stejnegeri. In: Alberts A (ed) West Indian iguanas: status survey and conservation action plan. IUCN SSC West Indian Iguana Specialist Group, Gland, pp 27–31
Wikelski M, Thom C (2000) Marine iguanas shrink to survive El Niño. Nature 403(6765):37
Williams J, Tieleman B, Shobrak M (1999) Lizard burrows provide thermal refugia for larks in the Arabian Desert. Condor 101(3):714–717
Williams ID, Polunin NVC, Hendrick VJ (2001) Limits to grazing by herbivorous fishes and the impact of low coral cover on macroalgal abundance on a coral reef in Belize. Mar Ecol Prog Ser 222:187–196. https://doi.org/10.3354/meps222187
Williams VS, Barrett PM, Purnell MA (2009) Quantitative analysis of dental microwear in hadrosaurid dinosaurs, and the implications for hypotheses of jaw mechanics and feeding. Proc Natl Acad Sci 106:11194–11199
Willians SH, Kay RF (2001) A comparative test of adaptive explanations for hypsodonty in ungulates and rodents. J Mamm Evol 8:207–229
Wilms T, Wagner P, Shobrak M (2010) Aspects of the ecology of the Arabian spiny-tailed lizard (Uromastyx aegyptia microlepis Blanford, 1875) at Mahazat as-Sayd protected area, Saudi Arabia. Salamandra 46(3):131–140
Wilson DE, Reeder DM (eds) (2005) Mammal species of the world: a taxonomic and geographic reference, 3rd edn. Johns Hopkins University Press, Baltimore, MD
Windley HR, Barron MC, Holland EP, Starrs D, Ruscoe WA, Foley WJ (2016) Foliar nutritional quality explains patchy browsing damage caused by an invasive mammal. PLoS One 11: e0155216
Wing SL, Hickey LJ, Swisher CC (1993) Implications of an exceptional fossil flora for late cretaceous vegetation. Nature 363:342–344
Winnard AL, Coulson G (2008) Sixteen years of eastern barred bandicoot Perameles gunnii reintroductions in Victoria: a review. Pac Conserv Biol 14:34–53
Wong V, Hickling GJ (1999) Assessment and management of hare impact on high-altitude vegetation. Sci Conserv 116:2–40
Wood AE (1947) Rodents – a study in evolution. Evolution 1:154–162
Wood AE (1965) Grades and clades among rodents. Evolution 19:115–130
Wood KA, Stillman RA, Clarke RT, Daunt F, O'Hare MT (2012) The impact of waterfowl herbivory on plant standing crop: a meta-analysis. Hydrobiologia 686:157–167
Wood KA, O'Hare MT, McDonald C, Searle KR, Daunt F, Stillman RA (2016) Herbivore regulation of plant abundance in aquatic ecosystems. Biol Rev 92:1128–1141
Woodruff DC (2017) Nuchal ligament reconstructions in diplodocid sauropods support horizontal neck feeding postures. Hist Biol 29:308–319
Woollard P, Vestjens WJM, Maclean L (1978) The ecology of the eastern water rat Hydromys chrysogaster at Griffith, Nsw: food and feeding habits. Wildl Res 5:59–73

Wrangham RW (1980) An ecological model of female-bonded primate groups. Behaviour 75:262–300
Wu Y, You HL, Li XQ (2017) Dinosaur-associated Poaceae epidermis and phytoliths from the early cretaceous of China. Natl Sci Rev. https://doi.org/10.1093/nsr/nwx145
Yeager CP, Kirkpatrick RC (1998) Asian colobine social structure: ecological and evolutionary constraints. Primates 39:147–155
Young MT, Rayfield EJ, Holliday CM et al (2012) Cranial biomechanics of *Diplodocus* (Dinosauria, Sauropoda): testing hypotheses of feeding behaviour in an extinct megaherbivore. Naturwissenschaften 99:637–643
Yu C, Pang XP, Wang Q et al (2017) Soil nutrient changes induced by the presence and intensity of plateau pika (Ochotona curzoniae) disturbances in the Qinghai-Tibet plateau, China. Ecol Eng 106:1–9. https://doi.org/10.1016/j.ecoleng.2017.05.029
Yugovic J (2015) Do ecosystems need top predators? A review of native predator-prey imbalances in south-East Australia. Vict Nat 132:4–11
Zhiheng L, Clarke JA (2016) The craniolingual morphology of waterfowl (Aves, Anseriformes) and its relationship with feeding mode revealed through contrastenhanced x-ray computed tomography and 2D morphometrics. Evol Biol 43:12–25
Zijlstra M, Loonen MJJE, Van Eerden M, Dubbeldam W (1991) The Oostvaardersplassen as a key moulting site for Greylag Geese Anser anser in western Europe. Wild 42:45–52
Zipperle AM, Coyer JA, Reise K, Stam W, Olsena JL (2010) Waterfowl grazing in autumn enhances spring seedling recruitment of intertidal Zostera noltii. Aquat Bot 93:202–205
Zosky KL, Wayne AF, Bryant KA, Calver MC, Scarff FR (2018) Diet of the critically endangered woylie (Bettongia penicillata ogilbyi) in South-Western Australia. Aust J Zool 65:302–312

Chapter 16
Browsers and Grazers Drive the Dynamics of Ecosystems

Iain J. Gordon and Herbert H. T. Prins

16.1 Introduction

Ever since plants evolved on land, vertebrate herbivores have been exploiting them for food. Plants in turn have evolved a range of adaptations to deter herbivores, both vertebrate and invertebrate. These defences, in part, reduce the nutritional value of plant material, through, for example, structural carbohydrates such as cellulose and lignin, which vertebrate digestive enzymes are not able to digest, and toxins such as tannins and glucosinolates. Vertebrate herbivores, across a range of taxa (see also **Gordon et al.** Chap. 15) have developed symbiotic relationships with microbes (housed in their gut) that are able to utilise these structural carbohydrates and detoxify plant secondary compounds, increasing the nutritional values of plants. This arms race between plants and vertebrate herbivores continues to this day.

More than any other vertebrate taxonomic Orders, the Ungulata (Perissodactyla and Artiodactyla), have exploited the nutritional opportunities offered by plant communities, mainly on land but even in the sea. Evolving from ancestral omnivorous taxa, ungulates have been part of the ecosystems of the globe for the past 54 M years before present (BP). It is not surprising, therefore, that they show adaptation to the environment nor that they have both direct and indirect impacts on the ecosystems in which they exist that are deep and broad ranging.

Compared with *The Ecology of Browsing and Grazing* (Gordon and Prins 2008), in *The Ecology of Browsing and Grazing II* we have slightly changed the emphasis to focus more heavily on the impact of browsing and grazing. However, our main

I. J. Gordon (✉)
James Cook University, Townsville, Australia
e-mail: Iain.gordon@jcu.edu.au

H. H. T. Prins
Animal Sciences Group, Wageningen University, Wageningen, The Netherlands
e-mail: Herbert.prins@wur.nl

I. J. Gordon, H. H. T. Prins (eds.), *The Ecology of Browsing and Grazing II*,
Ecological Studies 239, https://doi.org/10.1007/978-3-030-25865-8_16

aim is to draw together, and update, the leading research in understanding the mechanisms and processes which underlie the behaviour, distribution, movement and direct impacts on the vegetation of domestic and wild herbivores, and the dynamics of nutrients, plant species and the vegetation composition and associated fauna in terrestrial ecosystems, including the impacts of changes in management, environment and the climate. The Chapters in *The Ecology of Browsing and Grazing II* encompass fundamental and applied research aimed at crosscutting issues in ecology. This research demonstrates how an understanding of the processes of plant-herbivore interactions allows researchers to provide practical advice on the management of large herbivores and integrate production and conservation management objectives in terrestrial systems.

Let us start this concluding Chapter by stating that, 11 years on from our book on *The Ecology of Browsing and Grazing* (Gordon and Prins 2008), the foregoing Chapters demonstrate that the dichotomy between browse and grass dominated diets in large mammalian herbivores is still an extremely important heuristic for understanding the evolution, anatomy, physiology, behaviour, life history, population and community dynamics and composition of large mammalian herbivores. However, we were not just wanting to reconfirm the value of a theoretical approach to understanding the relationship between browsing and grazing large mammalian herbivores and the ecosystems in which they exist; that is sloppy science—we are after the nuggets of research that give new insights into the relationship. We will collate some example of this below.

16.2 Theory of Large Herbivore Impacts on Ecosystems

One of the issues that confronted us when we were reviewing the Chapters in this book is the fuzziness about the definitions as to what categorises small vs. medium vs. large vs. mega-herbivores. For example **Saarinen** Chap. 2 states “large herbivorous mammals, considered here as taxa ≥44 kg” *(i.e., 100 pounds)* and in **Rowan and Faith** Chap. 3 determine that “In the literature on late Quaternary extinctions, most scholars follow Martin (1967) in describing mammalian species ≥44 kg as ‘megafauna.’” Also in **Gordon et al.** Chap. 15, diverse authors, who deal with other herbivores than ungulates (e.g., reptiles or fish), use a range of boundary classes for these different categories of animal mass. However, in the ungulate ecological literature the term ‘megaherbivore’ is currently generally used for animals ≥1000 kg (Owen-Smith 1988). For the development of coherent ideas about how the world’s vertebrate assemblages or biomes are organized, it is important that scientists from different (sub-)disciplines easily comprehend each other’s insights and interpretations, and we thus call for the development of a communally acceptable set of boundary size classes (see Box 16.1).

Box 16.1

Because the modern sciences have embraced the SI-system, these class boundaries must be expressed in the metric system, but because these boundary values are not intrinsically determined by, say, molecular or physiological thresholds, they can be arbitrarily set at rounded-off values. Lastly, because many ecological processes seem to scale on a logarithmic scale (Peters 1983), we propose to use the following class boundaries: 0.01 kg, 0.1 kg, 1 kg, 10 kg, 100 kg and a 1000 kg. In other words, we propose:

- Megaherbivore is >1000 kg (log 3) (e.g., elephant, giraffe)
- Large herbivore is >100 kg but <1000 kg (between log 2 and log 3) (e.g., zebra, buffalo)
- Medium sized herbivore is >10 kg but <100 kg (between log 1 and log 2) (e.g., gazelle, roe deer)
- Small herbivore is >1 kg but <10 kg (between log 0 and log 1) (e.g., hares, oribi)
- Very small herbivore is >100 g but <1000 g (between log −1 and log 0) (e.g., pica, mole rats)
- Micro-herbivore is >10 g but <100 g (between log −2 and log −1) (e.g., voles, lemmings)

But more fundamentally: do we need these contradistinctions and do we need these categories? An important paper in this respect is that of Caughley and Krebs (1983), with the intriguing title '*Are big mammals simply little mammals writ large?*' in which they showed that the population processes of very small or even micro-mammals are fundamentally different from those of medium-sized and large herbivores, a conclusion echoed by the recent analysis of Hopcraft et al. (2012). However, in our craving for finding general patterns ('rules'), time-and-again we seek simple contrasts like 'large' vs. 'small' (or 'browser' vs. 'grazer'; 'foregut-fermenter' vs. 'hindgut-fermenter') and, in that respect, the work of Wolff (1997) is sobering: most population regulation is extrinsic and 'traits' such as female territoriality, or infanticide appear to be more important than density or body mass. And, of course, recent work (e.g., Illius and O'Connor 1999; Sullivan and Rohde 2002; Yatat et al. 2018) shows that most systems, and populations, are governed by non-equilibrium processes making distinctions based on body mass perhaps even less important at the scale of populations. Yet, at this stage of the development of thinking on ungulate ecology, we believe that 'body mass' is important for understanding of the impact of browsing and grazing on the habitats occupied by ungulates.

One of the great things about science is the way in which individual disciplines can bring insights that inform other related and unrelated disciplines (think ecology and human health via the microbiome; Lloyd-Price et al. 2016). For example, the inclusion of two Chapters (**Saarinen** Chap. 2; **Rowan and Faith** Chap. 3) in this book, that use advanced methods to assess the diets of fossil ungulates, demonstrates the importance of bringing this new information into our assessment of the ecology of extant ungulate species and populations. From these Chapters, it emerges that much of the past reconstruction of the evolution of grazing during the Miocene (23–5 My BP) and later, may prove to be less certain than is currently justified. This highlights that much of the literature prior to 2008 (Gordon and Prins 2008), and our understanding of the ecology of browsing and grazing, must be re-evaluated to test the validity of our present knowledge. That means that more recent literature that refers to that older literature also should also be re-evaluated—a massive task—as much of our present thinking carries the weight of older, perhaps dated, thinking and writing.

Fundamental to an understanding of the plant-herbivore relationships outlined in this book is the concept of 'niche'. This idea has been around for a long time in the ecological literature (Hutchinson 1959). Stated simply, "The joint action of such stressors determines an organism's ecological niche, which should be defined as the set of environmental conditions where population growth rate is greater than zero (where population growth rate is equal to $r = \log_e(Nt + 1/Nt)$)" (from Sibly and Hone 2002). Despite this formulation, many ecologists work on the assumption that a species' ecological niche is fixed (see Sexton et al. 2017 for review). However, a number of the Chapters in this book (e.g., **Codron et al.** Chap. 4) demonstrate that this assumption is incorrect, and that large mammalian herbivores have a *continuous niche* whereby they continuously adapt to the foodscape and thermal landscape (thermal ecology; **Boone** Chap. 8) in which they as an individual, population or species find themselves. The specific anatomical and, less so, physiological and behavioural "adaptations" that define a species will allow the individual or population to best derive nutrients from plants in their "ecological niche" (Codron and Clauss 2010; Damuth and Janis 2011), however, they can also feed on other vegetation components of the ecosystem, albeit, with lower nutritional efficiency if circumstances require. Indeed, in general, plant food is so marginal as compared to the dietary requirements of ungulates that they have to assemble diets from different plant species, different plant parts and different localities (*cf.* Van Langevelde and Prins 2008). Movement is perhaps the most important 'trait' that animals have; animals are not 'cabbages on feet'. Much of the theory in ecology has been formulated by plant ecologists (e.g., Clements 1936; Grime 1977), and some is of outstanding quality, but it is questionable how much of that theory is ultimately of use for analysing and predicting the ecology of the animals that eat those plants.

There is no doubt that ungulate species do have morphological, physiological and behavioural adaptations to the diets that they eat. This is particularly evident from the fossil record where the early browsing ungulates adapted to the increasing availability of grasses and grasslands. For example, in **Rowan and Faith** Chap. 3 state that, in the fossil record, "the skulls of browsing ungulates tend to have narrow muzzles, moderate-sized attachment surfaces for the masseter muscles, comparatively large attachment surfaces for the temporalis muscles and relatively shallow jaws holding low-crowned teeth, whereas the more derived grazers usually have wider muzzles, deeper jaws facilitating more high-crowned teeth, posteriorly-located orbits and larger attachment surfaces for the masseter muscles." However, simplification is perhaps impossible when it comes to understanding grazing and browsing. There are manifold anatomical, physiological and behavioural adaptations that large mammalian herbivores have to ingest, process, digest and extract nutrients. That is that the acquisition of nutrients from plants starts with ingestion (locating and biting food), there is then extensive processing in the mouth both before swallowing and after regurgitation (in ruminants), and finally there is fermentation in various sites in the digestive tract. In **Codron et al.** Chap. 4 analysed 155 variables across this set of processes and found that the relevance of many of these characteristics "is not always clear, and in fact dubious in many instances". Furthermore, as long ago as 2008 Clauss et al. (2008), found that the concepts propounded to link these traits of species to their diets, whilst useful, was often not supported by empirical/statistical evidence. For more thoughts on this see Box 16.2.

Box 16.2

Suppose we take the trait discussion further. And suppose the traits are independent and can take on different 'values'. For instance, 'jaw length' is the trait and 'short', 'medium', 'long' and 'very long' are to be taken as the values (i.e., 'levels'). Then we can construct the following set of 'traits' (to a large extent based on **Codron et al.** Chap. 4). We put in as a trait also 'thermal' on basis of the content of **Gordon et al.** Chap. 15 (the reptiles, fish vs. mammals and birds)), just like teeth (to cover the descendants of the dinosaurs that we now call 'birds' which lost their dentition but forage (still?) on foliage and grass):

1. **Incisor arch width** [*short, medium, wide, very wide*]
2. **Dentition** [*no mastication, superficial mastication, fine mastication*]
3. **Jaw length** [*short, medium, long, very long*]
4. **Stomach type:**

   (a) *Simple*
   (b) *Ruminant*
   (c) *Different (like pigs)*
   (d) *Hindgut*

4.b. If ruminant:

(i) rumen contents stratification (say: *low, medium, high*)
(ii) rumen fluid viscosity (say: *low, medium, high*)
(iii) the ratio of particle to fluid MRT (say: *low, medium, high*)
(iv) the cranial rumen pillar thickness (say: *low, medium, high*)
(v) the caudal rumen pillar thickness (say: *low, medium, high*)
(vi) the area of the intraruminal orifice (say: *low, medium, high*)
(vii) the area of the ruminoreticular orifice (say: *low, medium, high*)
(viii) the area of the reticuloomasal orifices (say: *low, medium, high*)
(ix) the height of the Papillae unguiculiformes (say: *low, medium, high*)
(x) the larger curvature of the abomasum (say: *low, medium, high*)

(continued)

Box 16.2 (continued)

4.c. If 'different' (like, e.g., pigs):

(i) 'pig-like'
(ii) 'camel-like'
(iii) 'hippo-like'
(iv) 'hare-like'

4.d. If hindgut fermenter:

(i) Caecum small
(ii) Caecum medium
(iii) Caecum large

5. **Throughput rate** (*slow, medium, fast*)
6. **Body size** (*small, medium, large, mega*)
7. **Palate** (*small, medium, large*)
8. **Liver mass** (*small, medium, large*)
9. **Diet 'trait':**

(a) *Browser*
(b) *Mixed-feeder*
(c) *Grazer*

10. **Thermal regulation:**

(a) poikilothermic
(b) homoeothermic

The mathematical solution for the number of combinations is: incisor∗dentition∗jaw∗stomach∗throughput∗body∗palate∗liver∗diet∗thermal $= 4*3*4*(1+[3^{10}]+4+3)*3*4*3*3*3*2 = 1{,}836{,}908{,}928$ in other words, 1.8 billion solutions to this part of an animal's Bauplan.

So the sum of all extant species and all extinct species is much lower than everything that can be thought to be possible. By a long shot, not all possible herbivore forms have evolved which may have been because of the 'channel-ling' of associated forms through evolved branches linking a past set of chronospecies to the next. The message is, do not look too closely for patterns if one only would try to describe the world in simple contradistinctions like 'hind-gut fermenters' vs. 'ruminants'. This is a humbling thought. If we would have added a trait as 'locomotion' with three levels (i.e., 'swimming', 'flying' and 'walking') then the unique series of 'life forms' would become 5.5 billion.

If it were simply a series of traits, with several levels (like, small, medium, large, mega), then the correct notation is x^k is that x is the number of levels, and k the number of traits. So, six traits with four levels is $4^6 = 4096$. If

(continued)

Box 16.2 (continued)
a finer gradation is made, then, say, with ten levels for six traits yields $10^6 = 1,000,000$ combinations. In other words, the finer one looks, the more one can revel in discerning uniqueness. Perhaps it would not be unreasonable to state that an animal is described by quite a number of traits (let us say, fifty) with quite some detail (say, six), then the total sum of possible combinations is an unfathomable $6^{50} = \sim 8 * 10^{38}$. Thus, the closer one looks and the more one understands, the more futile it becomes to have a few categories (e.g., 'browser', 'grazer') only.

Body size has been seen as a major explanatory variable in studies across animal taxa (Peters 1983; Prins and Olff 1998; Olff et al. 2002). Within the large herbivores, many studies have shown significant scaling of anatomical, physiological and behavioural variables with body size, and that species with different feeding styles have characteristic adaptations to the broad dietary niche they occupy (e.g., Gordon and Illius 1988; Illius and Gordon 1991; Gordon and Illius 1994). As **Venter et al.** (Chap. 5) highlight, there are, however, always exceptions to these rules. For example, red hartebeest (*Alcelaphus buselaphus*) and topi (*Damaliscus lunatus*), are both relatively large grazers, but have narrow muzzles and incisor arcades and are highly selective relative to their body mass (Venter et al. 2014; Murray and Brown 1993). Both species are adapted to feed on tall grasslands, selecting green shoots between less nutritious older leaves and stems, as compared to a short grass specialist such as the blue wildebeest (*Connochaetes taurinus*), with its broad, flat dentition and muzzles that can efficiently crop short swards with a high leaf content (*cf.* **Codron et al.** Chap. 4).

Although the body size scaling with ungulate species' life history and population demography has received a lot of attention in the past, **Kiffner and Lee** (Chap. 6) find only a very weak relationship between female body size and demographic traits, and that the grazer/browser dichotomy does not appear to add much in the way of explaining the significant variation in population densities across ungulate populations. Globally, the highest population densities occur in relatively small-bodied browsers (range: 20–233 kg; median: 45 kg), whereas the highest densities in grazers occur across a wider body mass range, and typically in larger species (range: 17–325 kg; median: 137.5 kg). They also found significant variation in within-species population densities—in grazers, within-species variability in density was generally negatively correlated with body mass, and above a body mass of ~ 100 kg, tends to be lower than for browsers. In browsers, the body mass–coefficient of variation relationship was "hump shaped" with highest variability in densities among browsing species of 200–600 kg body mass. They also found that among grazers, there was particularly high variability in densities of fallow deer (*Dama dama*), chital (*Axis axis*), southern reedbuck (*Redunca arundinum*), and African buffalo (*Syncerus caffer*). Browsers with highly variable densities were red deer

(*Cervus elaphus*), wild boar (*Sus scrofa*), bushbuck (*Tragelaphus scriptus*), and eland (*Taurotragus oryx*).

Linking these findings to the fossil record, suggests that we would expect the populations of large bodied species, and particularly grazers, to be at a higher extinction risk (Johnson 2002; Cardillo et al. 2005). Whilst, studies show that this is the case in the Cenozoic (e.g., Alroy 1999; Smith et al. 2018), it is not the case in the extinctions that occurred in the late Quaternary, leading to an ongoing debate over their underlying driver(s) (e.g., Koch and Barnosky 2006); one being the controversial hypothesis of overhunting by humans as they migrated into regions of the globe that housed large mammalian herbivores (Martin 1967; Sandom et al. 2014). However, recent analyses suggest that it was a combination of both climate change and human hunting pressure that led to these late Quaternary extinctions, and that the relativity of these drivers varied across the continents (Barnosky et al. 2004; Koch and Barnosky 2006; Stuart 2015; Prado and Alberdi 2017). **Boone** (Chap. 8) does not support the Pleistocene overkill hypothesis, and is of the opinion that climate shifts, such as the Younger Dryas, caused extinctions in many mammals such as Columbian Mammoth (*Mammuthus columbi*), and extinctions of mammals from their continents of origin, such as Bactrian camel (*Camelus bactrianus*), lion (*Panthera leo*), cheetah (*Acinonyx jubatus*) and perhaps Grevy zebra (*Equus gevyii*) in North America (Dalquest 1978; Byers 1997; Eisenmann and Kuznetsova 2004), which continued to co-exist with people in Eurasia and Africa. Indeed, climate has influenced the distribution of vegetation complexes (plant species assemblages) and those in turn influence the distribution and population dynamics of animal species such as ungulates (see the concept of Mammoth Steppe: Prins 1998; Guthrie 2001; Zimov et al. 2012). Across a number of places (e.g., in Florida or Patagonia) there is a significant mismatch in the timing of when large mammals went extinct in relation to the appearance of humans, in other places (e.g., Eurasia) many large mammals did not go extinct at all (*cf.* **Mishra et al.** Chap. 7) and in Africa, where hominids evolved, hardly any large mammal species went extinct. The debate continues, as exemplified by Chap. 7 in which **Mishra et al.** present evidence that not only the 'very large' ungulates went extinct during the transition from the Pleistocene to the Holocene, but also the very small bodied species. They speculate that this may have been due to the extremely low CO_2 levels during the hyper-arid Last Glacial Maximum that caused the near-extinction of plants with the C_3 photosynthetic pathway.

One fact that has emerged in recent years is that these late Quaternary extinctions had significant knock–on effects on the structure and function of ecosystems, that is still rippling through terrestrial ecosystems across the globe today (Malhi et al. 2016; Smith et al. 2016; Prado and Alberdi 2017 pp. 120 ff.; **Mishra et al.** Chap. 7). For example, Schweiger and Svenning (2018) showed that over the last ~ 50,000 years, the mean body size of fossil dung beetle assemblages gradually declined before abruptly crashing after the Pleistocene-Holocene boundary, which corresponds with the prehistoric mammal extinctions.

16.3 Observations on Research Methods and Approaches

16.3.1 Research Methods

In the past 10 years our understanding of ungulate ecology has benefited from significant improvements in methodological and analytical techniques. Of particular note are the developments in analyses and interpretation that have allowed researchers to gain a deeper understanding of the fossil record. Tooth wear (Walker et al. 1978), and stable isotope (Lee-Thorp and van der Merwe 1987), analyses have been available to palaeontologists for some time, but improvements in scanning electron microscopy have allowed palaeontologists to make more precise predictions about the dietary specialisations from fossil teeth (**Saarinen** Chap. 2; **Rowan and Faith** Chap. 3) including the findings on "structural fortifications of the cusps" of teeth. Alas, the application of the stable isotope technique is at this moment still limited by the requirement for the herbivore in question to be consuming C_4 grasses (i.e., the tropical and sub-tropical parts of the globe) vs. plants with a C_3 photosynthetic pathway. C_4 grasses did not become abundant until the Late Miocene (i.e., 10My BP; Cerling et al. 1997, 2015). In other sites and other species, the researcher will have to rely on the analysis of tooth wear (but have to make distinctions between micro-wear and meso-wear, noting that it now appears as though the last few meals that an animal took before it died better explains microwear than the long-term average diet [Rivals et al. 2015]), phytolite occurrence and other proxies. The earlier 'rule' that there was a simple one-on-one relation between brachydonty vs. hypsodonty and a browsing vs. a grazing diet respectively, seems be overwhelmed by the many exceptions that are now evident from modern research, also reported in this book (e.g., **Codron et al.** Chap. 4). This may necessitate a re-interpretation of the evidence that had been collated before, underlining that our previous book on *The Ecology of Browsing and Grazing* (Gordon and Prins 2008), did not represent the 'end of research into herbivore ecology' (*cf.* Fukuyama 2006).

A number of interesting developments are that giving a rich picture of paleontological ecosystems, firstly, the use of soil cores from paleolakes that tell the vegetation communities present in the neighbourhood (e.g., pollen analysis); secondly, the abundance of coprophilous fungi, *Sporormiella*, which is increasingly used as a measure of megaherbivore biomass on Quaternary landscapes (e.g., Gill et al. 2009, 2012); and finally, dung beetles (Order Coleoptera), which have evolved to consume herbivore faeces, often specialising on particular components of the large mammalian herbivore guild (Gunter et al. 2016). For example, the mean body size of fossil dung beetle assemblages has been shown to be was correlated with declines in megafaunal diversity over the last ~ 50,000 years (Schweiger and Svenning 2018), suggesting that dung beetle assemblages provide a proxy for megafaunal biomass and diversity paleo-landscapes.

Developments of analytical techniques in genetic and genomics is revolutionising the understanding of the relationship between DNA and animal morphology and physiology, the impact of evolutionary drivers at the levels of the individual, population and species, and the species composition of the microbiomes of herbivore guts and soils. This is the new frontier of research for the ecology of large mammalian herbivores and the ecosystems they inhabit. As **Sitters and Andriuzzi** (Chap. 9) emphasise, new theoretical and mechanistic approaches are needed to understand the stoichiometric impact of ungulates on soils and plants, and how this feeds back on the ungulates themselves. Livestock nutritionists are using genetic/genomic techniques to developing an increased understanding of the role that the microbiome plays in the digestion and detoxification of ingested plant material (e.g., Loor et al. 2015; Thomson 2016). Researchers studying wild large mammalian herbivores need to embrace these new techniques, or at least take cognisance of the recent developments, to both understand the species they are studying, but also help develop new theoretical, evolutionary frameworks from the broad range of dietary specialisations shown by these species, that can themselves inform livestock science and game management.

16.3.2 Data

We cannot underestimate the importance of data in providing insights into the ecology and impacts of browsing and grazing large mammalian herbivores. Of particular note is the value of the increasing number of long-term datasets on the dynamics of particular ungulate species, from across a range of populations, and also the increasing availability of data (**Kiffner and Lee** Chap. 6; **Mishra et al.** Chap. 7). The public availability of this data, as demonstrated in the provision of data from **Kiffner and Lee** Chap. 6 and **Mishra et al.** Chap. 7, will be of value to future researchers to develop and test theoretical frameworks arising from the patterns in nature. More researchers need to make their data freely available for others to access and analyse, rather than keep it under lock and key in their virtual filing cabinet.

One of the pitfalls into which many ecologists (including authors of Chapters in this book) have fallen into is to cherry-pick from the diversity of ungulate ecologies (individual, populations, species ([current day and historic]) to support their particular hypothesis. What is called for is a systematic review approach, such as that used in the medical sciences (e.g., the Cochrane Review Procedure; Higgins and Green 2011), to reduce the risk of bias in the analysis. Systematic reviews and meta-analyses can only be conducted when data repositories are filled and maintained. In our field of science, the best example is GenBank (https://www.ncbi.nlm.nih.gov/genbank/) which is well-curated and maintained. An other initiative that may change the theory and analysis of animal movement is MoveBank (https://www.movebank.org/). Search engines are now very advanced, and the first results are becoming available in which Artificial Intelligence is used to 'read' and analyse scientific texts. This will significantly increase the data upon which systematic reviews can be based.

16.3.3 Publications

If one goes through the list of references for this Chapter (see **References**), the reader may be struck by the fact that many papers we refer to were published in 'high-ranking journals' (such as Proceedings of the National Academy of Sciences, Science, Proceedings of the Royal Society, Nature). Perhaps that can be expected because the 'impact' of these journals is measured by the number of citations: these papers were influential. However, the reader may also notice that we also refer to 'non-glamorous journals' (Buzz Holling's seminal work on the 'disc-equation', that forms the basis for the functional response curves, was published in the *Canadian Entomologist*; Dale Guthrie's important review in the *Quaternary Science Reviews*; Tucker's formative work in the journal *Remote Sensing of Environment*; and John Wiens' important insights in *Annales Zoologici Fennici*). and to books from publishers not known to the general public (Peter Van Soest's very influential book was published by *O&B Books*; Reino Hofmann's famous one by *the East African Literature Bureau*; Francis Fukuyama's very influential book, not by a scientific publisher but by *Simon & Schuster*). The take home message here is that it is quite irrelevant in which journal one publishes—the important thing is that research should be published; modern electronic searching algorithms will find the work under the condition it can be found by a 'crawler'. This message cannot be repeated enough, and the systematic reviews that modern-day science of ecology needs so badly, to evaluate whether evidence supports our knowledge (= 'justified true belief'), or not, can only proceed if the conclusions and 'theories' are supported by hard and repeatable evidence. Judith Meyers so aptly wrote on the supposed understanding of population cycles "... I review several hypotheses that, although fascinating, have been rejected numerous times and should now be laid to rest" (Myers 2018). We came to the same conclusion when we analysed tens of studies in which hypotheses from animal ecology had been tested and rejected (Prins and Gordon 2014b): too often ecology does not demonstrate the principals of good scientific practice, but our conclusions were negatively received by some leading academics, who have vested interest in retaining the old ideas. Yet, we stick to our guns and call for putting to rest hypotheses that have been soundly rejected.

16.3.4 Models

Many of the authors in this book use models, be they conceptual, physical or digital, to help in understanding the ecosystem, and the part that herbivores play in its dynamics. They also use them to try to find general rules that explain patterns across varied landscapes and ecosystems. As our understanding of the relationships between large mammalian herbivores and the biotic and abiotic environment grows, the complexity of these models increases. For example, in **Kiffner and Lee** Chap. 6, state that "Advanced models to adequately depict ungulate population dynamics, . . . , need to (1) include biotic interactions (which could be additive, reciprocal, indirect, and interaction modifying) between resource availability, competition and facilitation, diseases, and predation; (2) incorporate spatiotemporal variation in abiotic factors, which determine resource availability, the relative strength of competition and facilitation, predation, and diseases in regulating herbivores of different body sizes; and (3) explicitly address possible feedback loops between abiotic factors, biotic interactions, and ungulate populations." With this complexity comes decreased generality as these models will be become so advanced, and, therefore, prone to contingency (see Prins and Gordon 2014b), that they are no longer universal. They most likely will be good at postdicting but not at predicting ecosystem dynamics. With the increase availability of data gathered from sensors, perhaps we would encourage the development of models that utilise Artificial Intelligence, neural networks and advanced diagnostics that will do the data mining and pattern analysis that result in "models" (likely black boxes) that are better than the human mind's mechanistic approach to modelling the forms and functions of the world around us. In much of science, we may have reached a time when 'hypothesis testing' becomes an outdated method. The old paradigm, that theories are formulated (either on the basis of deduction or on induction), and the validity of the theory is put to the test through analysing new data against a hypothesis derived from the theory, is losing creditability in a scientific landscape in which we increasingly become aware that our desire to believe in 'laws of nature' is not justified at the level beyond the physiology of the animals we study (Prins and Gordon 2014a, b). Indeed, oxygen transfer across a membrane can be understood by physics and chemistry: at the molecular level, that oxygen transfer is completely understandable, reproducible and predictable. Yet, at the assemblage ('community') level where hundreds of species and thousands of individuals interact in an often boundless part of the Earth, the development of the numerical abundance of all participants (i.e., number of species, number of individuals per population) over time is not blessed with that certainty. The world of the poet John Keats (1795–1821) who in his *Ode on a Grecian Urn* wrote

> Beauty is truth, truth beauty,—that is all
> Ye know on earth, and all ye need to know

(https://www.poetryfoundation.org/poems/44477/ode-on-a-grecian-urn) may perhaps still be part of the mental landscape of mathematicians, quantum physicists and law practitioners but is a paradise lost for ecologists who must predict the impact of changing causal factors and the management implications of the outcomes of their models.

Ecologists analyse patterns through space and time. A number of the Chapters in the book demonstrate the importance of spatial variation in the interaction between herbivores and their food supply. An interesting conceptual framework is the imagining of large mammalian grazers interacting with their preferred food supply in two dimensions (2D), whereas browsers interact in three dimensions (3D), because browse plants have a vertical structure that is of a dimension relevant to the scale of the herbivore (see also Gordon and Benvenutti 2006). Linking GPS technology with new sensing methods, such as Lidar, should allow researchers to create a digital map of herbivore's food supply and the ways that the animal interacts with its foodscape. Among others, **Van Langevelde et al.** (Chap. 10) point to the requirement for models of plant/herbivore/fire interactions to be spatially explicit, and temporally dynamic, if they are to have any relevant to the real world. For example, experimental work demonstrates the importance of flows of water through the landscape for vegetation establishment and Rietkerk et al. (2002a, b). Stigter and Van Langevelde (2004) and Van der Waal et al. (2009, 2011) shows unexpected effects of nutrients on seedling establishment of woody species, their competitive ability with grasses and the spatial patterns that develop as a consequence of these interactions. Much of that work was on the experimental (and thus small) scale, or in the elegant world of the simple mathematical equations (*cf.* Sinclair et al. 2006). Indeed, mathematics has a seductive beauty of its own, but **Sabo** (Chap. 11) notes that, even after decades of laboratory and field research into the impacts of ungulates on plants, useful generalizations across taxa and ecosystems remain largely elusive. The complexity of ungulate-vegetation dynamics demands continued, site-specific attention. 'Site-specific' means, indeed, that generalities beyond the level of the morphology and physiology of the multi-species assemblages of the ungulates that we study may forever be out of our grasp. That, however, is not necessarily a problem. Just as agricultural scientists understand much of the genetics, physiology, production, ecology, etc., of their crops and livestock, and can make very advanced forecasts about yields or disease effects, so ungulate ecologists can still study and discover untapped knowledge about 'their' animals and their impact. A major contrast between the agricultural scientists and ungulate ecologists, however, is that for every crop species there may be 10,000 scientists, and hundreds of thousands of farmers that put the ideas to the test, while for each of the 254 wild ungulate species (**Mishra et al.** Chap. 7) there are perhaps 10 full-time researchers while the number of 'experimental nature reserves' worldwide can perhaps counted on one hand. Even well-conducted experimental investigations on systematic hunting, and their numerical effects, are near-absent. A recent sobering experience, for example, has been the limited scientific evidence of science from the built-up of the populations of red deer (*Cervus elaphus*), roe deer (*Capreolus capreolus*), feral cattle and feral horses in the Oostvaardersplassen (in the Netherlands), and the recent culling of these animals. Decisions were taken by politicians and civil servants, and large mammal ecologists of Dutch universities were denied research permits year after year (but for the modelling studies on these animal populations see Groot-Bruinderink et al. 1999; Kramer et al. 2003; Kramer et al. 2006; Kramer et al. 2017).

An important new insight that is becoming more mainstream, also with this book, is the new knowledge that the three 'classical' functional response curves of Holling (1959), that describe the relationship between an individual's intake and the amount of food on offer, have to be increased with a fourth. This fourth functional response curve is hump-shaped and can be explained by the fact that plant forage quality declines with forage biomass and density (see **Venter et al**. Chap. 5 and references therein). The 'discovery' of the functional response curve Type IV sets herbivores firmly outside the realm of carnivores or granivores (where food quality is independent of food density): many of the mathematical models that have been developed thus for birds in the 1950s to 1990s may simply not be applicable to grazers (and perhaps browsers too). The application of the functional response curve Type IV to mathematical models, aimed to understand numerical abundance of herbivores, or the interaction between fire and vegetation mass, or the development of vegetation patterns (see **Van Langevelde et al.** Chap. 10 and references therein) is, therefore, critical to the further development of this field of plant-herbivore interactions.

The prediction of the future numerical abundance of a species, is perhaps The Holy Grail of animal ecology. Indeed, farmers want to know how many sheep they will be able to shear the wool off, or how many cattle can be sold off their ranch; hunters want to know how many red deer the glens will hold or how many wild boar (*Sus scrofus*) will there be in a country; conservationists want to know how the population of a species, brought back from extinction, will develop if released in a safe haven. However, the simplicity of the days of the Bureau for Animal Population's focus on understanding rodent cycles (see Elton 1962) are over, and even though Elton (op. cit.) observed "we can see that the investigation of single species populations, of predators and prey, the structure and interspersing of whole communities, and of foreign invaders into them, are all connected parts of the animal ecologist's field of study", Chitty's (1996) analysis of his own life-long pursuit of the understanding of animal cycles brings home the message that even he seems to have failed. Clutton–Brock and Coulson (2002) observed that the demographic mechanisms underlying observed population dynamics of ungulates are clearly complex, and suggest that there are strong context dependencies (see **Kiffner and Lee** Chap. 6; *cf.* Box 16.3). This dovetails with the conclusions that we reached some years ago, after studying many cases from Australia (Prins and Gordon 2014b). A recent review into the population cycles of vertebrates is sobering "Population cycles are one of nature's great mysteries. For almost a hundred years, innumerable studies have probed the causes of cyclic dynamics in snowshoe hares, voles and lemmings, forest Lepidoptera and grouse.... What determines variation in amplitude and periodicity of population outbreaks remains a mystery" (Myers 2018). Any modern ecologist should, we think, read Chitty's (1996) autobiography to study the true pitfalls of the science of ecology. General statements like "larger-sized species are mainly limited by food supply whereas the effect of predation may be most influential in relatively small species and in less productive environments" (**Kiffner and Lee** Chap. 6), therefore, must be treated with extreme caution, and the use of the qualifiers 'mainly' and 'most' must be understood as alarm bells. The recent research into animal population dynamics does not yet really incorporate the issues of the spatial dimensions of a (real) landscape, even though the first steps are being taken (see the spatially explicit model 'SAVANNA' of Mike Coughenour—e.g., Hille Ris-Lambers et al. 2001; Hilbers et al. 2015; or the spatially explicit model of Kramer et al. 2006). As **Kiffner and Lee** (Chap. 6) correctly observe, "at the core, theory of ungulate population dynamics needs to explicitly incorporate temporal and spatial aspects of environmental variation (*cf.* **Van Langevelde** Chap. 10; **Venter et al**. Chap. 5; **Sitters and Andriuzzi** Chap. 9; **Sabo** Chap. 11; **Katona and Coetsee**

Chap. 12). Forage availability determines individual body condition and, therefore, survival and reproduction in ungulates (**Boone** Chap. 8; Prins 1996; Parker et al. 2009), so resource availability ("bottom up regulation") is the ultimate causal factor determining population size and trajectory (Sinclair and Krebs 2002). Indeed, the observed variation in population densities of herbivores is primarily driven by primary production, which itself is mainly governed by soil fertility and precipitation. Thus, primary productivity is the main determinant of maximum density for a population". We think this is a valid assessment, but from this quote from **Kiffner and Lee** (Chap. 6) one must not deduce this is a linear relationship. It is not—the reality is more complicated, especially because of the interaction between woody species and grasses that leads to non-linearities (**Van Langevelde et al.** Chap. 10). Also, the fact that grass occurrence (on which more herbivore mass is founded: **Mishra et al.** Chap. 7) is generally observed at lower average annual precipitation levels than at higher rainfall regimes (Sankaran et al. 2005), and the observation that grass quality is higher at lower precipitation level (Prins and Olff 1998; Olff et al. 2002), make this apparent simple statement much richer in content than may be thought at first sight.

Box 16.3

Interdisciplinarity.

Science tends to operate within its disciplinary silos, for example, ecologists study the foraging behaviour of browsing and grazing ungulates, without including nutritional scientists who study how herbivores deal with toxins and remove nutrients from the ingested forages. The link with social scientists and economists, who study the cultural relationships of people with herbivores and the landscapes in which they exist, is even less common (Specht et al. 2015). In reality large mammalian herbivores live within complex socioecological systems, which means that their relationship with abiotic, biotic and human aspects of the systems is one where feedbacks occur, and so impacts reverberate through the system and come back to influence the herbivores themselves (Gordon 2019). As a consequence, the science that is needed to understand large mammalian herbivores and their impacts upon the systems in which they live requires interdisciplinarity. Interdisciplinarity, is not just about the disciplinary scientists working together (that is *multidisciplinarity*), it means "involving two or more different subjects or areas of knowledge" (Cambridge Dictionary 2019) or more specifically, the disciplines working together to develop theory and hypotheses about how the system works and the impacts of large mammalian herbivores within that system. A good example from this Book is the work of **Rowan and Faith** (Chap. 3) where a deep insight into the dynamics of ancient herbivores, and the vegetation communities in which they exist, can be drawn from modern large herbivore communities and the dung beetles that feed on their excrement and the fungi

(continued)

Box 16.3 (continued)
that live in their guts and which are excreted and then incorporated into the soil. In future of large mammalian herbivore research has to take an interdisciplinary approach as it is the intersection of disciplines that will forge new understanding of the system and how it can be managed.

From this, two general conclusions can be drawn: firstly, as the study of plant-herbivore interactions so critically hinges on understanding food quantity and its determinants, it is of real importance that (animal) ecologists measure that and appreciate keenly what their study animals really eat as well as the plants and their abundances that the herbivores do not eat (*cf.* Gordon 1989; Drescher et al. 2006; van Langevelde et al. 2008). It has been a tendency for many an ecologist to rely on satellite products, such as those produced by LandSat, or more recently by Galileo, but if one reads much of herbivore-plant interaction literature, even of the last 10–20 years, it appears that ecologists do not really take cognisance of the ever-better products that are available (see, e.g., Lu 2006), and assume that the simple Tucker-equation (NDVI, computed using the red and near infrared bands; see Tucker 1978, 1979) is still useful even though it is well known that these indices are of limited value since they saturate in dense vegetation (Mutanga and Skidmore 2004). Even then, the different algorithms used to interpret remotely sensed data from different products can yield vastly different results (see e.g., Gross et al. 2013). Recently, researchers have adopted LIDAR as a new technique (e.g., Zolkos et al. 2013); but at the basis of all these advanced techniques stays the laborious measurement of grass and browse production in the field, and too often real data are lacking. The same applies for the measurements of grass and browse quality: NIR-measurements look wonderful if one is interested in the N-content of forage, but they are not meant for specific samples (they work reasonable well for bulk samples). Yet, to understand the food of wild or domestic ungulates, one must collect the food the herbivore would eat (and not some random bulk sample), and carefully measure the NDF, ADF, cellulose, protein, digestibility, etc., etc. (see, e.g., Demment and Van Soest 1985; Van Soest 1982, 2018). There are no short cuts here, but, again, what we (the editors) observe is an ever-decreasing sets of studies in which these parameters are measured. Partly this may be explained because maintaining well-equipped laboratories is expensive, CITES-criteria stand squarely in the way of good science, and the funding structure for MSc or PhD students means that those that have field-based projects simply do not get enough time to do fieldwork which was not the case in the 1970s and 1980s.

But even with all this knowledge, the best ecologists may still be baffled by the fact that reintroductions of animals in their former range, where everything seems to match their requirements and where one would predict a thriving population with a maximum growth rate of the population when founders are reintroduced, still fail: this was observed twice now in the case of Grevy zebra (*Equus grevyi*) (Rubenstein et al. 2006). Also the failure of the Oribi (*Ourebia ourebi*) to re-establish itself in

Kruger National Park is difficult to understand (Prins and Olff 1998). For the ever-further decreasing 'rare antelopes' in the same ecosystem even Norman Owen-Smith, who has perhaps more knowledge on African herbivores than any other living ecologist, has been baffled as evidenced by a large number of papers and conference talks in which he attempted in a true Chitty'ean fashion to give explanation after explanation.

16.4 Adaption to Diet

Initially, in the Palaeocene and the early Eocene, the ungulate fauna comprised species of browsers and frugivores, with relatively rudimentary adaptations to their diets, possibly because at the time plant species had not yet adapted the complex chemical and morphological defences of browse species today. The stomach contents of the very early ancestors of the horse (*Eurohippus*), from the Middle Eocene (47 My BP), discovered in the Messel-pit (Germany), mainly comprised of fruit and browse (Koenigswald and Schaarschmidt 1983), and the same was true for an Eocene species of horse that also consumed flowers (Hellmund and Wilde 2009; Wilde and Hellmund 2010). These discoveries throw a little light on those ancient diets. Early evolution of a foregut-digestion system (perhaps later to include rumination itself) is likely to have been an adaptation to deal with alkaloids in fruits, which then became a preadaptation to tackle cellulose for which vertebrates have not evolved enzymes to tackle the intricately branched and fused sugar molecules. However, with the later increasing dominance of grasses (low in alkaloids but high in cellulose), in the late Miocene (12 to 5 MBP ago), the artiodactyls (e.g., cervids [deer] and bovids [antelopes incl. cattle and buffalo]) flourished. With this change in potential forage came a set of adaptations to the processing of plant material reflecting the significant differences in structural and chemical characteristics between the forage types, from the relative simple brachydont teeth to the more hypsodont teeth that characterise the teeth of many grazing species today. This adaptation, be it directly due to siliceous material in the tissues of graminoid plants, or due to soil adhering to the grasses at ground level, demonstrates the importance of processing the newly available plant material prior to digestion. Also, the rapid diversification of species of artiodactyls, with their adaptations of the gut (rumen, etc.) for fermenting and reprocessing ingested material, at this time, is another sign of the importance of grasses in the evolutionary story of large mammalian herbivory.

Large mammalian herbivores exist because of their symbiotic relationship with the microbial species that inhabit their guts, breaking down the structural tissues of plants and detoxifying the secondary compounds that plants have evolved to combat herbivory. The question of whether the plant defences are against large mammalian herbivores *per se* or to deal with herbivory by other taxa (e.g., insects or even fungi) is a moot point. The defences still limit the ability of large mammalian herbivores and other vertebrate herbivores (e.g., sections in **Gordon et al.** Chap. 15) to gain access to the energy and other nutrients held in plant material. The microscopic organisms inhabiting the gut of large mammalian herbivores deal with the plant material that mammalian digestive enzymes and detoxification pathways are not able to deal with. The microbes themselves provide nourishment and the by-products of their fermentation provide energy. One of the benefits for wild herbivore researchers is that, because of the importance of nutrition for livestock production, a great deal is known about the processes of digestion, and about the microbial communities (bacterial, archaeal, fungal and protozoan microbiome) with advances in molecular and genomics techniques. However, the diversity that occurs across the array of wild large herbivore species, and the fact that the major livestock species are grazers or are fed on grass rather than browse, we still have a lot to learn about how the microbiome differs between browsers and grazers and what the adaptive advantages are of these differences.

Across the adaptations that appear to be related to improvements in the ability to harvest and process different diets one of the key attributes of large mammalian herbivores appears to be the maintenance of dietary flexibility depending upon the local (spatial and temporal variable) circumstances (Rivals et al. 2015; Rivals and Lister 2016; Saarinen et al. 2016; Saarinen and Lister 2016; see **Saarinen** Chap. 2). This means that we should not look at the morphological and physiological adaptations as "straightjacketing" species into particular dietary niches but potentially allowing large mammalian herbivores advantages in diets dominated by plant material in general, be it grass or browse. Large mammalian herbivores are, therefore, forage specialists and retain the overall ability to gain nutritional value out of any forage available in the situation in which the individual, population or species finds itself. This exemplified in the quote from **Kiffner and Lee** (Chap. 6) "and even archetype browsers such as giraffes (*Giraffa camelopardalis*) may occasionally feed on grasses (Seeber et al. 2012)".

So, where does that leave us in the broad classification of ungulates into 3 main classes—i.e., 'browsers', 'grazers' and 'mixed feeders'?. The reader will have seen that many of the authors in the book use this classification in their analyses. It has been, and remains, a valuable filter through which to view the diversity of diets and adaptations of large mammalian herbivores across the globe (Gagnon and Chew 2000; Olff et al. 2002). However, a filter also obstructs what you are seeing and what is evident from the Chapters is the ability of large mammalian herbivores to adapt around their evolutionary Bauplan, demonstrating the importance of ecological adaptation as compared to evolutionary adaptation. In **Venter et al.** Chap. 5, provide a very interesting observation that grazers appear better able to handle greater variability in their diet, as compared to browsers. This may mean that grazers are more like mixed feeder (specialist) with consequent effects on their ability to adapt to contextual variation in the vegetation on offer. But perhaps the worst that can happen, is that we project the simple generalities of the trichotomy on a specific species without deeper understanding of its unique adaptations, its Bauplan, and its evolutionary history. Indeed, an African buffalo (*Syncerus caffer*) is a ruminant-grazer and so is the tiny oribi, as are a topi, a hartebeest, a wildebeest (cf. **Venter et al.** Chap. 5). But deep inside, these animals have many dissimilarities, especially in their guts (see Hofmann 1973).

In this book, we have willingly and knowingly used the simple trichotomy of 'grazer—mix feeder—browser', and we have, equally willingly and knowingly embraced the simple dichotomy 'ruminant—hind gut fermenter'. We think that while using that limited scope of trichotomy on the richness of the extant ungulates (some 240 species: **Mishra et al**. Chap. 7 plus the recently extinct 110-odd species), we may have gained insight in some broad patterning as evidenced by numerous Chapters in this book. Yet, it also became very evident that this trichotomy may hinder the further understanding of the richness of the adaptations of the ungulates (see Box 16.2). To consider the shear endless variation of plants, and their plurality of chemical compounds, as simply 'browse' or 'grass' denies that these life forms evolved in response to a myriad of past selection forces from the context in which they found themselves and from extant animal species and extinct grazing and browsing large herbivore species. As ungulate ecologists we feel trapped between a Scylla and a Charybdis—we need the simplicity to discover general patterns but we need autecology to understand the animals that we study (and love).

Much of our thinking has evolved around the startling insight that ruminants are so very different from hind-gut fermenters (Duncan et al. 1990). This dichotomy of approaches to fermenting plant material in the gut has yielded much understanding about the adaptations of ungulates. But we wish to make two important observations here—firstly, nearly all our knowledge about the physiology of ruminants is based on the excellent work of livestock scientists studying (in decreasing order of attention) cattle (*Bos taurus*), sheep (*Ovis aries*) and water buffalo (*Bubalus bubalis*). Not too much is known about goats (*Capra hircus*) (but see, for instance, the work of Mkhize et al. 2015). Of all those one hundred-odd wild ruminants, we know next-to-nothing about their physiology and proper digestion experiments have only been carried out on a handful of species by a handful of scientists (e.g., wildebeest, hartebeest and topi, Murray 1993; red deer, Arnold et al. 2015; white-tailed deer [*Odocoileus virginianus*], Bonin et al. 2016). For hind-gut fermenters, even less research has been conducted: only the horse has received much attention, but the donkey's (*Equus asinus*) physiology and digestion has been studied too (e.g., Izraely et al. 1989). Of the wild hindgut fermenters we may know a bit about the digestive efficiency of zebra (see for example Joubert 1977), but not much about that of rhinos, tapirs, elephant (**Codron et al.** Chap. 4). There is, therefore, a tremendous dearth of proper experimental digestion studies, and yet this is the cornerstone of our thinking about ungulate adaptations! Secondly, the purported dichotomy of 'ruminant' vs. 'hind-gut fermenter' ignores the fact that there are many other digestive systems outside this contrast, but, alas, we know very little about the digestive efficiency or digestion system of, say, hippos, Bactrian camels, or dugongs, let alone about it of the enormous array of extinct forms such as the giant klipdas-forms that had radiated in Africa before the advent of the modern Artiodactyla and Perissodactyla. And, yes, we can look at the simple dichotomy of hypsodont cheek teeth vs. brachydont dentition and deduce dietary adaptations from that contrast; but, again, several Chapters in this book make clear that our insights should take more recent research into consideration: palaeontologists have unearthed many other tooth forms, and any field ecologist holding the jaw of a study species in her hands may wonder at the exact classification of the molars of a sheep (grazer), roe deer(browser), goat (mixed feeder), fallow deer (*Dama dama*; grazer), red deer (mixed feeder), sambar (*Rusa unicolor*; browser), or Himalayan thar (*Hemitragus jemlahicus*; mixed feeder). And how to classify a horse (grazer) of which the winter diet proves to be mainly the roots of *Urtica* nettles in the Netherlands (Kramer et al. 2006), Burchell's zebra that feed to a large extent on browse in Welgevonden Game Reserve (South Africa) during the dry season (B. Schroder, forthcoming), cattle that take more browse than grass in long dry times of the year in Zambia (Rees 1974)?

Consequences of Environmental Variation for Browsers and Grazers
Paleontological and ecological research demonstrates that ungulate population size, species distribution and community composition are broadly determined by the resources available to them (**Rowan and Faith** Chap. 3; **Kiffner and Lee** Chap. 6; **Mishra et al.** Chap. 7). In the fossil record, with the expansion of grassland, there has been a flourishing of species of large mammalian herbivore adapted to this forage type. This has resulted in a shift in the community composition of the ungulates over time away from a dominance of frugivores and browsers from the Palaeocene and early Eocene until the Miocene and a dominance of grazing species from the Miocene until the present day. With the increase dominance of grass as food, also fire became more widespread (Bond et al. 2004). In other words, food availability and food quality started to be modulated by fire from the very first days that large mammalian grazers became important on Earth, highlighting the significance of work presented by **Van Langevelde et al.** (Chap. 10) and **Smit and Coetsee** (Chap. 13).

The dynamics of extant species of browsers and grazers show a complex of direct and indirect effects of contextual factors affecting population abundance and distribution. Bottom-up effects of vegetation resource availability (through primary productivity driven by soils and weather) are clearly evident across a range of populations, taxa and circumstances, but top down-effects of predators, disease, fire, water availability and land use change are also impacting population dynamics. This shows the clear socioecological contextualisation of large mammalian herbivores and their population dynamics and distribution should be seen through the 'systems lens' rather than in isolation. This is nowhere more evident than their interaction with humans through direct and indirect effects including climate change (Box 16.3; Gordon 2019 and this Chapter's section **Management of browsers and grazers and the ecosystems they inhabit**).

In **Boone** Chap. 8 takes us away from a focus on ungulate food to demonstrate the importance of weather and climate in defining the selection pressures that determine the taxonomic diversity, geographic ranges, physiology and behaviours of ungulates. Climatic constraints are evident in the biogeography of ungulates, such as there being more than 80 bovid species alone in Africa, but just two species of ungulates breeding in the far northern Polar Regions (i.e., muskox, *Ovibos moschatus* and caribou/reindeer, *Rangifer tarandus*). It is not surprising, therefore, that large mammalian herbivores seek out ways to remain in their thermal comfort zone through movement, or other behaviours, but, we would expect that, through their endothermic metabolism, coat and water conserving adaptations, large mammalian herbivores would have a broad tolerance of thermal condition (see for instance Shrestha et al. 2014). This we see from the human movement of species such as red deer into areas well away from their native range (e.g., into Australia: Davis et al. 2016) where the thermal environment is substantially different. Again, as with diet, we see the great ecological adaptation potential of ungulate species.

In ecological time there is spatial and temporal variation in the food available to the large mammalian herbivore. Ungulates have adopted two main strategies to cope with this spatiotemporal variability in food resources. Firstly, populations of red deer and impala (*Aepyceros melampus*), remain spatially located across the seasons adjusting their feeding strategies, with changes in forage availability (mainly feed on grasses during the growing season and increase their intake of woody vegetation during winter or dry seasons; Meissner et al. 1996; Verheyden-Tixier et al. 2008). Secondly, other populations e.g., bison (*Bison bison*) and wildebeest, retain their dietary specialisation and migrate to track areas of higher forage quantity and quality (**Venter et al.** Chap. 5; Merkle et al. 2014). The increasing evidence from ungulate population studies across the globe (**Kiffner and Lee** Chap. 6), demonstrates that no specific demographic rate is central in governing all population dynamics, and that environmental variation in resource availability and predation directly and indirectly affect vital rates. In addition, there is increasing evidence that ungulates can flexibly adjust reproductive tactics (influencing population growth) in response to environmental variation, for example in droughts or high rainfall periods (Gamelon et al. 2017). Therefore, population ecologists are less inclined to seek out single factors to explain population dynamics and are embracing more complex models, which include resource availability, predation, droughts, fire, land use change, etc., which interact in multiple complex ways (e.g., **Van Langevelde et al.** Chap. 10). Again, this brings us back to the issue of the spatial and temporal variation in the determinants of ungulate population dynamics (Boyce et al. 2006; Hempson et al. 2015), and the inclusion of space and time in concepts, experiments and models. We risk continuing to be drawn to mathematical simplicity of the models of the likes of May and Lotka-Volterra (Boyce 2000), when the world is complex (and on continents often nearly unbounded) and animals interact with it in complex ways. As we observed before, mathematics has a seductive beauty of its own, and many of us are drawn by the simplicity of the ethereal world of interacting formulas and the patterns they generate. But we may be duped into believing that when we see a pattern developing on our computer screen that, because it is an analogy with a pattern we deduced from, say, census data of a population, that the factors that generated the patterns on our screen are identical to the causal factors that determined the dynamics of our study population. Yet, reasoning through analogy is perhaps the weakest of all reasoning systems. In that respect, *The Ecology of Browsing and Grazing II* is a celebration of "humility" because we still know so little but, at the same time, this book should serve as a clarion call for the next generation of ungulate scientists. Indeed, because we know so little, after nearly 100 years of dedicated ungulate research, the world lies open for more knowledge to be unearthed. As a side remark we observe that societies spend way more money to discover whether there is life on

Mars or whether seeds may be able to germinate on the Moon, than to understand the ecology of (wild) ungulates and the life support system that they share with us on Earth.

This brings us to climate change, we hark back to **Boone** (Chap. 8) who reminds us of the important of the thermal environment for the distribution and behaviour of individuals, populations and species of ungulates. Individuals and populations can be significantly affected by single events, for example harsh weather, sudden storms, heat waves, etc., that are independent of density. Individuals that are in poor condition are generally more affected by these events (Loison et al. 1999), but on occasion, a large portion of the population can die (e.g., Mitchell et al. 2015). The ability of ungulates to cope with these events will be determined, in part, by the spatial distribution of the resources in the environment, their ability to exploit this spatial variation (see also **Management of browsers and grazers and the ecosystems they inhabit**) and the frequency of events. **Boone** (Chap. 8) concludes that populations of ungulates, in semi-arid and arid areas, are most at risk, with primary production linked to precipitation that can be highly variable. This in turn leads to populations in these areas showing non-equilibrium population dynamics because the frequency with which these events occur (mainly droughts) keeps populations below their "carrying capacity" (Bowyer et al. 2014). For more on the effects of 'climate change' see also the section on **The future for browsers and grazers and the ecosystems they inhabit** in this Chapter.

16.5 The Assemblage of Grazers and Browsers and the Habitat they Use

Neither browsers nor grazers live in isolation. Individuals live in populations, and species co-inhabit parts of the Earth in local assemblages. We refrain from using the word 'community' as much as possible for two reasons. First is the connotation with an ordered set of families living in a structured village: indeed, the word 'community' was coined at a time when ecologists shared a worldview that was about orderliness more than we believe today. At times, communities of plants and animals were even viewed as superorganisms in which individual species had their 'allotted roles'. This was taken to its most extreme in the so-called 'French-Swiss School' of vegetation science. The second reason is that we have come to appreciate the fact that the distributions of all species of ungulates, and their predators, and their parasites are not totally spatially congruent. This means that in one place, say, leopard (*Panthera pardus*), roe deer, red deer, wild boar, Usuri brown bear (*Ursus arctos lasiotis*) and Asian black bear (*Ursus thibetanus)* co-occur, but in other places tiger (*Panthera tigris*) may form a part of the set, and roe deer or black bear do not occur and in yet others both species and the Usuri brown bear are replaced by Brown bear (*U. a. ursus*). In addition, species go locally extinct: during the era of Blue Babe (36,000 YBP) (Guthrie 2013) the steppe bison (*Bison priscus*), scimitar cat

(*Homotherium latidens*), lion (*Panthera leo*), wolf (*Canis lupus*), red deer, Dall sheep (*Ovis dallii*) and reindeer formed an assemblage, but since that time the lion disappeared from Alaska and the steppe bison hybridized with the aurochs and became the European bison (*Bison bonasus*) and the scimitar cat went extinct. So, assemblages are formed and reshaped over time. Thus, **Mishra et al.** (Chap. 7) and **Kiffner and Lee** (Chap. 6) show that species do not live in isolation. Ultimately, these assemblages are limited by their regional species pools of extant species or even the continental ones (see **Mishra et al**. Chap. 7; *cf*. Prins and Olff 1998). A big issue in this respect is whether one can discover structuring forces (factors) in these assemblages similar to those inspired by Santa Rosalia (Hutchinson 1959; Wiens 1982), or whether the edifice is merely constructed like the San Marco (Gould and Lewontin 1979). The thinking of **Mishra et al**. (Chap. 7; but also Olff et al. 2002; Ahrestani et al. 2016) firmly harks back to Hutchinson (Hutchinson and MacArthur 1959), but the issue is not yet resolved. This should be high on the research agenda of plant-herbivore ecologists because certainty about this issue has important implications for restoration ecology (Prins and Olff 1998; Venter et al. 2014), and data should be collected before wild ungulates go locally extinct and we, therefore, cannot ascertain what the "intact assemblage" looked like. A random case in point: did Murchison's Falls National Park (Uganda) 'originally' contain Burchell's zebra or not? Presently, its grasslands are used by African elephant, African buffalo, hartebeest, topi, Uganda kob (*Kob kob*) and oribi—but is that the 'natural assemblage'? It looks strange without zebra. Perhaps it is not the time to put an altar to Santa Rosalia in the San Marco.

16.6 Impact of Browsers and Grazers on Ecosystems

In *The Ecology of Browsing and Grazing* (Gordon and Prins 2008), we emphasised the role that large mammalian herbivores play in shaping assemblages of plants, and their associated fauna, through both the direct effects of browsing and grazing, and indirect effects of e.g., trampling, nutrient redistribution. Significant evidence continues to accumulate of the keystone role that species of large herbivores play in ecosystem dynamics (**Sitters and Andriuzzi** Chap. 9; **Sabo** Chap. 11; **Katona and Coetsee** Chap. 12; see also Prins and Van Oeveren 2014). The advent of new techniques in soil microbiology (see also **Research methods and approaches** above) has allowed for the better understanding of the community composition and dynamics of the soil microbiome and role that these organisms play in nutrient dynamics, which affect vegetation both below and above ground. However, it is clear from a number of the Chapters (e.g., **Sitters and Andriuzzi** Chap. 9) that the systems interactions between large mammalian herbivores and the ecosystems in which they exist are complex, dynamics and context specific. For example, **Sitters and Andriuzzi** (Chap. 9) highlight the two-way interactions between soils and large mammalian herbivores—it has been known for some time that soil nutrients impact herbivore population densities and community structure via their effects on vegetation community and plant productivity, nutrient composition and biomass. In turn, mainly through their faeces and urine (the composition of which are, in part, determined by the diets of the herbivores) large mammalian herbivores affect the spatial distribution and nutrients available for plant growth and also below-ground macro-invertebrates (e.g., earthworms, termites and dung beetles [see above]; Howison et al. 2017), which in turn could explain the observed hysteresis in the system (Searle et al. 2009). Whilst, it is, therefore, expected that grazers and browsers would have differing impact on below-ground processes, at this stage the evidence is weak and contradictory (**Sitters and Andriuzzi** Chap. 9), as are the general effect of large mammalian herbivores on e.g., nitrogen cycling (**Sitters and Andriuzzi** Chap. 9). This suggests that the factors influencing below-ground microbial, faunal communities and ecological processes are complex, likely to be interactive and most probably context specific. We also note the conclusion from du Toit and Olff (2014) that plant traits determine nutrient cycling rates in savannas rather than the community composition and density of large mammalian herbivores. This may be true, but **Sitters and Andriuzzi** (Chap. 9) are justifiably less certain in their conclusions and put many a question mark about general conclusions. They also observe that recent studies have highlighted more complex interactions; in African savannas, the formation of nutrient hotspots, preferred by ungulates—e.g., grazing lawns (see for references their Chapter)—may be either facilitated or hindered by soil fauna. **Sitters and Andriuzzi** (Chap. 9) do not look at plant traits (as one would expect a plant ecologist or herbivore-plant specialist to do) but state that evidence has been accruing that interactions of ungulates with other ecosystem engineers, such as earthworms, termites, ants, and dung beetles, are a major contributor to the patchiness of grazed ecosystems (*cf.* Howison et al. 2017). Finally, **Sitters and Andriuzzi** (Chap. 9) point out that several alternative mechanisms have been proposed to explain the accelerating *or* decelerating effects of mammalian ungulates on nutrient cycling (synthesized in Table 9.1). When present, these mechanisms will, together, determine the net effect ungulates have on nutrient cycling; but, as they point out, empirical studies have found positive, negative, or mixed effects (references in their Chapter). In other words, a 'unifying theory' seems to be still far over the horizon and perhaps the real conclusion must be that there is no general rule, and the context-dependency and contingency are overriding (*cf.* Prins and Gordon 2014b). In other words, it is a mess from a scientific point of view. We have no doubt that the next 10 years will produce individual studies (hopefully experimental rather than pattern analyses), reviews (ideally bases upon e.g., the Cochrane Review procedure [see above]) and dynamic models of below-ground food webs and their functional attributes. Hopefully, many more studies will be carried out on forbs too, since they form an important part of the herb layer, and they are not woody species. Indeed, many browsing species feed on forbs, and the dynamic interaction between grasses and forbs (especially over a transition from a drought to a wet period) is to a very large *terra incognita*. And, we also have to carry out future work, taking scale effects much more seriously (e.g., Van der Waal et al. 2009, 2011).

For large plant-herbivore interactions, there may be value in the browser/grazer dichotomy because the food plant categories vary in their response to damage, as a result of differences in their lifespan, size, and biomass allocation as well as physiological integration between plant parts (Haukioja and Koricheva 2000). Grasses, and grass-like plants, characteristically can regrow rapidly after grazing events because the growing point (the meristem) is at the bottom of the leaf of the plant; for browse plants (including forbs) the growing point (meristem) is nearly always at the end of the shoot and so a browsing event can be catastrophic for the plant. Again, spatially explicit and temporally dynamic models of herbivore-plant interactions may help in understanding, interpreting and predicting the how changes in e.g., herbivores' ability to freely move about the landscape, affect plant populations and vegetation communities. A word of caution though, the more deeply we research into herbivore-plant interactions, the more complex our interpretation and the generality described above has exceptions (Kirby 2001; Moser and Schütz 2006).

In *The Ecology of Browsing and Grazing II* the indirect role that large mammalian play in fire dynamics is much more evident than in *The Ecology of Browsing and Grazing* (Gordon and Prins 2008). There is a simple, systems elegance to the empirical and modelling work showing the difference that browsing and grazing have on grass and tree dynamics (**Smit and Coetsee** Chap. 13; **Van Langevelde et al.** Chap. 10). **Kiffner and Lee,** in Chap. 6, observe that the global distribution of tree cover is mainly affected by climate, but at intermediate precipitation and mild seasonality, fire is the main force differentiating savannas from forests (**Langevelde et al**. Chap. 10; **Smit and Coetsee** Chap. 13). Without fire, closed forests could double in extent (Bond et al. 2004; Bond and Scott 2010). In arid and semi-arid regions, woody cover is limited by the interaction between precipitation, fire, and herbivory. Competition with grasses also limits the recruitment of woody vegetation (Van der Waal et al. 2011). In areas with average rainfall of greater than 650 mm per annum, savannas may transform to forests (and vice versa) following perturbations (Sankaran et al. 2005; Murphy and Bowman 2012; below 650 mm of rainfall they are more 'stable'). The presence or absence of ungulates can also affect vegetation structure and quantity both directly and indirectly, but, as **Van Langevelde et al.** (Chap. 10) point out, the interactive effects within an assemblage of grazers, or between grazers and browsers, have not yet been explored in these models (*cf.* **Kiffner and Lee** Chap. 6; **Mishra et al.** Chap. 7).

Fundamental to the models of **Van Langevelde et al.** (Chap. 10) is the so-called 'two-layer hypothesis', originally formulated by Brian Walker (for references see **Van Langevelde et al.** Chap. 10). The idea behind it is elegant and easy to imagine: grasses have access to shallow ground water and trees to deep ground water. Assuming this makes it easy to comprehend how the two life-forms can co-exist, and it led to elegant modelling (references in **Van Langevelde et al.** Chap. 10). The outcome of the models make sense, and as so often, are then used to develop further ideas. Models, however, do not provide evidence, even if the outcome results in 'things' that make sense (see section in this Chapter '**On models'**) and should not be taken as proof. **Van Langevelde et al.** (Chap. 10) evaluate the validity of the 'two layer hypothesis' and conclude that there is much evidence against its generality. This proves that ideas can be very stimulating for the development of science but also that the ideas should not be mistaken for facts: ultimately, it is the evidence that must result in knowledge which is justified by facts. An intriguing outcome of **Van Langevelde et al.**'s modelling exercise is that the catastrophic behaviour of the model (the 'instability' of the transition between grassland and woodland) seems to disappear when the modellers change animal population dynamics from static to dynamic (and, therefore, more realistic). In other words, models are very good tools for investigating one's assumptions, dangerous for making predictions and often fallacious for providing a certain foundation to build the edifice of knowledge upon.

Also **Smit and Coetsee** Chap. 13 provides important pointers for how the science of herbivore-plant interactions could proceed, and the issue of non-linearity between causes and effects is also highlighted here. They observe that "historically, studies on disturbances such as fire and herbivory have primarily focused on their effects separately. Although many studies continue in this way, it is increasingly appreciated that the interactive effects of these two processes are often different to the sum of the separate main effects. These interactive effects can be so strong that it has been argued that herbivory and fire should be viewed as a single disturbance in ecosystems that evolved with both processes active. Furthermore, while the effects of these disturbances were traditionally considered to be linear and unidirectional, current understanding explicitly highlights non-linearity, time-lags, possibility of multiple stable-states and several feedbacks". With the increased knowledge of the interaction between factors, and the realisation that there are many factors to consider in the ecology of herbivore-plant interactions (*cf.* Box 16.2), the simple 'truths' of the 1980s may disappear altogether. Perhaps the once novel 'stable limit cycles' of a genius such as Bob May were merely the result of too large time steps (imposed by limitations in computing power) in the models of that era. **Smit and Coetsee** (Chap. 13) even observe that there may be "evidence that herbivory (sometimes interacting with climate and sometimes with anthropogenic management), contributed to, and interacted with, historic fire regimes, dynamically shaping vegetation patterns over time". If this is true, then predictability will fade away even though we may be increasingly able to postdict what happened in the past: contingency becomes overwhelming. But for ever-better postdicting, we may have more-and-more need of inter- and transdisciplinarity (Specht et al. 2015; cf. Boxes 16.3 and 16.4) and new Artificial Intelligence tools to analyse the very rich and multi-layered data sources. The simple classifications of 'browser' vs. 'grazer' may stand in the way of the mining of information that we will need to tackle the issues at stake, because not all herbivory effects are similar, and herbivore traits (e.g., body mass and feeding guild: **Kiffner and Lee** Chap. 6; **Mishra et al.** Chap. 7) clearly influence the resulting herbaceous dynamics. The conclusion of **Smit and Coetsee** (Chap. 13) is, like those in other Chapters, that uncertainty about the outcomes is rife "in mixed tree/grass and grazer/browser systems, increased grazing pressure can reduce fire frequency and intensity directly through fuel removal and compositional and structural changes to the herbaceous layer. Increased browsing may increase fire indirectly through reducing woody recruitment and decreasing the competitive effect of trees on grasses. However, there may also be other feedbacks working in opposing directions, and it is sometimes hard to ascertain the net effect."

Box 16.4

Transdisciplinarity.

In contrast to multidisciplinarity and interdisciplinarity, *transdisciplinarity* enables "inputs and scoping across scientific and non-scientific stakeholder communities and facilitating a systemic way of addressing a challenge. This includes initiatives that support the capacity building required for the successful transdisciplinary formulation and implementation of research actions" (Belmont Forum 2019). The evidence to date suggests that research is not being brought into policy and management (Knight et al. 2008). If the science of large herbivores and their impacts is to be translated into policy and management then the people who will be developing and implementing policy and management actions have to be involved in the research programmes from the outset. This gives them "skin in the game" and means that the issues faced by policy and managers in their professional lives are at the fore in the development of the research programmes (Gordon et al. 2014).

The link with palaeontology (**Rowan and Faith** Chap. 3), shows that human extirpation of wild ungulate species of browsers and grazers has had a profound impact on fire dynamics, and the consequent composition of trees and grass in ecosystems across the globe. The relevance of this for modern day ecosystems cannot be underestimated (see **Van Langevelde et al**. Chap. 10; **Smit and Coetsee** Chap. 13). The removal of both grazing and browsing, and its suppression of 2D and 3D vegetation biomass, have made many ecosystems more fire prone. This is particularly the case close to human conurbations, in peri-urban landscapes, in the developed world, which wild herbivores tend to avoid and where farming has been abandoned. Here removal of both domestic and wild herbivores is linked to increases in fire frequency and intensity (Barbero et al. 1990). Reinstating browsing and grazing, to provide ecosystem services, through fire suppression is seem as a possible mechanism to reduce insurance liabilities associated with these catastrophic fires (Dallimer et al. in review).

Whilst evident today, it is clear from fossil record that humans (Hominids) have had a significant impact on species of large mammalian herbivores for millennia. Given the highly skewed size selectivity of the Late Quaternary extinctions (**Mishra et al**. Chap. 7) and the timing of megafaunal disappearances across continents tracking the chronology of human dispersals out of Africa, and across the globe, during the Late Quaternary, it seems probable that humans have played a role in these extinctions (**Rowan and Faith** Chap. 3), although this is likely to have interacted with climate change (**Rowan and Faith** Chap. 3; **Boone** Chap. 8), with the effect depending upon the continental context (**Rowan and Faith** Chap. 3; **Mishra et al**. Chap. 7). What is becoming clear is that the Late Quaternary extinctions, where a significant portion of the large mammalian herbivore fauna was lost, fundamentally altered the structure and function of ecosystems and has consequences for today's vegetation and faunal communities and fire dynamics

(**Saarinen** Chap. 2; **Smit and Coetsee** Chap. 13; Schweiger and Svenning 2018). This portents the potential impact that the current mega-extinction event is likely to have on the globes ecosystems for aeons to come (see also Gordon and Prins 2008). This may become important for ecosystem restoration and rewilding (see elsewhere in this chapter) but also for conservation itself (*cf.* Rubenstein et al. 2006). The facilitation role of very large herbivores and mega-herbivores may have been lost when in the Palaearctic mammoths, mastodons, woolly rhinos and steppe bison, or when in South America gomphoteres and giant sloths went extinct (*cf.* Janzen and Martin 1982). Indeed, with the imminent extinction of the yak (*Bos grunniens*) their keystone role (Prins and Van Oeveren 2014) will be lost. Likewise, when megaherbivore keystone species arrive after an absence of some centuries and exert their original role again, conservationists who love hole-breeding birds or picturesque trees rise to arms and call for the shooting of African elephant that are elsewhere on the brink of extinction or went locally extinct already (for details, see the debates on Kruger National Park). Similarly, the interaction between fire and herbivores may change substantially if there are particular herbivores or not present in an ecosystem (*cf.* **Rowan and Faith** Chap. 3 on fire coals and archaeology; **Van Langevelde et al.** Chap. 7; **Smit and Coetsee** Chap. 13). Yet, most of understanding of the impact by different sized browsers and grazers is so rudimentary, that meaningfully modelling the impact of differential extinction in assemblages in real habitats is, at this moment, beyond the ability of herbivore-plant scientists. It is a clear challenge though for the coming 10 years. Indeed, it appears too easy to write about 'homogenization of communities' or 'cascading effects on trophic interactions' but it appears that the science of herbivore-plant interactions merely scratches the surface. This means that we, the editors, are fully confident that much can be done the coming years that is exciting and useful.

16.7 The Future for Browsers and Grazers and the Ecosystems they Inhabit

Given the historic and present day evidence for humans creating global and local (population) extinctions of large mammalian herbivores, and the accumulating evidence for the role that these keystone species play in ecosystem dynamics it is no wonder that large mammalian herbivores feature strongly in "rewilding" debates and projects (e.g., Martin 2005; Navarro and Pereira 2015; Van Maanen and Convery 2016) The research summarised in the Chapters of this book shows the depth of understanding of large mammalian herbivore impacts on ecosystems and provides the evidence base upon which rewilding management decisions can be made. The Chapters also provide information from which managers can assess the dietary characteristics that will meet their needs for broader ecosystem management. One of the keys here, is the apparent flexibility in diets of individuals, and, more broadly, those across populations within species; a similar flexibility seems to emerge from studies on 'thermal niches' of many ungulate species and of their predators. This means that the current local context need not be a barrier to the introduction or reintroduction of species of large mammalian herbivores in rewilding projects. More controversially, however, the breadth of dietary adaptations in livestock species means that they also can play a role in ecosystem engineering when the local circumstances mean that wild herbivores are not suitable (see **Management of browsers and grazers and the ecosystems they inhabit**). Before we get carried away by the tide of rewilding, however, it is clear from the Chapters in the book that context is all important in determining the dynamics of complex ecosystems. Just because something works in one place, does not mean that the same management intervention (and we should not forget that rewilding is a management intervention) will work elsewhere even though they may appear to be very similar in their biotic and abiotic characteristics.

That said, some changes are happening due to changes in e.g., global CO_2 concentrations, leading to changes in the competitive interactions between browse and grass species, and their relationships with fire—in savannas browse plants are generally more efficient users of CO_2 for growth (when they are carbon limited) than grasses, and browse plants are suppressed by frequent high intensity fires fuelled by grass biomass (**Van Langevelde et al.** Chap. 10; Brown and Carter 1998; Van Auken 2000; Groen et al. 2017). This may explain the world-wide trend to ever more woody species as compared to grasses, i.e., bush encroachment (see Gordon and Prins 2008). Whether these insights from savanna systems apply in other ecosystems across the globe, for example the taiga in the Palearctic, is yet to be determined and could have major implications for global circulatory models that are being used to make climate change predictions. However, a very important observation was made by Petty and Werner (2010), namely, "have your historical [animal census] data correct before you make sweeping statements about the importance of potential explanatory variables such as CO_2" [we paraphrase]. **Sabo** in Chap. 11 draws attention to an interesting phenomenon, namely, dense layers of understory vegetation that impede tree regeneration called recalcitrant layers. These layers appear to develop after a period of high density of ungulates occupying a particular place. She observed that these dense layers of rhizomatous ferns, ericaceous shrubs, or graminoids block regeneration, or vegetation development to shrub and forest communities. If this is a general phenomenon, then it would be a nice example of a knock-on effect of ungulates on a habitat in which ungulate density (possibly with a certain threshold) can cause habitats to develop along different trajectories and in which hysteresis may be an important consideration (*cf.* Rietkerk et al. 1996). We think that state-and-transition models deserve more attention in vegetation ecology: for example, different densities of ungulates, during particular states, due to chance effects (like disease or drought), may trigger different cascading trajectories (and thus different sets of states—such as even-aged stands of forest [cf. Prins and Van der Jeugd 1993]) of vegetation development. Indeed, this may cause, like fire (**Smit and Coetsee** Chap. 13), patchiness and heterogeneity in the landscape (**Fynn et al.** Chap. 14) which of quintessential importance for herbivores to survive in the landscape allowing them to move and collect a diversity of food. Each 'state' or 'patch' individually will not be able to meet the ungulate's requirements but collectively they can (Illius and O'Connor 1999; Prins and van Langevelde 2008). An important feature of ecosystems having alternative stable states is the prediction that they are vulnerable for abrupt discontinuous changes, so called 'catastrophic shifts' (see **Van Langevelde et al.** Chap. 10 and references therein; Rietkerk et al. 1996). If this type of catastrophe-modelling (founded by René Thom and popularised by Christopher Zeeman; see Zeeman 1976, 1977) is as useful as the recent upsurge in models suggest (see e.g., the beautiful mathematics of Egbert H. van Nes e.g., Van Nes and Scheffer 2007), then it would be good if this were to become more integrated into modelling, and hopefully understanding of herbivore-plant interactions. A word of caution here, though: catastrophe theory (*sensu* Thom and Zeeman) does not deal with 'catastrophe' in the sense of 'disaster' or 'calamity' but in the sense of 'revolutionary' or 'unforeseen' and 'sudden'. The changes of one state to the other may be sudden but that does not mean that the outcomes are disastrous: for instance, Mills (2004) did not find significant differences in bird assemblages in burnt and unburnt areas of Kruger National Park, South Africa.

However, one should be acutely aware that the savanna tree to grass ratio is determined by a range of factors, including rainfall, soil type, herbivory, fire occurrence and fire intensity (**Van Langevelde et al.** Chap. 10 and references therein). Savanna tree-grass dynamics are of particular interest in ecology, as sudden shifts can occur between a low biomass system state (i.e., tree-grass co-dominance) and a high biomass system state (i.e., tree dominance). Intriguingly, tree-savanna dynamics has had much more attention in tropical, mainly African, savannas; **Van Langevelde et al.** observe that as the two competing life forms trees and grasses co-occur under many environmental conditions in savanna ecosystems, savannas provide an ideal ecosystem to investigate several ecologically important processes such as competition and facilitation (Van Langevelde et al. 2010). So even though the outcome of the models (in for instance **Van Langevelde et al.** Chap. 10) is already quite complicated and contains surprises, which leads the science of herbivore-plant interactions away from the simplicity of the 1950s to 1990s (see section **On models** in this Chapter), one must realise that the simplicity of the dichotomy between 'trees' vs. 'grass' hides many unknown and poorly understood interactions. The work on spinesence in trees of Tomlinson (e.g., Tomlinson et al. 2013, 2016, 2018) clearly shows that plant responses differ largely between species (e.g., between evergreen and deciduous tree species), which has not been included in modelling studies so far. These different responses imply that the model results may only be valid for a select group of plant species.

We believe that much still has to be gained using modelling to explore vegetation dynamics, not only of savannas, but of many other ecosystems. Indeed, savannas are defined as stable associations of grasses and trees but much of the Palearctic is covered in taiga, birch woodlands or open oak (*Quercus*) woodlands with grass understory which basically is the same formation yet their dynamics are hardly studied in depth that savannas have been studied.

16.8 Management of Browsers and Grazers and the Ecosystems they Inhabit

The human population of the globe currently sits at around 7.5 Billion and is likely to increase to 9 to 11 Billion in the next 30 years (United Nations 2017). Much of this increase has happened in the developing world, and Asia and Africa are likely to be the major contributors to population growth by 2050 (United Nations 2017). The growing population in the developing world is also getting richer, with Gross Domestic Product increasing in lower- and middle-income countries by 180% in the past 10 years (World Bank 2019). As well as moving people out of poverty, this increase in wealth is associated with changing dietary habits, as people shift from a predominately grain-based diet to one that includes increasing amounts of meat, initially poultry but then pigs, and red meat from goats, sheep and cattle (Delgado et al. 1999; Alexandratos and Bruinsma 2012).

Although grain is increasingly used to provide the feed required for these livestock, land use change is also associated with increased livestock production. For example, it is projected that the land area under permanent pasture in lower- and middle-income countries (particularly in South America and sub-Saharan Africa) will expand by ~ 320 Mha by 2030 (from 2010; Wirsenius et al. 2010), with increases in livestock numbers of between 40% and 50% for buffalo and cattle and 30–45% for sheep and goats (Bruinsma 2003). In contrast, as compared to livestock (including pets) that make up 66% of the total weight of land mammals today, wildlife species (including elephants, whales, etc.) make up less than 3% (Bar-On et al. 2018) i.e., per unit weight of wildlife, livestock are 14 times more abundant). This has profound effects on the way in which the vegetation is shaped by the impact of large mammalian herbivores.

Livestock production now occupies over a quarter of the land surface area of the globe (Robinson et al. 2014). Obviously, conversion of land to support livestock production occurs at the expense of land that is currently occupied by forests and other native vegetation (Foley et al. 2005). This change in land use puts pressure on wildlife populations, through either direct effects (e.g., persecution) or indirect effects (e.g., land clearing, fencing). These effects (impacts) are likely to increase as the demand, for particularly, red meat increases. The extent to which they herald the decline in wildlife populations depends upon the nature of the interactions; as discussed below.

The agricultural intensification that has been prevalent since the start of the agricultural revolution in the 1800s has not passed by livestock production. The approach has been to have greater control over the nutrition of livestock by reducing the diversity of forage species that typifies extensive grazing systems to the ultimate level of single species pastures or at best inclusion of a legume into the grass swards, primarily to maintain fertility of the soil but also to add more protein to the diets. This "man knows best" approach takes away choice and in rotational grazing systems (as compared with continuous systems) livestock are also discouraged from choosing the parts of the grass plants they eat (i.e., leaf, stem or pseudostems).

Apart from buffalo, cattle and the equids (horses and donkeys), sheep, goats and the camelids (camels, llamas and alpacas) are mixed feeders, adapted to a diverse diet including grasses and browse. This removal of dietary choice is likely to reduce the resilience of grazing systems in the face of the vagaries of weather without substantial inputs of energy, water and fertiliser. As **Fynn et al**. (Chap. 14) emphasise, the human management of ungulate density (stocking rate) and the duration of herbivory, especially in livestock systems, appears to critical determinant of the heterogeneity and sustainability of forage resources. The concept of functional heterogeneity (inherent [e.g., soils] and induced [e.g., grazing/browsing, fire, rainfall]) variation in forage quantity and quality) may become a general theme guiding their management, across scales or ecosystems. Yet, there appears still to be a lot conjecture about which there has to be a note of caution; in the South African conservation context, the argument to manage for 'heterogeneity' may be less based upon a true understanding of the reasons why heterogeneity is important and more because of the mantra created by certain charismatic individuals who expound the message (see du Toit et al. 2013). We cannot help thinking that the message appears to be: "wow, there is heterogeneity *ergo* that must be important *ergo* because heterogeneity is important *ergo* we must create/maintain heterogeneity". However, **Van Langevelde et al**. Chap. 10 suggest that when pastoralists (or *pars pro toto*, any herbivore population) respond quickly to decreases in forage availability and slowly to vegetation regeneration, the stability of savannas increases, and the risk on vegetation collapse decreases. This may have bearing on this heterogeneity discussion, but we think that the fuller understanding of what 'heterogeneity' implies (see what can be learned from 'the Kruger Experience' [du Toit et al. 2013]), is a fruitful field of science in the coming years. But of course the response of herbivore populations depends upon more than just forage availability, for example the addition of water accessibility, vegetation structure and predation risk. Another issue is that at a longer time scale, the forage base of the ungulates may shift towards more browse and less grass (Gordon and Prins 2008; **Smit and Coetsee** Chap. 13). Perhaps this is due to the ever-increasing CO_2-levels or changing rainfall patterns. In itself, these changes have been around for hundreds of thousands of years and—by definition—the presently extant ungulates (**Mishra et al.** Chap. 7) have been able to cope with these changes so far. In contrast to the wild ungulates in the rich countries (Gordon and Prins 2008), those elsewhere have not found 'an answer' to the changes in land use and local eradication (Gordon 2009). Much attention is given to the climate change debate, but, as Richard Leakey once observed, "climate change is a 300-years problem but land use intensification and eradication are with us now". The grazers and browsers that are subject of this book, live in a socio-ecological system (Biggs et al. 2015; Box 16.4), meaning that humans play a major role in determining the outcome of interactions between components of the system. Thus, a focus on the interactions between livestock and wildlife can miss the perceptions and behaviours of, for example, farmers or wildlife managers, who will determine management actions that impact on those interactions. As researchers, we cannot, therefore, approach the issue of the future of wildlife in livestock production dominated landscapes through reductionism-based research alone; we have to see it within the context of the whole socio-ecological system, yet, simultaneously, we know that most progress was made in the agricultural sciences through reductionism and carefully conducted experiments. In other words, in an approach based on inter- and transdisciplinarity there must always be place for 'hard-nosed reductionism' too simply prevent the risk of losing touch with reality (see Box 16.4). In this balancing act we see increasing need for the application of Artificial Intelligence to help us maintain the overview in the transdisciplinary approach but also to discern patterns in the ever-increasing richness of our data bases: indeed, the reductionist approach may be changed in revolutionary ways in the coming decades.

The foregoing Chapters demonstrate the vibrancy of the fields of research into browsing and grazing larger mammalian herbivores, their history, their role in the world and their important for ecosystem dynamics. They also make clear that there is still a great deal to be learnt; it is upon the pillars of knowledge outlined in *The Ecology of Browsing and Grazing* (Gordon and Prins 2008) and now *The Ecology of Browsing and Grazing II* (Gordon and Prins 2019) that future generations of researchers will seek to better understand the whats, whys and the wherefores of the interactions between herbivores and the ecosystems in which they live. Given the vital importance of large mammalian herbivores to those ecosystems in, and also the role they play in providing ecosystem services to humanity (an issue that requires more research focus in the future), researchers must seek partnership with policy and management practitioners in delivering evidence-based solutions for the future management and conservation of these amazing creatures (Box 16.4), in a world that is changing before our eyes. But researchers should not forget that these ungulates are made of flesh and blood, that they graze and browse in real landscapes, and that there is a profound need for hard-core ungulate ecologists with a broad set of skills and deep understanding of 'their' animals. As a bonus, we, and all other ungulate ecologists, get to see, feel and understand some of the most beautiful creatures that share our planet.

References

Ahrestani FS, Heitkönig IMA, Matsubayashi H, Prins HHT (2016) Grazing and browsing by large herbivores in South and Southeast Asia. In: Ahrestani FS, Sankaran M (eds) The ecology of large herbivores in South and Southeast Asia. Springer, Dordrecht, pp 99–120

Alexandratos N, Bruinsma J (2012) World agriculture towards 2030/2050: the 2012 revision. ESA Working paper No. 12-03. Food and Agriculture Organization, Rome, Italy

Alroy J (1999) Putting North America's end-Pleistocene megafaunal extinction in context. In: MacPhee RDE, Sues HD (eds) Extinctions in near time. Springer, Boston, MA, pp 105–143

Arnold W, Beiglböck C, Burmester M, Guschlbauer M, Lengauer A, Schröder B, Wilkens MR, Breves G (2015) Contrary seasonal changes of rates of nutrient uptake, organ mass, and voluntary food intake in red deer (*Cervus elaphus*). Am J Phys Heart Circ Phys 309:R277–R285

Barbero M, Bonin G, Loisel R, Quézel P (1990) Changes and disturbances of forest ecosystems caused by human activities in the western part of the Mediterranean basin. Vegetatio 87:151–173

Barnosky AD, Koch PL, Feranec RS et al (2004) Assessing the causes of Late Pleistocene extinctions on the continents. Science 306:70–75
Bar-On YM, Phillips R, Milo R (2018) The biomass distribution on earth. Proc Natl Acad Sci 115:6506–6511
Belmont Forum (2019) http://www.belmontforum.org/
Biggs R, Schlüter M, Schoon ML (2015) Principles for building resilience: sustaining ecosystem services in social-ecological systems. Cambridge University Press, Cambridge
Bond WJ, Scott AC (2010) Fire and the spread of flowering plants in the Cretaceous. New Phytol 188:1137–1150
Bond WJ, Dickinson KJ, Mark AF (2004) What limits the spread of fire-dependent vegetation? Evidence from geographic variation of serotiny in a New Zealand shrub. Glob Ecol Biogeogr 13 (2):115–127
Bonin M, Tremblay JP, Côté SD (2016) Contributions of digestive plasticity to the ability of white-tailed deer to cope with a low-quality diet. J Mammal 97:1406–1413
Bowyer RT, Bleich VC, Stewart KM, Whiting JC, Monteith KL (2014) Density dependence in ungulates: a review of causes, and concepts with some clarifications. Calif Fish Game 100:550–572
Boyce MS (2000) Modeling predator-prey dynamics. In: Boitani L, Fuller TK (eds) Research techniques in animal ecology - controversies and consequences. Columbia University Press, New York, pp 253–287
Boyce MS, Haridas CV, Lee CT, the NCEAS Stochastic Demography Working Group (2006) Demography in an increasingly variable world. Trends Ecol Evol 21:141–148
Brown JR, Carter J (1998) Spatial and temporal patterns of exotic shrub invasion in an Australian tropical grassland. Landsc Ecol 13:93–102
Bruinsma J (ed) (2003) World agriculture: towards 2015/2030. An FAO Perspective. Food and Agriculture Organization/Earthscan Publications, Rome/London
Byers JA (1997) American pronghorn: social adaptations and the ghosts of predators past. University of Chicago Press, Chicago
Cambridge Dictionary (2019) https://dictionary.cambridge.org/dictionary/english/interdisciplinary
Cardillo M, Mace GM, Jones KE, Bielby J, Bininda-Emonds OR, Sechrest W, Orme CD, Purvis A (2005) Multiple causes of high extinction risk in large mammal species. Science 309:1239–1241
Caughley G, Krebs CJ (1983) Are big mammals simply little mammals writ large? Oecologia 59:7–17
Cerling TE, Harris JM, MacFadden BJ, Leakey MG, Quade J, Eisenmann V, Ehleringer JR (1997) Global vegetation change through the Miocene/Pliocene boundary. Nature 389:153–158
Cerling TE, Andanje SA, Blumenthal SA, Brown FH, Chritz KL, Harris JM, Hart JA, Kirera FM, Kaleme P, Leakey LN, Leakey MG (2015) Dietary changes of large herbivores in the Turkana Basin, Kenya from 4 to 1 Ma. Proc Natl Acad Sci 112:11467–11472
Chitty D (1996) Do lemmings commit suicide? Beautiful hypotheses and ugly facts. Oxford University Press, New York
Clauss M, Kaiser TM, Hummel J (2008) The morphophysiological adaptations of browsing and grazing mammals. In: Gordon IJ, Prins HHT (eds) The ecology of browsing and grazing. Springer, Heidelberg, pp 149–178
Clements FE (1936) Nature and structure of the climax. J Ecol 24:252–284
Clutton–Brock TH, Coulson T (2002) Comparative ungulate dynamics: the devil is in the detail. Philos Trans R Soc London B: Biol Sci 357:1285–1298
Codron D, Clauss M (2010) Rumen physiology constrains diet niche: linking digestive physiology and food selection across wild ruminant species. Can J Zool 88:1129–1138
Dallimer D, Gordon IJ, Martin-Ortega J, Paavola J, Afionis S, Rendon O, Bark R (in review) The insurance value of ecosystems: taking stock of what we know so far. Ecol Econ
Dalquest W (1978) Phylogeny of American horses of Blancan and Pleistocene age. Ann Zool Fenn 15:191–199

Damuth J, Janis CM (2011) On the relationship between hypsodonty and feeding ecology in ungulate mammals, and its utility in palaeoecology. Biol Rev 86:733–758
Davis NE, Bennett A, Forsyth DM, Bowman DM, Lefroy EC, Wood SW, Woolnough AP, West P, Hampton JO, Johnson CN (2016) A systematic review of the impacts and management of introduced deer (family Cervidae) in Australia. Wildl Res 43:515–532
Delgado C, Rosegrant M, Steinfeld H, Ehui S, Courbois C (1999) Livestock to 2020: the next food revolution. International Food Policy Research Institute, Washington, DC
Demment MW, Van Soest PJ (1985) A nutritional explanation for body-size patterns of ruminant and nonruminant herbivores. Am Nat 125:641–672
Drescher M, Heitkönig IMA, Raats JG, Prins HHT (2006) The role of grass stems as structural foraging deterrents and their effects on the foraging behaviour of cattle. Appl Anim Behav Sci 101:10–26
du Toit JT, Olff H (2014) Generalities in grazing and browsing ecology: using across-guild comparisons to control contingencies. Oecologia 174:1075–1083
du Toit JT, Rogers KH, Biggs HC (eds) (2013) The Kruger experience: ecology and management of savanna heterogeneity. Island Press, New York
Duncan P, Foose TJ, Gordon IJ, Gakahu CG, Lloyd M (1990) Comparative nutrient extraction from forages by grazing bovids and equids: a test of the nutritional model of equid/bovid competition and coexistence. Oecologia 84:411–418
Eisenmann V, Kuznetsova T (2004) Early Pleistocene equids (Mammalia, Perissodactyla) of Nalaikha, Mongolia, and the emergence of modern Equus Linnaeus, 1758. Geodiversitas 26 (3):535–561
Elton CS (1962) The first 30 years of the Bureau of Animal Population. In the Elton Archive (transcribed & edited by C.M. Pond 2014). https://ora.ox.ac.uk/objects/uuid:89c5e479-6003-45ba-bd78-8a8a12858bf1/download_file?file_format=pdf&safe_filename=Elton%2BArchive%2BElton%2Blecture%2Bon%2B30%2Byears%2Bof%2BBureau%2Bof%2BAnimal%2BPopulation.pdf&type_of_work=Dataset. Accessed 12 Feb 2019
Foley JA, DeFries R, Asner GP, Barford C, Bonan G, Carpenter SR, Chapin FS, Coe MT, Daily GC, Gibbs HK, Helkowski JH (2005) Global consequences of land use. Science 309:570–574
Fukuyama F (2006) The end of history and the last man. Simon and Schuster, New York
Gagnon M, Chew AE (2000) Dietary preferences in extant African Bovidae. J Mammal 81:490–511
Gamelon M, Foccardi S, Baubet E, Brandt S, Franzetti B, Rochni F, Venner S, Sæther B-E, Gaillard J-M (2017) Reproductive allocation in pulsed resource environments: a comparative study in two populations of wild boar. Oecologia 183:1065–1076
Gill JL, Williams JW, Jackson ST, Lininger KB, Robinson GS (2009) Pleistocene megafaunal collapse, novel plant communities, and enhanced fire regimes in North America. Science 326 (5956):1100–1103
Gill JL, Williams JW, Jackson ST, Donnelly JP, Schellinger GC (2012) Climatic and megaherbivory controls on late-glacial vegetation dynamics: a new, high-resolution, multi-proxy record from Silver Lake, Ohio. Quat Sci Rev 34:66–80
Gordon IJ (1989) Vegetation community selection by ungulates on the Isle of Rhum III. Determinants of vegetation community selection. J Appl Ecol 26:65–79
Gordon IJ (2009) What is the future for wild, large herbivores in human-modified landscapes? Wildl Biol 15:1–9
Gordon IJ (2019) Review: livestock production increasingly influences wildlife across the globe. Animal 12:s372–s382
Gordon IJ, Benvenutti M (2006) Food in 3D: how ruminant livestock interact with sown sward architecture at the bite scale. In: Bell V (ed) Feeding in domestic vertebrates: from structure to behaviour. CABI Publishing, Wallingford, pp 263–277
Gordon IJ, Illius AW (1988) Incisor arcade structure and diet selection in ruminants. Funct Ecol 2:15–22

Gordon IJ, Illius AW (1994) The functional significance of the browser-grazer dichotomy in African ruminants. Oecologia 98:167–175
Gordon IJ, Prins HHT (2008) The ecology of browsing and grazing. Springer, Berlin
Gordon IJ, Prins HHT (2019) The ecology of browsing and grazing II. Springer, New York
Gordon IJ, Evans DM, Garner TWJ, Katzner T, Gompper ME, Altwegg R, Branch TA, Johnson JA, Pettorelli N (2014) Enhancing communication between conservation biologists and conservation practitioners: letter from the conservation front line. Anim Conserv 17:1–2
Gould SJ, Lewontin RC (1979) The spandrels of San Marco and the Panglossian paradigm: a critique of the adaptationist programme. Proc R Soc Lond B 205:581–598
Grime JP (1977) Evidence for existence of three primary strategies in plants and its relevance to ecological and evolutionary theory. Am Nat 111:1169–1194
Groen TA, van de Vijver CADM, van Langevelde F (2017) Do spatially homogenising and heterogenising processes affect transitions between alternative stable states? Ecol Model 365:119–128
Groot-Bruinderink G, Lammertsma DR, Kramer K, Wijdeven S, Baveco JM, Kuiters AT, Cornelissen P, Vulink JT, Prins HHT, van Wieren SE, de Roder F (1999) Dynamische interacties tussen hoefdieren en vegetatie in de Oostvaardersplassen. IBN-DLO, Report no. 436, Wageningen (the Netherlands)
Gross D, Dubois G, Pekel JF, Mayaux P, Holmgren M, Prins HHT, Rondinini C, Boitani L (2013) Monitoring land cover changes in African protected areas in the 21st century. Eco Inform 14:31–37
Gunter NL, Weir TA, Slipinksi A, Bocak L, Cameron SL (2016) If dung beetles (Scarabaeidae: Scarabaeinae) arose in association with dinosaurs, did they also suffer a mass co-extinction at the K-Pg boundary? PLoS One 11(5):e0153570
Guthrie RD (2001) Origin and causes of the mammoth steppe: a story of cloud cover, woolly mammal tooth pits, buckles, and inside-out Beringia. Quat Sci Rev 20:549–574
Guthrie RD (2013) Frozen Fauna of the mammoth steppe: the story of blue babe. University of Chicago Press, Chicago
Haukioja E, Koricheva J (2000) Tolerance to herbivory in woody vs. herbaceous plants. Evol Ecol 14:551–562
Hellmund M, Wilde V (2009) Der "Mageninhalt" von Propalaeotherium isselanum aus dem Geiseltal (Sachsen-Anhalt, Deutschland). Hercynia-Ökologie und Umwelt in Mitteleuropa 42:167–175
Hempson GP, Illius AW, Hendricks HH, Bond WJ, Vetter S (2015) Herbivore population regulation and resource heterogeneity in a stochastic environment. Ecology 96:2170–2180
Higgins JPT, Green S (eds) (2011) Cochrane handbook for systematic reviews of interventions version 5.1.0 [updated March 2011]. The Cochrane Collaboration, 2011. Available from http://handbook.cochrane.org
Hilbers JP, van Langevelde F, Prins HHT, Grant CC, Peel MJ, Coughenour MB, de Knegt HJ, Slotow R, Smit IP, Kiker GA, Boer WF (2015) Modelling elephant-mediated cascading effects of water point closure. Ecol Appl 25:402–415
Hille Ris-Lambers R, Rietkerk M, van den Bosch F, Prins HHT, de Kroon H (2001) Vegetation pattern formation in semi-arid grazing systems. Ecology 82:50–61
Hofmann RR (1973) The ruminant stomach. East African Literature Bureau, Nairobi
Holling CS (1959) Some characteristics of simple types of predation and parasitism. Can Entomol 91:385–398
Hopcraft JG, Anderson TM, Pérez-Vila S, Mayemba E, Olff H (2012) Body size and the division of niche space: food and predation differentially shape the distribution of Serengeti grazers. J Anim Ecol 81:201–213
Howison RA, Olff H, van de Koppel J, Smit C (2017) Biotically driven vegetation mosaics in grazing ecosystems: the battle between bioturbation and biocompaction. Ecol Monogr 87:363–378
Hutchinson GE (1959) Homage to Santa Rosalia or why are there so many kinds of animals? Am Nat 93:145–159

Hutchinson GE, MacArthur RH (1959) A theoretical ecological model of size distributions among species of animals. Am Nat 93(869):117–125
Illius AW, Gordon IJ (1991) Prediction of intake and digestion in ruminants by a model of rumen kinetics integrating animal size and plant characteristics. J Agr Sci 116:145–157
Illius A, O'Connor T (1999) On the relevance of nonequilibrium concepts to arid and semiarid grazing systems. Ecol Appl 9:798–813
Izraely H, Choshniak I, Shkolnik A, Stevens CE, Demment MW (1989) Factors determining the digestive efficiency of the domesticated donkey (*Equus asinus asinus*). Q J Exp Physiol 74:1–6
Janzen DH, Martin PS (1982) Neotropical anachronisms: the fruits the gomphotheres ate. Science 215:19–27
Johnson CN (2002) Determinants of loss of mammal species during the Late Quaternary 'mega-fauna' extinctions: life history and ecology, but not body size. Proc R Soc B 269:2221–2227
Joubert E (1977) Preliminary observations on the digestive and renal efficiency of Hartmann's zebra *Equus zebra hartmannae*. Modoqua 10:119–121
Kirby KJ (2001) The impact of deer on the ground flora of British broadleaved woodland. Forestry 74:219–229
Knight AT, Cowling RM, Rouget M, Balmford A, Lombard AT, Campbell BM (2008) Knowing but not doing: selecting priority conservation areas and the research–implementation gap. Conserv Biol 22:610–617
Koch PL, Barnosky AD (2006) Late Quaternary extinctions: state of the debate. Annual Rev Ecol Evol Syst 37:215–250
Koenigswald W v, Schaarschmidt F (1983) Ein Urpferd aus Messel, das Weinbeeren fraB. Natur Mus 113:79–84
Kramer K, Groen TA, van Wieren SV (2003) The interacting effects of ungulates and fire on forest dynamics: an analysis using the model FORSPACE. For Ecol Manag 181:205–222
Kramer K, Groot-Bruinderink G, Prins HHT (2006) Spatial interactions between ungulate herbivory and forest management. For Ecol Manag 226:238–247
Kramer K, Cornelissen P, Groot-Bruinderink G, Kuiters AT, Lammertsma DR, Vulink JT, van Wieren SE, Prins HHT (2017) Effects of weather variability and geese on population dynamics of large herbivores creating opportunities for wood-pasture cycles. In: Cornelissen P (ed) Large herbivores as a driving force of woodland-grassland cycles. Wageningen University, Wageningen, pp 109–123
Lee-Thorp J, van der Merwe NJ (1987) Carbon isotope analysis of fossil bone apatite. S Afr J Sci 83:712–715
Lloyd-Price J, Abu-Ali G, Huttenhower C (2016) The healthy human microbiome. Genome Med 8:51
Loison A, Langvatn R, Solberg EJ (1999) Body mass and winter mortality in red deer calves: disentangling sex and climate effects. Ecography 22:20–30
Loor JJ, Vailati-Riboni M, McCann JC, Zhou Z, Bionaz M (2015) Triennial lactation symposium: Nutrigenomics in livestock: systems biology meets nutrition 1. J Anim Sci 93:5554–5574
Lu D (2006) The potential and challenge of remote sensing-based biomass estimation. Int J Remote Sens 27:1297–1328
Malhi Y, Doughty CE, Galetti M et al (2016) Megafauna and ecosystem function from the Pleistocene to the Anthropocene. Proc Natl Acad Sci U S A 113:838–846
Martin PS (1967) Prehistoric overkill. In: Martin PS, Wright HEJ (eds) Pleistocene extinctions: the search for a cause. Yale University Press, New Haven, pp 75–120
Martin PS (2005) Twilight of the mammoths: ice age extinctions and the rewilding of America. University of California Press, Berkeley, CA
Meissner HH, Pieterse E, Potgieter JHJ (1996) Seasonal food selection and intake by male impala *Aepyceros melampus* in two habitats. S Afr J Wildl Res 26:56–63
Merkle JA, Fortin D, Morales JM (2014) A memory-based foraging tactic reveals an adaptive mechanism for restricted space use. Ecol Lett 17:924–931

Mills MSL (2004) Bird community responses to savanna fires: should managers be concerned? S Afr J Wildl Res 34:1–11
Mitchell CD, Channey R, Aho K, Kie JG, Bowyer RT (2015) Density of Dall's sheep in Alaska: effects of predator harvest? Mammal Res 60:21–28
Mkhize NR, Heitkönig IM, Scogings PF, Dziba LE, Prins HHT, de Boer WF (2015) Condensed tannins reduce browsing and increase grazing time of free-ranging goats in semi-arid savannas. Appl Anim Behav Sci 169:33–37
Moser B, Schütz M (2006) Tolerance of understory plants subject to herbivory by roe deer. Oikos 114:311–321
Murphy BP, Bowman DMJS (2012) What controls the distribution of tropical forest and savanna? Ecol Lett 15:748–758
Murray MG (1993) Comparative nutrition of wildebeest, hartebeest and topi in the Serengeti. Afr J Ecol 31:172–177
Murray MG, Brown D (1993) Niche separation of grazing ungulates in the Serengeti: an experimental test. J Anim Ecol 62:380–389
Mutanga O, Skidmore AK (2004) Narrow band vegetation indices overcome the saturation problem in biomass estimation. Int J Remote Sens 25:3999–4014
Myers JH (2018) Population cycles: generalities, exceptions and remaining mysteries. Proc R Soc B Biol Sci 285(1875):20172841
Navarro LM, Pereira HM (2015) Rewilding abandoned landscapes in Europe. In: Navarro LM, Pereira HM (eds) Rewilding European landscapes. Springer, Cham, pp 3–23
Olff H, Ritchie ME, Prins HHT (2002) Global environmental controls of diversity in large herbivores. Nature 415:901–904
Owen-Smith RN (1988) Megaherbivores: the influence of very large body size on ecology. Cambridge University Press, Cambridge
Parker KL, Barboza PS, Gillingham MP (2009) Nutrition integrates environmental responses of ungulates. Funct Ecol 23:57–69. https://doi.org/10.1111/j.1365-2435.2009.01528.x
Peters RH (1983) The ecological implications of body size. Cambridge University Press, Cambridge
Petty AM, Werner PA (2010) How many buffalo does it take to change a savanna? A response to Bowman et al. (2008). J Biogeogr 37:193–195
Prado JL, Alberdi MT (2017) Fossil horses of South America. Springer, Cham
Prins HHT (1996) Ecology and behaviour of the African buffalo: social inequality and decision making. Chapman & Hall, London. (now Springer Science & Business Media)
Prins HHT (1998) Origins and development of grassland communities in northwestern Europe. In: WallisDeVries MF, Vries MFW, Bakker JP, Bakker JP, van Wieren SE (eds) Grazing and conservation management. Springer, Berlin, pp 55–103
Prins HHT, Gordon IJ (eds) (2014a) Invasion biology and ecological theory. Insights from a continent in transformation. Cambridge University Press, Cambridge
Prins HHT, Gordon IJ (2014b) A critique of ecological theory and a salute to natural history. In: Prins HHT, Gordon IJ (eds) Invasion biology and ecological theory: insights from a continent in transformation. Cambridge University Press, Cambridge, pp 497–516
Prins HHT, Olff H (1998) Species-richness of African grazer assemblages: towards a functional explanation. In: Newberry DM, Prins HHT, Brown ND (eds) Dynamics of tropical communities: 37th symposium of the British Ecological Society. Cambridge University Press, Cambridge, pp 449–490
Prins HHT, Van der Jeugd HP (1993) Herbivore population crashes and woodland structure in East Africa. J Ecol 81:305–314
Prins HHT, van Langevelde F (2008) Assembling a diet from different places. In: Prins HHT, van Langevelde F (eds) Resource ecology. Springer, Dordrecht, pp 139–155
Prins HHT, Van Oeveren H (2014) Bovini as keystone species and landscape architects. In: Melletti M, Burton J (eds) Ecology, evolution and behaviour of wild cattle. Cambridge University Press, Cambridge, pp 21–29

Rees WA (1974) Preliminary studies into bush utilization by cattle in Zambia. J Appl Ecol 11:207–214
Rietkerk M, Ketner P, Stroosnijder L, Prins HHT (1996) Sahelian rangeland development; a catastrophe? J Range Manag 49:512–519
Rietkerk M, Boerlijst M, van Langevelde F, HilleRisLambers R, van de Koppel J, Kumar L, Prins HHT, de Roos A (2002a) Self-organization of vegetation in arid ecosystems. Am Nat 160:524–530
Rietkerk M, Ouedraogo T, Kumar L, Sanou S, van Langevelde F, Kiema A, van de Koppel J, van Andel J, Hearne J, Skidmore AK, de Ridder N, Stroosnijder L, Prins HHT (2002b) Fine-scale spatial distribution of plants and resources on a sandy soil in the Sahel. Plant Soil 239:69–77
Rivals F, Lister AM (2016) Dietary flexibility and niche partitioning of large herbivores through the Pleistocene of Britain. Quat Sci Rev 146:116–133
Rivals F, Prignano L, Semprebon GM, Lozano S (2015) A tool for determining duration of mortality events in archaeological assemblages using extant ungulate microwear. Sci Rep 5:17330
Robinson TP, Wint GW, Conchedda G, Van Boeckel TP, Ercoli V, Palamara E, Cinardi G, D'Aietti L, Hay SI, Gilbert M (2014) Mapping the global distribution of livestock. PLoS One 9:e96084
Rubenstein DR, Rubenstein DI, Sherman PW, Gavin TA (2006) Pleistocene Park: does re-wilding North America represent sound conservation for the 21st century? Biol Conserv 132:232–238
Saarinen J, Lister AM (2016) Dental mesowear reflects local vegetation and niche separation in Pleistocene proboscideans from Britain. J Quat Sci 31:799–808
Saarinen J, Eronen J, Fortelius M, Seppä H, Lister AM (2016) Patterns of diet and body mass of large ungulates from the Pleistocene of Western Europe, and their relation to vegetation. Palaeontol Electron 19.3.32A:1–58. palaeo-electronica.org/content/2016/1567-pleistocene-mammal-ecometrics
Sandom C, Faurby S, Sandel B et al (2014) Global late Quaternary megafauna extinctions linked to humans, not climate change. Proc R Soc B 281:20133254
Sankaran M, Hanan NP, Scholes RJ, Ratnam J, Augustine DJ, Cade BS, Gignoux J, Higgins SI, Le Roux X, Ludwig F, Ardö J, Banyikwa F, Bronn A, Bucini G, Caylor K, Coughenour M, Diouf A, Ekaya W, Feral CJ, Zambatis N (2005) Determinants of woody cover in African savannas. Nature 438:846–849
Schweiger AH, Svenning JC (2018) Down-sizing of dung beetle assemblages over the last 53000 years is consistent with a dominant effect of megafauna losses. Oikos 127:1–8
Searle KR, Gordon IJ, Stokes CJ (2009) Hysteresis responses to grazing in a semi-arid rangeland. Rangel Ecol Manage 62:136–144
Seeber P, Ndlovu HT, Duncan P, Ganswindt A (2012) Grazing behaviour of the giraffe in Hwange National Park, Zimbabwe. Afr J Ecol 50:247–250
Sexton JP, Montiel J, Shay JE, Stephens MR, Slatyer RA (2017) Evolution of ecological niche breadth. Annu Rev Ecol Evol Syst 48:183–206
Shrestha AK, Van Wieren SE, Van Langevelde F, Fuller A, Hetem RS, Meyer L, De Bie S, Prins HHT (2014) Larger antelopes are sensitive to heat stress throughout all seasons but smaller antelopes only during summer in an African semi-arid environment. Int J Biometeorol 58:41–49
Sibly RM, Hone J (2002) Population growth rate and its determinants: an overview. Philos Trans R Soc London B: Biol Sci 357:1153–1170
Sinclair ARE, Krebs CJ (2002) Complex numerical responses to top–down and bottom–up processes in vertebrate populations. Philos Trans R Soc London B: Biol Sci 357:1221–1231
Sinclair N, Pimm D, Higginson W (eds) (2006) Mathematics and the aesthetic: new approaches to an ancient affinity. Springer, New York
Smith FA, Doughty CE, Malhi Y, Svenning JC, Terborgh J (2016) Megafauna in the earth system. Ecography 39:99–108
Smith FA, Smith RE, Lyons SK, Payne JL (2018) Body size downgrading of mammals over the late Quaternary. Science 360:310–313

Specht A, Gordon IJ, Groves RH, Lambers H, Phinn SR (2015) Catalysing transdisciplinary synthesis in ecosystem science and management. Sci Total Environ 534:1–3

Stigter JD, van Langevelde F (2004) Optimal harvesting in a two-species model under critical de-pensation. The case of optimal harvesting in semi-arid grazing systems. Ecol Model 179:153–161

Stuart AJ (2015) Late Quaternary megafaunal extinctions on the continents: a short review. Geol J 50:338–363

Sullivan S, Rohde R (2002) On non-equilibrium in arid and semi-arid grazing systems. J Biogeogr 29:1595–1618

Thomson JM (2016) Impacts of environment on gene expression and epigenetic modification in grazing animals. J Anim Sci 94(Suppl 6):63–73

Tomlinson KW, van Langevelde F, Ward D, Bongers F, da Silva DA, Prins HHT, de Bie S, Sterck FJ (2013) Deciduous and evergreen trees differ in juvenile biomass allometries because of differences in allocation to root storage. Ann Bot 112:575–587

Tomlinson KW, van Langevelde F, Ward D, Prins HHT, de Bie S, Vosman B, Sampaio EV, Sterck FJ (2016) Defence against vertebrate herbivores trades off into architectural and low nutrient strategies amongst savanna *Fabaceae* species. Oikos 125:126–136

Tomlinson KW, Sterck FJ, Barbosa ER, de Bie S, Prins HHT, van Langevelde F (2018) Seedling growth of savanna tree species from three continents under grass competition and nutrient limitation in a greenhouse experiment. J Ecol. https://doi.org/10.1111/1365-2745.13085

Tucker CJ (1978) Post senescent grass canopy remote sensing. Remote Sens Environ 7:203–210

Tucker CJ (1979) Red and photographic infrared linear combinations for monitoring vegetation. Remote Sens Environ 8:127–150

United Nations (2017) World population prospects: the 2017 revision. Population Division, Population Estimates and Projections Section, United Nations Department of Economic and Social Affairs, New York, USA. http://esa.un.org/unpd/wpp/index.htm. Retrieved 16 February 2019

Van Auken OW (2000) Shrub invasions of North American semiarid grasslands. Annu Rev Ecol Syst 31:197–215

Van der Waal C, De Kroon H, De Boer WF, Heitkönig IMA, Skidmore AK, De Knegt HJ, Van Langevelde F, Van Wieren SE, Grant RC, Page BR, Slotow R, Kohi EM, Mwakiwa E, Prins HHT (2009) Water and nutrients alter herbaceous competitive effects on tree seedlings in a semi-arid savanna. J Ecol 97:430–439

Van der Waal C, de Kroon H, Heitkönig IMA, Skidmore AK, van Langevelde F, de Boer WF, Slotow R, Grant RC, Peel MP, Kohi EM, de Knegt HJ, Prins HHT (2011) Scale of nutrient patchiness mediates resource partitioning between trees and grasses in a semi-arid savanna. J Ecol 2011:1124–1133

Van Langevelde F, Prins HHT (2008) Introduction to resource ecology. In: Prins HHT, van Langevelde F (eds) Resource ecology. Springer, Dordrecht, pp 1–6

Van Langevelde F, Drescher M, Heitkönig IMA, Prins HHT (2008) Instantaneous intake rate of herbivores as function of forage quality and mass: effects on facilitative and competitive interactions. Ecol Model 213:273–284

Van Langevelde F, Tomlinson K, Barbosa ER, de Bie S, Prins HHT, Higgins SI (2010) Understanding tree-grass coexistence and impacts of disturbance and resource variability in savannas. In: Hill MJ, Hanan NP (eds) Ecosystem function in savannas: measurement and modelling at landscape to global scales. CRC Press, Boca Raton, pp 257–271

Van Maanen E, Convery I (2016) Rewilding: the realisation and reality of a new challenge for nature in the twenty-first century. In: Convery I, Davis P (eds) Changing perceptions of nature. Boydell & Brewer, Suffolk, pp 303–319

Van Nes EH, Scheffer M (2007) Slow recovery from perturbations as a generic indicator of a nearby catastrophic shift. Am Nat 169:738–747

Van Soest PJ (1982) Nutritional ecology of the ruminant. 0 & B Books, Corvallis, OR

Van Soest PJ (2018) Nutritional ecology of the ruminant. Cornell University Press, Ithaca

Venter JA, Prins HHT, Balfour DA, Slotow R (2014) Reconstructing grazer assemblages for protected area restoration. PLoS One 9(3):e90900

Verheyden-Tixier H, Renaud P-C, Morellet N, Jamot J, Besle J-M, Dumont B (2008) Selection for nutrients by red deer hinds feeding on a mixed forest edge. Oecologia 156:715–726

Walker A, Hoeck HN, Perez L (1978) Microwear of mammalian teeth as an indicator of diet. Science 201:908–910

Wiens JA (1982) On size ratios and sequences in ecological communities: are there no rules? Ann Zool Fenn 19:297–308

Wilde V, Hellmund M (2010) First record of gut contents from a middle Eocene equid from the Geiseltal near Halle (Saale), Sachsen-Anhalt, Central Germany. Paleobiodivers Paleoenviron 2:153–162

Wirsenius S, Azar C, Berndes G (2010) How much land is needed for global food production under scenarios of dietary changes and livestock productivity increases in 2030? Agric Syst 103:621–638

Wolff OJ (1997) Population regulation in mammals: an evolutionary perspective. J Anim Ecol 66:1–13

World Bank (2019) World bank data for low and middle income countries. https://data.worldbank.org/income-level/low-and-middle-income. Retrieved 16 February 2019

Yatat V, Tchuinté A, Dumont Y, Couteron P (2018) A tribute to the use of minimalistic spatially-implicit models of savanna vegetation dynamics to address broad spatial scales in spite of scarce data. Biomath 7:1812167

Zeeman EC (1976) Catastrophe theory. Sci Am 234:65–83

Zeeman EC (1977) Catastrophe theory: selected papers, 1972–1977. Addison-Wesley, Oxford

Zimov SA, Zimov NS, Tikhonov AN, Chapin FS (2012) Mammoth steppe: a high-productivity phenomenon. Quat Sci Rev 57:26–45

Zolkos SG, Goetz SJ, Dubayah R (2013) A meta-analysis of terrestrial aboveground biomass estimation using lidar remote sensing. Remote Sens Environ 128:289–298

Index

I. J. Gordon, H. H. T. Prins (eds.), *The Ecology of Browsing and Grazing II*, Ecological Studies 239, https://doi.org/10.1007/978-3-030-25865-8

W

X

www.ingramcontent.com/pod-product-compliance
Ingram Content Group UK Ltd.
Pitfield, Milton Keynes, MK11 3LW, UK
UKHW021839270726
14058UKWH00002B/246
* 9 7 8 3 0 3 0 2 5 8 6 7 2 *